Environmental Hydrology

Environmental Hydrology

Edited by **William Sobol**

SYRAWOOD
PUBLISHING HOUSE
New York

Published by Syrawood Publishing House,
750 Third Avenue, 9th Floor,
New York, NY 10017, USA
www.syrawoodpublishinghouse.com

Environmental Hydrology
Edited by William Sobol

International Standard Book Number: 978-1-68286-303-9 (Hardback)

Printed in China.

Contents

Preface

Environmental hydrology integrates the concepts of hydrology with environmental planning. The applications of this field spread across agriculture, geology, climatology, etc. Waste management, global warming, hydrologic cycle, etc. are some of the topics that have been extensively covered in this book. With its detailed analyses and data, the text will prove immensely beneficial to professionals and students involved in this area at various levels. It will also help new researchers by foregrounding their knowledge in environmental hydrology.

This book is the end result of constructive efforts and intensive research done by experts in this field. The aim of this book is to enlighten the readers with recent information in this area of research. The information provided in this profound book would serve as a valuable reference to students and researchers in this field.

At the end, I would like to thank all the authors for devoting their precious time and providing their valuable contributions to this book. I would also like to express my gratitude to my fellow colleagues who encouraged me throughout the process.

Editor

Study of specific power input and mass transfer coefficient in aeration of volatile organic compounds

M. K. Verma and R. K. Tyagi

Department of Mechanical and Automation Engineering, Amity School of Engineering and Technology, Amity University, Noida (U.P), India.

The aeration of volatile organic compounds (VOC) is based on the mass transfer rate of VOCs from the liquid phase to the gaseous phase. Principle of air stripping involves the mass transfer of volatile organic contaminant from water to air. The system can be easily upgraded to strip greater amount of VOCs with relatively small increase in capital cost. Aeration tanks strip volatile compounds by bubbling air into tank through which contaminated water flows. The present study shows power input in VOC by using advanced designed gas dispersing radial impeller that is, Rushton, CD-6 and BT-6, the factors which affect the design of aeration tank baffles and the impeller used for best possible result. This study also shows the power input of VOCs at different height of submergence. A comparison of the result from experiment that shows the power input of VOCs during aeration is a function of mass transfer coefficient of air and it increases with the increase of submergence of height, by supply of air, and increase in impeller speed. Best power input result can be obtained at the 2/3rd height of the submergence of the total depth of water for impeller used.

Key words: Power input, gas dispersion impellers, Rushton impeller, CD-6 impeller, BT-6 impeller.

INTRODUCTION

Aeration is the most important and indispensable operation for the treatment of wastewater. The main purpose of aeration is to dissolve the oxygen into the water. In order to improve the dissolution of oxygen in water, the common methods include bubbling and surface aeration. For bubbling type of aeration, oxygen dissolves in the water during the rising of air bubbles in liquid phase. For surface aeration type, a vertical axis is designed to induce updraft flows through a pumping action, causing a rapid change in the air–water interface to facilitate dissolution of the air. Both types of aeration use the air–water contact to achieve oxygen dissolution (Cheng et al., 2003).

A number of volatile organic compounds (VOCs) are found in the wastewater and some of these VOCs are very harmful for the human health as well as environment. Although there are several methods which are being used for the removal of VOCs, but the air stripping process are the low energy usage, low preventive and maintenance cost and high efficiency process which can stripes VOCs from wastewater. The present study consists of aeration system for the removal of VOCs. The aeration of VOC is based on the mass transfer rate of VOCs from the liquid phase to the gaseous phase. Principle of air stripping involves the mass transfer of volatile organic contaminant from water to air (Verma and Tyagi, 2012). Aeration is the process of bringing water and air in close contact in order to remove dissolved gases, such as carbon dioxide, and to oxidize dissolved metals such as iron. It can also be used to remove volatile organic chemicals (VOC) present in the water. Aeration is often the first major process at the treatment plant. During aeration, constituents are removed or modified before they can interfere with the treatment processes. Aeration is an important step in the treatment of sewage by bio oxidation and in much industrial fermentation (Sathiyamoorthy et al., 2012).

Bhuyar et al. (2009) found out the effect of a curved-blade-surface mechanical aerator for oxidation, which is used to treat municipal and domestic sewage. Aeration experiments were conducted to study the design characteristics of curved blade surface mechanical aerator. The paper critically examines six different configurations of aerators, which were developed, fabricated and tested in the laboratory for its various dynamic parameters, such as diameter of aerators (D), speed (N) and immersion depth (h).

Oxygen transfer rate and the corresponding power requirement to operate the rotor are vital for design and scale-up of surface aerators. It has been demonstrated that energy can be saved substantially if the aeration tanks are run at relatively higher input powers. It is also demonstrated that smaller sized tanks are more energy conservative and economical when compared to big sized tanks, while aerating the same volume of water, and at the same time by maintaining a constant input power in all the tanks irrespective of their size. Oxygen transfer characteristics for three sizes of equipment correlate well and permit scaling of fermentation systems in size were studied (John, 2003).

Oxygen transfer, the process by which oxygen is transferred from gaseous to liquid phase is a vital part of the waste-water treatment process (Metcalf and Eddy, 2001). Due to low solubility of oxygen and consequent low rate of oxygen transfer, sufficient oxygen to meet the requirement of aerobic waste does not enter through normal surface air-water interface. To transfer the large quantities of oxygen that is needed, additional interfaces are created by employing aeration process. The creation of additional interfaces enhances the rate of oxygen transfer so that the dissolved oxygen level gets raised.

The three basic categories of aeration methods (Thakre et al., 2008) are:

1) Surface or mechanical aeration method, which increases interfacial area by spraying water droplet into the air.
2) Diffused aeration method, which release air bubbles beneath the surface of water.
3) Combined and turbine aeration methods, which introduced large air bubble into water and reduce their sizes mechanically

Out of these three, the mechanical surface aerators are widely used because they offer better efficiency as well as convenience in operation and maintenance (Rao and Kumar, 2007). Further, oxygen transfer rate from gas to liquid phase is dependent on various factors for given method of aeration such as dynamic variables like speed, mixing intensity and turbulence, geometrical parameters like size and number of blades, depth of flow etc and physicochemical properties of the liquid.

Objective

Objective of present work is to show specific power input required for the different types of impellers at different speed of rotation and at different height of submerge. This work also shows that mass transfer coefficient depends upon specific power input.

EXPERIMENTAL PROCEDURE

Experimental procedure and setup used are the same as mentioned in Verma and Tyagi (2012) which consist of cylindrical tank with a diameter of 30 cm and height of 35 cm with total working volume of 24.74 L, and working volume of water is 16 L with water height in tank is 22.6 cm. Verma and Tyagi (2012) described in detail VOC`s for combination of impellers and height of submergence and the result shows removal rate of different VOC`s used. This article intend to specify power input and mass transfer coefficient for different impellers at different height of submergence.

Estimation of K_La

In aeration, open vessel were used that allows the sufficient air volume to avoid significant gas phase saturation above the liquid surface, which allow us to assume $C_s = 0$ (Chu-Chin et al., 1993). Therefore from the equation (Bhuyar, 2009), the liquid concentration changes over time were used to estimate the mass transfer coefficient K_La using the equation below.

$$\ln\left(\frac{C_t}{C_o}\right) = -K_{La}(t - t_0) \qquad (1)$$

Estimation of specific power input

$P/V = 2.05E\text{-}06\ (N)^{3.11}$

Materials used

Chemical used in this study are acetone and dichloromethane. Table 1 shows some physical and chemical properties of these chemicals:

Impeller used in the study

In this study, gas dispersion impellers that is, Rushton or D-6 and hollow blade impellers that is, CD-6 and BT-6 are used which is shown in Table 2 and Figure 1 (Verma and Tyagi, 2012).

RESULTS

Relation between specific power input (P/V) versus mass

Table 1. Physical and chemical properties of VOCs used.

Property	Value (Dichloromethane)	Value (Acetone)
Molecular weight	84.9 g	58.08 g
Aspect	Liquid	Liquid
Melting point	- 94.9°C	-95.35°C
Boiling point	39 - 40°C at 1.013 Pa	-94.7°C at 1 Pa
Density	1.33 g/ml at 20°C	0.78996 g/ml at 20°C
Vapor pressure	475 mm of Hg at 20°C	181.72 mm of Hg at 20°C
Water solubility	13.7 g/L at 20°C	11.7 g/L at 20°C

Table 2. Description of the impeller used.

Impeller	Geometrical characteristic of the impeller			
	D (m)	L	W	D_D
D-6	0.1	D/4	D/5	7D/10
CD-6	0.1	D/4	D/5	7D/10
BT-6	0.11	D/4	2D/13	2D/3

transfer co-efficient (KLa) at 250, 300, and 350 rpm for VOC1 and VOC2 when Rushton impeller was used. The maximum value is 0.185 at 350 rpm and VOC1, minimum value is 0.118 at 250 rpm and VOC2 Rushton impeller at height H1(Figure 2).

Relation between specific power input (P/V) versus mass transfer co-efficient (KLa) at 250, 300, and 350 rpm for VOC1 and VOC2 when CD6 impeller was used. The maximum value is 0.220 at 350 rpm and VOC2, minimum value is 0.154 at 250 rpm and VOC1 CD6 impeller at height H1 (Figure 3).

Relation between specific power input (P/V) versus mass transfer co-efficient (KLa) at 250, 300, and 350 rpm for VOC1 and VOC2 when BT6 impeller was used. The maximum value is 0.323 at 350 rpm and VOC1, minimum value is 0.246 at 250 rpm and VOC1 BT6 impeller at height H1 (Figure 4).

Relation between specific power input (P/V) versus mass transfer co-efficient (KLa) at 250, 300, and 350 rpm for VOC1 and VOC2 when Rushton impeller was used. The maximum value is 0.187 at 350 rpm and VOC1, minimum value is 0.132 at 250 rpm and VOC2 Rushton impeller at height H2 (Figure 5).

Relation between specific power input (P/V) versus mass transfer co-efficient (KLa) at 250, 300, and 350 rpm for VOC1 and VOC2 when CD6 impeller was used. The maximum value is 0.246 at 350 rpm and VOC1, minimum value is 0.192 at 250 rpm and VOC1 CD6 impeller at height H2 (Figure 6).

Relation between specific power input (P/V) versus mass transfer co-efficient (KLa) at 250, 300, and 350 rpm for VOC1 and VOC2 when BT6 impeller was used. The maximum value is 0.346 at 350 rpm and VOC1, minimum

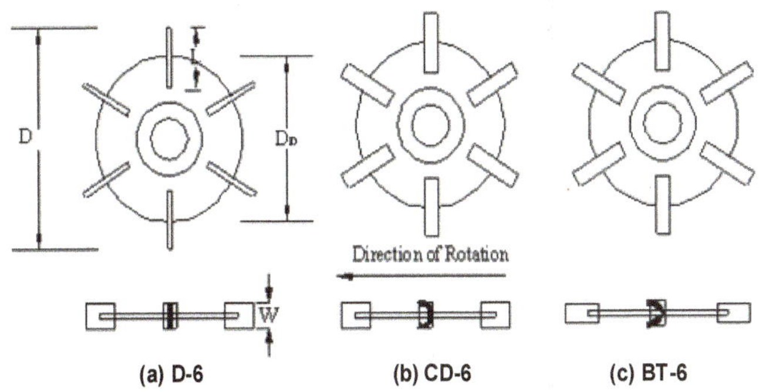

Figure 1. Impeller used in the study.

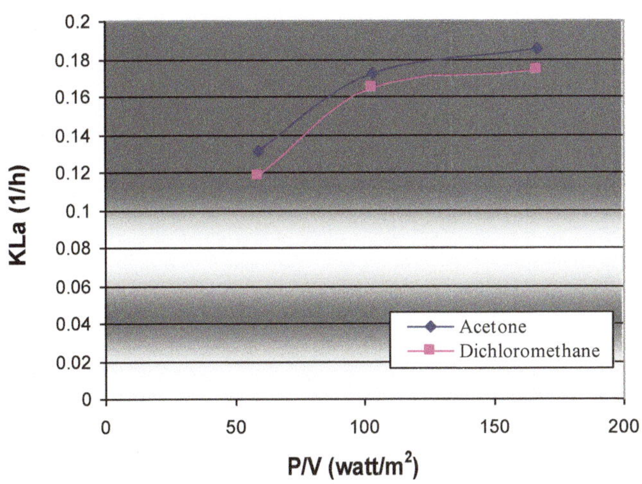

Figure 2. Plot of mass transfer co-efficient (KLa) versus specific power input (P/V) by Rushton impeller at height H1.

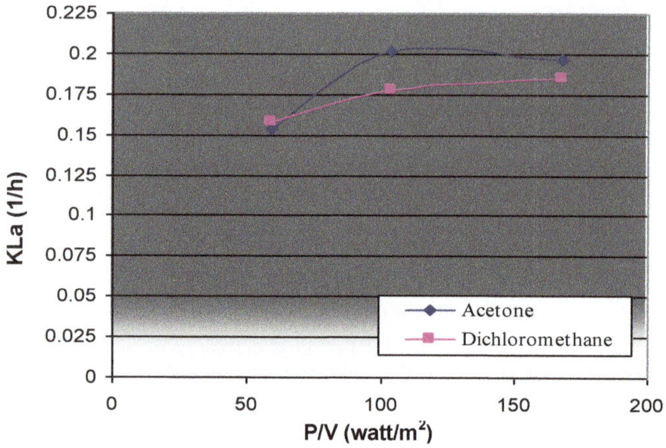

Figure 3. Plot of mass transfer co-efficient (KLa) versus specific power input (P/V) by CD6 impeller at height H1.

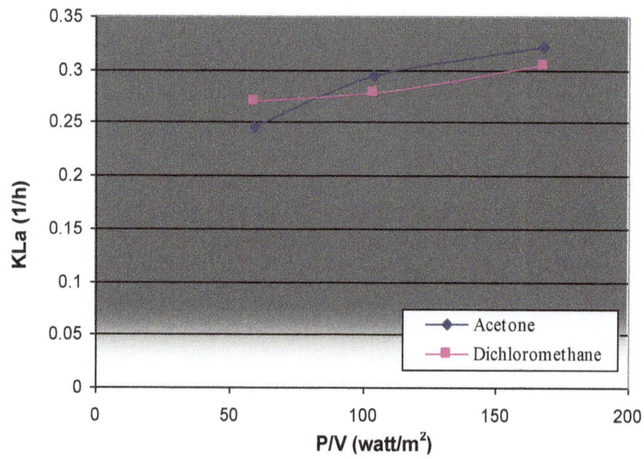

Figure 4. Plot of mass transfer co-efficient (KLa) versus specific power input (P/V) by BT6 impeller at height H1.

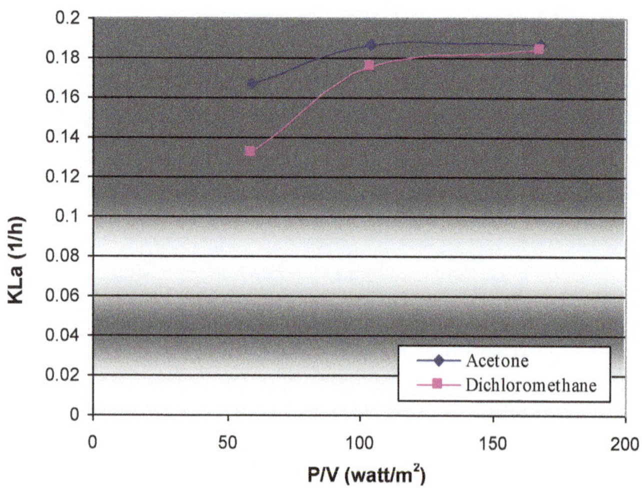

Figure 5. Plot of mass transfer co-efficient (KLa) versus specific power input (P/V) by Rushton impeller at height H2.

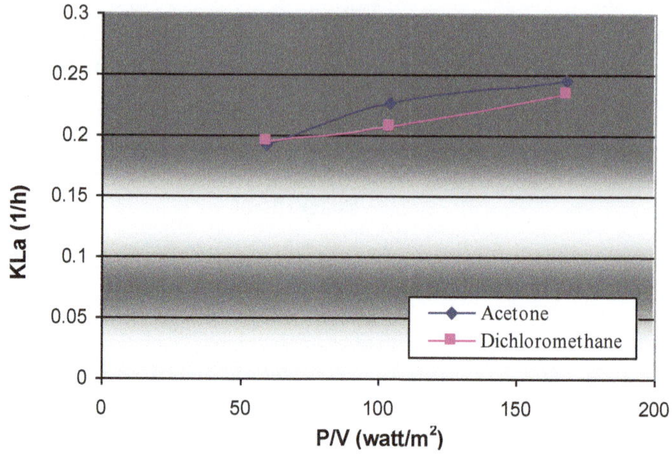

Figure 6. Plot of mass transfer co-efficient (KLa) versus specific power input (P/V) by CD6 impeller at height H2.

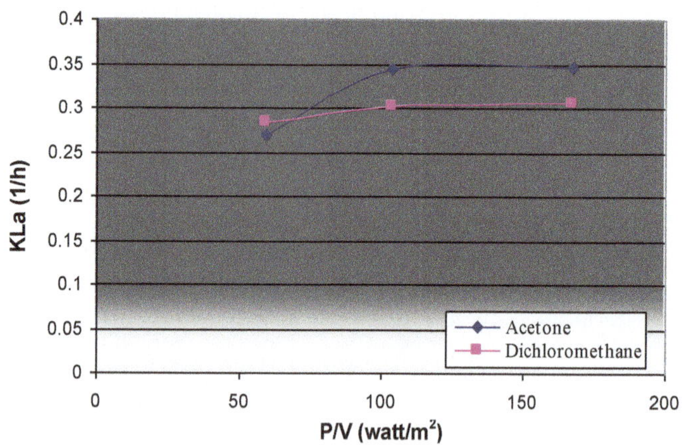

Figure 7. Plot of mass transfer co-efficient (KLa) versus specific power input (P/V) by BT6 impeller at height H2

value is 0.269 at 250 rpm and VOC1 BT6 impeller at height H2 (Figure 7).

Relation between specific power input (P/V) versus rotation with the help of bar chart at 250, 300, and 350 rpm for VOC1 and VOC2 when Rushton, BT6, CD6 impeller was used. The maximum value is obtained at 350 rpm and minimum value is obtained at 250 rpm (Figure 8).

DISCUSSION

Mass transfer coefficient depends upon specific power input (as specific power input proportional to impeller speed) graph clearly shows that it increases with the increase in specific power input at both height of submergence H1 and H2. These graph shows that mass transfer coefficient increases with the increase in specific power input, and its value changes more frequently for

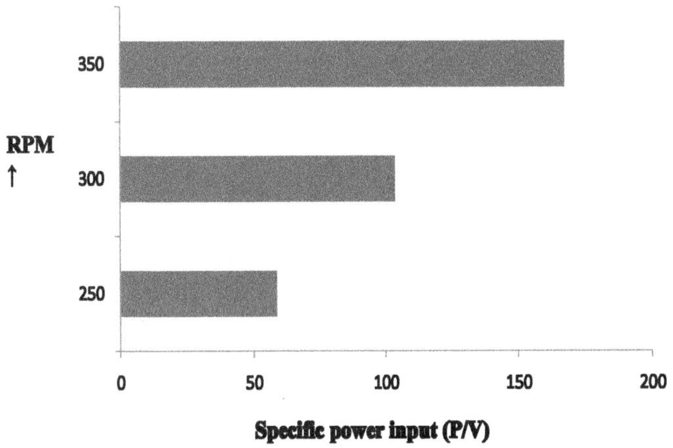

Figure 8. Plot of Specific power input (P/V) versus rotation in RPM for Rushton, CD6 and BT6 impeller

BT- impeller, and it's having lower value for Rushton impeller. The value of K_La changes with increase in height of submergence. The result obtained from model is fairly new, which was calculated experimentally. The result found are based on prototype model and the value of parameters such as specific power input, mass transfer coefficient are within the range of literature cited in this research article.

NOMENCLATURE

H, Height of the tank; H1=H/3; H2=2H/3; $\mathbf{C_s}$**,** gas phase concentration in equilibrium; $\mathbf{C_L}$**,** liquid phase concentration in equilibrium; $\mathbf{K_La}$**,** mass transfer coefficient, 1/h; **B,** tank baffle width, (m); **T,** tank diameter, (m); $\mathbf{C_S}$**,** Sparger clearance, (m); **D,** impeller diameter, (m); $\mathbf{D_D}$**,** impeller central disk diameter, (m); **W,** impeller blade width, (m); **L,** impeller blade length, (m).

REFERENCES

Bhuyar LB (2009). Design characteristics of curved blade aerator w.r.t aeration efficiency and overall oxygen transfer coefficient and comparison with CFD modeling. IJEST l(1):1-15.

Cheng JH, Hsu YC, Chen YF, Lin CC (2003). Application of gas-inducing reactor to obtain high oxygen dissolution in aeration process. Water Res. 37(1):2919-2928.

Chin H, Kyoung H, Stenstrom SR, Michael K (1993). Estimating emissions of 20 VOCs (diffused aeration and surface aeration). JEE (ASCE) 1(5):119-126.

John JW (2003). Paddle wheel aerator design, fisheries series. Timely Info. Agric. Nat. Resour. 1(3).

Metcalf and Eddy Inc. (2001). Waste water engineering: Treatment disposal and reuse. 3rd ed. Tata McGraw Hill, New Delhi India, pp. 1200.

Rao AR, Kumar B (2007). Neural modeling of square surface aerators. JEE 125(3):215-223.

Sathiyamoorthy M, Sanghavi BJ, Tumsa Z, Shemsedin A, Birhane H (2012). Increasing aeration efficiency by redesigning the aerator for a tertiary treatment plant in a nitrogenous fertilizer plant. IJMIE 2(4):6.

Thakre SB, Bhuyar LB, Deshmukh SJ (2008). Effect of different configurations of mechanical aerators on oxygen transfer and aeration efficiency with respect to power consumption. IJMSSE 2(2):100-108.

Verma MK, Tyagi RK (2012). Aeration of volatile organic compounds by using gas dispersion impellers. JMER 4(5):213-224.

Scouring control of the eroded stilling basin: Case study, Namrood Dam

H. Taebi and M. Fathi-Moghadam*

[1]Graduate Student, School of Water Sciences Engineering, Shahid Chamran University, Ahwaz, Iran.
[2]School of Water Sciences Engineering, Shahid Chamran University, Ahwaz, Iran.

Local scouring immediately downstream of stilling basin is unavoidable local velocity exceeds incipient motion velocity. This phenomenon causes a difference level between concrete floor of basin and river bed. In such circumstances a riprap is normally used to prevent the development of scouring hole. In this study, hydraulic model of Namrood dam is used to estimate stone size that is able to resist scouring as result of basin turbulent outlet flow. Tail water (y_t), the difference in level between basin floor and river bed (Z) and critical depth of flow in stilling basin (y_c) are required to be measured in order to develop a mathematical model for estimation of scouring. Results confirm considerable reduction of the stone size with increase of (Z) and (y_t) for same flow discharges, while stone size increases with flow discharge.

Key words: Riprap protection, stilling basin, hydraulic jump, stability number.

INTRODUCTION

Scouring in downstream of a stilling basin will gradually degrade the river bed resulting to a difference in level with concrete of the basin floor. Hence, downstream of basin and the whole structure is always vulnerable to scouring and the potential threats caused by scouring. Figures 1 and 2 show pictures of Jafarabad diverging-regulating dam in south west of Iran before and after dam operation, respectively. As it is shown in Figure 2, the scouring has caused 2.5 m difference in level of the basin and river bed.

Many studies have been conducted in the past for estimation of scouring in channels and downstream of the hydraulic structure when the structure bed and the down-stream channel bed are level. Peterka (Chanson, 2008) presented a method to estimate the stone size for riprap to protect downstream of stilling basin against scouring. The flow velocity near the bed was key parameter for estimation of stone size. Since the nature of flow near bed and stones is not clear, the method contains many uncertainties in practice. Rice and Kadavy (1991, 1992) conducted experiments to formulate scouring down-stream of straight drop spillways and riprap design for SAF stilling basins as a function critical flow depth and drop height. Shafai-bajestan and Albertson (1993) carried out several studies on designing riprap for outlet flow from pipes and developed an equation for estimating minimum stone diameter for a stable riprap. Lauchlan and Melville (2001) studied riprap protection at bridge piers. Farhoudi and valizadegan (2004) conducted an experimental work on a physical model of stilling basin leveled with channel bed in order to develop criteria for protection of downstream against scouring. Having Froude Number, riprap diameter that could resist against the outflow can be determined. More studies on grade control of scouring downstream of structures can be found in Maynord (1991) and D'Agos-tino and Ferro (2004).

The purpose of this study is to investigate effects of tail water and difference in level between floor of stilling basin and downstream channel bed on incipient motion of ripraps downstream of basins. The Namrud dam stilling basin model with eroded downstream was used to develop a mathematical model for estimation of stable stone size of riprap.

*Corresponding author. E-mail: fathi49@gmail.com.

Dimensional analysis

Size of stable riprap downstream of stilling basin is a

Figure 1. Jafarabad dam - Before scouring

Figure 2. Jafarabad dam - After scouring.

function of flow characteristics, stone material characteristic and river characteristic as follows:

A. Flow characteristics: critical flow depth (y_c), flow velocity (v), water specific density (ρ_w), dynamic viscosity (μ), and gravity (g).

B. Riprap characteristics: stone size (D_s), stone specific density (ρ_s), geometric variance of distributing riprap (σ_g), static angle of riprap particles (ϕ), shape factor (F_s).

C. River characteristics: bed slop(s), bed roughness (n), level difference between basin and downstream bed (Z), river width (B), water depth downstream of stilling basin (y_t), spillway height (H_p), stilling basin length (L_B).

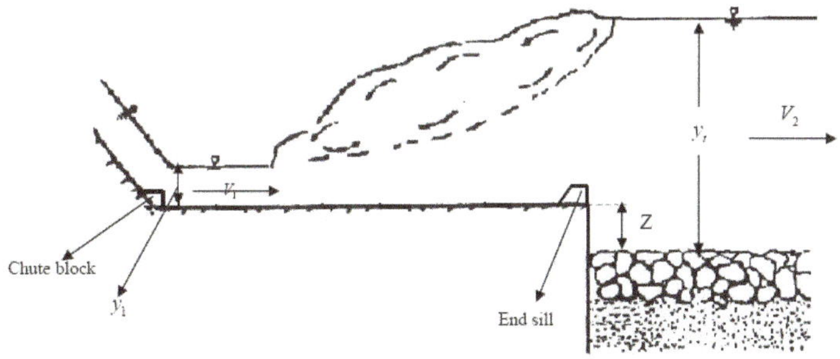

Figure 3. Profile of stilling basin flow and downstream riprap.

The influential parameters on riprap stable downstream of degraded stilling basin are show in Figure 3 and summarized as follow;

$$f(y_c, V, \upsilon, \rho_w, g, D_s, \rho_s, \sigma_g, \phi, F_s, n, Z, y_t, S, B, H_p, L_B) = 0$$
(1)

Since n (manning roughness coefficient) is a function of stone size, it is ignored; with respect to trivial change in riprap size and to the unifying hypothesis, the same happens to static angle (ϕ) and shape factor of sediment particles (F_s). In present study, bed width downstream of basin is kept constant in all the experiments and bed slop is supposed to be zero; there for, (B) and (S) are ignored in experiments.

For all tested ripraps, geometric variance (σ_g) of particles were calculated to be less than 1.066, thus ripraps used in the study are unified and (σ_g) is cancelled out from analysis.

In the experimental model basin length and spillway level are constant, so the effect of (H_P) and (L_B) can also be ignored.

Since stilling basin floor and downstream bed are not at the same level and flow characteristic are not same within basin and downstream, y_c (calculated based on flow discharge at sub-critical flow inside of basin) is applied as flow depth in basin and y_t as flow in downstream of basin.

$$f(y_c, V, \rho_w, \upsilon, D_s, \rho_s, g, Z, y_t) = 0 \quad (2)$$

Following dimensionless parameters can be produced from Equation 2,

$$f\left(\frac{D_s}{y_c}, \frac{D_s}{y_t}, \frac{D_s}{Z}, \frac{Z}{y_t}, \frac{Z}{y_c}, \frac{y_c}{y_t}, Fr, S.N, Re\right) = 0 \quad (3)$$

Where;

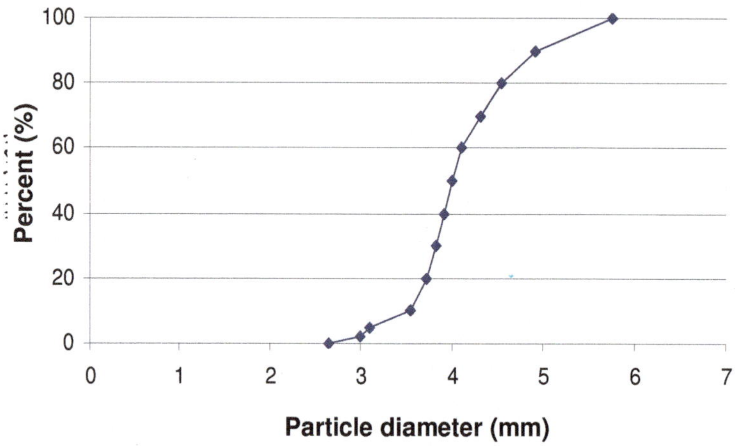

Figure 4. Particle size distribution of riprap material.

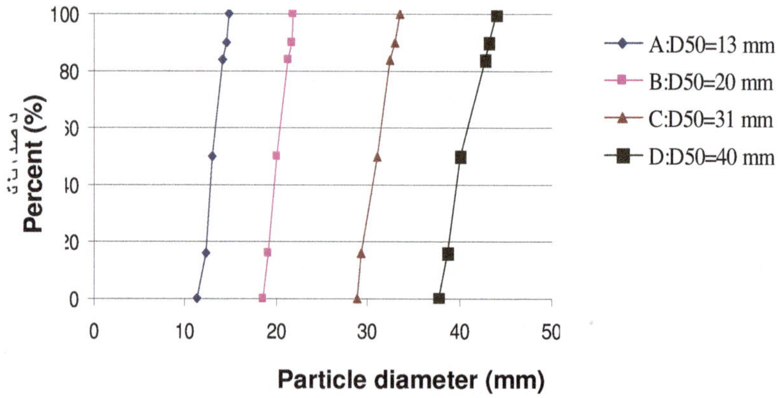

Figure 5. Particle size distribution of bed material.

$$S.N=\frac{V}{\sqrt{(G_s-1)gD_s}}$$

is Density Froude Number (which is called Stability number in literature) and Gs = g. $\square_s$. The seventh and eighth dimensionless parameters stand for Froude Number (Fr) and Reynolds Number (Re), respectively. Since S.N takes car of gravity effect as well, (Fr) is ignored in farther analysis. All flows passing on riprap are turbulent, (Re) is deleted and final form of the equation is:

$$S.N=f\left(\frac{D_s}{y_c},\frac{D_s}{y_t},\frac{D_s}{Z},\frac{Z}{y_t},\frac{Z}{y_c},\frac{y_c}{y_t}\right) \qquad (4)$$

MATERIAL AND METHODS

To estimate the diameter appropriate for riprap in eroded downstream of stilling basin, the experiments were carried out on a hydraulic model of Namrood dam constructed with scale of 1/40 in Water Research Institute of Ministry of Power of Iran. The characteristic and parts of model are as follows;

- Approach channel
- Ogee
- Chute cannel
- Stilling basin (USBR type II)
- Downstream area
- Check gate
- Bed and riprap material

The bed material were chosen from gravels with a diameter of D_{50} = 4.2 mm in model (0.17 m in prototype). The average diameters of the ripraps were 13, 20, 31, 40 mm in the model, fixed on the top layer of bed material. The particles size distribution of riprap materials and bed materials are shown in Figures 4 and 5.

A Nixon probe micro-propeller was used to measure average flow velocity of downstream of stilling basin the flow velocity was measured in middle direction of the flow, in two depths of 0.2 and 0.8 from water level; then these measures were used to calculate the average flow velocity. 0.8 from water level; then these measures were used to calculate the average flow velocity The experiments were conducted for level differences (Z) of 2.5, 5, 7.5 and

Table 1. S.N correlation and mean square error with individual parameters.

Mean square error *(MSE)*	Correlation coefficient (%)	Independent variable	Dependent variable
0.14	7.8	Z/y_c	$S.N$
0.345	19.1	D_s/Z	$S.N$
0.567	31.4	Z/y_t	$S.N$
1.613	89.5	D_s/y_c	$S.N$
0.563	31.3	y_c/y_t	$S.N$
1.532	85.0	D_s/y_t	$S.N$

Table 2. S.N correlation and mean square error with two parameters.

Mean square error *(MSE)*	Correlation coefficient (%)	Independent variable	Dependent variable
0.795	88.2	$Z/y_t, D_s/y_t$	$S.N$
0.776	86.1	$Z/y_t, y_c/y_t$	$S.N$
0.815	90.5	$Z/y_t, D_s/y_c$	$S.N$
0.831	92.2	$y_c/y_t, D_s/y_c$	$S.N$
0.813	90.2	$D_s/y_t, D_s/y_c$	$S.N$
0.771	85.6	$D_s/y_t, y_c/y_t$	$S.N$

10 cm. So the ratio of z/yt were measured as 2/3, 0.5, 0.4, 1/3, 0.25 and 0.2 respectively. For each (Z), the downstream water depth was increased (decreasing z/yt ratio) to get the best classic jump; for Z=2.5 cm, the experiments were repeated for_z/yt = 0.15, and also for Z = 5 cm, there was an experiment for z/yt = 0.15. The experiments were conducted for classic hydraulic jump while hydraulic jump was adjusted so that end of jump exactly corresponds the beginning of riprap downstream of stilling basin then downstream water depth was recorded and z/yt ratio was calculated.

RESULTS AND DISCUSSION

A linear regression process was conducted to evaluate the resulted dimensionless parameters in Equation 4. Table 1 shows result of correlation of individual parameters with S.N. According to the results, z/yt and Ds/yt did not show a significant relation with S.N, that they were deleted from multiple regression analysis.

A multiple regression for two parameters in Table 2 reveals some improvement in MSE and R^2 coefficients. Moreover, result of correlation with three parameters shows a slightly improvement in Table 3, while no improvement was recognized for involvement of more parameters. However, results of most significant correlation of S.N with one, two, and three parameters are summarized in Table 4 for ease in practice. Among all equations in Table 4, Equation 8 is the most useful one due to (a) having variable of Z, and (b) appearance of the stone size (Ds) just in left hand side of the equation. Moreover, based on available parameters in practice, other equations can be used with a trial and error procedure to estimate the required stone size for riprap. The single parameter equations can be just used to get rough idea of scouring when Z is not known.

However, results of most significant correlation of S.N with one, two, and three parameters are summarized in Table 4 for ease in practice. Among all equations in Table 4, Equation 8 is the most useful one due to (a) having variable of Z, and (b) appearance of the stone size (Ds) just in left hand side of the equation. Moreover, based on

Table 3. S.N correlation and mean square error with three parameters.

Mean square error *(MSE)*	Correlation coefficient (%)	Independent variable	Dependent variable
0.541	90.0	$D_s/y_t, y_c/y_t, Z/y_t$	$S.N$
0.557	92.7	$D_s/y_c, y_c/y_t, Z/y_t$	$S.N$
0.554	90.5	$D_s/y_c, D_s/y_t, Z/y_t$	$S.N$
0.564	93.9	$D_s/y_c, D_s/y_t, y_c/y_t$	$S.N$
0.423	93.9	$D_s/y_c, D_s/y_t, y_c/y_t, Z/y_t$	$S.N$

Table 4. Linear regression equations for estimation of S.N.

(MSE)	R^2 *(%)*	Formula	Equation number	Number of independent variable
1.613	89.5	$S.N = 1.252 - 0.723\,(\frac{D_s}{y_c})$	(5)	1
1.532	85.0	$S.N = 1.203 - 1.291(\frac{D_s}{y_t})$	(6)	1
0.795	88.2	$S.N = 1.234 - 0.153\,(\frac{Z}{y_t}) - 1.171\,(\frac{D_s}{y_t})$	(7)	2
0.776	86.1	$S.N = 1.673 - 0.597\,(\frac{Z}{y_t}) - 1.105\,(\frac{y_c}{y_t})$	(8)	2
0.815	90.5	$S.N = 1.241 - 0.792\,(\frac{D_s}{y_c}) + 0.105\,(\frac{Z}{y_t})$	(9)	2
0.831	92.2	$S.N = 1.343 - 0.662\,(\frac{D_s}{y_c}) - 0.263\,(\frac{y_c}{y_t})$	(10)	2
0.813	90.2	$S.N = 1.243 - 0.536\,(\frac{D_s}{y_c}) - 0.361\,(\frac{D_s}{y_t})$	(11)	2
0.771	85.6	$S.N = 1.154 - 1.386\,(\frac{D_s}{y_t}) + 0.146\,(\frac{y_c}{y_t})$	(12)	2
0.564	93.9	$S.N = 1.503 - 1.178(\frac{D_s}{y_c}) - 0.642(\frac{y_c}{y_t}) + 1.169(\frac{D_s}{y_t})$	(13)	3

available parameters in practice, other equations can be used with a trial and error procedure to estimate the required stone size for riprap. The single parameter equations can be just used to get rough idea of scouring when Z is not known.

Figure 6 illustrates graphical relation of S.N with z/yc for number of dimensionless degradation parameters, and Figure 7 shows variation the stone size with critical water depth for same number of degradations. Both figures confirm a minor nonlinearity or relationship between parameters. Figure 6 confirms reduction of stone size with increase of (Z) and tail water (y_t) for same flow discharges, while Figure 7 demonstrates increase of stone size with discharge. Alternatively, Figures 6 and 7 can be used to estimate the required stone sizes based on the available parameters. Since estimation of velocity over riprap for calculation of S.N contains many uncertainties, application of Figure 7 has advantage over Figure 7.

Comparison of the results with previous studies: Previous studies by large have investigated incipient motion and scouring for the stilling basins leveled with the downstream channel bed, and have suggested different coefficients of (a) and (m) for following equation.

$$\frac{V}{[\,g(G_s - 1)D_s\,]^{0.5}} = a(\frac{y_t}{D_s})^m \qquad (14)$$

In order to show effect of degrading level (Z) on scouring and S.N, experimental results of this study are correlated

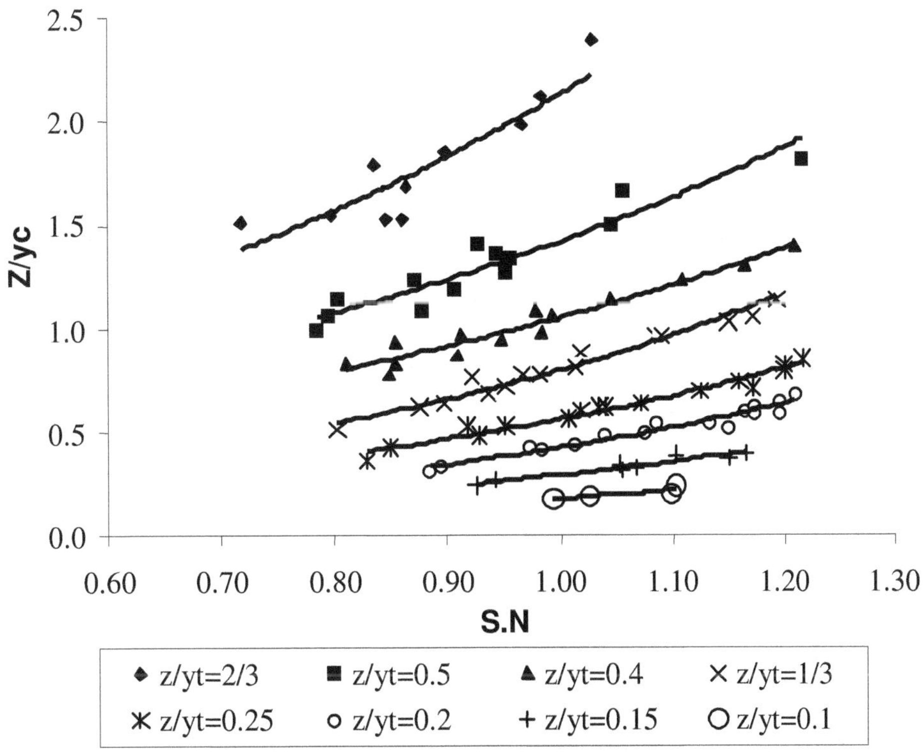

Figure 6. Relation of S.N with z/yc for variable ratio of_z/yt.

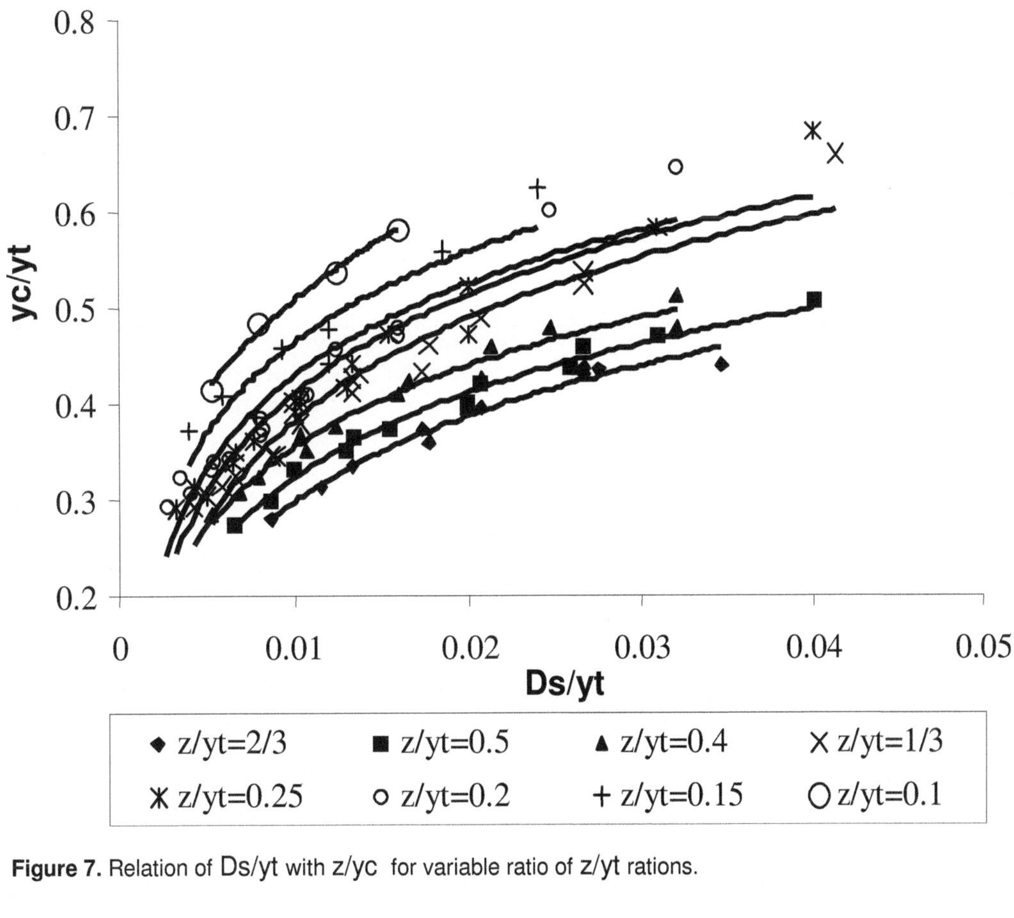

Figure 7. Relation of Ds/yt with z/yc for variable ratio of z/yt rations.

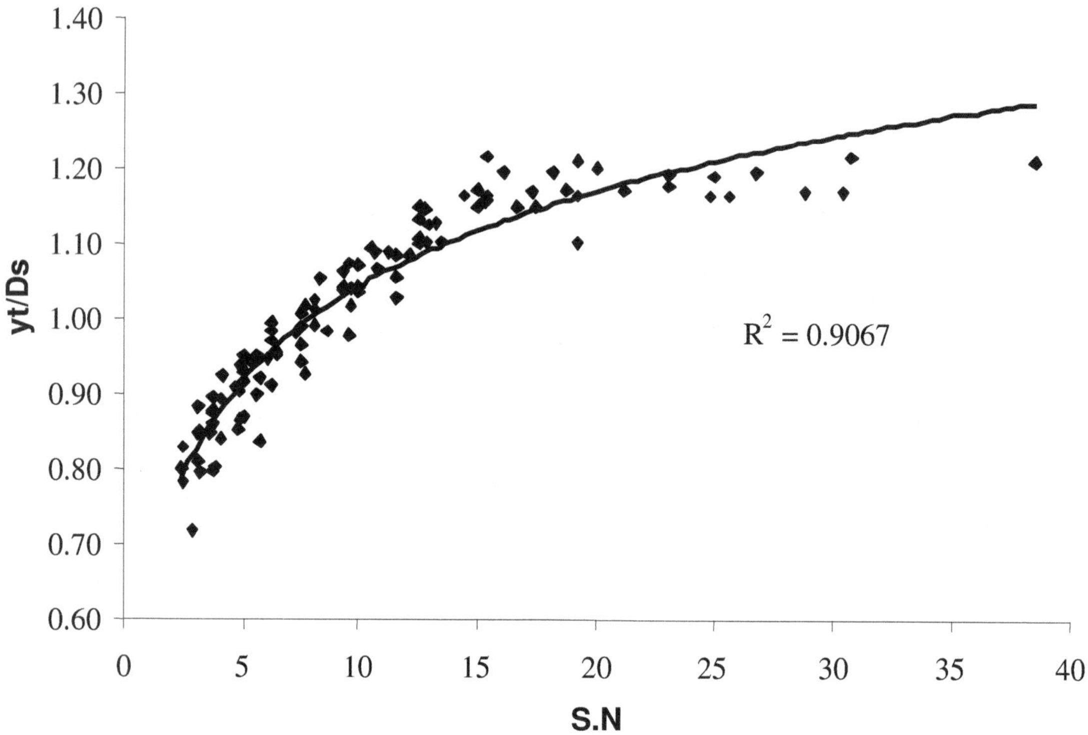

Figure 8. y_t/Ds verses S.N for all degrading levels (Z).

based on same parameters in Equation 14 (that is, Z affects the correlation indirectly) and following coefficients for (a) and (m) were calculated.

$$S.N = 0.68\left(\frac{y_t}{D_s} \right)^{0.18} \qquad (15)$$

The graphical illustration of Equation (15) is shown in Figure 8. For a particular riprap, decrease of S.N (that is, decrease of flow velocity) with tail water is observed in Figure 8. The suggested coefficients of (a) and (m) for the leveled stilling basins from previous studies in Brown and Clyde (1989) are compared with calculated coefficients of this study in Table 5. A considerable variation of coefficients with previous study confirms significant effect of degrading on increase of S.N. in this study. This means that scouring is reduced with increase of (Z) for particular tail water.

Table 5. (a) and (m) coefficients for Equation

Researchers	*m*	*a*
Isbash	0	1.7
Straub	0.17	1.49
Neill	0.1	1.58
Bogardi	0.095	1.7
Maynord	0.1	3.33
Current Study	0.18	0.68

Conclusion

Results of this study reveal considerable effect of degradation on scouring of channel bed downstream of the stilling basins. This finding disqualifies application of previous methods and equations which by large have been developed for condition of leveled basin floor with the downstream channel bed. A linear relationship proposed for scouring and estimation of size for stable riprap. Alternatively, nonlinear relations of the involved dimensionless parameters are presented graphically which can be used to estimate size of riprap based on available filed information. Since estimation of velocity near bed in S.N is difficult in practice, Figure 7 can be used to estimate the stable stone size based on critical flow depth. However, to some extends results reveal a decrease of scouring as degradation level (Z) increases. In general, previous findings such as increase of scouring with decrease of tail water and increase of discharged are confirmed in this study.

ACKNOWLEDGEMENT

The authors would like to acknowledge Shahid Chamran University of Ahwaz, Iran for financial support of the research (Grant No. 696) and the Research Institute of the Ministry of Power of Iran for facilitation of the experiments.

REFERENCE

Agostino D, Ferro V (2004). Scour on Alluvial Bed Downstream of Grade-Control Structures J. Hydraulic Eng. 130 (1): 24-37.

Brown SA, Clyde ES (1989). Design of RipRap Revetment. Hydraulic Engineering Circular No, 11, Federal Highway Administration. McLean, Va., USA p156.

Chanson H (2008). Hydraulics of Open Channel Flow: An Introduction – Basic Principles, Sediment Motion, Hydraulic Modeling, Design of Hydraulic Structures, 2nd ed., Butterworth Heinemann pub., Oxford, UK, p 650.

Farhoudi J, Valizadeghan E, (2004). "Bed protection criterion downstream of stilling basins", Proc. 9th ISRS conference, China.

Lauchlan CS, Melville BW (2001). Riprap Protection at Bridge Piers J. Hydraulic Eng. 127 (5): 412-418.

Maynord ST, (1991). "Flow resistance of riprap", J. Hydraulic Eng., ASCE, 117 (6).

Rice CE, Kadavy KC (1991). "Riprap design downstream of straight drop spillways", Trans. ASCE 34(4): 7-12.

Rice CE, Kadavy KC (1992). "Riprap design for SAF stilling basins", Trans. ASCE, 35 (6): 1817-1825.

Shafai-Bajestan M, Albertson ML (1993). "Riprap criteria below pipe outlet", J. Hydraulic Eng., ASCE 119 (2): 181-200.

3

Integration of HEC-RAS and geographical information system (GIS) in the hydrological study of peak flow response to deforestation on a small watershed in Malaysia

Ali M. F., Kamarudin A. F., Khalid K. and Abd Rahman N. F.

Flood-Marine Excellence Center, IIESM, Faculty of Civil Engineering University Teknologi MARA, Malaysia.

Land use change is one of the most serious environmental and hydrological problems nowadays. It has become a global issue including Malaysia due to the greenhouse effect. Land use changes cause a significant adjustment of the hydrological balance of a watershed mainly by eliminating evapotranspiration, redistributing and modifying the amount of rainfall that reach the ground. A 20 years historical data were used to represent a geographical and hydrological data for a Kinta watershed, in the state of Perak. The geographical data were processed using geographical information system (GIS) to show the land use change in the study area. The hydrological data were analyzed through regression analysis, flow duration curve, and soil conservation service (SCS) method for analyzing runoff for land use change. The regression analyses have found significant increase of r^2 value from 3.1% between year 1990 to 1996, 7.5% between year 1997 to 2000, and 13.2% between the years of 2001 to 2006. The flow duration curve has selected streamflow data event on December 28 to 29, 1996 on the Kinta River at outlet of Kinta watershed result which has shown volume of streamflow that was 2.209×10^6 m^3. Meanwhile the direct runoff was 8.979 mm from volume divided by area of watershed 246 km^2. The ϕ index calculated was 28.144 mm/h and $\phi\Delta t$ was 28.144 mm. Land use analysis shown result for runoff depth and S, potential max retention followed by CN for each year. The result showed increase of runoff depth and decrease of S, potential maximum retention. These findings have proved that the changes in land use in terms of the increasing deforestation activities or increasing new development will give a crucial impact to the hydrology response.

Key words: Land use change, flow duration curve, soil conservation service (SCS) method, runoff depth, geographical information system (GIS).

INTRODUCTION

Land cover which remains constant for long periods will reveal equilibrium between various human activities, natural hazards and climate. In some areas, land cover undergoes progressive to drastic changes more frequently or over a larger expanse. The changes usually relate to economic and social response such as agricultural development, timber harvesting or urbanization and followed by the changes influenced by watershed hydrology (Savary et al., 2009). Scanlon et al. (2007) revealed deforestation for agricultural or urban

Figure 1. The hydrology station in Tanjung Rambutan.

purposes that can affect local water balance. Those influences still remain difficult to be quantified until the recent developments in decision support systems based on geographical information system (GIS), remote sensing, and distributed hydrological models. These developments have led to practical and useful tools due to hydrological responses of watershed (Fohrer et al., 2001). In addition to the natural heterogeneity in the physical characteristics of the catchment, the distribution and disposition of different land uses may hugely influence catchment response by modifying the connectivity and continuity of drainage lines (Schulze, 2000). Removal of vegetation in general causes reduction of interception diminishing transpiration and boost up soil water content. Harvesting maneuver locally decrease infiltration rates due to soil compaction and may increase transit to the stream network of water conveyed in road ditches and ruts. Removal of vegetative cover causes changing of the hydrological cycle, mainly by eliminating evapotranspiration and redistributing and modifying the amount of rainfall that reaches the ground, thus increasing the surface runoff (Pizarro et al., 2006). As removing vegetation cover causes important changes in the hydrological balance of a watershed, afforestation and reforestation generate opposite effects on flows, although the magnitude of the response is highly variable and unpredictable (Hibbert, 1967).

Runoff can be an important component of the water balance (Lal, 1991). For hydraulic designs on very small watersheds, a complete hydrograph of runoff is not constantly compulsory. As a result, a number of methods for approximating a design discharge is the maximum value of the flood runoff hydrograph has been developed (David, 2006). Knebl et al. (2005) was successfully modeling the San Antonio river basin for flood study by the using of Hydraulic Engineering Centers-River Analysis System (HEC-RAS). The model is capable of being used to calculate one-dimensional steady and unsteady flow. Shamsi (2005) highlighted that GIS applications have the potential to enhance the management of water, wastewater, and stormwater systems. Dangermond (2010) agreed with the functions of the GIS for hardware, software and data integration.

Objectives

The study covers three main objectives. The first objective is to determine the runoff characteristics of the annual hydrograph of the stream. Secondly, this research analyzes the annual peak flows between pre and post development in a study area. The last but not least is to assess the influence of land use changes on the hydrological regime for a small watershed.

MATERIALS AND METHODS

Study area

The study has been carried out in a small watershed at Kinta, Perak. The area located in Tanjung Rambutan near to Korbu Mountain which has deforestation activities. The main river in the area is Kinta River. The major land use change is the dam construction of Kinta River. A commissioning date of dam construction of Kinta River was issued on 9 March, 2004 and the completion date was 29 June, 2009. The dam was located an approximately 12 km from the Ipoh centre to coordinate of 4° 40' 2" N and 101° 13' 5"' E. A set of hydrology data was obtained from the Department of Irrigation and Drainage (DID) hydrology station located in Tanjung Rambutan and coordinates of station point is at (4°67' N and 101° 16' E) as shown in Figure 1.

Data sets

Data of discharge stream flow, rainfall and water level for 20 years historical data from the year 1990 to 2009 were gained from the Department of Irrigation and Drainage (DID). Sets of soil and land use data for the year 1990, 1997, 2000 and 2006 were obtained from the Department of Agriculture (DOA). A topographical map of the study area was obtained from the Department of Surveying and Mapping (JUPEM). Figure 2 shows the classification of soil types in 2006 meanwhile Figure 3a, 3b, 3c and 3d are the land used maps with the sequences of year 1990, 1997, 2000 and 2006.

Approach

This research is a computer based study which involves two crucial data sets; a hydrological and geographical data. Geographical data sets will go through the digitizing processes using GIS tools before it is ready to embed in the HEC-RAS model.

The next step is a daily hydrological simulation using HEC-RAS. Regression analysis is used to detect a relation between the values of two or more variables, of which at least one is subject to random variation, and to test whether such a relation, either assumed or calculated, is statistically significant. It is a tool for detecting relations between hydrologic parameters in different places, between the parameters of rainfall and discharge. The analyses

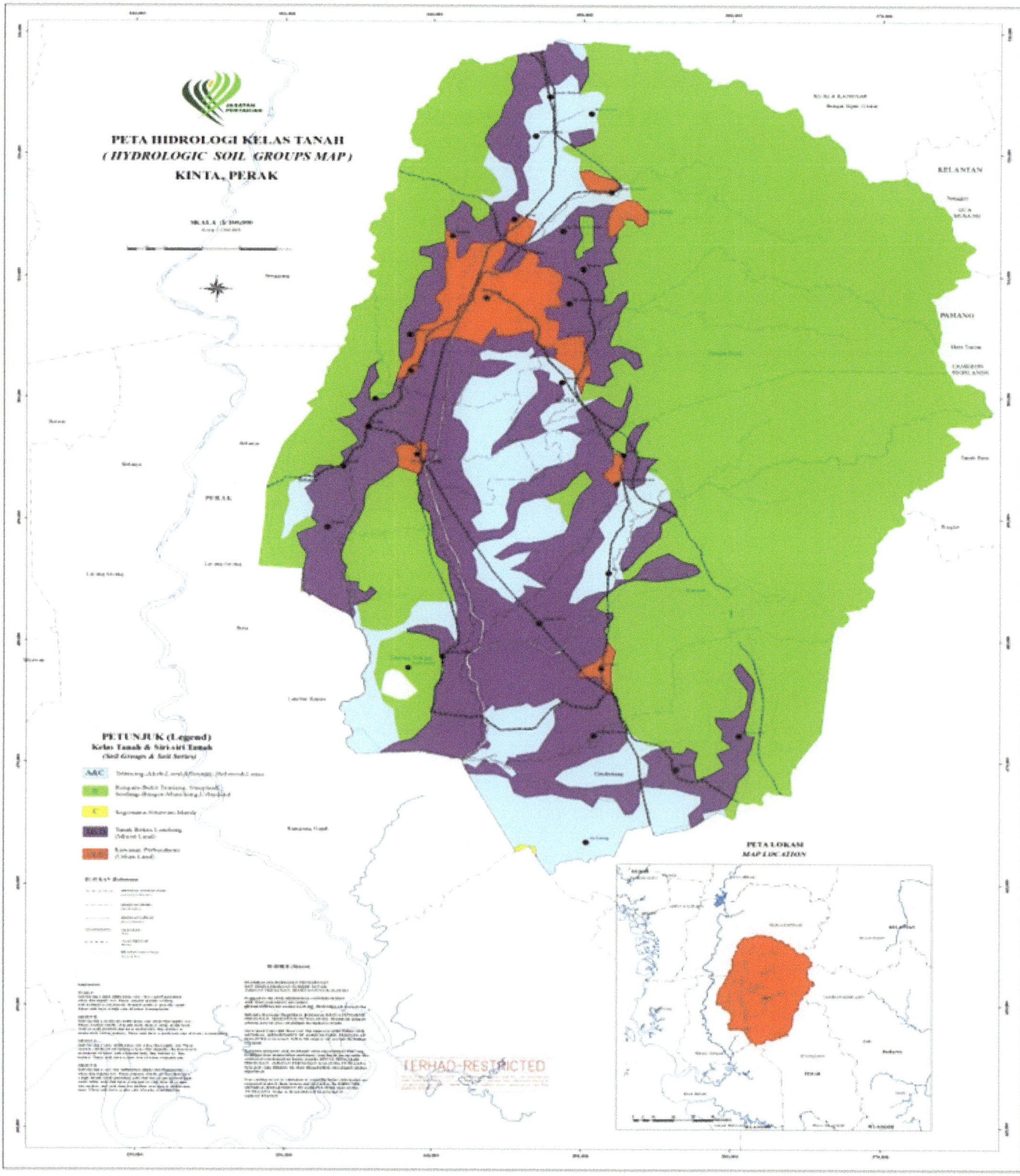

Figure 2. The types of soil at Kinta watershed on 2006.

established are between year 1990 to 1996, 1997 to 2000, and 2001 to 2006.

The flow duration curve is a plot that shows the percentage of time that flows in a stream is likely equal or exceeds some specified

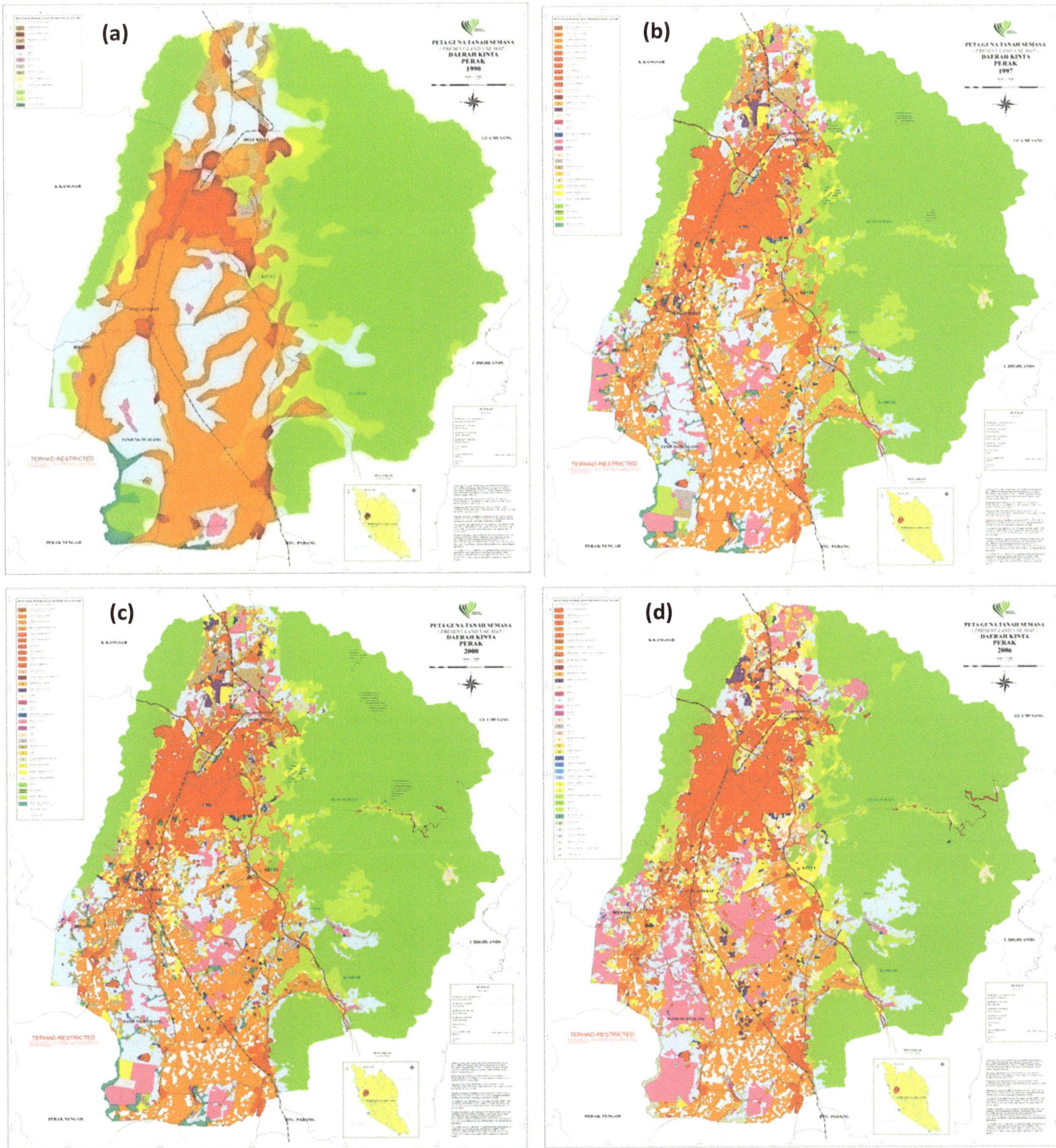

Figure 3. (a) Land used in 1990, (b) Land used in1997, (c) Land used in 2000, (d) Land used in 2006.

value of interest. For example, it can be used to show the percentage of time river flow expected to exceed a design flow of some specified value. On the contrary, it can also show the discharge of the stream that occurs or exceeded in certain percentage of the time. A flow duration curve is a plot of discharge versus percent of time that a particular discharge was equaled or exceeded. The excess rainfall hyetograph determined from the hyetograph in one of two ways, on whether streamflow data event on December 28 to 29, 1996 on the Kinta River at outlet of Kinta watershed. Baseflow has been separated from the streamflow to

Table 1. Land use classification in year of 1990, 1997, 2000, and 2006.

No	Land use type	1990 (km^2)	1997 (km^2)	2000 (km^2)	2006 (km^2)
1	Aquaculture	*	3.153	3.256	3.628
2	Development	88.154	180.221	171.822	201.845
3	Diversified crops	28.110	51.782	43.198	39.309
4	Ex-Tin Mining	*	279.656	129.155	190.753
5	Ex-Quarry	*	*	*	9.612
6	Forest	833.757	812.620	810.362	799.951
7	Lake/Pond	*	72.700	87.563	80.864
8	Limestone Hill	*	*	*	10.543
9	Market Gardening	*	9.461	10.984	14.309
10	Mangrove	*	*	18.359	17.188
11	Mixed Horticulture	36.425	46.003	47.528	49.432
12	Neglected Grassland	19.908	66.568	68.509	78.147
13	Newly Cleared land	1.441	9.800	0.112	16.91
14	Oil Palm	10.814	77.242	123.458	158.357
15	Orchard	6.437	12.935	8.717	16.601
16	Rubber	305.587	207.536	207.976	147.006
17	Secondary Forest/Scrub	152.51	117.755	104.863	93.456
18	Tin Mining	455.820	20.355	118.445	10.543
	Total area	**1938.963**	**1967.787**	**1955.368**	**1938.454**

produce the direct runoff hydrograph in order to determine excess rainfall hyetograph. The parameter of infiltration used in this study was ϕ index (rainfall index). The ϕ index is a constant rate of abstractions (mm/h) that will yield an excess rainfall hyetograph (ERH) with a total depth equal to the depth of direct runoff over the Kinta watershed. The value of ϕ is determined by picking a time interval length, Δt judging the number of intervals M of rainfall that actually contribute to direct runoff, subtracting $\phi\Delta t$ from the observed rainfall in each interval, and adjusting the value of ϕ and M as necessary so that the depths of direct runoff and excess rainfall.

In the soil conservation service (SCS) method, the amount of the total runoff is affected by soil type, land use or plant cover, cover treatment, and the hydrological characteristic of the soil within the watershed boundary. Quantitative measurements for the runoff affected by these factors are suggested in the SCS method. The relationship between the total precipitation and the total runoff is suggested as shown in Equation 1.

$$Q= [(P-0.2S)2/(P+0.8S)] \quad (1)$$

Where:
Q = the total runoff in millimeters (mm)
P = the total precipitation in millimeters (mm)
S = the potential maximum retention of a site

The terms S and I_a are both affected by the character of the land cover and soil condition. Thus, the relationship is verified in the following equation:

$$I_a = 0.2S \quad (2)$$

S, the potential maximum retention is a function of such variables as soil, plant cover, the amount of impervious area and runoff curve number (CN) is made after a detailed analysis to model these effects for various soil types and land cover complexes – hydrologic land cover complexes. The CN for a given set of surface cover and soil condition is used to estimate the value of S (SCS, 1986).

$$S= 1000/CN - 10 \quad (3)$$

Classification area of land use has done by using GIS tools. The processes started with a conversion of the different land use of the area via polygon shape in shape file and followed by a calculating of a land use area using a geometry calculator as shown in Table 1. The information area of land use was extracted from the attributes. Further investigation has used SCS method for estimating total runoff (Table 2).

RESULTS AND DISCUSSION

Between 1990 and 2006, land use type of oil palm and development in the Kinta catchment exhibited progressive changes as highlighted in Table 1.

In 1990, 26.11% of the catchment was under cultivated activity, 66.51% of forest, 5.94% of development area and 1.44% of pasture as further summarized in Table 2. In 1997 the development activity has raised to be double compared with 1990. Pasture has risen to 4.88%. Others land use has reduced in cultivated by 0.67% and forest reduced by 11.63 to 54.88%. During the period 1997 to 2000 as shown in Figure 3c, there is a notable decline in development by 10.75%, pasture 4.30% and forest continue to decrease by 57.30%. The cultivated area is increased by 27.65%. In 2006, a cultivated area decreased to 26.31% and the forest remains a momentum of decrease by 55.31%. Development and pasture have risen to 12.50 and 5.88%, respectively.

Development activities will reflect the changes in streamflow rate on the watershed. Figure 4 shows the

Table 2. Total runoff curve number for year 1990, 1997, 2000 and 2006.

Land use	Hydrology soil group		
	%	CN	Product
	Year 1990		
Cultivated Land	26.11	82	2141.02
Forest Land	66.51	66	4389.66
Development	5.94	92	546.48
Pasture	1.44	79	80.44
Total	**100**		**7157.6**
	Year 1997		
Cultivated Land	26.11	82	2141.02
Forest Land	66.51	66	4389.66
Development	5.94	92	546.48
Pasture	1.44	79	80.44
Total	**100**		**7157.6**
	Year 2000		
Cultivated Land	25.44	82	2086.08
Forest Land	58.44	66	3857.04
Development	11.32	92	1041.44
Pasture	4.80	79	379.2
Total	**100**		**7363.76**
	Year 2006		
Cultivated Land	26.31	82	2157.42
Forest Land	55.31	66	3650.46
Development	12.50	92	1150
Pasture	5.88	79	4645.52
Total	**100**		**7422.4**

scatterplot rainfall versus the streamflow for three interval development. The overall trend of the 16 years duration of study shows a fair consistency with a slight positive trend of the streamflow. For the year 1990 to 1996 it has indicated a value of r^2 by 3.1% and equal P-value at 0.107. Intermediate period (Figure 4b) showed positive increased of r^2 by 7.5% and 0.06 in P-value. The value of r^2 has further increased from 2001 to 2006 by 13.2% but the P-value is decreased by 0.002. Although the reason of the increase of streamflow is uncertain, these data suggested that it is related to the increase in the percentage of developed land.

Excess rainfall and direct runoff

The excess rainfall hyetograph was determined from the hyetograph in one of two ways, on streamflow data event on December 28 to 29, 1996 on the Kinta River at outlet of Kinta watershed. A baseflow has been separated from streamflow to produce the direct runoff hydrograph in order to determine excess rainfall hyetograph. A constant baseflow rate of 3 m^3/s was selected in the study. The result in Figure 5 shows the volume of streamflow to be 2.209×10^6 m^3. Meanwhile the direct runoff is 8.979 mm from volume divided by area of watershed of 246 km^2. The calculated ϕ index is 28.144 mm/h and $\phi\Delta t$ is 28.144 mm.

The result shows increasing runoff depth and decreasing S, potential maximum retention. These results proved that the changes in land use in terms of increasing deforestation activities or increasing new development give a crucial impact to the hydrology response.

Estimating total runoff depth

The result of runoff depth and potential maximum retention (S) after considering the CN value for each year is shown in Table 3. The result shows the increasing runoff depth and the decreasing S, potential maximum retention. In the year 1990 to 1997, results showed the increasing runoff depth by 4.7% and the decreasing S by 10.89%. Results on the second interval from year 1997 to 2000 showed runoff depth increasing by 0.02 mm or 0.35% and S was reduced to 0.03 mm or 0.84%. The increasing trend of runoff depth was also observed on the last interval of 0.05 mm and S has decreased by 0.08 mm or 2.3%. These results proved that the changes in land use in terms of the increasing deforestation activities or increasing new development will give a crucial impact to the hydrology response.

Flow duration – curve

The effect of varying the time unit is not the same for all streams. A daily streamflow of Kinta River observed was almost uniform and the daily curves were also nearly identical. Once a smooth curve was drawn, the mean discharge for the selected day time unit was used as the discharge was grouped into the class interval from 1990 to 1999. Frequency distribution for class interval was plotted as number of days within a specified discharge interval against discharge interval to determine the number of occurrences for each class. The result in Figure 6a shows the highest number of occurrences that was in the class interval 2.66 to 5.33 m^3/s. The data then was combined by plotting a curve of the cumulated days of the 20 equal classes against the discharge for the entire period of study as in Figure 6b and followed by the flow duration curve for 20 equal classes as in Figure 6c. The flow duration curve as shown in Figure 6d was a cumulative frequency curve that shows the percent of time during which specified discharges were equaled or exceeded in period 1990 to 1999. In this study, on the year 1990 to 1999, the daily mean flow of Kinta River was at least 30 m^3/s during 90% of the time. The flow duration

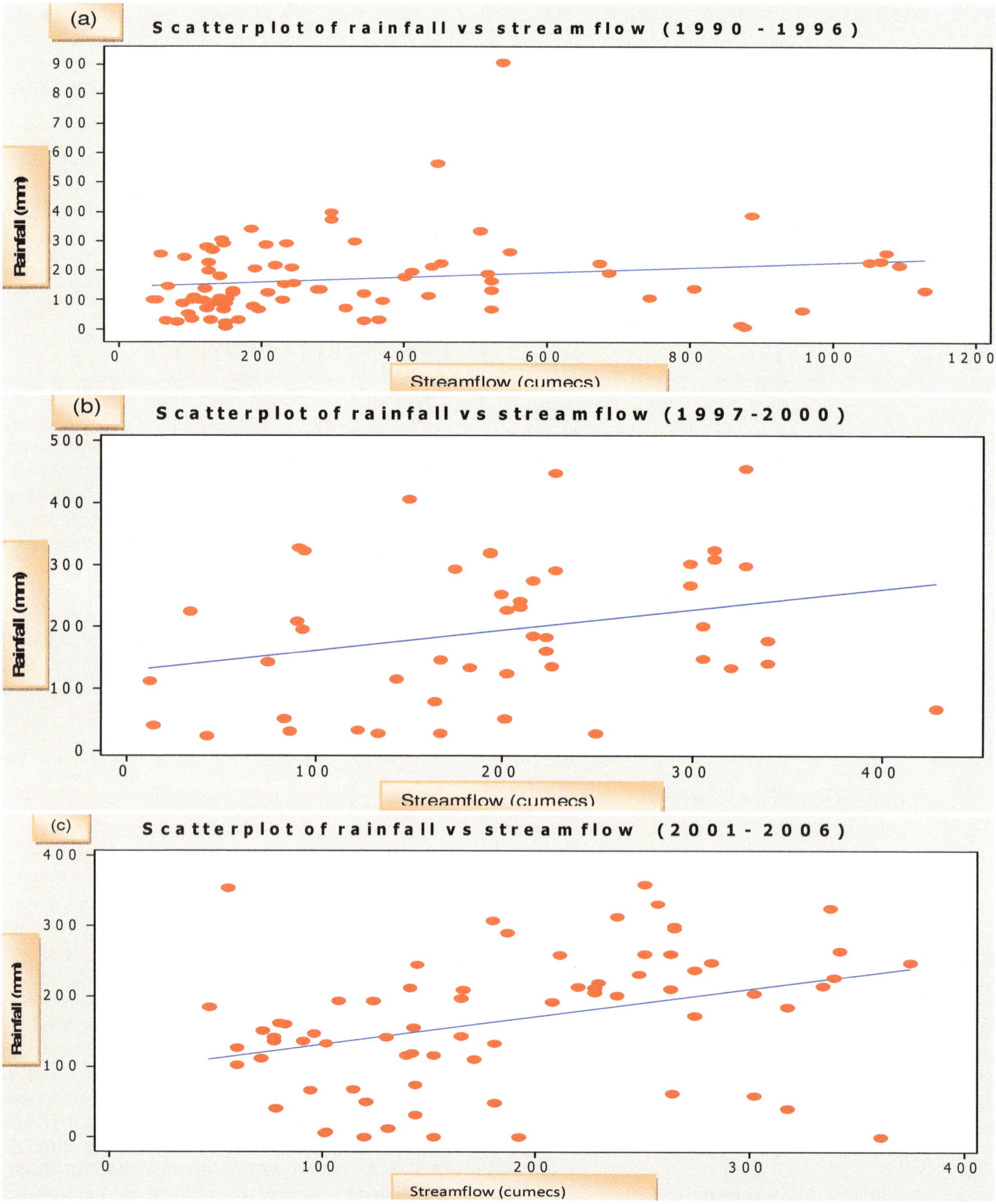

Figure 4. Scatterplot rainfall vs streamflow between year (a) 1990 to 1996 (b) 1997 to 2000 and (c) 2001 to 2006.

curve is the integral of the frequency diagram. Simpler concept of the flow duration curve is that, it is another way of presenting streamflow data combining in one curve and the flow characteristic of a stream throughout the range of discharge. This research has selected a daily flow as a time unit. The most important use of such

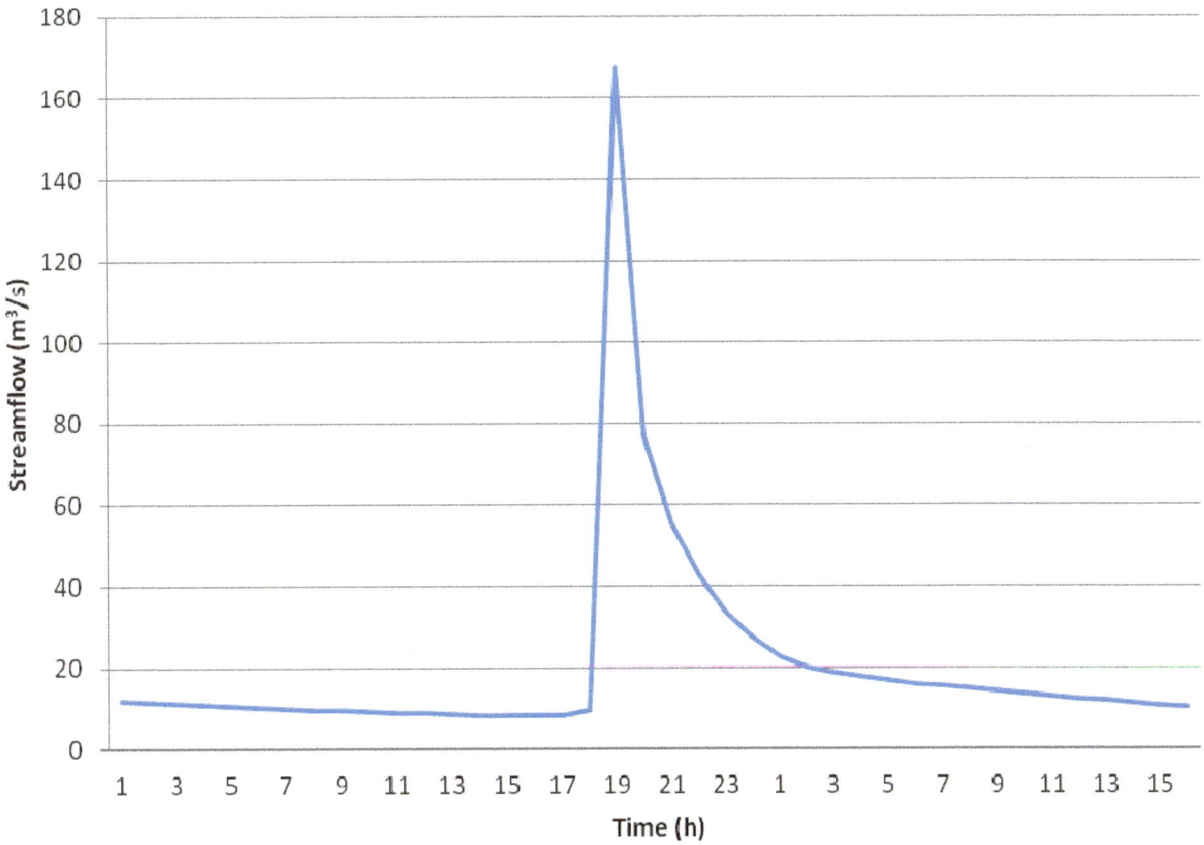

Figure 5. Streamflow for the storm event of December 28 to 29, 1996, on Kinta River.

Table 3. The effects of Land use change on Runoff Depth (mm), Curve Number and Potential Maximum Retention.

Parameter	Year			
	1990	1997	2000	2006
Runoff Depth (mm)	5.51	5.77	5.79	5.84
CN	71.58	73.64	73.78	74.22
S (Potential max retention)	3.97	3.58	3.55	3.47

duration curves is to show the characteristics of the flow. The details of the variations in flow are obscured of the time unit is long. For most of the streams, the monthly discharges are unsatisfactory for showing the variation in flow, although duration curve of annual mean discharges data exists but their range in variation is comparatively small and only a few values are available for defining the curve. A study by Foster (1934) in North Carolina showed differences as great as 35% between a duration curve based on monthly mean discharges and one based on daily mean discharges. Weekly mean discharge have been used in the study (Saville et al., 1933) for certain relatively stable streams. The selected daily time unit was used as the discharge that was grouped into the class interval from 2000 to 2006. Figure 7a shows class interval of 2.53 to 5.07 m^3/s for the highest number of occurrences. The class interval was provided with 20 equal distribution points of the curve. The data were combined by plotting a curve of the cumulated days against the discharge for the entire period (2000 to 2006) that was computed by one set of class intervals (Figure 7b). The flow duration curve for 20 equal classes is shown in Figure 7c. The flow duration curve as in Figure 7d was a cumulative frequency curve that shows the percent of time during which the specified discharges were equaled or exceeded in period 2000 to 2006. In this study, a daily mean flow of the Kinta River in the year of 2000 to 2006 was at least 15 m^3/s during 90% of the time. Figure 7d showed data computed from daily discharge in the period 2000 until 2006.

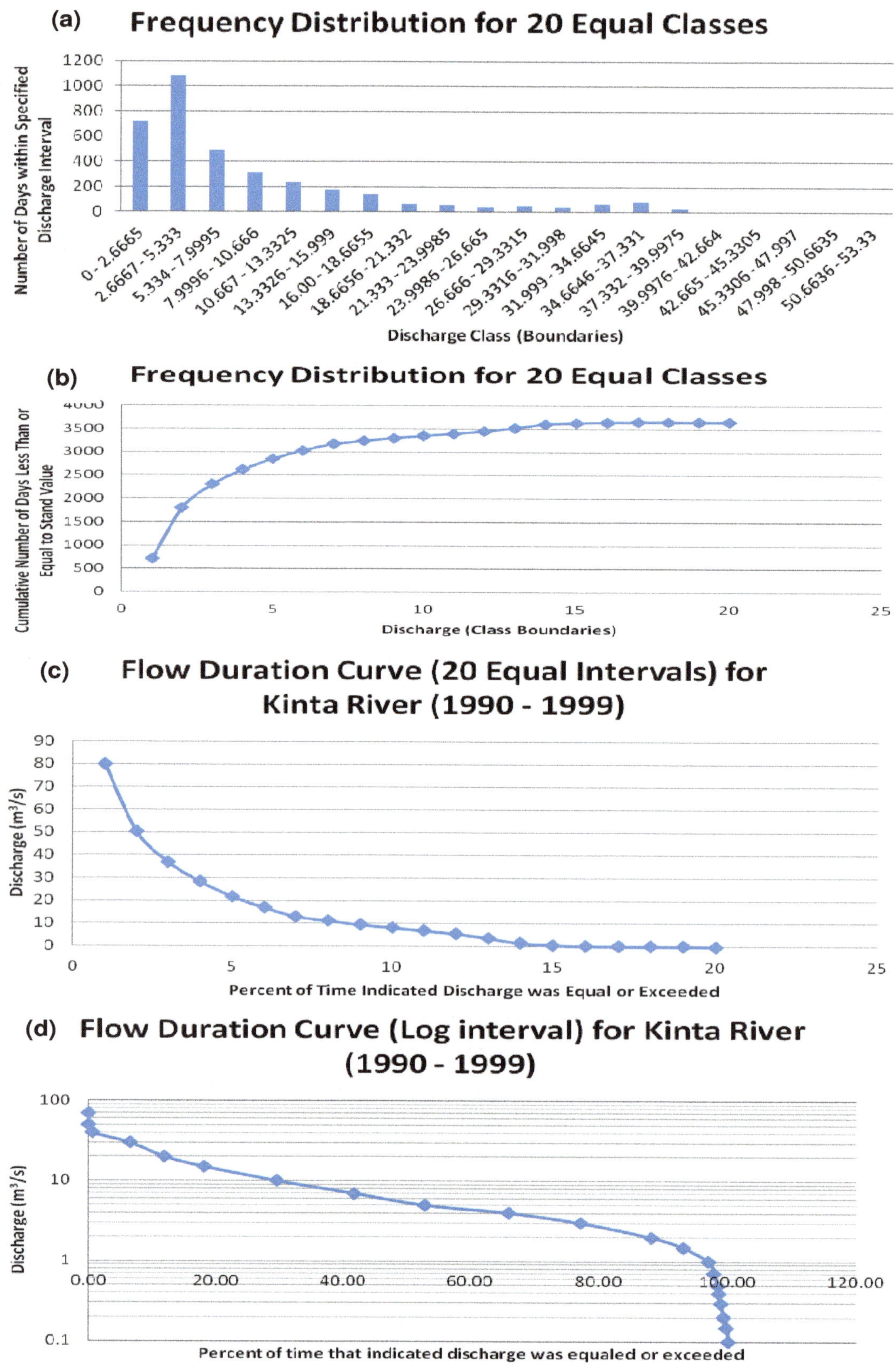

Figure 6. (a) Number of occurrence for each class for year 1990 to 1999, (b) Cumulative number of days for year 1990 to 1999, (c) Percent of time indicated discharge was equal or exceeded for year 1990 to 1999, (d) Percent of time indicated discharge was equal or exceeded for year 1990 to 1999.

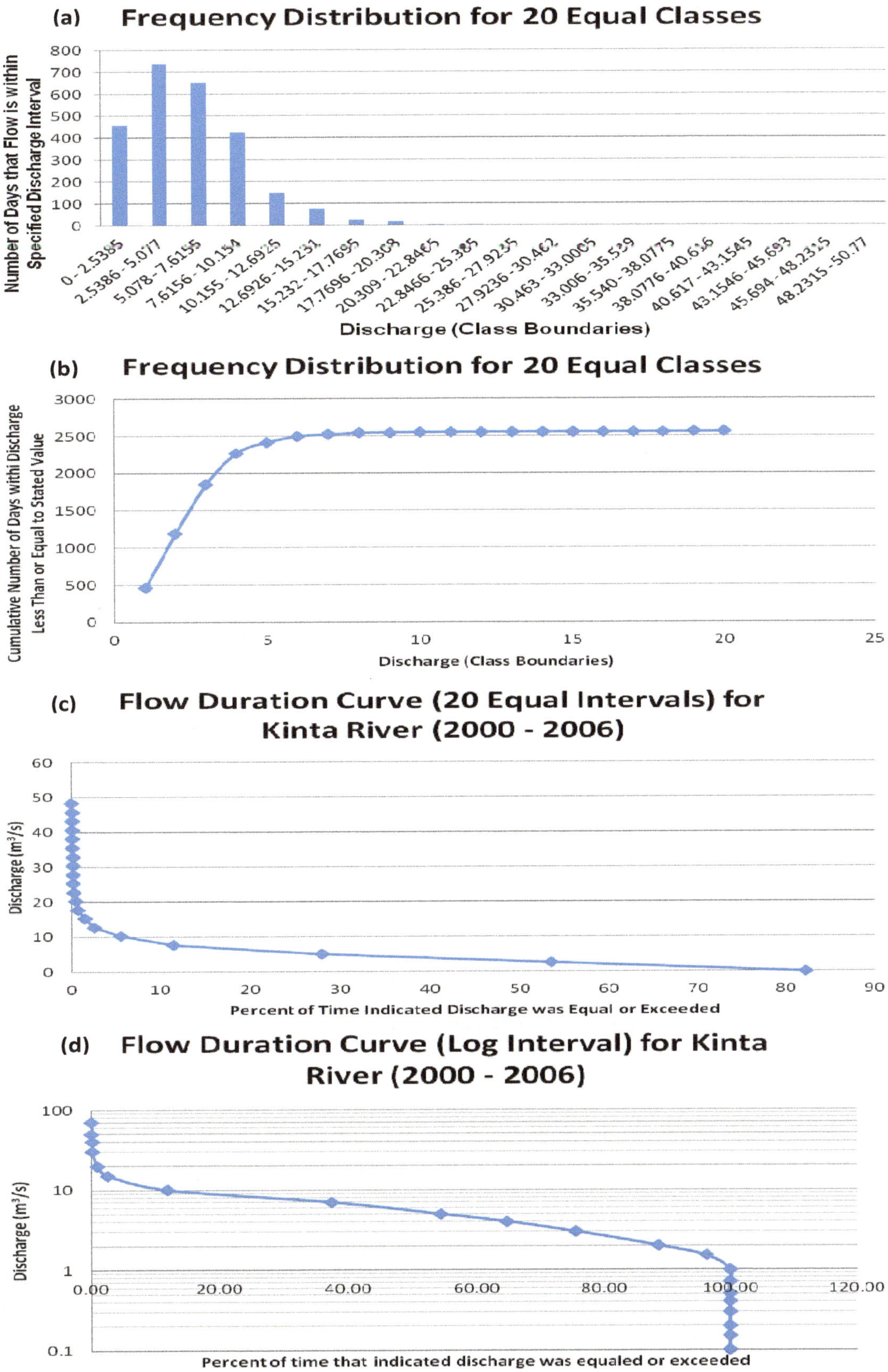

Figure 7. (a) Number of occurrence for each class for year 2000 to 2006, (b) Cumulative no. of days for year 2000 to 2006, (c) Percent of time indicated discharge was equal or exceeded for year 2000 to 2006, (d) Percent of time indicated discharge was equal or exceeded for year 2000 to 2006.

Conclusion

Contributions of land use change to that of major hydrological components in the Kinta watershed were evaluated using a combination of total runoff curve number and regression analyses. Major changes of land use on reducing forest area and increasing the development activities lead to increase of CN value followed by the increasing surface runoff and decreasing the potential maximum retention. This trend shows the changes of land use responses to hydrology regimes. Therefore, the precipitation was not the major contributing factor, thus when precipitation condition was kept constant, the runoff variations were mainly coming from land use change of watersheds or land use periods, which also need to be further studied.

In this study, the annual hydrograph analyses, the previous 20-years' hydrology data and the trend or the runoff characteristic of annual hydrograph has been verified as perennial stream. The relationship between precipitation and discharge rate affecting each other were not proven. A few factors need to be considered to make the conclusion more accurate and trustworthy. The insufficient data at station point of precipitation is the main concerning issue. The collecting station point of precipitation needs more than one station data and determining the precipitation will flow into Kinta River directly or flow to tributary before entering the Kinta River. The other factor that needs to be considered is the shape of the basin itself. The profile of the basin influences the time taken for water from the remote parts of the catchment to arrive at the outlet. Thus the occurrence of the peak and hence the shape of the hydrograph are exaggerated by the basin shape. In addition, size of basin is also another factor. Small basins act differently from the large ones in terms of the relative importance of various phases of the runoff phenomenon. In small catchments the overland flow phases are main over the channel flow. Therefore, the land use and intensity of rainfall have a significant role in the peak flood. On large basins these effects are covered up as the channel flow phase is more predominant. The basin slope is important in small catchments where the overland flow is relatively more imperative. In such cases the steeper slope of the catchment outcome in larger peak discharges. The slopes of the main river control the velocity of flow in the channel. As the recession limb of the hydrograph correspond to the depletion of storage, the stream channel slope will have a pronounced effect on this part of the hydrograph. Large stream slopes increase depletion of storage and hence results in steeper recession limbs of hydrographs. This would clearly result in a smaller time base. This study is focusing on the effects of deforestation activity that is related to land use. The vegetation and forests raise the infiltration rate and storage capacities of the soils. Additionally, they cause considerable retardants to the overland flow. Thus the vegetal cover decreases the peak flow. This effect is generally very pronounced in small catchment of area less than 150 km^2. Furthermore, the effect of the vegetal cover is prominent in small storms. In general, for two catchments of equal area, other factors being identical, the peak discharge is superior for a catchment that has a lower density of forest cover. Of the range of factors that control the peak discharge, probably the only factor that can be manipulated is land use, thus it represents the only practical means of exercising long-term natural control over the flood hydrograph of a catchment.

ACKNOWLEDGEMENTS

The project was supported by the Exploratory Research Grant Scheme (ERGS), Ministry of Higher Education, and Universiti Teknologi MARA (UiTM), Malaysia. Additional thanks to the Department of Irrigation and Drainage, Department of Agriculture, and Department of Surveying and Mapping, Malaysia for providing climatic and geospatial data.

REFERENCES

Dangermond J (2010). Geographical information system. Environmental System Research Institute, Redland, CA. Retrieved from http://www.esri.com.

David BT (2006). The rational method, Civil Engineering Department, Texas Technology University (In PDF).

Fohrer N, Haverkamp S, Eckhardt K, Frede H (2001). Hydrologic response to land use changes on the catchment scale. Physics and Chemistry of the Earth, Part, B: Hydrol. Oceans Atmos. 26:577-582.

Foster HA (1934). Duration curves: Am. Soc. Civil Engrs. Trans. 99:1213-1267.

Hibbert A (1967). Forest treatment effects on water yield. In: Proceedings of a National Science Foundation Advanced Science Seminar, International Symposium on Forest Hydrology. Pergamon Press, Estados Unidos, pp. 527-543.

Knebl MR, Yang ZL, Hutchison K, Maidment DR (2005). Regional scale flood modeling using NEXRAD rainfall, GIS, and HEC-HMS/RAS: A case study for the San Antonio River Basin Summer 2002 storm event. J. Environ. Manage. 75(4):325-336.

Lal R (1991). Current research on crop water balance and implications for the future. In: Sivakumar MVK, Wallace JS, Renard C, Giroux C (Editors), Soil water balance in the Sudano-Sahelian Zone, Proceedings of an International Workshop, Niamey, Niger. IAHS Press. Inst. Hydrol. Wallingford, UK pp. 31-44.

Pizarro R, Araya S, Jordan C, Farias C, Flores JP, Bro PB (2006). The effects of changes in vegetative cover on river flows in the Purapel river basin of central Chile. J. Hydrol. 327:249-257.

Savary S, Rousseau AN, Quilbé R (2009). Assessing the effects of historical land cover changes on runoff and low flows using remote sensing and hydrological modeling. J. Hyd. Eng. 14(6):575-587.

Saville T, Watson JD (1933). An investigation of flow-drtition characteristics of North Carolina streams. Am. Geophys. Union Trans. pp. 406-425.

Scanlon BR, Jolly I, Sophocleous M, Zhang L (2007). "Global impacts of conversion from natural to agricultural ecosystems on water resources: Quantity versus quality." Water Resour. Res. 43(3):1-18.

Schulze R (2000). Transcending scales of space and time in impact studies of climate and climate change on agrohydrological responses. Agric. Ecosyst. Environ. 82:185-212.

Shamsi UM (2005). GIS applications for water, wastewater, and stormwater system. CRC Press, 2000 N.W. Corporate Blvd., Boca Raton, Florida 33431.

4

Simulation of interaction of side weir overflows with bed-load transport and bed morphology in a channel (SSIIM2.0)

Samane Arvandi[1]*, Amir Khosrojerdi[2], Mohammad Rostami[3] and Hosein Baser[4]

[1]Department of Water Science, Islamic Azad University, Science and Research Branch, Tehran, Iran.
[2]Soil Conservation and Watershed Management Research Institute, Tehran, Iran.
[3]Department of Water Engineering, Amirkabir University of Technology, Tehran, Iran.
[4]Master of Water Engineering, Amirkabir University of Technology

To describe the complexity of flows accurately around side spillways, a 3D numerical flow model has been tested. In this study 3D numerical model SSIIM2.0 is used to simulate unsteady flows in side spillways in a schematic test channel. Side channel spillways are one of the types of outlet works at dams with wide applications in irrigation, drainage systems, in water and wastewater facilities. Overflow side, known as cross overflow and overflow sections as a free overflow and flow diversion devices are used in hydraulic engineering. They are made in the side channels or rivers to overflow part of discharge's crown at the top of the main level. When accidental drop of water happen, sediment transport capacity in the main channel of investment and deposits alluvium for deposition in downstream reduced. Reduce levels, provides back water, additional expansion and contraction. So the height of the energy loss of the side spillway and overflow discharge increases. Design discharge to offset increases in the flow of sediment transport. The present confrontation with the overflow stream bed morphology and bed load in a regular channel experiments conducted has not been studied. Results showed that increase in spillway length turbulence in the flow is reduced and morphological changes decrease. Intensity of erosion and sedimentation over time has been increased, in the early minutes of calculations, the rate changes over time in bed was significantly reduced. The results demonstrate the ability of numerical models SSIIM2.0, to simulate the flow in the side spillway.

Key words: Overflow side, cross flow, bed morphology, sediment deposited, side spillway, 3D CFD model, – ε turbulence model.

INTRODUCTION

Considering the widespread destruction of the side channels offset overflows and floods that has been occurred, correction the method is necessary. So the design of such structures must be done correctly to reduce construction cost, repair and the annual reconstruction plans flood control. Construction significantly would reduce, If design of spillways in diversion of water spreading systems done according to engineering principles. Degradation that has occurred in the channel system of flood peripherals overflows or hydraulic conductivity of this system is the main motivation for this scheme. Using a mathematical model of the experimental model designed to achieve a basis consistent with the circumstances of this kind of structures can also reduce costs, increase correct understanding of its performance and increased design accuracy for each type of geometry.

Forecast during the transfer process within the main channel and side weir can be both experimental and computational research. Nowadays, numerical methods in computer calculations are important as an efficient tool in the design and reducing construction and engineering costs. In this study numerical model, SSIIM2.0, was used to simulate side spillway and sedimentation in the channel.

Fluid-sediment interactions control river channel forms and processes. Analysis of spatial hydraulic patterns and the resulting boundary shear stress are required to aid understanding of river system behavior. In this paper, the hydraulic processes active in a side channel spillways simulated using a three-dimensional computational fluid dynamics (SSIIM2.0) model.

Methods employed for the prescription of model boundary conditions are outlined. Model calculations are assessed using comparisons with field observations acquired over a range of flows. Simulations are then used to illustrate flow structures.

Side weirs, also known as a lateral weirs, and overflow dams are free overflow regulation and diversion devices commonly encountered in hydraulic engineering. They are set into the side of a channel or river allowing spilling a part of the discharge over their crest when the surface of the flow in the main-channel exceeds a certain level.

The lateral loss of water is reducing the sediment transport capacity in the main-channel yielding to aggradations and the formation of a local sediment deposit in the downstream weir alignment.

The reduced cross section generates backwater effects and additional contraction and expansion losses. As a consequence, the head over the side weir rises and the side overflow discharge as well. The design discharge to be diverted over the weir is increased by this flow-sediment transport interaction.

Since the interaction of side overflow with bed-load and bed morphology in a channel has not been studied so far, systematic tests have been performed.

Talebbeydokhti et al. (2006) stated that resistance to flow is an important and primary parameter in the determination of water surface elevation. A variety of bed forms, especially dunes, have a sensible effect on total roughness. Because of the complexity of bed form development, previous methods differ drastically from each other in predicting dune bed forms. In this paper, laboratory experiments were conducted to investigate the geometry of dunes in a sand-bed channel and its influence on total channel resistance. The experiments were performed in a flume in the hydraulic laboratory of Shiraz University using sand particles. Simple relations were sought for dune dimensions via some dimensional parameters, and previous methods were compared to each other in light of this new data. Thus, a new boundary condition has been introduced which has always been a free overall in the previous investigations.

The effective crest length has been taken into account introducing a correction factor for the true crest length. The main-channel geometry in experiments restricted to subcritical flow has been rectangular and the weirs crests have been sharp (7 experiments) and broad crested (15 experiments). The discharge coefficient was assumed to be a function of the approach Froude number (Fr_1). For the broad crested weir, an empirical being a function of $y1 - \omega D$ and crest width was added to the c_D-relation. The value $y1 - \omega D$ was invariably maintained greater than 10.00 cm to eliminate effects of viscosity and surface tension thatmight be important at small heads. With respect to the constant specific energy approach, a different between the up- and downstream weir corner of less than 2% has been observed. Thus, this assumption has been concluded to be reasonable. The investigations resulted in a design procedure to determine the discharge to be passed into a branch channel.

Imanshoar et al. (2012) stated that subsurface erosion in river banks and its details, in spite of its occurrence in various parts of the world has rarely been paid attention to by researchers. In this paper, quantitative concept of the subsurface bank erosion has been investigated for vertical banks. Vertical banks were simulated experimentally by considering a sandy erodible layer overlaid by clayey one under uniformly distributed constant overhead pressure. Results of the experiments indicated that rate of sandy layer erosion is decreased by an increase in overburden; likewise, substituting 20% of coarse (3.5 mm) sand layer bed material by fine material (1.4 mm) may lead to a decrease in erosion rate by one-third. This signifies the importance of the bed material composition effect on sandy layers erosion due to subsurface erosion in river banks.

Yalin and da Silva (2001) chose the dimensionless chezy friction factor (c) to establish a relation for the grain roughness. The main input parameters are the grain size Renlds number (Re*) and the relative flow depth (y/d). The resistance due to form roughness is expressed in terms of the bed form length and steepness.

In addition an experimental and numerical study investigating the interaction of a side overflow with a mobile bed has been performed at the libratory of Hydraulic Constructions (LCH) by Teiller (2000).

Willey et al. (2010) researched that a series of relatively small floods caused extensive rock erosion, approximately 5000 m^3, in the unlined section of the spillway channel at Googong Dam. A range of protective remedial works were installed during the 1980s with varying success. The most recent phase of work commenced in 2006 with a review of the spillway's performance, assessment of future erosion potential and a comparison of remedial works options. The detailed design was developed for the preferred option, comprising the retro-fitting of a concrete-lined chute, the raising and extension of the spillway chute walls and strengthening of other existing components. Construction is currently underway by the Bulk water Alliance and is due for completion in late 2010. This paper will present details of the history of this project including the initial assessment, review of rock erosion potential, options comparison and the detailed design.

Chiew (1991) stated that none of the classification diagrams refer to non-uniform bed material. For this reason, a classification method for bed features in non-uniform sediments has been proposed.

Interaction between surface and tectonic processes plays a key role in the structural evolution, kinematics, and exhumation of rocks in orogenic wedges. The deformation patterns observed in analog models show that strain partitioning has a strong impact on the vertical component of displacement of tectonic units, which in return favors erosion in domains of important uplift. Partitioning is controlled by tectonic processes and by climate-dependent surface processes, including erosion and sedimentation. The effects of partitioning include localization of deformed domains, exhumation above areas of deep underplating, and steady-state maintenance of wedges for long time periods. Simple models illustrate well how the morphostructural evolution of mountain belts is determined by these complex interactions (Malavieille, 2009).

Resuming, the literature review indicates that at the current stage of research lateral overflow on fixed bed conditions is well studied. The same accounts for expressions referring to the side weir discharge coefficient, sediment transport and bed morphology. Almost no investigations deal with the interaction of lateral overflow, sediment transport and bed morphology as a combined problem and no integral approach relating them to each other have been developed yet.

Numerical model description

The numerical model used in this study is the CFD code called SSIIM (Olsen, 2002). SSIIM2.0 is an abbreviation for Sediment Simulation in Intakes with Multiblock option. It solves the Reynolds-averaged Navier-Stokes equations with the two equation k-e turbulence closure in three dimensions to compute the water flow using the finite volume approach as discretisation method. SSIIM is based on the solution of the Navier-Stokes equations, with the k-e model.

This gives the water velocity and turbulence field which is used for solving the convection-diffusion equation for the sediment concentration. The model simulates water and sediment movement in a complex three-dimensional geometry. The model has a graphical user interface with pre and post processors. The program uses a standard graphical user interface for OS/2 and runs only on this operating system.

This paper shows several examples where the program has been used. The initial motivation for making SSIIM2.0 was the limited possibilities of determining the flow of finer sediment particles in a complex geometry.

Equations governing this phenomenon

Bed load and suspended load sediment transport is usually divided into two groups. Suspended load can be determined by the equation (Convection-diffusion) for sediment concentration (C). Generally transport equation for sediment concentrations case is according to the Equation 1.

$$\frac{\partial c}{\partial t} + U_i \frac{\partial c}{\partial x_i} + W \frac{\partial c}{\partial z} = \frac{\partial}{\partial x_i}\left(\Gamma \frac{\partial c}{\partial x_i}\right) \qquad (1)$$

W Fall velocity of sediment particles, U Flow rate, X Dimension and Γ is total confusion and diffusion coefficient of molecular diffusion coefficient. Value of Γ as the coefficient of viscosity ν_t and obtained by turbulence model.

$$\Gamma = \frac{\nu_T}{S_c}$$

Number S_c is Schmidt an integer value is between 0.7 And 1 (The default value SSIIM2.0, is 1). To calculate the bed load in the default software SSIIM2.0 the empirical formula of bed at Van Rhine (Van Rijn) Used is as follows.

$$\frac{q_b}{d^{1.5}\sqrt{\frac{(\rho_s-\rho)}{\rho}g}} = 0.053\frac{\left[\frac{\tau-\tau_c}{\tau_c}\right]^{1.5}}{d^{0.3}\left[\frac{(\rho_s-\rho)g}{\rho\nu^2}\right]^{0.1}}$$

q_b Bed load discharge per unit width, d Average particle diameter (D_{50}), τ Bed shear stress, τ_c Critical shear stress for sediment movement (Shields diagram is obtained) and ρ, ρ_s : The mass per unit volume of water and sediment grains are.

MATERIALS AND METHODS

Laboratory flume

The experiments have been conducted in a recirculating rectangular prismatic glass-sided open-channel main flume being 40.00 m long, 2.00 m wide and 1.20 m deep. The flume slope was horizontal; the requested bottom slope of the mobile bed has been achieved by adjusting the sediment layer. The main flume was subdivided longitudinally into two separate channels. The first channel, 20.00 m long and 1.50 m wide represents the actual testing facility with the mobile bed. The second one, 0.47 m wide, constitutes a lateral channel enabling to evacuate the laterally diverted discharge.

The side weir was located on the right channel bank 5.00 m or 40

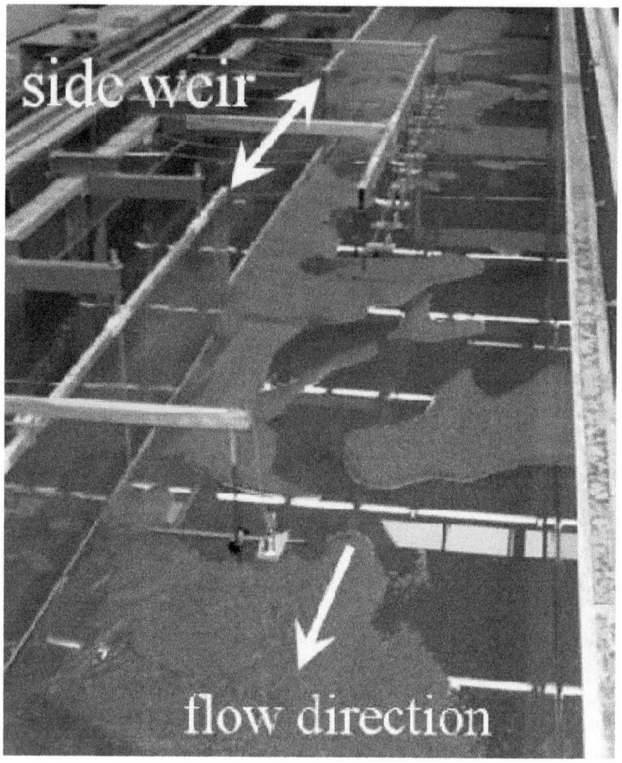

Figure 1. Laboratory setup with main channel and mobile bed, side weir and evacuation channel.

flow depths from the main channel inlet (test series B). The crest was horizontal and rectangular with a crest width of 0.025 m. the up-and downstream weir corner consisted of semi-circle profiles.

At the end of the mobile bed reach a plate has been installed to fix the sediments. The collection of bed material transported out of the main channel was attained by the arrangement of three restitution basins at the channel outlet.

A group of test have been carried out in a 20.00 m long, 1.50 m wide and 1.20 m high rectangular flume and consisted of a 3.00 m long side weir, that they were simulated in SSIIM2.0.

The overall flow regime has been subcritical. The average initial bottom slope was 0.21%. The mobile bed was characterized by a median particle size of d_{50}=0.72 mm. During the simulation the water surface, the 2D-velocity field, the side overflows discharge and sediment supply was measured.

NUMERICAL MODELING AND ANALYSIS RESULTS

The software SSIIM2.0 field network solutions, has been done in the editor of network software. In order to reduce computation time, network size has compacted in spillway due to velocity gradient. For the dimensions of a network 12 * 31 * 276 used (From left to right, respectively, the number of grid lines in the x, y and z). The network created by network software in SSIIM2.0 is as shown in Figure 1. Sediment transport roughness for calculation was about d_{50} = 0.00072. Figures 2 and 3 showed Channel shape and its spillway.

The final shape of the substrate with one side weir (condition B) has been showed in Figure 4. This study presents the development and comparison performed in the numerical model SSIIM2.0 and a prototype. This study examined the model results with respect to those observed in the field as shown in Table 1) in order to determine whether the numerical model (SSIIM2.0) is able to predict velocity distribution in the study reach.

Results of the experiments have been shown on the graph (Figure 5). According to Figure 5 discharge with sediment and without sediment is close together. The difference between them was about 10%, so there is a little difference and it's ignored.

Figure 6 showed the view of mesh network, and Figure 7 showed surface bed vectors.

Conclusions

In this study systematic experimental flume has been simulated in SSIIM2.0. According to the result, water and sedimentation simulation in SSIIM2.0 can show the situation of the flume in any characteristics.

For determining the water flow, sediment transport, bed deposition and its pattern SSIIM2.0 can be used effectively at all stages of project. Care is required to prepare grid and control file. It is strongly recommended

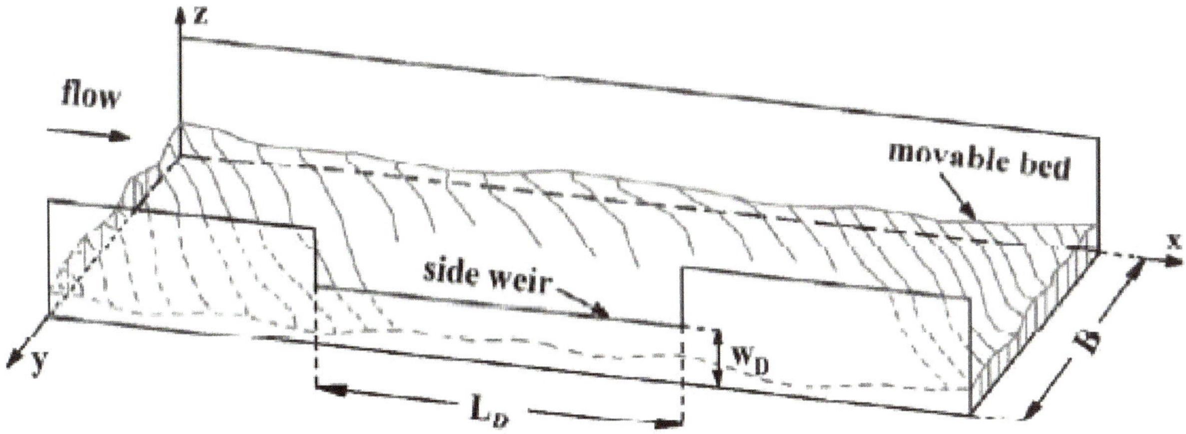

Figure 2. Definition sketch of experimental setup for one side weir.

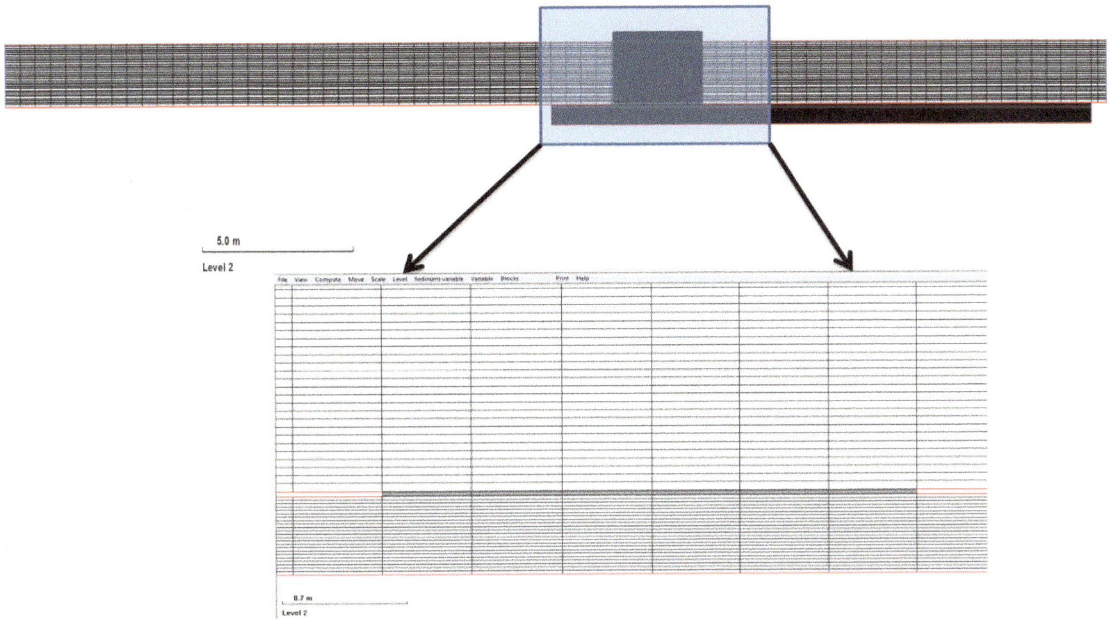

Figure 3. View of the mesh network in SSIIM2.0.

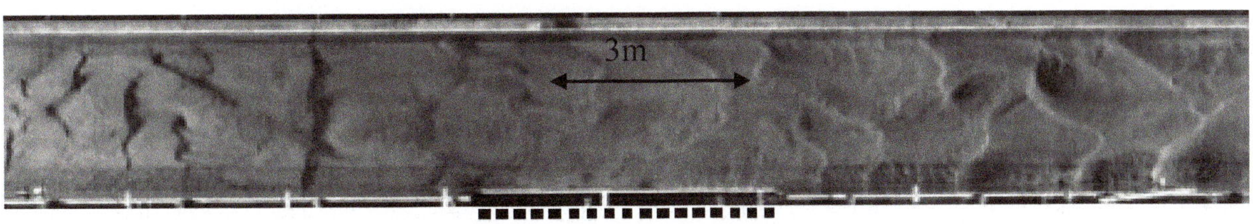

Figure 4. The final shape of the substrate with one side weir (condition B).

for further study SSIIM has great potential.

According to the results obtained by this study, the following conclusions are reached:

1. The rate of erosion and sedimentation over time has been an increasing trend, so that in the early minutes of the start of the calculations, changes in the bed has had

Table 1. channel characteristics in B(with one side weir).

No. of experiment	No. of weirs (n_D)	Length of weir crest (L_D(m))	Sill heigh t(w_D (m))	Bottom Slope (S_0 (%))	Upstream discharge(Q_1 (l/s))	Sediment supply ($Q_{sb/in}$) (kg/min)
B01	1	3.00	0.10	0.2	131	8.70
B02	1	3.00	0.10	0.2	181	17.73
B03	1	3.00	0.10	0.4	177	9.10
B04	1	3.00	0.10	0.1	98	9.67
B05	1	3.00	0.10	0.2	144	16.72
B06	1	3.00	0.10	0.3	148	17.61

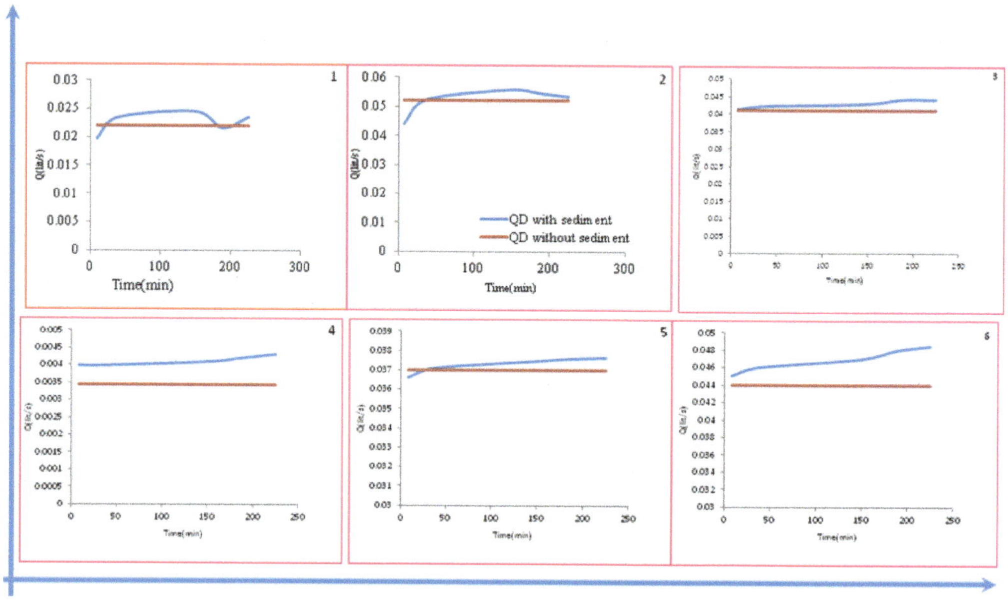

Figure 5. View of the morphological changes in flow.

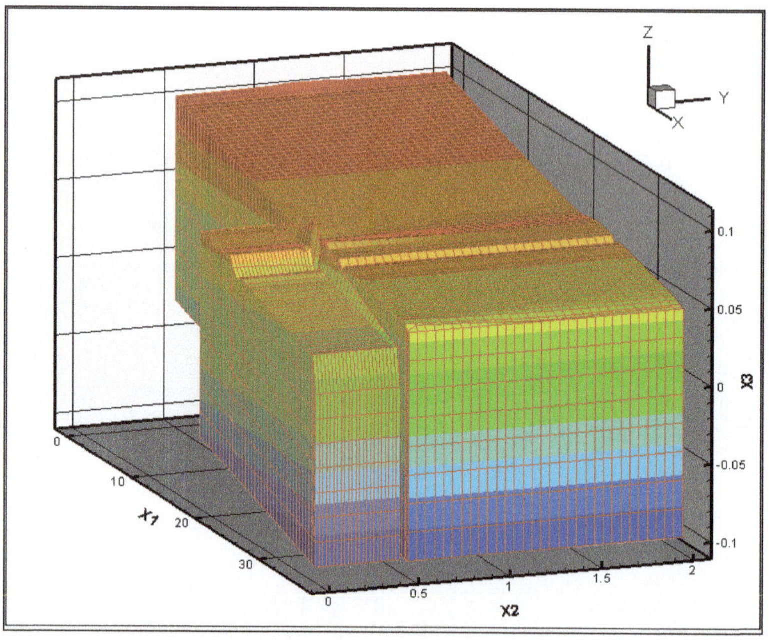

Figure 6. View of the mesh network in Tecplot360.

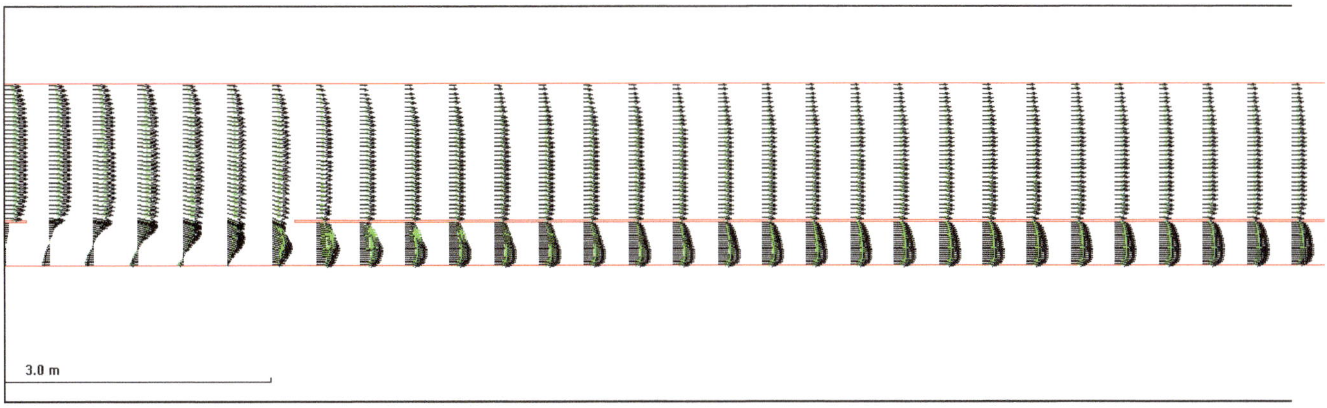

Figure 7. Surface bed vectors.

a reduced rate of change over time.

2. A good relation was observed between the measured and computed values of velocity at the study reach in three dimensions.

3. The SSIIM is one of the useful tools to predict the velocity distributions in three dimensions which gave good idea about the behavior of the flow velocities.

REFERENCES

Chiew YM (1991). Bed features in non-uniform sediments, J. Hydraul. Engr. 117(1):116-121.

Imanshoar FMRM, Tabatabai HY, Rostamipoor M (2012). Experimental Study of Subsurface Erosion in River Banks. World Academy of Science and Engineering Technology, pp. 61.

Malavieille J (2009). Impact of erosion, sedimentation, and structural heritage on the structure and kinematics of orogenic wedges: Analog models and case studies, Université Montpellier 2, CNRS UMR 5243, Géosciences Montpellier, 34095 Montpellier cedex 5, France; and International Laboratory (LIA), "ADEPT," CNRS-NSC, France-Taiwan. DOI: 10.1130/GSATG48A.1.

Nils R, Olsen B (2003). Three-Dimensional CFD Modeling of Self-Forming Meandering Channel. NTNU. S.P.Andersensvei5.N-7491 Trondheim. Norway. Dept. Hydr. Environ. Engrg. 129(5):366-372.

Talebbeydokhti N, Hekmatzadeh AA, Rakhshandehroo GR (2006). Experimental Modeling Of Dune Bed Form In A Sand-Bed Channel, Iran. J. Sci. Technol. Shiraz Univ., Trans. B, Eng. 30(B4).

Willy J, Buchanan J P, O'Rourk M, Greentree B (2010). Googong Dam Spillway- Remedial works to repair extensive rock erosion. http://www.ghd.com/australia/

Yalin MS, da Silva AMF (2001). Fluvial processes, IAHR Monograph, Delft, The Netherlands.

5

Application of synthetic unit hydrograph methods to construct storm hydrographs

B. F. Sule[1] and S. A. Alabi[2]

[1]Department of Civil Engineering, University of Ilorin, Ilorin, Nigeria.
[2]Lower Niger River Basin Development Authority, Ilorin, Nigeria.

Synthetic unit hydrograph methods were used to generate unit hydrographs for the Awun River Basin in Kwara State, Nigeria. The synthetic methods used were those of Snyder's, Soil Conservation Service (SCS), and Gray's. The study watershed has a maximum relief of 183 m with an area of 954 km^2 and a slope of 0.15%. The unit hydrograph peak flows for the methods employed ranged from 100.15 to 318.65 m^3/s while the times to peak ranged from 15.82 to 62.93 h. For the storm hydrograph development, the design frequency or return period of 25 year, 24 h storm hydrographs have peak flows ranging from 4565.83 to 11277.93 m^3/s while the times to peak ranged from 23.73 to 62.93 h. For the 100 year, 24 h storm hydrographs the peak flows ranged from 6177.92 to 15155.08 m^3/s while the times to peak ranged from 23.73 to 62.93 h. The statistical evaluation carried out on the design storm hydrograph flows indicated that there were significant differences in the methods employed. Generally, the three methods employed have been found useful in one way or the other, but Snyder's and SCS methods have distinct features and utilize most major unit hydrograph characteristics and watershed parameters in generation of unit hydrographs. The generation of these unit hydrographs was found to give some useful parameters of runoff such as peak flow rates and time to peak which are normally used in hydraulic structures design and general flood studies.

Key words: Hydrographs, watershed, peak flow, design storm.

INTRODUCTION

In many parts of the world, rainfall and runoff data are seldom adequate to determine a unit hydrograph of a basin or watershed. This situation is common in Nigeria due to lack of gauging stations along most of the rivers and streams. Generally, basic stream flow and rainfall data are not available for planning and designing water management facilities and other hydraulic structures in undeveloped watershed. However, techniques have been evolved that allow generation of synthetic unit hydrograph. This includes Snyder's method, Soil Conservation Service (SCS) method, and Gray's method. Straub et al. (2000) simply defined unit hydrograph as a discharge time graph (hydrograph) of a unit volume of direct runoff resulting from a spatially uniformly distributed effective precipitation with a uniform intensity over a given duration. Bedient and Huber (2002) defined unit hydrograph as basin outflow resulting from 1.0 inch of direct runoff generated uniformly over the drainage area at a uniform rainfall rate during a specified period of rainfall duration. The unit hydrograph is essentially a hydrological tool for predicting flood peak discharges and determining the direct runoff response to rainfall. Viessman et al. (1989) defined a watershed as a land area that contributes surface runoff to any point of interest.

The unit hydrograph can be developed for both gauged

and ungauged basin. For gauged basins, unit hydrographs can be derived from observed data, by measuring the concurrent rainfall and runoff amounts for the storms. For ungauged basins, some synthetic methods are used to determine the unit hydrographs. In most watersheds in Nigeria, there is considerable lack of data with regard to rainfall and river discharges. However, when enough data or concurrent observations of precipitation and streamflow are not available in a given watershed, a synthetic unit hydrograph could be developed. Synthetic unit hydrograph methods are based on theoretical or empirical formulas relating hydrograph peak flow and timing to watershed characteristics (Bedient and Huber; 2002).

The watershed or basin characteristics have been described by many researchers in the hydrological literature. Mustapha and Yusuf (2012) described some of the important basin characteristics. They include basin area, stream order, stream lengths, stream density, basin slope and others. Mustapha and Yusuf (2012) described procedures and mathematical formulas for determining these basin characteristics. Synthetic unit hydrograph methods are popular and play an important role in urban storm water drainage design. The synthetic unit hydrograph methods have also been adopted to some basins in Turkey where rainfall and runoff data are seldom adequate. Also, Straub et al. (2000) developed synthetic unit hydrographs for small rural watersheds in Illinois. Runoff hydrographs were generated from flow data and unit hydrographs (UH) were obtained for 1 and 2 h duration in Midnapore and Bankura districts of West Bengal state in India (Jena and Tiwari, 2006). UH parameters such as time to peak (t_p), time base (t_b), and peak discharge were modeled with geomorphologic parameters of the watershed such as channel parameters as well as basin parameters.

In this study, the main objectives include: collection and determination of basin physiographic characteristics, determination of peak runoff using unit hydrographs, convolution of 24 h rainfall at selected return period with the unit hydrograph and estimation of design flood for the watershed. Such estimates of design storms are useful for routing the flood through a dam proposed for construction on the river. The routed floods will be used to determine dimensions of spillway and the length of the stilling basin to be provided below the proposed dam.

MATERIALS AND METHODS

The study area is Awun basin, which is a small watershed in Kwara State, Nigeria. Figure 1 shows map of Nigeria and network of rivers including the catchment areas of River Awun. The entire Awun basin is located between Latitudes 8°28'00" North and 9°00'00"North and Longitudes 4°30'00"East and 4°45'00"East. Sule (2003) described a river basin to be the most appropriate scale of management of water resources. Based on the stream order concept, the Awun River can be classified as the highest order stream of the watershed or basin. This is because it is the main stream channel that carries the flow from the entire tributaries area upstream of River Niger. The Awun River and some other rivers are shown in Figure 2. The topographic map of Awun basin was digitized on a computer using a software known as Global Mapper and watershed characteristics were obtained from the computer using engineering softwares such as AutoCAD 2002 and Land Developer. Geographic Information Systems (GIS) allows topographic data and flow networks to be addressed more accurately than was possible with paper maps (Bedient and Huber, 2002).

The characteristics of the watershed obtained are summarized in Table 1. The watershed has an area of 954 km^2, watershed slope of 0.15%, average channel slope of 0.12% (Figure 3), maximum relief of 183 m, main river length of 80.23 km, and length along the main channel from the outlet to a channel point nearest the watershed centroid as 42.29 km. Each of these characteristics has special relevance in hydrology and plays a significant role in the development of a unit hydrograph for the watershed. The textural class of the soil in the watershed is sandy loam, and it belongs to Hydrologic Soil Group B (HSG B), with an Antecedent Moisture Condition II (AMC II). These characteristics were used to determine the curve number for the watershed as shown in Table 2.

Development of synthetic unit hydrograph

The three methods that were used in the generation of unit hydrograph for the watershed includes Snyder's, SCS, and Gray's method.

Development of unit hydrograph by Snyder's method

The Snyder's method was used to compute the unit hydrograph characteristics such as lag time or basin lag, unit-hydrograph duration, peak discharge, time base or base period, and hydrograph time widths at 50 and 75% of peak flow. Determination of all these parameters allows for the development of unit hydrographs. Snyder considered the shape and area of the basin and gave the following empirical equations after analyzing a large number of hydrographs from drainage basins of areas from 25 to 25000 km^2 (Arora, 2004).

Lag time or basin lag: The lag time was defined as the time from the center of mass of effective rainfall to the peak rate of flow (Viessman et al., 1989). The basin lag is given by:

$$t_p = C_t \, (L L_c)^{0.3} \tag{1}$$

where t_p = the basin lag (hours), C_t = a coefficient which depends upon the characteristics of the basin, L = length of the main stream of the catchment (km), L_c = distance from the basin outlet to a point on the stream which is nearest to the centroid of the area of the basin(km).

Unit- hydrograph duration: The duration of rainfall excess for Snyder's synthetic unit- hydrograph development is a function of lag time. The unit duration of the storm was given as follows (Arora, 2004).

$$t_r = \frac{t_p}{5.5} \tag{2}$$

Where t_r = the unit duration of the storm (hours), t_p = the basin lag (hours). If the unit hydrograph of another duration $t_{r'}$ is required, Equation (1) for the basin lag is modified as follows (Arora, 2004).

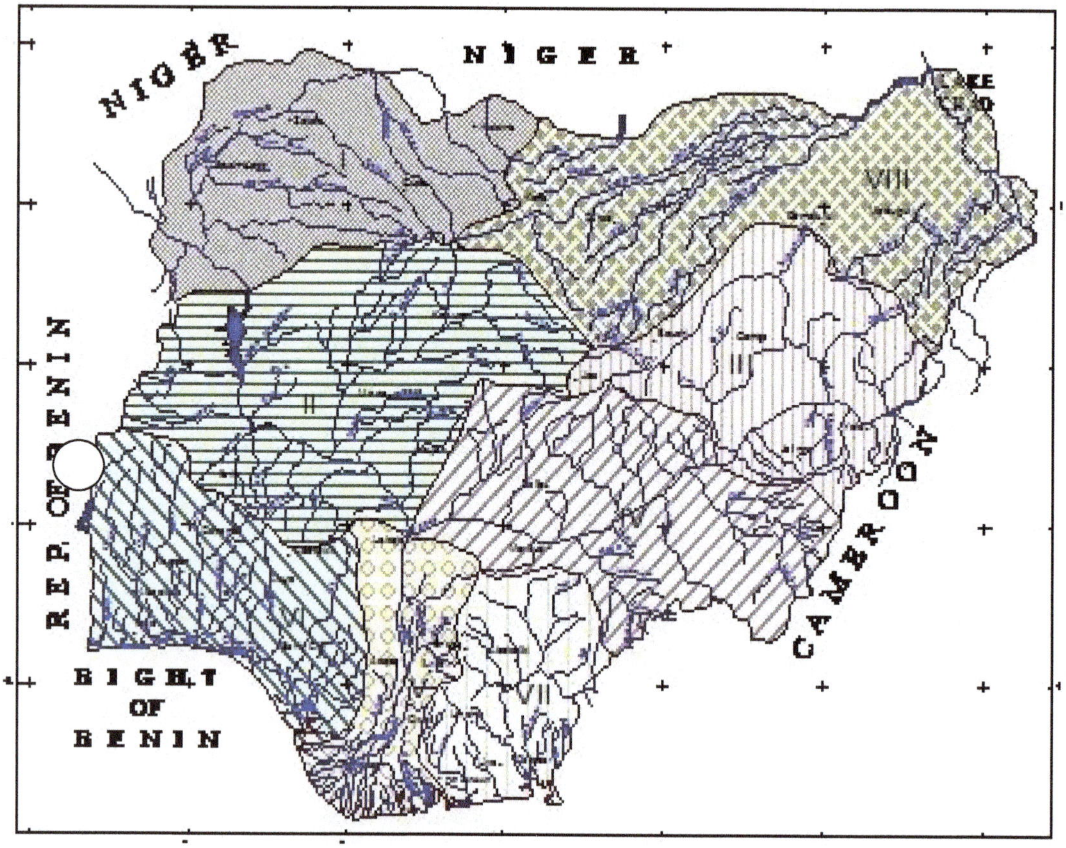

Figure 1. Map of Nigeria showing major rivers and catchment of River Awun circled.

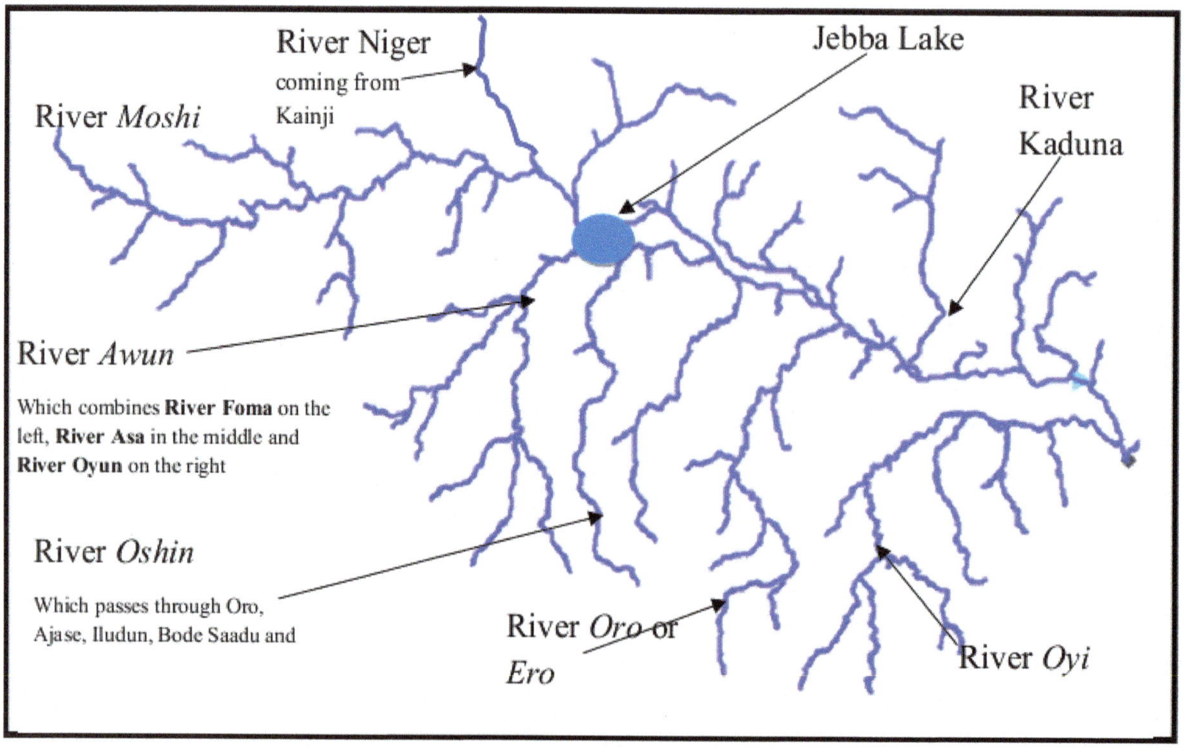

Figure 2. River Awun and other rivers dischrging into the River Niger in Nigeria.

Table 1. Summary of the results obtained for the watershed characteristics/parameters.

Watershed characteristics/parameters	Values obtained
Watershed area (A)	954 km^2
Watershed slope (S)	0.15%
Average channel slope (S_c)	0.12%
Maximum Relief (MR)	183 m
Length of main river (L)	80.23 km
Length along the main channel from the outlet to a channel point nearest the watershed centroid (L_c)	42.29 km
Weighted Curve Number, CN	76
Potential maximum retention, S	80.213 mm

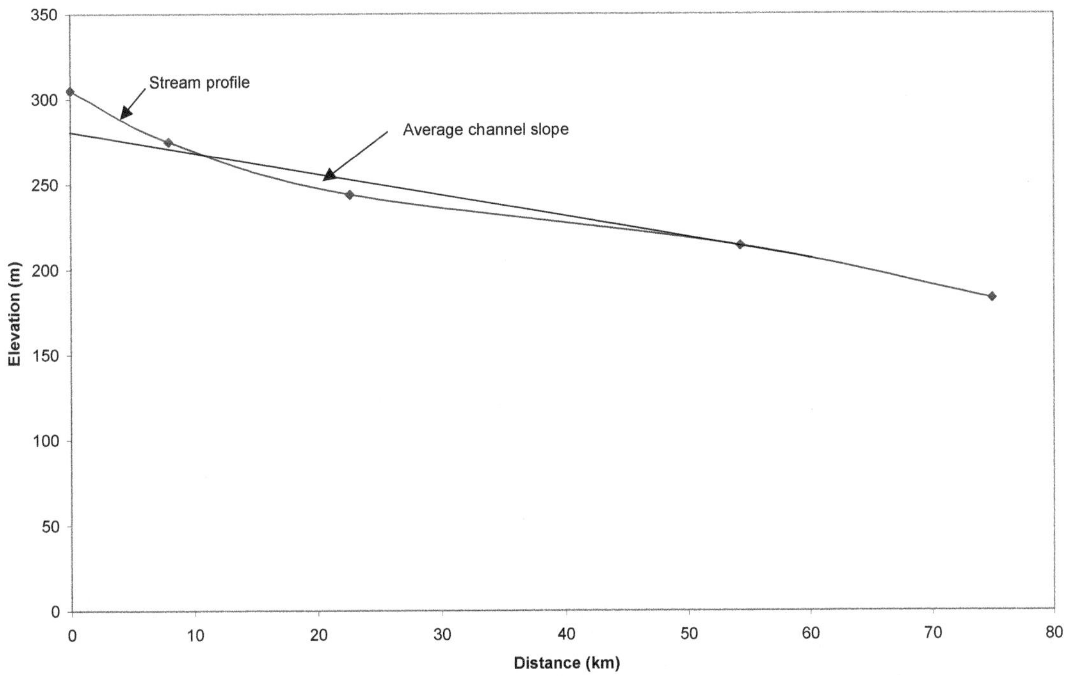

Figure 3. Elevation- distance from head of stream.

Table 2. Land use and runoff curve number for the watershed.

Sub areas	Land use	Area (km^2)	Curve Number CN
A	Residential	205	98
B	Streets and roads	174	85
C	Cultivated land	296	75
D	Wood or forest land	279	55

$$t_{p'} = t_p + \left(\frac{t_{r'} - t_r}{4}\right) \quad (3)$$

Where: $t_{p'}$ = *the basin lag for a storm of duration* $t_{r'}$

Peak discharge: Peak discharge is the highest volume of runoff over the basin. It is a function of the hydrographic time relation parameters. The determination and knowledge of peak discharge is very crucial to hydraulic designs and flood characteristics in basins (Ifabiyi, 2004).

$$Q_p = \frac{2.78 C_p A}{t_p} \qquad (4)$$

The peak discharge is given by the equation below (Arora, 2004):

$$q_p = \frac{Q_P}{A} = \frac{2.78 C_P}{t_p} \qquad (5)$$

where: Q_p =the peak discharge (m^3/s), C_p = the coefficient which depends upon the retention and storage characteristics of the basin (Values of C_p varies from 0.3 to 0.93). A = area of the basin (km^2); t_p = the basin lag (hours).

Also, the peak discharge per unit area is given by:
If an X-hr unit hydrograph is required or desired, equation (4) for the peak discharge is modified as follows:

$$Q_{p'} = \frac{2.78 C_p A}{t_{p'}} \qquad (6)$$

and

$$q_{p'} = \frac{2.78 C_p}{t_{p'}} \qquad (7)$$

Time base or base period: The time base of a hydrograph is the time from which the concentration curve (rising portion of a hydrograph) begins until the direct runoff component reaches zero. The base period (T) of the unit hydrograph is given by:

$$T = 3 + 3(t_p / 24) \qquad (8)$$

where: T = the base period (days), t_p= the basin lag (hours).

Equation (8) above can be modified as follows:

$$T' = 3 + 3(\frac{t_{p'}}{24}) \qquad (9)$$

Hydrograph time widths at 50 and 75% of peak flow: As a general rule of thumb, the time width at W_{50} and W_{75} ordinates should be proportioned each side of the peak in a ratio of 1:2 with the short time side on the left of the synthetic unit- hydrograph peak (Viessman et al., 1989). U.S. Army Corps of Engineers gave the following expressions for W_{50} and W_{75} (Arora, 2004; Mustafa and Yusuf, 2012).

$$W_{50} = \frac{5.9}{(q_{p'})^{1.08}} \qquad (10)$$

$$W_{75} = \frac{W_{50}}{1.75} = \frac{3.4}{(q_{p'})^{1.08}} \qquad (11)$$

Development of unit hydrograph by SCS method

The SCS method is a method developed by the soil conservation service for constructing synthetic unit hydrographs which is based on a dimensionless hydrograph, and which relates ratios of time to ratios of flow. This dimensionless graph is the result of an analysis of a large number of natural unit hydrographs from drainage areas ranging widely in size and geographic locations. The method requires only the determination of the time to peak and the peak discharge. The peak discharge can be expressed as follows (Viessman et al., 1989).

$$q_p = \frac{484 A}{t_p} \qquad (12)$$

where q_p = peak discharge (ft^3/s); A = drainage area (mi^2) and t_p = the time to peak (hour). Time to peak is the time it takes a stream of water to build up to it peak. It is important in flood prediction and basin management and controlled by basin length, length of mainstream, slope and others.

$$t_p = \frac{D}{2} + t_L \qquad (13)$$

The time to peak is given by:

where: t_p = the time to peak (hour);D = the duration of rainfall (hour); t_L = the lag time (hour)

The lag time can be described by the equation below:

$$t_L = 0.6 t_c \qquad (14)$$

where: t_c = the time of concentration (hours).

The time of concentration can be defined as the time required, with uniform rainfall, for 100% of a tract of land to contribute to the direct runoff at the outlet (Viessman et al., 1989; Viessman and Lewis, 2008; Wurbs and James, 2010). The time of concentration can be expressed by the equation below:

$$t_c = 0.0195\, L^{0.77} S^{-0.385} \qquad (15)$$

where: t_c = Time of concentration (min); L = Length of main river (m); S = the watershed gradient or slope (m/m).The watershed slope can be described by the expression below:

$$S = \frac{\textit{Difference in elevation along the flow path}}{\textit{Maximum length of flow}} \qquad (16)$$

The duration of rainfall can also be expressed as:

$$D = 0.133 t_c \qquad (17)$$

where: D = the duration of rainfall (hour); t_c = time of concentration (hour).

Development of unit hydrograph by Gray's method

The Gray's method is a synthetic unit hydrograph method that is based on dimensionalizing the incomplete gamma distribution in its generation of unit hydrograph. The method requires the determination of some important characteristics of the watershed such as main stream length, channel slope, area, period of rise and others. These parameters allows for the computation of discharge ordinates for the unit hydrograph at times equal to ¼ intervals of the period of rise. The incomplete gamma distribution is:

$$Q_{t/P_R} = \frac{25.0(\gamma')^q}{\Gamma(q)}(e^{-\gamma' t/P_R})(\frac{t}{P_R})^{q-1} \quad (18)$$

Where $Q_{t/PR}$ =percent flow in $0.25P_R$ at any given t/P_R value, q and γ = shape and scale parameters, respectively. Γ (q) = the gamma function of q which is equal to (q-1)!, e = the base of the natural logarithm, P_R = the period of rise (min), t = time (min)

The relationship for γ' is defined as:

$$\gamma' = \gamma P_R \quad (19)$$

and

$$q = 1 + \gamma' \quad (20)$$

Correlations with physiographic characteristics of the watershed can be developed to get the values of both P_R and γ. The storage factor P_R / γ' has been linked with watershed parameter L/S_c, where L is the length of main channel of the watershed in miles. The average channel slope, S_c is achieved by plotting the elevation points along the main river channel on the map against the distance from head of stream. The step by step solution procedure is available in Viessman et al. (1989).

Development of design storm hydrographs

The real importance of the unit hydrograph approach is the development of storm hydrographs due to an actual rainfall event over the watershed. Design storm hydrographs for selected recurrence interval (25 and 100 year) were developed for the three methods through convolution (adding and lagging procedures), with Ilorin rainfall data taken from Ogunlela et al. (1995). The procedure of deriving a storm hydrograph from a multi period of rainfall excess is called hydrograph convolution (Bedient and Huber, 2002). It involves multiplying the unit hydrograph ordinates U_n by incremental rainfall excess P_n, adding and lagging in a sequence to produce a resulting storm hydrograph. The SCS type II curve was used to divide the different rainfall data into successive equal short time events (4 h) and the SCS-Curve Number method was used to estimate the cumulative rainfall excess. The incremental rainfall excess is obtained by subtracting sequentially, the rainfall excess from the previous time events. The equation that applies to the SCS- Curve Number method is:

$$Q = \frac{(P - I_a)^2}{(P - I_a) + S}$$

$$(Q = 0 \ if \ P \le 0.2S) \quad (21)$$

Q = cumulative rainfall excess (inches), P = cumulative precipitation (inches), I_a = initial abstraction = 0.2S.

and

$$S = \frac{1000}{CN} - 10$$

S= potential maximum retention after runoff begins (inches)
CN= SCS Curve Number

Statistical evaluation of different methods of storm hydrograph development

A statistical analysis known as randomized complete block design (RCBD) (Oyejola, 2003) was used to evaluate the different methods of storm hydrograph development for the two return periods of 25, 24 and 100 years, 24 h. The different methods are represented by Treatments (T_1, T_2, and T_3) while the return periods are represented by Blocks (B_1 and B_2). An analysis of variance table (ANOVA Table) for the RCBD was constructed for the statistical analysis by calculating some parameters such as degree of freedom, sum of squares, mean squares, and F-Ratio.

RESULTS AND DISCUSSION

The summary of unit hydrograph and storm hydrograph peak flows and times to peak, for the three methods employed are presented in Tables 3 and 4 respectively. It could be seen from Table 3 that the unit hydrograph peak flows for the methods employed ranged from 100.15 to 318.65 m^3/s while the times to peak ranged from 15.82 to 62.93 h. From Table 4, it could be seen that the design frequency or return period of 25year, 24 h storm hydrographs have peak flows ranging from 4565.83 to 11277.93 m^3/s while the times to peak ranged from 23.73 to 62.93 h. For the 100 year, 24 h storm hydrographs, the peak flows ranged from 6177.92 to 15155.08 m^3/s while the times to peak also ranged from 23.73 to 62.93 h. From Tables 3 and 4, it could be observed that the lowest value of peak flows was found in Gray's method while the highest value was obtained in the SCS method. Figures 4 to 6 shows the unit hydrographs for the three methods which are used for the development of design storm hydrographs through convolution method.

The design storm hydrograph flows obtained from the different methods were statistically evaluated using the RCBD to determine if there were significant differences in the methods. The results shown in Tables 5 and 6 indicated that there were significant differences in the methods. This can be confirmed by using the F-ratio to test whether the different methods have the same effect. The F-ratio has an F-distribution on numerator degree of freedom (df1=2) and denominator degree of freedom (df2=2). The critical value is the number the test statistic must exceed to reject the test. $F_{cr}(0.05,2,2)=19.00$. From Table 6, $F=46.16>F_{cr,}$ hence the results are significant at 5% significant level, and the test is accepted. It can then be inferred that there is strong evidence that the treatment methods differ. Generally, it could be seen that the efficiency of each method depends to some extent on the main parameters of the watershed likewise the duration

Table 3. Summary of unit hydrograph peak flows and times to peak for the methods employed.

Method	Q_p (m^3/s)	t_p (h)
Snyder's	156.70	17
SCS	318.65	15.82
Gray's	100.15	62.93

Table 4. Storm hydrograph peak flows and times to peak for the methods employed.

Method	Frequency	Q_p (m^3/s)	t_p (h)
Gray's	25 years, 24 h	4565.83	62.93
	100 years, 24 h	6177.92	62.93
Snyder's	25 years, 24 h	6504.48	33
	100 years, 24 h	8717.19	33
SCS	25 years, 24 h	11277.93	23.73
	100 years, 24 h	15155.08	23.73

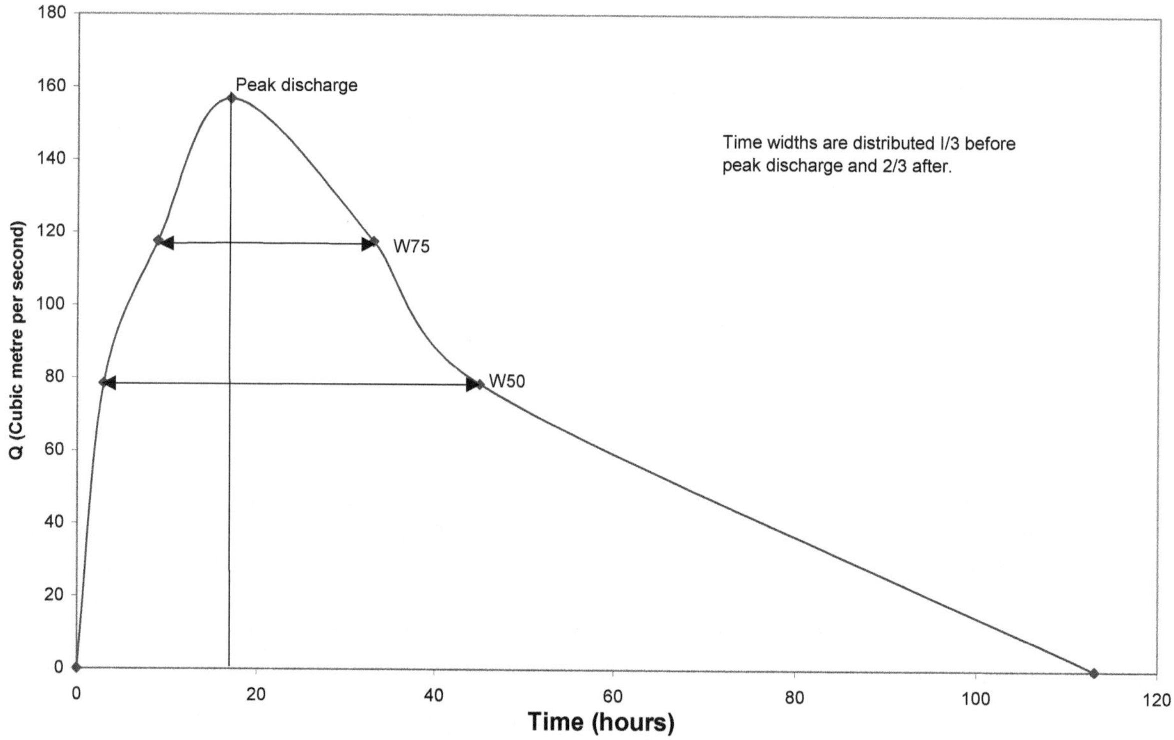

Figure 4. A sketch of unit hydrograph using Snyders method.

of unit hydrograph is dependent upon the parameters used in the equation specific to the method.

Salami et al. (2009) reported similar results on the use of synthetic unit hydrograph to generate ordinates for the development of design storm hydrographs for the catchment of eight selected rivers located in the South West Nigeria. Unit hydrographs were developed based on Snyder, Soil Conservation Service (SCS) and Gray's

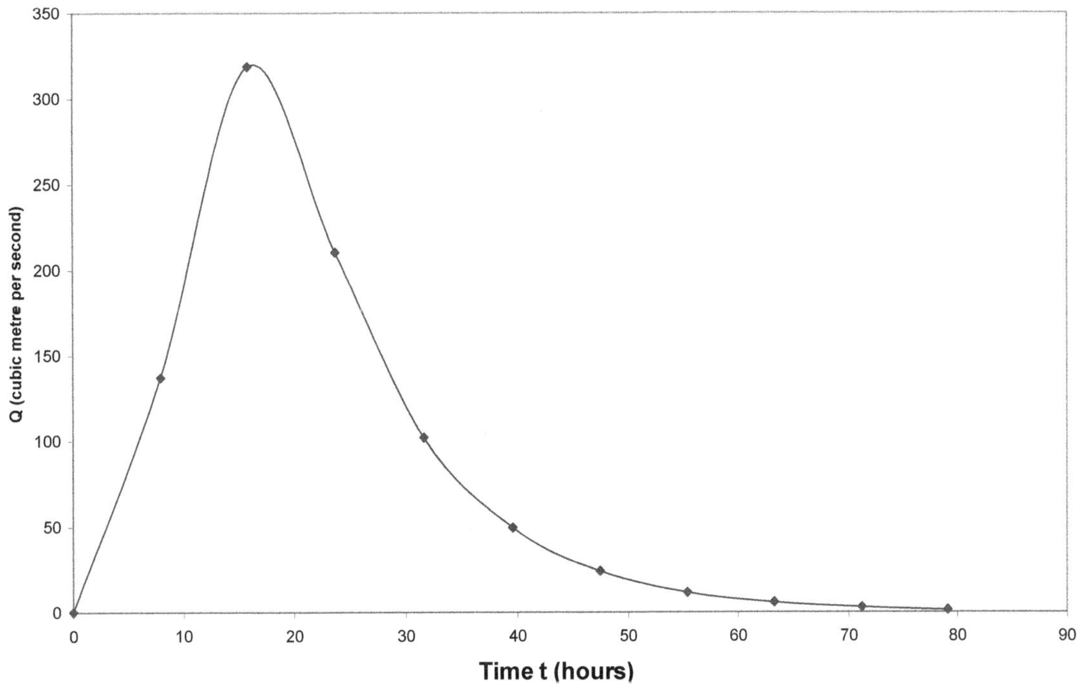

Figure 5. A sketch of unit hydrograph using SCS method.

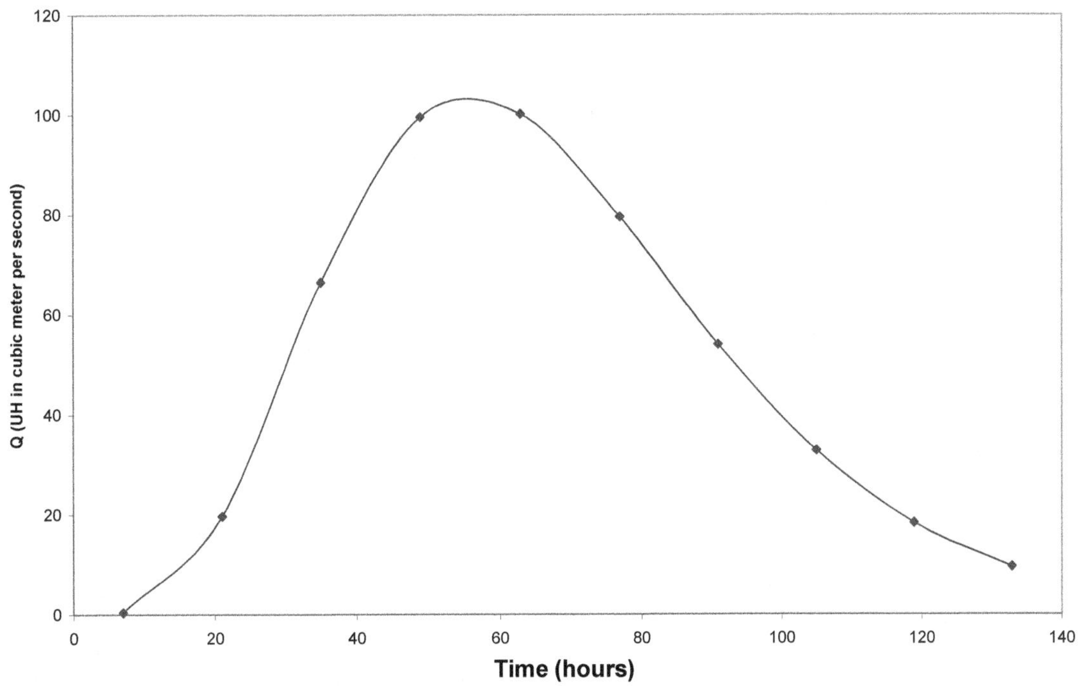

Figure 6. A sketch of unit hydrograph using Grays method.

methods. The peak storm hydrograph flows obtained based on the unit hydrograph ordinate determined by Snyder's for 20, 50, 100, 200 and 500years return periods varied from 112.63 to 13364.30 m^3/s, while those based on the SCS method varied from 304.43 to 6466.84 m^3/s and those based on Gray's varied from 398.06 to 2607.42 m^3/s for the eight watersheds. The analysis showed that the values of peak flows obtained by Gray's

Table 5. Table of observations.

Methods	Treatment	Return periods (Blocks)		Total
		25 years, 24 h	100 years, 24 h	
		B1	B2	
Gray's	T_1	2255.68	3047.81	5303.49
Snyder's	T_2	4717.09	6360.96	11078.05
SCS	T_3	3072.10	4156.82	7228.92
Total		10044.87	13565.59	23610.46

Table 6. ANOVA Table for Randomized Complete Block Design (RCBD).

Source of variation	d.f	SS	MS	F-Ratio
Treatment	2	8644770.9	4322385.5	46.16
Block	1	2065911.5	2065911.5	22.06
Error	2	187286.4	93643.2	
Total	5	10897968.8	6481940.2	

and SCS methods for five watershed were relatively close, while the values of peak flows obtained by Gray's and Snyder's methods for two watershed were relatively close and the values of peak flows obtained by Snyder's and SCS methods were relatively close for only one watershed. Salami et al. (2009) concluded that SCS method can be used to estimate ordinates required for the development of peak storm hydrograph of different return periods of different rivers as it was done in the present study.

Conclusions

Based on the results obtained so far for the ungauged watershed, it could be seen that the generation of unit hydrograph through synthetic methods has been found useful and effective. The statistical evaluation of the storm hydrograph flows obtained in this study from the three methods employed have indicated that there were significant differences in the methods. Though all the three methods employed have been found useful in one way or the other, but Snyder's and SCS method have been considered distinct and more important since they both utilize most major unit hydrograph characteristics and watershed characteristics in the generation of unit hydrographs. These two methods were found simple, requiring only an easy determination of watershed and land use characteristics.

REFERENCES

Arora KR (2004). Irrigation, Water Power and Water Resources Engineering, Standard Publishers Distributors, Delhi, pp. 96-99.

Bedient BP, Huber CW (2002). Hydrology and Floodplain Analysis. Prentice- Hall, Upper Saddle River, United States of America.

Ifabiyi IP (2004). The Response of Runoff and its Components to Basin Parameters in the Upper Kaduna Catchment of Nigeria. Ph.D Thesis, Department of Geography, University of Ilorin, Ilorin, Nigeria.

Jena SK, Tiwari KN (2006) Modeling synthetic unit hydrograph parameters with geomorphologic parameters of watersheds. J. of Hydrol. 319(1–4):1–14

Mustafa S, Yusuf MI (2012). A Textbook of Hydrology and Water Resources, Revised Edition, Topsmerit Page Publishing Co., Abuja, Nigeria

Ogunlela AO, Nwa EU, Jatto BO (1995). Runoff Prediction for Major Catchments in Ilorin City, Nigeria. Paper presented at the 9th Afro-Asian Regional Conference of the International Commission on Irrigation and Drainage (ICID), Algiers, Algeria.

Oyejola BA (2003). Design and Analysis of Experiments for Biology andAgriculture Students. Olad Publishers, Ilorin, Kwara State, Nigeria, pp. 29-51.

Salami AW, Bilewu SO, Ayansola AM, Oritola, SF (2009) Evaluation of synthetic unit hydrograph methods for the development of design storm hydrographs for Rivers in South-West, Nigeria. J. Am. Sci. 5(4):23-32.

SCS (2000), Soil Conservation Service. Design Hydrographs. U.S. Department of Agriculture, Washington, D.C.

Straub DT, Melching SC, Kocher EK (2000). Equations for Estimating Clark Unit- Hydrograph Parameters for Small Rural Watersheds in Illinois. U.S Department of the Interior U.S Geological Survey, Water-Resources Investigations Report 00-4184.

Sule BF (2003). Water Security: Now and the Future, Sixty-fifth Inaugural Lecture of University of Ilorin, Ilorin, Nigeria.

Viessman W, Knapp JW, Lewis GL (1989). Introduction to Hydrology, Harper and Row Publishers, New York.

Viessman WJr, Lewis GL (2008). Introduction to Hydrology, Prentice-Hall of India Private Ltd, New Delhi.

Wurbs RA, James WP (2010). Water Resources Engineering, PHI Learning Private Ltd, New Delhi.

6

Classification of transmissivity magnitude and variation in calcarious soft rocks of Bhaskar Rao Kunta Watershed, Nalgonda District, India

K. Srinivasa Reddy

ICAR-Central Research Institute for Dryland Agriculture, Hyderabd-59, India.

Pumping test was used for the appraisal and evaluation of groundwater potential and design of well. Pumping tests of the calcarious sedimentary rocks of Bhaskar Rao kunta watershed area were carried out for twenty five selected bore wells for quantitative understanding of the groundwater for crop water requirement and groundwater use efficiency. The tests were carried out independently for short duration under constant rate conditions. The acquisition of the drawdown data was interpreted by Jacob straight line method. The results of transmissivity vary from 2.67 to 236.9 m^2/day with mean of 37 m^2/day, whereas the specific capacity varies from 5.47 to 451.63 m^3/d/m with a mean of 76 m^3/d/m. Spatial variation of transmissivity values was further analyzed using statistical testing and Krasny's classification systems; from the results of the statistical testing, 72% of the wells were under covered background transmissivity anomalies; 12% under positive anomalies; 8%, negative anomalies and remaining 8%, positive extreme anomalies. From the results of Krasny's classification system, 12% of the wells was under high magnitude (withdrawals of lesser regional importance), 40% of wells was under intermediate magnitude (withdrawals for local water supply), 48% was under low magnitude (smaller withdrawals for local water supply) and 100% of the wells was under covered moderate variations (fairly heterogeneous hydrogeological environment). Spatial variation of transmissivity magnitude and variation was identified as best useful in management practices.

Key words: Transmissivity, magnitude, variation, statistical testing, Krasny's classification.

INTRODUCTION

Quantitative understanding of most problems in hydrogeology (Ramakrishna, 1998), determination and evaluation of aquifer parameters of transmissivity and storage coefficient from aquifer test data is a continual field research (Birpinar, 2003); it is field–scale prediction (Illman and Tartakovsky, 2006) and integral part of assessment and management of groundwater study (Sarwade et al., 2007; Mayooran et al., 2011; Sudher Kumar et al., 2012). Generally, there are two types of pumping tests evaluation methods for determining aquifer parameters: (i) drawdown and (ii) recovery. In the present study, only drawdown test was conducted.

The present study was done in a semi-arid region of Bhaskar Rao kunta watershed area (40 km^2), which is a

Figure 1. Location map of the study area.

purely remote and tribal area in the Nalgonda District. The Government Organization of Andhrapradesh State Irrigation Development Corporation (APSIDC) was incited to help each selected group of economically backward farmers by giving them a single bore well with a nominal subsidy to share the groundwater for irrigation use only. After the completion of bore wells drilling in the area, randomly pumping tests were conducted to check the availability of quantitative understanding of the groundwater used for a few number of bore wells. It is valuable and useful for crop water requirement and enhances agriculture productivity and sustainable development of the farmers.

Study area

Semi-arid region of Bhaskar Rao kunta watershed geographically lies between Northern latitudes from 16° 42' 25" to 16° 37' 58" and Eastern longitudes from 79° 28' 15" to 79° 32' 30" of the Krishna Lower Basin. The watershed elevation ranges between 80 and 140 m above the mean sea level, with slightly undulating terrain to moderate slopes (2 to 3%); its annual normal rain fall is 737 mm. The average maximum and minimum temperature is 40 and 28°C, respectively. The drainage system has dendritic to sub-dendritic pattern, governed by regional slope; its homogenous lithology and relief are exhibited by 146 streams (1st, 2nd and 3rd order streams), which curve and contribute to the flow of mostly dry stream except for seasonal run-off. Soils are covered with red sandy and black clay.

Geology

The study area, part of the Kurnool group of Palnadu sub basin (Upper Proterozoic period of Vindyan rocks), is partially covered by Srisailam succession of Kadapa Super Group (Pre-Cambrian period of Archaean rocks) (Figure 1). Srisailam sub basin rocks are exposed with Quartzites; the Quartzites are inter bedded with thin siltstone units and are usually thick bedded, with dense and fine to medium grain. Palnadu sub-basin rocks are exposed with Calcareous sedimentary rocks of quartzites, shales and flaggy-massive limestones (Geology and Mineral Resources, 2006). General sequence of sub-surface strata is encountered in the top soil, weathered/semi weathered layered shale inter bedded with quartzite.

Hydrogeology

In the study area, groundwater occurs mainly along the bedding planes, cleavages, solution channels, cavernous formations and joints. Aquifers often have different hydraulic heads, caused by various surface topographic undulations or cap rock structures. Aquifers are under confined to semi-confined conditions with shallow to deep zones. The shallow aquifer depth and thickness range between 30 to 40 m and 5 to 25 m (Kotturu, Kalvakatta Villages), respectively. Deep aquifer depth and thickness range between 40 to 60 m and up to 60 m (Banjaranagar Thanda, Gonina Thanda, Champla Thanda, Ham Thanda and JK Thanda) respectively. It has been found that, most of the aquifer zones are encountered within 40 to 60 m depth. The depth of open wells ranges from 5 to 20 m, whereas the bore wells are about 60 m deep. An average yield is 448 m^3/day (Table 1). Due to fluctuations of the groundwater in the monsoon pattern, static water levels were changed in depth from 1 to 7 m below ground level.

MATERIALS AND METHODS

Acquisition of data

In the study area, twenty five pumping tests data were collected and carried out in selected bore wells (Figure 2). The average depth of bore wells is 60 m below ground level. Submersible pump of 7.5 HP is lowered to a depth of 40 m. Static water level and drawdown were recorded with automatic water level indicator. During duration pumping test for 300 min, to ensure uninterrupted

Table 1. Results of aquifer parameters of twenty five bore wells.

Well No.	Pumping duration (min)	Static water levels (m)	Drawdown (m) Min.	Drawdown (m) Max.	Discharge (m^3/day)	Specific capacity (m^3/d/m)	Transmissivity (m^2/day)
1	300	5	16	29.8	130	4.36	3.65
2	300	1	2.9	4.9	726	148.16	69.98
3	300	5.8	13.3	26.95	473	17.55	16.98
4	300	6	4.25	11.4	453	39.81	16.59
5	300	1	2.5	29.6	162	5.47	2.67
6	300	1	2	13.9	466	33.53	12.91
7	300	0.2	2.6	14.55	602	41.37	16.22
8	300	6	1.15	2.85	971	451.63	236.9
9	300	7.15	1.35	6.75	773	114.52	54.17
10	300	1.5	7.65	15.76	456	29	9.73
11	300	1.3	1.4	2.3	869	377.83	186.15
12	300	6.8	9.45	22.58	359	15.9	7.07
13	300	0.2	18.6	35	272	7.77	5.41
14	300	6.2	12.3	28.3	188	6.76	6.53
15	300	2.1	1.6	4.55	727	199.18	102.39
16	300	4	16.2	31	250	8.06	4.82
17	300	6	0.15	13.5	227	16.81	5.21
18	300	1.2	2.3	5.4	512	94.81	52.13
19	300	0.2	2.5	6.6	455	68.94	26.01
20	300	3.5	0.2	12.7	557	43.86	8.43
21	300	7	8.2	25.2	235	9.36	10.26
22	300	7	0.8	12	173	14.42	6.14
23	300	6	14.1	24	225	9.38	5.87
24	300	6	0.6	7.6	413	54.34	20.43
25	300	6.2	1.7	13.3	216	17	5.31
Minimum		0.20	0.15	2.30	130.00	5.47	2.67
Maximum		7.15	18.60	29.6	971.00	451.63	236.90
Mean		3.89	5.33	15.45	448.33	76.06	37.01
Standard Deviation		2.67	5.66	9.93	233.54	115.76	59.55

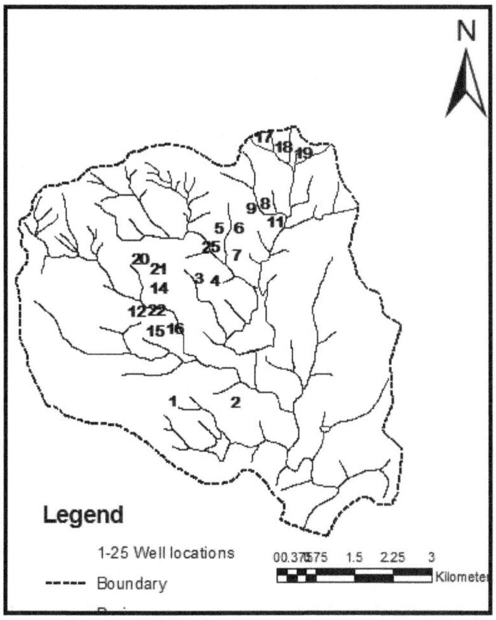

Figure 2. Pump testing locations in the study area.

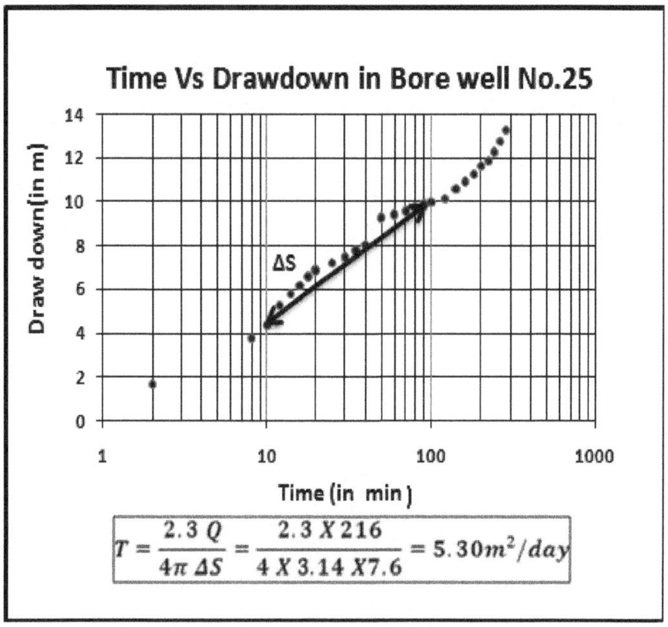

Figure 3. Data Interpretation on log-log graph in Jacob method.

Table 2. Transmissivity analysis based on transmissivity index 'Y' classification.

Classification	Description	Range of T index 'Y'
Negative extreme anomalies	Less than [mean – (2 * standard deviation)]	Less than 4.13
Negative anomalies	Between (mean – standard deviation) and [mean – (2 * standard deviation)]	4.13-4.66
Background transmissivity	Between (mean – standard deviation) and (mean + standard deviation)	4.66-5.74
Positive anomalies	Between (mean + standard deviation) and [mean + (2 * standard deviation)]	5.74-6.28
Positive extreme anomalies	Greater than [mean + (2 * standard deviation)]	Above 6.28

power supply a stand by 15 KV diesel generator was used. 200 L drum was used for measuring discharge. In the present study, a single well drawdown test was adopted, as an aquifer pre-test to determine an optimal pumping rate of the wells and neighboring wells could not be used as observation wells due to field limitations. In single well test drawdown well is influenced by well losses as well as well-bore storage. Pumping test data were statistically analyzed by MS Excel-2007 version.

Data interpretation

The pumping test data were interpreted by Jacob straight line method (Jacob, 1950). To determine transmissivity, on the semi-log paper the values of 'drawdown' were plotted against the corresponding values of 'time' and a straight line was drawn through the plotted points; and to determine the slope of the straight line '∆S' the drawdown difference was plotted over one log cycle of time (Figure 3). The transmissivity is determined from the following equation:

$$T(\text{Transmissivity}) = \frac{2.3Q}{4\pi\Delta s}$$

Where, T=Transmissivity in m^2/day, Q=Pumping well discharge in m^3, ∆s = Slope of time vs drawdown plot.

RESULTS AND DISCUSSION

Spatial analysis of transmissivity

The transmissivity analysis is carried out using two methods. One method is based on descriptive statistical testing by identifying transmissivity and anomalies and the other method is based on a classification scheme introduced by Krasny in 1993.

Statistical testing

In this approach, all the transmissivity values collected are pooled in a particular region using Transmissivity index 'Y'. The relationship between transmissivity (T) and logarithmic transmissivity index (Y) is;

$$T(m^2/day) = 10^{Y-8.96} X\ 86400 \quad (1)$$

Found by Jetal and Krasny in 1968, it is used to calculate the logarithmic transmissivity index (Y) from transmissivity (T) values. The above stated equation can be modified as, Logarithmic Transmissivity Index,

$$Y = log\left[\frac{T}{86400}\right] + 8.96 \quad (2)$$

Where T – Transmissivity in m^2 / day

The logarithmic transmissivity index (Y) values are calculated using the modified equation and the calculations are tabulated in Table 2. The mean value of transmissivity index is obtained as 5.2 and the standard deviation of transmissivity index is 0.54. By using these two values, the classification is found as given in Table 4. From the results of the statistical testing, 72% of the wells are under covered background transmissivity anomalies; 12%, under positive anomalies; 8%, negative anomalies and remaining 8%, positive extreme anomalies. The values outside this interval are considered as positive or negative anomalies. The positive anomalies represent prospective zones for groundwater exploration relatively compared to the areas of background transmissivity and the negative anomalies represent less favorable zones. The areas with extreme positive anomalies are highly suitable for local water.

Krasny's classification

Jiri Krasny (1993) proposed a transmissivity classification system based on transmissivity and standard deviation of transmissivity index magnitude and variation of the transmissivity and transmissivity index values. The methods of classifications for transmissivity magnitude and variation are tabulated in Table 3. By using the classification given in Table 3, the groundwater supply potential was identified as given in Table 4, and the standard deviation value of 0.54 observed in transmissivity index 'Y' represents a moderate transmissivity variation and fairly heterogeneous hydro geological environment. From the results of classification of magnitude, 12% of the wells are under high magnitude (Withdrawals of lesser regional importance),

Table 3. Krasny's classification of transmissivity of magnitude and variation.

Classification of T Magnitude				Classification of T Variation			
Coefficient of T (m^2/d)	Class of T magnitude	Designation of T magnitude	Groundwater supply potential	Standard deviation of T Index (Y)	Class of T Variation	Designation of T Variation	Hydro geological environment
> 1000	I	Very high	Withdrawals of great regional importance	< 0.2	a	Insignificant	Homogeneous
1000 - 100	II	High	Withdrawals of lesser regional importance	0.2 – 0.4	b	Small	Slightly heterogeneous
100 - 10	III	Intermediate	Withdrawals for local water supply (small communities and plants)	0.4 – 0.6	c	Moderate	Fairly heterogeneous
10 - 1	IV	Low	Smaller withdrawals for local water supply (private consumption)	0.6 – 0.8	d	Large	Considerably heterogeneous
1 - 0.1	V	Very low	Withdrawals for local water supply with limited consumption	0.8 – 1.0	e	Very large	Very heterogeneous
< 0.1	VI	Imperceptible	Sources for local water supply are difficult	> 1.0	f	Extremely large	Extremely heterogeneous

Table 4. Results of summary statistics of transmissivity index (Y), Krasny's magnitude and variation.

Well No.	T (m^2/day)	T Index 'Y'	Results of T index 'Y'	Results of 'T' magnitude	Results of 'T' variation
1	3.65	4.59	Negative anomalies	Smaller withdrawals for local water supply	Extremely heterogeneous
2	69.98	5.87	Positive anomalies	Withdrawals for local water supply	Extremely heterogeneous
3	16.98	5.25	Background transmissivity	Withdrawals for local water supply	Extremely heterogeneous
4	16.59	5.24	Background transmissivity	Withdrawals for local water supply	Extremely heterogeneous
5	2.67	4.45	Negative anomalies	Smaller withdrawals for local water supply	Extremely heterogeneous
6	12.91	5.13	Background transmissivity	Withdrawals for local water supply	Extremely heterogeneous
7	16.22	5.23	Background transmissivity	Withdrawals for local water supply	Extremely heterogeneous
8	236.9	6.40	Positive extreme anomalies	Withdrawals of lesser regional importance	Extremely heterogeneous
9	54.17	5.76	Positive anomalies	Withdrawals for local water supply	Extremely heterogeneous
10	9.73	5.01	Background transmissivity	Smaller withdrawals for local water supply	Extremely heterogeneous
11	186.15	6.29	Positive extreme anomalies	Withdrawals of lesser regional importance	Extremely heterogeneous
12	7.07	4.87	Background transmissivity	Smaller withdrawals for local water supply	Extremely heterogeneous
13	5.41	4.76	Background transmissivity	Smaller withdrawals for local water supply	Extremely heterogeneous
14	6.53	4.84	Background transmissivity	Smaller withdrawals for local water supply	Extremely heterogeneous
15	102.39	6.03	Positive anomalies	Withdrawals of lesser regional importance	Extremely heterogeneous
16	4.82	4.71	Background transmissivity	Smaller withdrawals for local water supply	Extremely heterogeneous
17	5.21	4.74	Background transmissivity	Smaller withdrawals for local water supply	Extremely heterogeneous
18	52.13	5.74	Background transmissivity	Withdrawals for local water supply	Extremely heterogeneous
19	26.01	5.44	Background transmissivity	Withdrawals for local water supply	Extremely heterogeneous
20	8.43	4.95	Background transmissivity	Smaller withdrawals for local water supply	Extremely heterogeneous
21	10.26	5.03	Background transmissivity	Withdrawals for local water supply	Extremely heterogeneous
22	6.14	4.81	Background transmissivity	Smaller withdrawals for local water supply	Extremely heterogeneous
23	5.87	4.79	Background transmissivity	Smaller withdrawals for local water supply	Extremely heterogeneous
24	20.43	5.33	Background transmissivity	Withdrawals for local water supply	Extremely heterogeneous
25	5.31	4.75	Background transmissivity	Smaller withdrawals for local water supply	Extremely heterogeneous

40% of wells are under intermediate magnitude (withdrawals for local water supply), 48% are under low magnitude (smaller withdrawals for local water supply) and 100% of the wells are under covered moderate variations (fairly heterogeneous hydrogeological environment).

CONCLUSION AND RECOMMANDATIONS

The study has shown that the transmissivity varies from 2.67 to 236.9 m^2/day (with a mean of 37 m^2/day), whereas the specific capacity varies from 5.47 to 451.63 m^3/dd/m (with a mean of 76 m^3/d/m). Spatial variation of transmissivity values was further analyzed using statistical testing and Krasny's classification systems. From the results of the statistical testing, 72% of the wells are under covered background transmissivity anomalies; the remaining 12, 8 and 8%, under positive, negative and positive extreme anomalies, respectively. From the results of Krasny's classification system, 48, 40 and 12% of the wells aree under covered low magnitude (smaller withdrawals for local water supply), intermediate magnitude (withdrawals for local water supply) and high magnitude (withdrawals of lesser regional important) respectively. 100% of the wells are under covered moderate variations (fairly heterogeneous hydrogeological environment). Spatial variation of transmissivity magnitude and variation were identified as best useful in management practices and sustainable development of groundwater.

A much more appropriate quantitative understanding of the groundwater for a long time pumping test and observation is well required.

ACKNOWLEDGEMENT

The author is thankful to Executive Engineer, Andhrapradesh State Irrigation Development Corporation (APSIDC), Miryalguda Division, Nalgonda Districts, for collection and utilization of the data.

Conflict of Interests

The author(s) have not declared any conflict of interests.

REFERENCES

Birpinar ME (2003). Aquifer parameter identification and interpretation with different analytical methods. Water SA, 29 (3):251-256.

Geology and mineral resources of Andhra Pradesh (2006). Geological survey of India, India.

Illman WA, Tartakovsky DM (2006). Asymptotic analysis of cross-hole hydraulic tests in fractured granite. Ground Water 44(4):555-563.

Jacob CE (1950). Flow of groundwater in engineering hydraulics: New York, John Wiley, pp. 321-386.

Krasny J (1993). 'Classification of Transmissivity Magnitude and Variation", Vol. 31, No. 2-GROUND WATER

Mayooran S, Manarathna SP, Gogulan N, Rajapakse RLHL (2011). An Aquifer Characteristic Analysis For Identifying Groundwater Resource Development Alternatives In The Wet Zone Of Sri Lanka. Civil Engineering Research for Industry, Department of Civil Engineering – University of Moratuwa.

Ramakrishna S (1998). Text book of groundwater. 1st Edition, K.J.Graphars, Chennai.

Sarwade DV, Singh VS, Puranik SC, Mondal NC (2007). Comparative study of analytical and numerical methods for estimation of aquifers parameters: A case study in basaltic terrain. J. Geol. Soc. India 70(6):1039-1046.

Sudher Kumar M, Srinivas CR, Srinivasa RK, Mondal NC (2012). Evaluation of hydrogeological properties of fractured granitic aquifer in a micro watershed of Kandukuru -Southern India. Int. J. Water Res. Environ. Eng. 4(12):386-396.

7

Estimation and mapping of groundwater characteristics in Greater Wad-Medani Locality, Gezira State, Sudan

Sami O. E.[1], Mustafa Y. M.[2], Shamseddin M. A.[3] and Hilmi H. S.[4]

Faculty of Agriculture, Alzaiem Alazhari University, Khartoum, Sudan.

Groundwater dominates the domestic water supply (85%) in the Gezira State, Central Sudan. Groundwater resources in Gezira are basically represented by two water-bearing geological formations: the Nubian Sandstone formation overlain by the Gezira formation and shallow quaternary to recent deposits in minor streams. This study aims at identifying the pattern distribution of selected aquifer characteristics of the Gezira geological formation. It applies empirical Driscoll's equation to generate the aquifer's transmissivity. Geospatial techniques in the Geographic Information System (GIS) accompanied with Digital Elevation Model (DEM) datasets were used in interpolation and map transmissivity. The obtained transmissivity of 7.44 to 875 m^2 a day were found to be conformable with those conventionally based values of 28 - 824 m^2 a day. The spatial variation in the water table was extracted using the DEM data, resulting in a reliable potentiometric map (water table map) for the Gezira formation as compared with that traditionally produced. Potentiometric map was used effectively for mapping trends in the water levels and flow directions. The study showed a powerful example of using GIS and DEM techniques in studying groundwater aquifer characteristics, specifically the geospatially dependent variables.

Key words: Groundwater, aquifer characteristics, spatial distribution, DEM.

INTRODUCTION

Geospatial sciences have become increasingly vital in managing natural resources, especially groundwater resources. With climatic changes and their detrimental impacts on all components of the hydrologic cycle, there is a persistent need to employ cost effective techniques in capturing, storing, processing, analyzing and displaying hydrologic and hydrogeologic data to assist decision making and planning to develop robust conservation and sustainable management practices. Integrating different themes of surface and groundwater in a geospatial context will provide an opportunity to understand the various processes pertaining to groundwater occurrence and development. For instance, the output of the geospatial analysis, whether it is aquifer parameters (that is, transmissivity) or water quality problems (saline zones, polluted zones, etc.) could be further utilized as inputs to develop groundwater flow conceptual and numerical models for deeply studying groundwater issues (Goodchild, 1992; Koch, 1997; Brodie, 2002; Gachet, 2006; Muley, 2008).

The number of functioning water wells is estimated at 2000m deep and shallow wells. Each of these formations has its own depositional characteristics, hydrogeological flow patterns, recharge and boundary conditions (GSWC,

2011); thus they should be differently managed. A wide range of groundwater studies were carried out in the Gezira region, Central Sudan, especially on pollution of urban centers, saline zones, and water barren zones of the basement complex as well as the potential of the aquifers (Saeed, 1974; Magboul, 1992; Elzein, 2007). However, the lack of persistent data has greatly reduced sound management decisions, reflected on occasional failures of drilling successfully freshwater wells in some areas, total drawdown of yet some other wells and miss-pinpointing of well sites. The advance techniques of remote sensing and GIS could help a lot in managing groundwater resources in a low cost manner (Gachet, 2006; Muley, 2008).

The main objective of this study is to understand groundwater aquifer characteristics in the Gezira state, Sudan. The specific objectives were to estimate transmissivity and water table depth using geospatial analysis techniques in GIS with remote sensing technique. The study area, lies within the administrative boundaries of Greater Wad-Madani locality Gezira, Sudan (Figure 1). The area lies between 14° 24' 33.25 N, 33° 25' 47.07 E, 14° 07' 22.56 N and 33° 50' 31".41 E, 14° 35' 21.09 N and 33° 24' 47.73 E and 14° 19' 59.82N and 33° 32' 51.57 E. The area occupies about 180 km^2.

The study area generally has flat slope of 10 cm per km. Some depressions represent relicts of proto-Blue Nile system spreading in the central part of the state. However, it was bounded by isolated hills in the eastern, western and southern direction. The Blue Nile River and its tributaries are the main surface water hydrological features in the area, reacting with groundwater within a distance of 2 to 4 km. The state is characterized by a short rainy season (July – September) and high rain fall variability (30%). The soil is vertisols where cracks developed due to change in soil moisture. The rock sequence in the study area is generally comprised of the Per- Cambrian basement complex rocks; the Cretaceous Nubian Sandstone; Quaternary Gezira Formation, Pleistocene buried channel deposits and recent alluvial and superficial deposits (Andrew, 1948; Whiteman, 1971; Vail, 1982). Groundwater in Gezira State is generally found in three aquifers, viz. Gezira, Nubian and Basement aquifers. (Ahmed, 2009) indicated that 90% of Gezira water supply is from the Gezira aquifer, while about 10% is from Nubian sandstone aquifers. Water in the Gezira aquifer occurs under the water table and semi confined conditions. The Blue Nile is the main sources of groundwater recharge for this aquifer The Vertisol soil has negative impact on groundwater recharge as it does not support percolation of water downwards because of its very low infiltration rate. Water hyacinths have demonstrated a great potential for purification of wastewater through physical, chemical and biological mechanisms. Artificial wetlands have been used for secondary treatment and for specific tertiary treatment such as removal of nitrogen and bacteria Successful case studies indicate that wetlands significantly reduce organic matter, suspended solids (SS), pathogens, heavy metals and excessive nutrients such as nitrogen, phosphorus and heavy metals from wastewater. The deviation in chemical, biological, and physical characteristics among wetland ecosystems and complications in understanding and predicting the efficiency of such systems, have motivated the development of artificial wetland systems (Barrie, 2002). As a result, for over 50 years, natural and artificial wetlands have been engineered for wastewater treatment particularly for small and medium sized communities and isolated areas in Europe and the USA where over 700 artificial wetlands have been constructed.

MATERIALS AND METHODS

The required wells and boreholes data were collected from the databases of the (GWC) and Water Management and Irrigation Institute (WMII), University of Gezira. The collected data included specific capacity (m^2/day), discharge (m^3/h), static water level (m), dynamic water level (m) and drawdown (m). The conventional fragmented available data were extracted from the documents of Gezira Water Corporation (GWC).

The transmissivity characteristics of the Gezira formation unconfined aquifer were estimated using specific capacity following the empirical formula of Driscoll (1986) as follows:

$$T=1.042\ (Q/Sw) \text{ for unconfined aquifer} \quad (1)$$

Where, T is transmissivity ($m^2\ day^{-1}$), Q is constant discharge rate ($m^3\ day^{-1}$) and Sw is drawdown in the pumped well after 1 day (m).

The inherent GIS capability in analyzing and interpolating vector and raster datasets was used to draw up aquifer characteristics maps, that is, transmissivity distribution, water table, directions of flow and drainage system. Survey data were collected using the Global Positioning System, (GPS), which were combined with the table attributes of boreholes data. DEM of Shuttle Radar Topography Mission (SRTM) with 90 m resolution was used.

Digital base map of Gezira State was prepared and compiled. The interpolation technique was used to infer distribution of the estimated transmissivity values. These inferred values were calibrated against calculated pumping test data by the analytical methods of Jacob and Theis's Recovery method:

$$\frac{Q}{s_w} = \frac{T}{0.183 \log\left(\frac{2.25Tt}{r_w^2 S}\right)} \quad (2)$$

where the term (Q/Sw) represents the specific capacity of a well, Q denotes the constant discharge rate (L3 T-1), rw is the radius of pumped well (L) and S denotes the storativity (unitless).

Well head elevations derived from the DEMs and water level elevations (static water level, SWL) were used to produce the potentiometric/water table map. The result is the z_value field (pot_surf) that is used to draw the potentiometric map:

$$\text{Pot\_surf} = \text{Raster value} - \text{SWL} \quad (3)$$

Where SWL is the static water level (m); Raster value is the elevation value extracted from the DEM grid (m) and Pot_surf is the elevation to the water table (m). Drainage systems are essential elements to construct the hydrogeologic potential of the renewable aquifers of the study area. The study results were validated using the previous studies mentioned above.

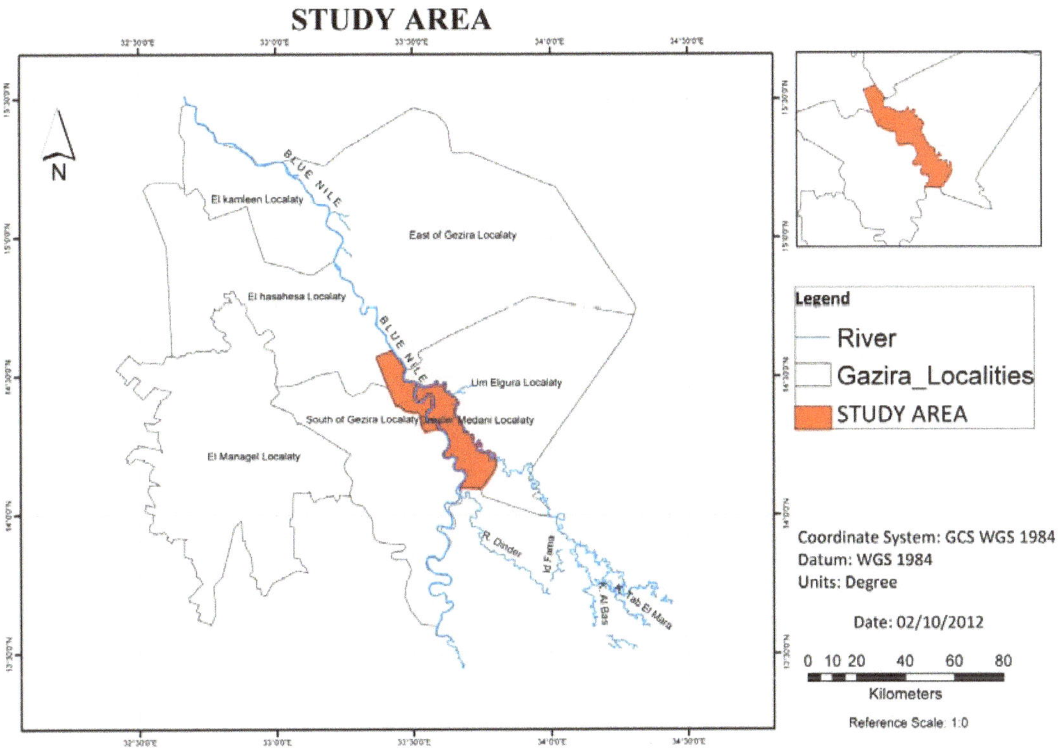

Figure 1. Study area location in Central Sudan.

RESULTS AND DISCUSSION

The data was collected from several wells distributed within and around the study area. Figure 2 shows the distribution of the wells. The transmissivity values for the Gezira aquifer estimated are shown in Figure 3. The average transmissivity was found to be 360.79 m^2/day. In the southern part transmissivity value was found to be as high as 2841 m^2/day. This odd value is attributed to problems in data quality since the drawdown was found to be inconsistent with the general gradations of drawdown values.

The transmissivity values based on the analytical solutions of Theis and Cooper-Jacob were 28 and 824 m^2/day minimum and maximum values respectively, and the average 368 m^2/day, as reported by Elkrai (2004) which is almost the same transmissivity values obtained from the empirical specific capacity calculations as shown in Table 1. The variation in transmissivity values in Gezira formation is attributed to the nature of the sediments depositional cycle of the geological formation. The intercalations of sediments in rapid changes may result in varied hydraulic conductivity and anisotropic conditions. The analysis of hydraulic conductivity of the Gezira formation resulted in values ranged between 34 and 95 m/day.

The aquifer's data were further analyzed using GIS techniques to detect the distribution of hydraulic parameters. The distribution of transmissivity values can be seen in Figure 3. The trend and pattern of the transmissivity distribution is conformant with the hydrogeological knowledge base of the area. Gezira formation has a rather moderate transmissivity values having higher values at the south and gradually getting flat and similar values at the middle area, Figure 4.

Digital Elevation Model of the study area is used to produce a potentiometric surface water table which defines the level to which water rises in a well under unconfined aquifer of the Gezira formation as seen in Figure 5. Potentiometric head data was used to map trends in the water levels, Figure 5 and directions of decreasing flow, Figure 5. The highest water table levels were found in the southern part of the study area gradually trending in the northern and north-western directions. This distribution is fairly conformable with the generally known hydrogeology of the area. Caution should be practiced as to the integrity of potentiometric maps drawn by interpolation methods which is subject to hypothesis in the first place and depends on the resolution of the DEM (90 m in this study). The trend and direction of flow of groundwater is safely indicated. The drainage pattern in Figure 5, was found conformant with the general observed hydrology of the area which reveals the potential of recharge to the unconfined aquifers from the Blue Nile River. A further comprehensive analysis based on the watershed scale to delineate net flow

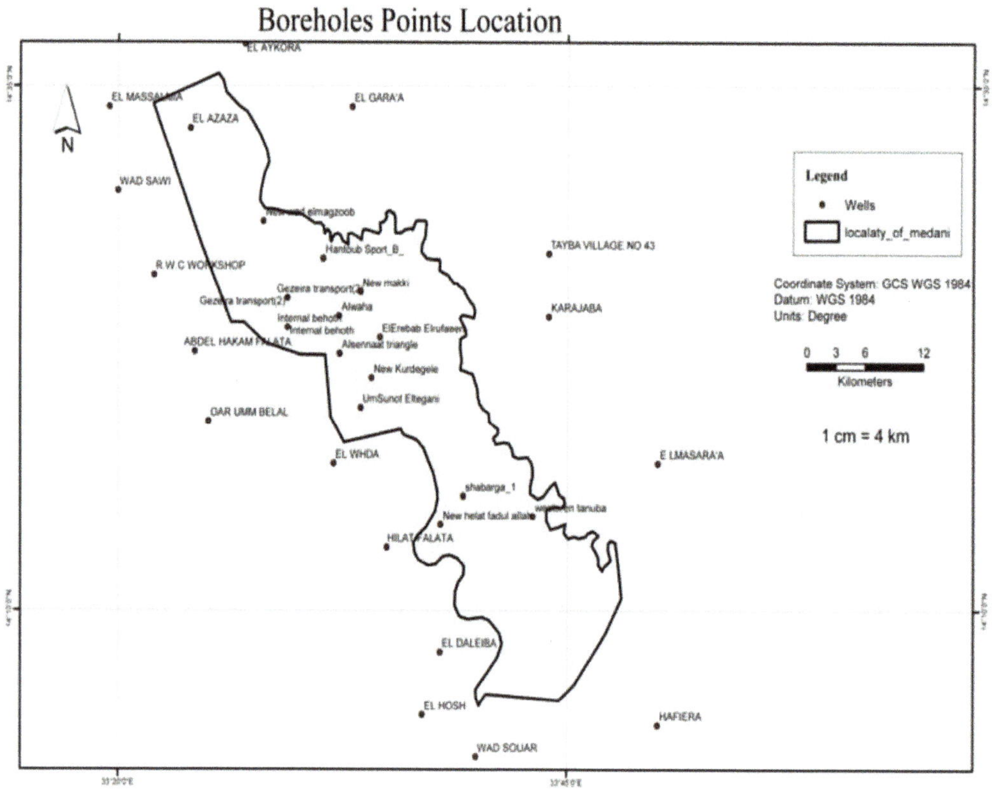

Figure 2. Location of the studied wells.

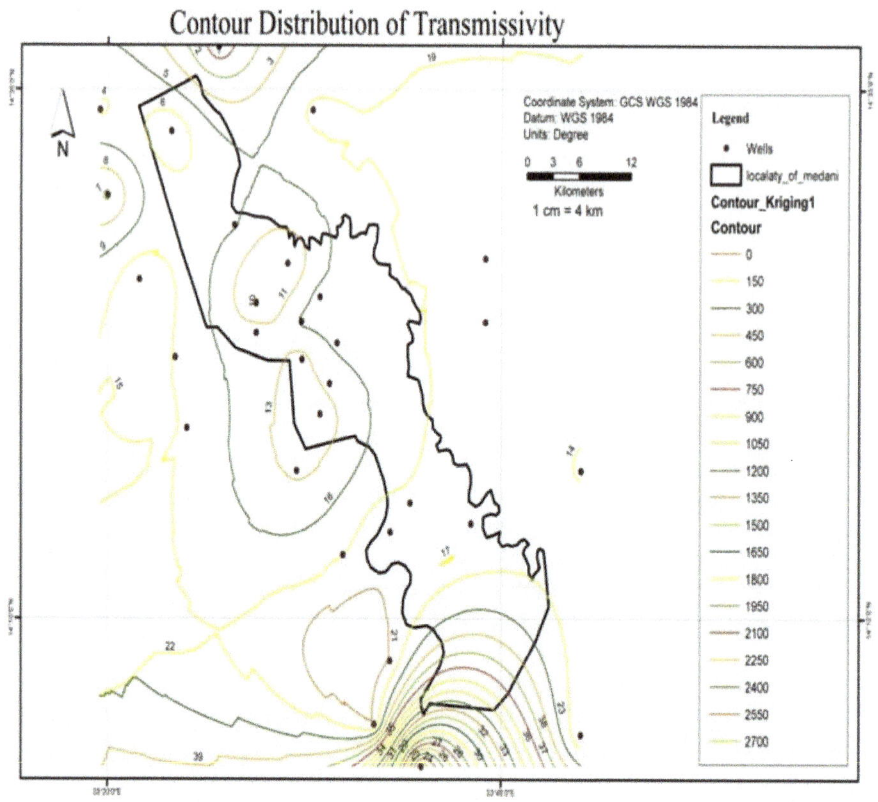

Figure 3. Transmissivity distribution.

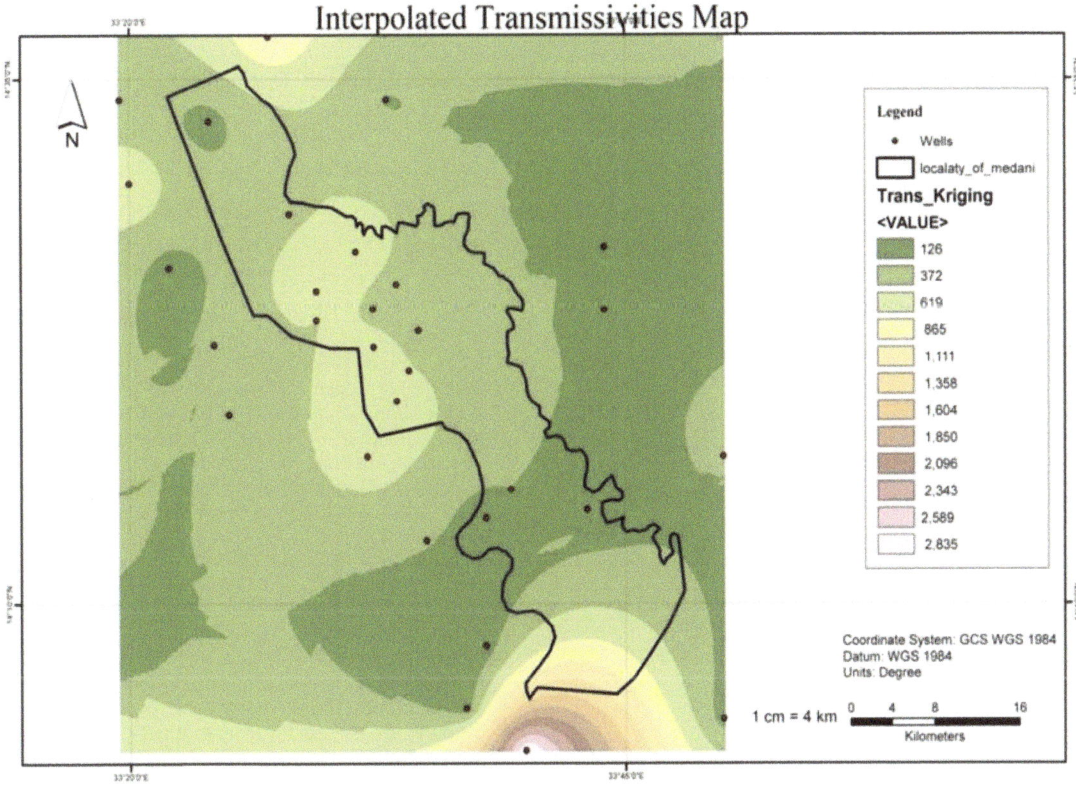

Figure 4. Spatial distribution of the transmissivity.

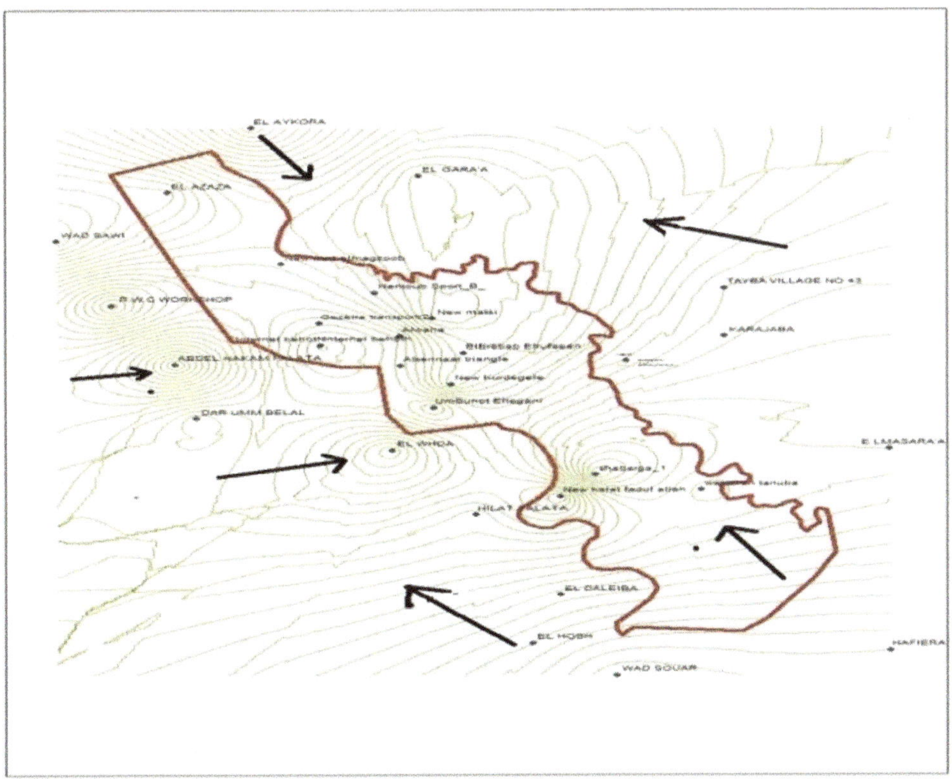

Figure 5. Directions of decreasing water table level depths.

Table 1. Transmissivity calculated from Pumping Test Data.

Method	Transmissivity (m^2/day)			Hydraulic Conductivity (m/day)		
	Min.	Max.	Mean	Min.	Max.	Mean
Jacob's	28	824	368.7	0.411	91.55	34.03
Theis's Recovery	66.5	643.2	355.2	0.977	71.47	26.56

*Source: Elkrai (2004).

Volumes and stream orders is needed.

Conclusion

Groundwater resources in Gezira State are basically represented by two water-bearing geological formations: the Nubian Sandstone formation overlain by the Gezira formation and shallow quaternary to recent deposits in minor streams. This study aims at identifying the pattern distribution of selected aquifer characteristics of the Gezira geological formation. The study applied empirical Driscoll's equation to generate the aquifer's transmissivity. Geospatial techniques in the Geographic Information System accompanied with Digital Elevation Model datasets were used in interpolation and map transmissivity. The obtained transmissivity of 7.44 to 875 m^2 a day were found to be conformable with those conventionally based values of 28 - 824 m^2 a day. The spatial variation in the water table was extracted using the DEM data, resulting in a reliable potentiometric map for the Gezira formation as compared with that traditionally produced. Potentiometric map was used effectively for mapping trends in the water levels and flow directions. The study showed a powerful example of using GIS and DEM techniques in studying groundwater aquifer characteristics, specifically the geospatially dependent variables. Further comprehensive analysis based on the watershed scale to delineate net flow volumes and stream orders is needed.

REFERENCES

Ahmed NM (2009). Development of a Computerized Information System For Domestic Water Management- Case Study: Greater Wad Medani Locality, Gezira, State Gezira University. Sudan. WMII, gezera university. Wad Medani Ph.D.

Andrew G (1948). Geology of the Sudan. In: Tothil, J. D. (ed.) Agriculture in the Sudan, London. Oxford Univ. Press, pp. 84-128.

Brodie RS (2002). Putting Groundwater on The Map: A Status Report On Hydrogeological Mapping In Australia. Bureau of Rural Sciences. International Groundwater Conference. Darwin, Northern Territory. Australia.

Driscoll FG (1986). Groundwater and Wells USA, Johnson Division, St. Paul, Minnesota.

Elkrai A, Kheir O, Shu L, Zhenchun H (2004). "Hydrogeology of the Northern Gezira Area, Central Sudan." J. Spatial Hydrol. 4(2).

Elzein EA (2007). Geoelectrical and Hydrogeological Characteristics of the Groundwater Aquifers in the Gezira Area Central Sudan. Khartoum, Sudan. Ph.D.

Gachet A (2006). Water Exploration in Darfur-Sudan- Sub-agreement #:K K6110, Contract funded by USAID, Operated by Radar Technologies France (RTF).

GSWC (2011). Gezira Water and Sanitation Strategic Plan. Khartoum, Public Water Corporation, Ministry of Water Resources. Sudan.

Koch M, Mather PM (1997). "Lineament mapping for groundwater resource assessment: a comparison of digital Synthetic Aperture Radar (SAR) imagery and stereoscopic Large Format Camera (LFC) photographs in the Red Sea Hills, Sudan." Int. J. Rem. Sens. USA 18:1465-1482.

Goodchild MF (1992). "Geographical information science". Int. J. Geogr. Info. Sys. 6(1):3-45.

Magboul AB (1992). Hydrogeology of the Northern Gezira area. Khartoum,Sudan, Kartoum. M.Sc thesis.

Muley RB, Babar M, Atkore SM, Ghute BB (2008). Application Of Geology And Remote Sensing In The Groundwater Potential Zones In Drought Prone Areas of Parbhani District, Maharashtra, Department of Geology, Dnyanopasak College, India.

Saeed TM (1974). Geological and Hydrogeological studies of Khartoum and north Gezira area, Cairo University, Eygpt. Ph.D.

Vail JR (1982). Geology of the central Sudan: The Land Between Two Niles. Balkema, Rotterdam.

Whiteman AJ (1971). The geology of the Sudan Republic London, Oxford, Clarendon Press. UK.

8

Numerical simulation of groundwater of large-scale irrigation in Northern China: Case of Baoyang Irrigation Area, Shaanxi, China

HUSSEIN I. Ahmed, Junmin LIU and ZHANG –Yinqin

Civil Engineering Department, College of Water Resources and Architectural Engineering, Northwest A & F University, Yangling, Shaanxi Province, P. R. CHINA.

The numerical simulation of groundwater for the single aquifer layer system of Baoyang irrigation area was performed. Visual MODFLOW was carried out to estimate the sustainability of the groundwater resource and to delineate suitable plan for future groundwater use. The conceptual model was built by analyzing the hydrogeological data, aquifer properties, pumping wells, head observations; etc. A groundwater model for Baoyang Irrigation Aquifer was developed, calibrated, and validated using observed groundwater levels for the period of 2006 to 2009. The reliability of the model is tested by long series historical groundwater monitoring data and the model is then used to predict the impact of groundwater exploitation until 2048, during this period; all parameters and well exploitation of 2009 were maintained constants. The water balance was checked for differences between the volume of water that is leaving and entering into the system. Calibration model was performed by adjusting the input parameters; the optimization results of parameters were calibrated automatically during the model processing using the Parameter Estimation Package "PEST". The results show that the computed water heads fit into with perfectly the measured data, which indicate that the conceptual model and the parameters used in the model can reflect the actual physical system of the study domain. Also the results found that the water table in the aquifer is slightly falling down due to continuous withdrawal of water. Finally, the model applications were reported.

Key words: Numerical simulation model, finite difference method, groundwater, MODFLOW, Baoyang irrigation, Shaanxi, Northern China.

INTRODUCTION

The Baojixia Yangmaowan irrigation district (hereinafter referred to as Baoyang irrigation). It is located in Xianyang district west of GuangZhong Plain; Shaanxi Province; a part of loess plateau, and an important part of loess platform. Extending from the steep valley of the Wei River in the west to the Jing River in the east. It is a unique landform of Loess Plateau in Northern China, which is also relatively short in water resources. In this area; the surface water resources cannot meet the growing demand of water for the development of domestic, industrial supply, crop irrigation, transport, recreation, sport and commercial fisheries, power generation and so on. Therefore, scientific and effective groundwater utilization becomes increasingly important. In addition, the loess has good water-retaining property and forms a good water-rich aquifer, which makes it possible to use and develop groundwater. For these reasons, even disregarding the potential threats due to the climate change, this situation appears as one of the biggest challenges of the current era. If the actual trends of development, and population growth does not change. So, this study focused on Numerical Simulation of Groundwater, using Visual MODFLOW Model. The requirement, to develop the numerical modeling of groundwater, to ensure appropriate response and sustainable of the aquifer and make decisions in perspective, the current and historic water use and management in to the existing knowledge of the hydrologic cycle in the area. Compare between the observation and estimation levels of groundwater, and calibrate parameters to show the optimization results, etc. This may help in water resources planning and management and may lead to improvements and managing in water resource, in the present and future.

A groundwater model is any computational method that represents an approximation of an underground water system (Anderson and Woessner, 1992). Generally, mathematical model is the common in groundwater - surface water analysis; it is the best method to develop the main equation of ground water, such as Darcy's law and the partial differential equation of ground water flow in two or three-dimension, the solution of partial differential equation for transient three- dimensional groundwater; requires for the use of a numerical method, such as a finite difference method; Visual MODFLOW is a famous computer program that simulates groundwater using finite difference solution; developed by the U.S. Geological Survey (McDonald and Harbaugh, 1988). A mathematical model (e.g. analytical model and Numerical model) describes the physical processes and boundaries of a groundwater system using one or more governing equations. Numerical models are usually solved by a computer and more computationally demanding than analytical models, (Barnett et al., 2012). The numerical models are widely used for the analysis of basin-scale groundwater hydrology and to quantify the water exchange between different aquifer levels (McDonald and Harbaugh, 1984). Among the numerical methods for solving groundwater flow and transport equations: finite difference method (FDM) and finite volume method (FVM) seem to be more popular for their relative simplicity and ease of implementation (Ataie et al., 1999; Moldrup et al., 1996). The finite difference method (FDM) was applied in this study.

The water resources of the rainfall, runoff of Weihe River and Qi shui he River through five irrigations channels (streams) as main source gravity irrigation is used (Yacor, 2003), and groundwater as the subsidiary (Zhang and Wei, 2007). Agricultural irrigation is the main water utilization. Groundwater is also used for irrigation, in addition to a little living and production water. In 1975, the two schemes (Baojixia and Yang mao irrihation area) were merged to form the Baoyang irrigation scheme.

Modeling software selection

After hydrogeological characterization of the site has been completed and the conceptual model developed, computer model software is selected (Visual MODFLOW-2000), which is a computer program that numerically solves the three-dimensional groundwater-flow equation for a porous medium using the finite-difference method. It is being extensively used worldwide to carry out research in the field of groundwater resource management. The advantages of MODFLOW include numerous facilities for data preparation, easy exchange of data in standard form, extended worldwide experience, continuous development, availability of source code, and relatively low price and forth. A full description of the capabilities of MODFLOW can be found in (McDonald et al., 1988, 2000).

Many hydrological studies and projects have been carried out on this aquifer system. For instance, by using FEFLOW and GIS as a platform, a numerical simulation model fort Typical Irrigation has been established by (Liu, 1991). The high degree of dispersion of groundwater level has been published by (Hao and Junmin, 2011) in his M.Sc Dissertation. Chen (2011) found that the water consumption can generally be satisfied, but there are many potential problems. In the groundwater quantity, there is no water shortage, but the growth of groundwater capacity has slowed down considerably, and decreases in the last two year. A simplified multi-objective optimization model for optimal allocation of water resources in Baojixia; up-tableland irrigation district has been established by Hong et al. (2009). The current model described in this study was conceptualized by identification of the objective, data collection and screening, conceptual model, mathematical model (computer code) was selected; model was designed and run, and model calibration with certain parameters. Finally, model sensitivity, uncertainty, validating,and predictive for future forecasting was reported.

Study area

The study area; Long: 107° 50' ~108° 50' E, Lat: 34° 12' ~34° 35' N; of the Baojixia Yangmaowan irrigation district (hereinafter referred to as Baoyang irrigation). It located in Xianyang district in the west of GuangZhong Plain; Shaanxi Province, a part of loess plateau, and an important part of loess platform, extending from the steep valley of the Wei River in the west to the Jing River in the east. Its is a unique landform of Loess Plateau in Northern China, which is also relatively short in water resources (Figure 1). It is one of the biggest irrigation districts in Shaanxi and an important agricultural production area in China. Climatically, this area belongs to continental monsoon climate, the scheme situated in semi-arid and sub-humid region, which is featured with four clear seasons with rains in spring and autumn, hot in summer and cold in winter. This region was characterized by variant landforms. In general, it is high in the south and north, and low in the center. Moreover, this area is also declining from the west to the east in elevation. Topographically, the area is situated at an elevation in the Guanzhong Basin in the south as 400 to 800 m a.s.l with open forest vegetation (Zhao et al., 2003). The total area is about 1758.26 km^2 (175826 ha), and the cultivate area (agricultural acreage) is about $12*10^4$ ha (1200 km^2); (Liu et al., 2001).

MATERIALS AND METHODS

To carry out this simulation, there are 6 gauging stations of rainfall, and two stations for meteorological data within the area were selected, with record data ranged from 1981 to 2009 years in length. As well as more than 60 observation wells were selected. The gauging stations covered most area. In addition, the data belong to the streams (channels), Aquifer properties, hydraulic conductivity, and forth were gained from College of Water Resources and Architectural Engineering, Northwest A & F University and previous studies (Liu et al., 1991, 2001). A mathematical formulation that consist of appropriate differential equations for the system, and their solutions, was used for simulation. The steps in mathematical modeling are: Examination of physical problems and understanding the physical behavior of system in relation to cause and effect to formulate a conceptual model of how the system operates (Domenico, 1972). Translation of physical problem into mathematical terms, making appropriate simplifying assumptions, developing governing equations, Calibration and valiadation for reliability, and application of the mathematical results in terms of physical problems that require solution.

The reference evapotranspiration "ETo" in the area was calculated using meteorological data of two stations within the area; with monthly record data, which ranged from 1981 to 2009 in length (Table 1), this period was chosen preliminary to develop groundwater of the area as the part of study plan of Numerical simulation of groundwater.

Model design and input parameters

A conceptual model was created to represent the aquifer system beneath the Baoyang aquifer. Model design includes all parameters that are used in the development and calibration. The input parameters involve: model grid size and spacing; the model was assumed as single layer, and divided in to 100 rows, 151 columns, the total number of cells is 15100 cells, the length unit is meter and time unit is day (Fingure 2). Lithological data along with top and bottom elevations of the Aquifer layer were gained from the Google earth and literature reviews (Liu and Yang, 1989). Figure 3 presents the top and bottom elevations as contour lines. The thickness of the Aquifer is varying between 80 to 260 m depth. Boundary conditions consists of Stream flow (Channels irrigation); streams data were gained from College of Water Resources -Northwest A&F University-, the initial recharge was assumed as percentage of rainfall, while the evapotranspiration was calculated from meteorological data using the FAO-56 application of the Penman-Monteith equation (Allen et al.,1998). ET_O calculator software was applied. Aquifer properties including horizontal, vertical hydraulic conductivities, specific yield (specific storage for confined aquifer) were collected from previous studies (Liu et al., 1991, 2001) and the global standard values. The optimizations values of recharge, hydraulic conductivity, and the specific yield are adjusted automatically by PEST method during the calibration process until the measured groundwater levels match the calculated values. During calibration the primary storage coefficient is 0.037 and 0.0246 is estimated by model, hydraulic conductivity along rows is 7.68, the estimated is 5.012. There are 5 artificial perennially Irrigation Channel (like small streams) in the area were coded as 1, 2, 3, 4, and 5 (Figure 2). The stream data consists stream stage and are difference from time to time according to the discharge. Some information about the stream that are used in the model are presented in Table 2. The stream conductance were calculated automatically by MODFLOW; using the formula described below.

$$C = \frac{L \times W \times K_z}{T.el - B.el} \quad (1)$$

Where; C: is the Streambed conductance, L: is the reach length of stream line in each grid cell, W: is the Stream width in each grid cell, K_Z: is the Streambed hydraulic conductivity, T.el: is the streambed top elevation, and B.el: is the Stream bottom elevation.

The irrigation requirement in the area

In Baoyang irrigation area, the irrigation requirement has been calculated directly by the law of irrigation quota. This water-saving irrigation system was established according to actual situation in irrigation district. The major crops grown in the area are: winter wheat, maize, grain, cotton, oil and other cash crops. Many researchers, (Zhang and Wei, 2007; ZHANG and Li, 2008) had used it and so on. Researchers used the irrigation efficiency 0.552. By computing, the total irrigation requirement was 322 and 502 million m^3 in normal and dry year respectively. Based on analyzing climatic material of Baoji gorge irrigation district from 1981 to 2003, result shows that the range of perennial average winter wheat water requirement in the district is 459.5~ 559. 6 mm, the range of maize water requirement is 454.6~ 528. 3 mm, the range of cotton water requirement is 609.2~ 681. 3 mm. The spatial distribution of crop water requirement for these three kinds of crop has similar regulation, (LI et al., 2010). According to 2009 total water consumption is about 1021.5 Mm^3, for irrigation, animals and industrial. 31% (314.67 Mm^3) of them from groundwater withdrawn per year and about (752.73 Mm^3) is withdrawn for irrigation purposes 27% (203.75 Mm^3) from groundwater. Likewise, according to 2010 about 725.46 Mm^3 is withdrawn for irrigation purposes, 27.7% (200.88 Mm^3) from groundwater.

The flow equation and finite differential equation

Groundwater modeling begins with a conceptual understanding of the physical problem. The next step in modeling is translating the physical system into mathematical terms, various numerical solution techniques have been established after (Freeze and Witherspoon, 1966, 1967; Bredehoeft and Pinder, 1973). In general, the results are the familiar groundwater flow equation and transport equations. Among the most approaches were used in groundwater modeling; Finite Difference Method, Finite Element Method, and Analytical Element Method. All techniques have their own advantages and

Figure 1. Study area location-the numbers within the map refer to the wells and the blue lines refer to streams

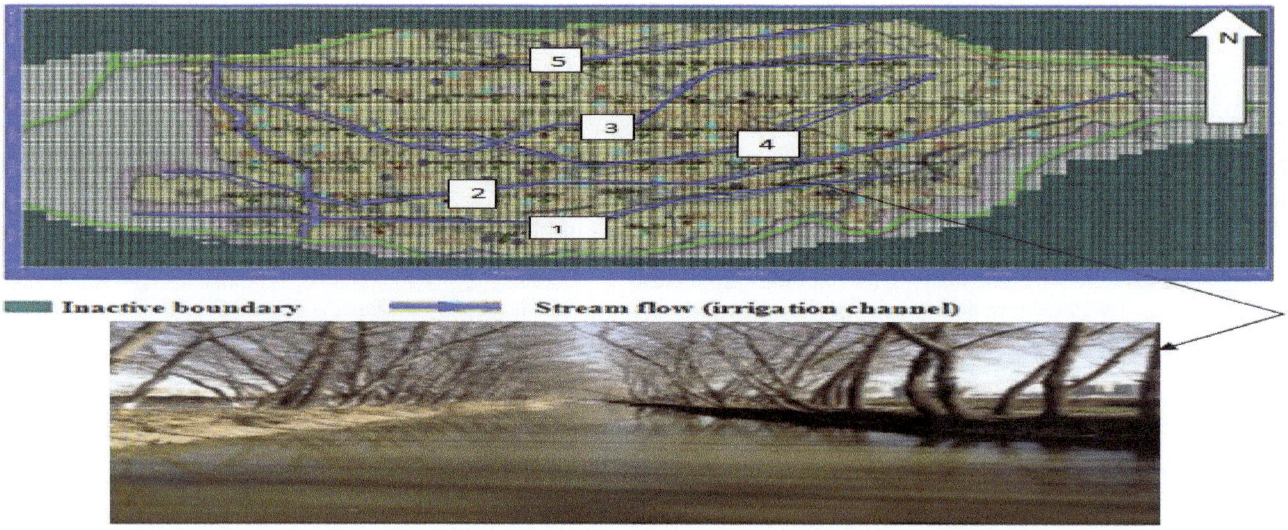

Figure 2. Map showing location of the stream flow and grid lines and size in the area

Table 1. Meteorological data of two stations (Xing and Wuging, 1981-2009).

Type of data	Period	Length of data (years)
Monthly mean relative humidity %	1981-2009	29
Monthly mean wind speed (m/s)	1981-2009	29
Monthly mean air temperature (^{o}C)	1981-2009	29
Monthly mean air pressure (Kpa)	1981-2009	29
Sunlight (Hours)	1981-2009	29

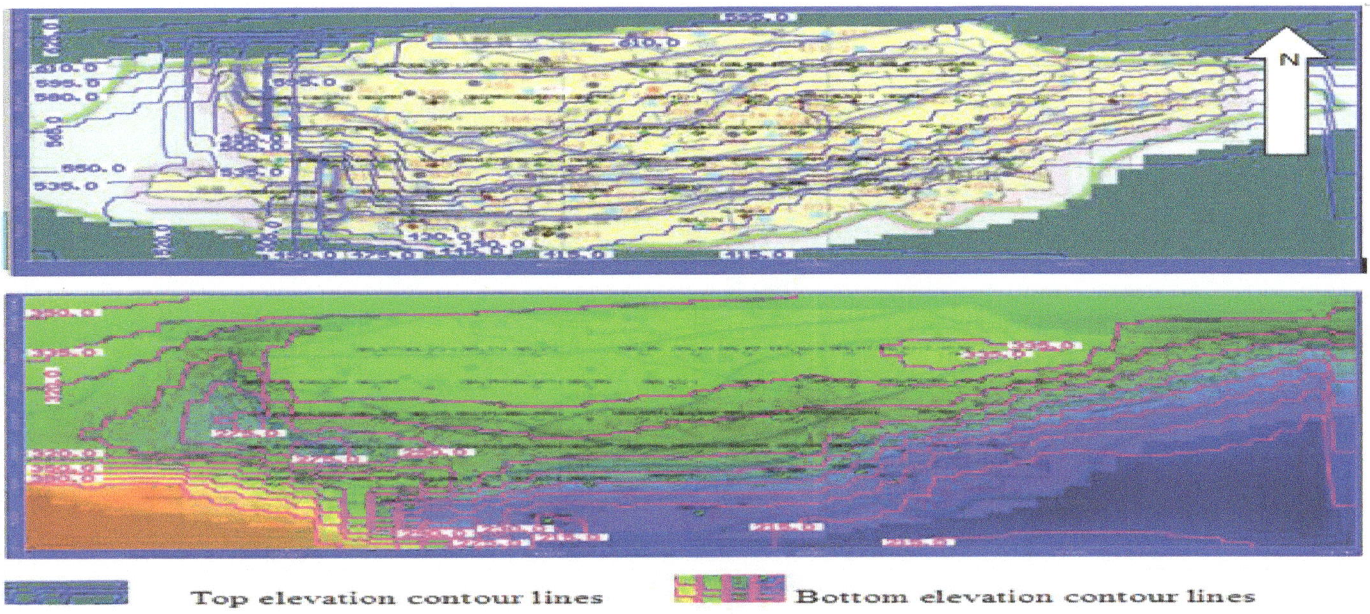

Figure 3. Map presents the Top and Bottom elevations contour lines.

Table 2. Presentes the streams parameters and conductances that calculated by V.MODFLOW

Name of stream (Chanal)	Code	Width (m)	Initial flow (m^3/s)	K_z (m/day)	Calculated conductance (m^2/day)	
					Min	Max
Wei qu	1	6.5	44	0.022	2.574	60.47
Wei gao gan gu	2	5.5	25.3	0.022	0.439	57.69
Zong dong gan qu	3	6	25	0.022	3.506	51.52
Wei zong gan qu	4	5.6	12.5	0.022	0.1055	68.58
Yangmao wan gan qu	5	6.5	20	0.022	5.84	65.4

disadvantages with respect to availability, costs, user friendliness,applicability, and user background. In this study Finite Difference Method was carried (Remson et al., 1971). This method has been preceded by replacing the derivatives in the differential equations by finite difference approximations (LeVeque, 2004; Luka, 2008). These approximations are algebraic in form, and the solutions are related to grid points.

Water balance method for groundwater evaluation

The water balance is merely a detailed statement of the law of conservation of matter, which states simply that matter can neither be created nor destroyed, but can only be changed from one state or location to another. Water balance also means the balanced relationship between mean annual values of groundwater recharges and discharges in the evaluated area (Liu and Jin, 2004; Chen, 2010). In this paper, evaluated area is the agricultural irrigated region as mentioned earlier. The groundwater balance evaluation in 2008 and 2009 are taken as an example. The mass (water) balance for shallow aquifer in the area can be determined as: Inflow - Outflow = Change in storage; the equation can be written as:

$$W_{str\,l} + W_{rech} + W_{sto} - (W_{str\ out} + ET + W_{well}) = \pm\Delta S \quad (2)$$

Where; $W_{str\ l}$ and $W_{str\ out}$ are the stream (Irrigation channels) Leakage in and out respectively. W_{sto} : is the water storage, W_{rech} is the water recharge to the aquifer, ET is the total evapotranspiration, and W_{well} : is the amount of water withdrawn from the wells, and ΔS is the change of storage. The calculating terms of recharge and evapotranspiration are mentioned above.

MODFLOW - model calibration and parameter estimation

Model calibration is the process of adjusting the input parameters and boundary conditions of a model to achieve a close fit to observed data in a real groundwater system (Ye, 2011). After the first run of a model, model results may differ from field measurements, this is expected because modeling is just a simplification of reality and approximations and computational

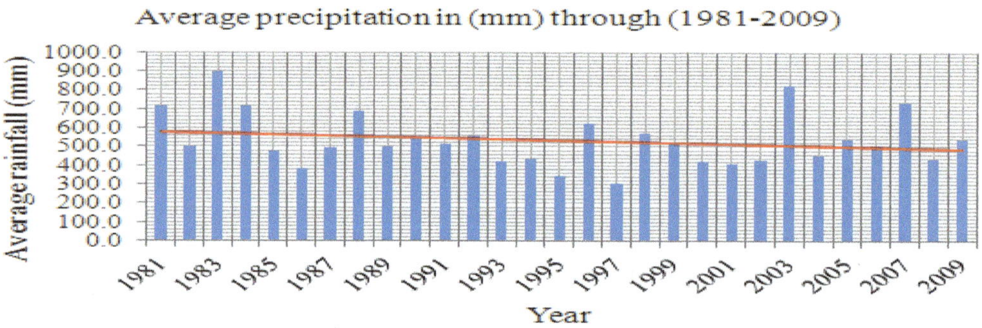

Figure 4. The average precipitation for six stations through (1981-2009).

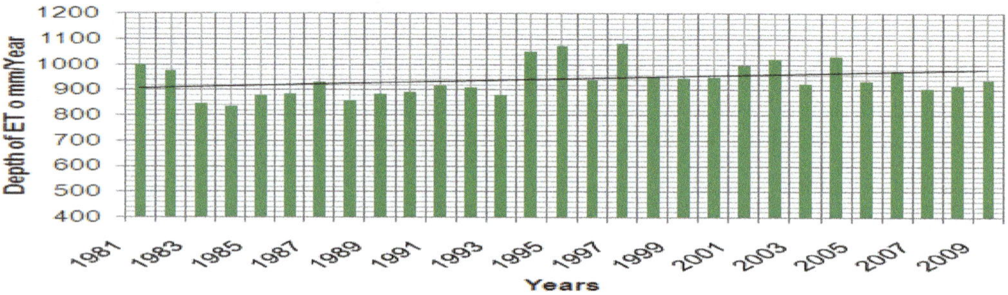

Figure 5. The mean annual reference evapotranspiration (1981-2009)

errors are inevitable. Calibration can be done manually or automatically. Software like Parameter Estimation Package "PEST" (Doherty et al., 1994) and UCODE (Poeter and Hill, 1998) can be used for automatic calibration. In this research the PEST is carried out to calibrate the parameters (hydraulic conductivity, specific yield and groundwater recharge). Drawdown for each grid cell is calculated as the difference between the initial (starting) head value and the calculated head value, at each grid cell. To illustrate the calibration process of a groundwater model, consider the groundwater head measurements (h_{obs}) i at the observation point i, the simulated head at the same point is (h_{sim}) i. The root mean square error of the residual is given by:

$$RMSE = \left[\frac{1}{n}\sum_{i=1}^{n}(h_{obs} - h_{sim})^2\right]^{1/2} \quad (3)$$

RESULTS AND DISCUSSION

Statistic analysis of rainfall

The precipitation data from a number of stations can be used to depict the isohyetal line, or isohyets, which joints all points that receive the same amount of precipitation. In this paper, the result of statistic analysis found that the maximum and minimum annual precipitation of Baoyang irrigation district through (1981 to 2009) according to available data are ranged between 895 mm in 1983, and 303.4 mm, in 1997 (Figure 4). The average of rainfall such as: maximum, minimum and mean is 603.4, 475.9, and 534.9 mm. The standard deviation (SD), Variation (VAR), Skew and the Coefficient of variance (CV) were found as 47.6, 2802, 0.2 and 0.01 respectively. Generally, the trend line was slightly decreasing by the resent years, which indicate that the rainfall is decreasing recently. While the Evapotranspiration was calculated using the FAO-56 application of the Penman-Monteith equation is visualized in (Figure 5). the trend line indicated that the mean annual reference evapotranspiratin was increased by resent years, this may due to increasing the temperature and climate changes as general.

Groundwater-level fluctuations

The groundwater of the Baoyang irrigation Aquifer System generally occurs under unconfined conditions. So, the water table is not stable and fluctuates according to wet and dry seasons and also according to the withdrawal rate of groundwater. Many wells were used to monitor the water level fluctuations of the aquifer. These wells have total depths of about 20 to 185 m. The water levels were presented (Figure 8a and b) as example, the results found that in some well there are no change in head during predictive period, but as general, the groundwater level was decreasing by resent years, and will be rose in autumn when precipitation and streams stage were high, and declined during winter; spring and summer when precipitation and streams stage were low and water demand for agriculture were increasing. The water table level is falling (see cross-sections in Figure 6) during 30 days, 360 days and 720 days [calibration period (2008 to 2009)]. Clearly, the water table is slightly falling down due to continuous withdrawal of water. The maximum and average drawdown during the calibration and predictive periods were presented in Table 2. The result also found that the movement of Groundwater is from north to south, meanwhile to the east and west. The comparison between the simulated and the observed head, as well as groundwater direction (Figure 7).

Model calibration, validation and predictive

Model validation is the next step after calibration and defined as the comparison of model results with numerical data independently derived from laboratory experiments or observations of the environment (ASTM, 1984). The term "validation" is not completely true when used in groundwater modeling (Oreskes et al., 1994) asserted it is impossible to validate a numerical model because modeling is only approximation of reality. The objective of model validation is to check that if the calibrated model works well on any dataset. In this research the calibration and validation process were performed by comparing results, which yielded from the Visual MODFLOW model with recording data for years 2008 to 2009 as calibration period and 2006 to 2007 as the validation period. Registered historical data from monitoring locations in the study area were used for these purposes. A statistical analysis was did automatically by Visual MODFLOW. Statistics used for this purpose includes; the correlation coefficient (R^2), Root Mean Square Error (RMSE), Normalized Root Mean Square Error (N.RMSE) and Recharge. Uncertainty in groundwater modeling is inevitable for a number of reasons,evaluated computing confidence intervals at a 0.05 significance level was carried in this model., according to (Baalousha and Köngeter, 2006).

A model was used to predict some future forecasting scenarios of groundwater simulation of Baoyang irrigating area; the predictive was carried for 40 years (that is, 2009 to 2048) during this period, all parameters and well exploitation of 2009 were maintained constants. The results of this simulation can be used to study the behavior of Baoyang irrigation aquifer system versus continuous exploitation. The maximum and average drawdown of some well-selected as examples- were scheduled in Table 3. The model may also be used to evaluate different remediation alternatives. However, errors and uncertainties in a groundwater flow analysis and solute transport analysis make any model prediction no better than an approximation. For this reason, all model predictions should be expressed as a range of possible outcomes that reflect the assumptions involved and uncertainty in model input data and parameter values.

The calibrated and predictive model can be used to derive components of the groundwater budget and to estimate the response of the regional system to new stresses, such as increased groundwater withdrawals due to increasing demand of water for the development of industry, agriculture, and daily life. These results may lead to sufficient enough information, and Water-resource managers can use these informations to make informed decisions when planning for future groundwater development. The uncertainty associated with inaccuracies in the groundwater-flow model is carried forward to the model applications.

The statistics summary of model calibration and validation

For the ground-water level calibration and validation, the results found that the minimum and maximum Standard Error of Estimation (SEE) of all heads is ranged between 1 and 2.365 m for calibration and 1.011 to 1.065 for validation, root mean square error (RMSE) of the observation wells is 11.17 m to 19.5 m, and 7.878 m to 11.241 m for calibration and validation respectively. The minimum and maximum normalize root mean square error (N.RMS %) is 6 to 10.48, and 4.235 to 6.043 for calibration and validation respectively. Generally, the results showed that the values are increased during the time period (2008 to 2009), and decreasing during (2006 to 2007) this may be due to the difference in conditions or for reasons that are mentioned by Oreskes et al. (1994). The correlation coefficient (R^2) is ranged between 0.98 and 0.93, the average values of SEE, RMSE, N.RMSE and R^2 are 1.8 m, 16.2 m, 8.7 and 0.9% respectively (Table 4). The water levels, a spatial comparison between the measured and simulated head of the observations wells were mentioned in Figure 8a and b. Multiple ground-water-flow systems are not well simulated by a single layer (layer 1) where all properties; ground-water levels, and flows are average values. As a result, there may be little differences between measured and simulated head in some wells. The time interval of the model during calibration and validation process is divided as monthly.

Sensitivity analysis

Sensitivity analysis refers to the assessment of the dependence between model inputs and outputs, and specifically, how the latter respond to perturbations in the former. Sensitivity analysis is important for calibration,

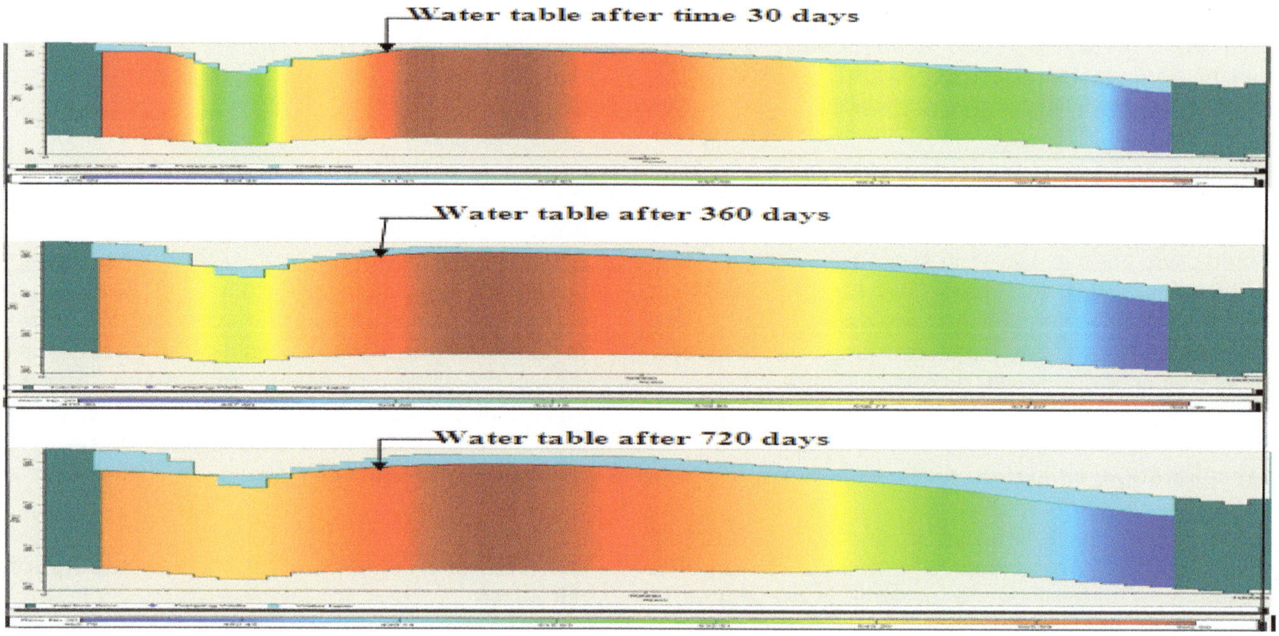

Figure 6. Cross-sections presente the falling in water table during the calibration period (2008-2009).

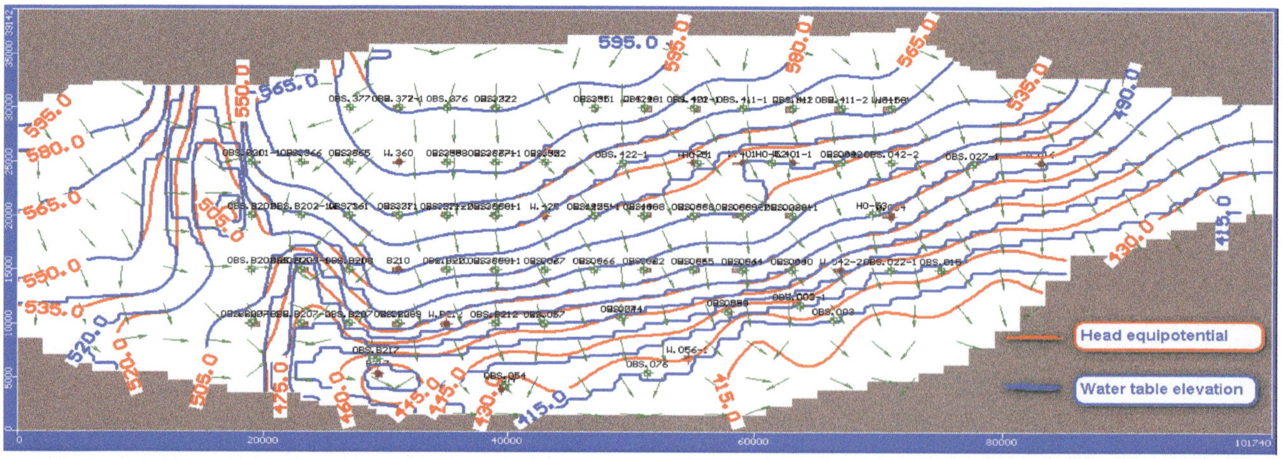

Figure 7. Map showing the velocity direction of groundwater in the domain and comparison between the simulated and the observed head, the observed is appear as red ,while the blue is the calculated.

optimization, risk assessment and data collection. In regional groundwater models, there is a large number of uncertain parameter, coping with these uncertainties is time-consuming and requires considerable effort. Sensitivity analysis indicates which parameter(s) have greater influence on the output. Parameters with high influence on model output should get the most attention in the calibration process and data collection. In this research, the optimization results of parameters (hydraulic conductivities, specific yield and recharge) were calibrated during the model processing. The results showed that the horizontal hydraulic conductivity "K_x" and specific yield "S_y" have the large values of sensitivity, so that, they have high influence in the model (Table 5). The initial values of recharge and optimization results were presented in Figure 9. The magnitude of recharge rates varied by node which ranged from 0.035 to 17.29 mm/year; and 0.035 to 11.105 mm/year; estimated and observed respectively, and the average values of recharge rates is 4.62 mm/year (observed) and 7.89 mm/year

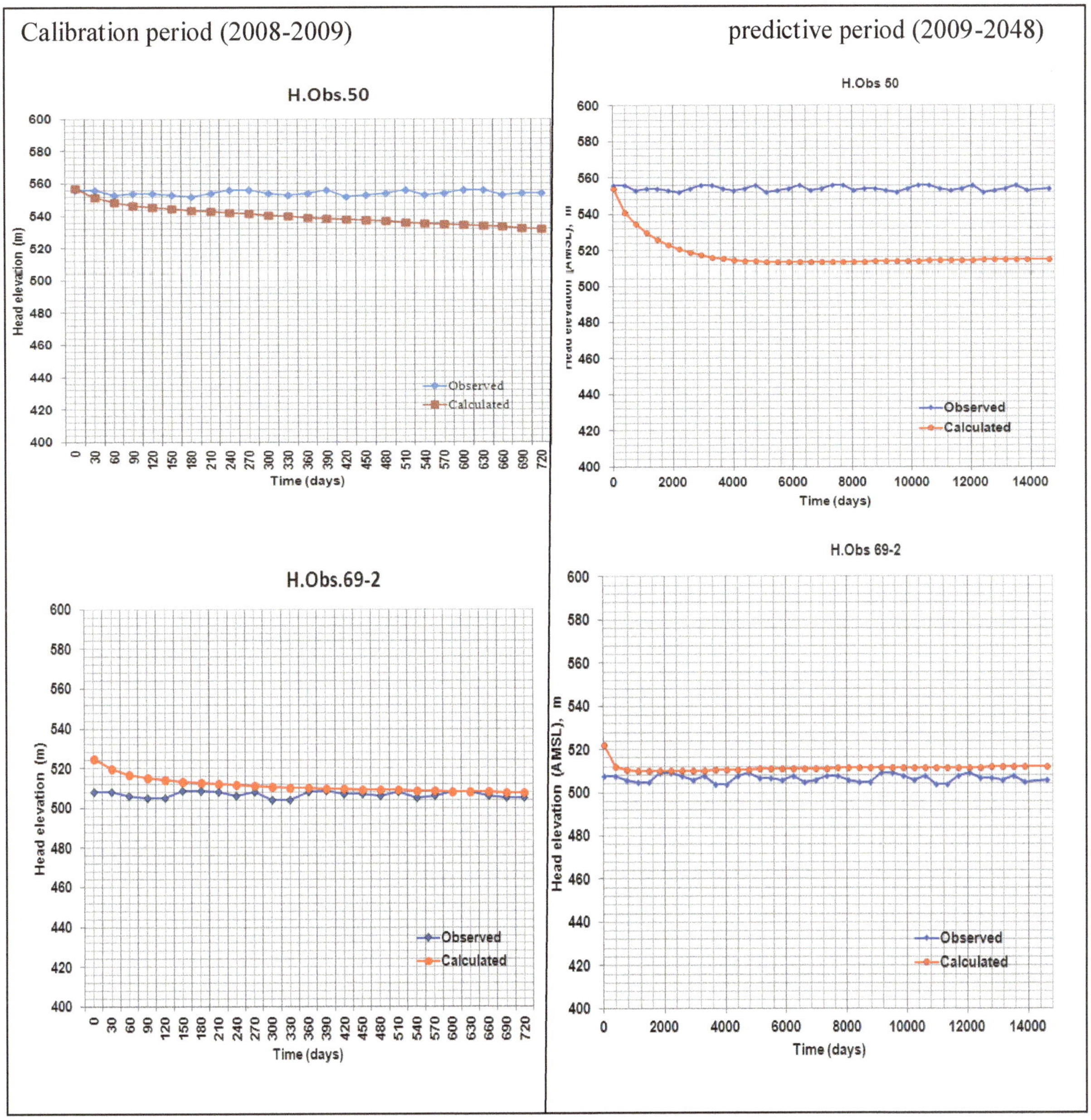

Figure 8a. Illustrated the head elevation during the calibration and predictive period for head observations .H.obs.50 and H.Obs.69-2

as estimated for the calibration period (2008 to 2009), the time interval divided as monthly.

Water balance of the system

Finally, the water balance was checked as differences between the volume of water leaving and entering intothe system. An approximate water cumulative volume during the study period of calibrated model (2008 to 2009) results by Visual MODFLOW, there are three major sources of inflow to the ground-water flow system; recharge from precipitation, stream leakage and groundwater flow to the storage. As cumulative volume, recharge accounts for about 6.38% (0.444 × 10^8 m^3). Channels (stream leakage) accounts for about 3.82% (0.266 × 10^8 m^3), and groundwater flow (storage inflow) accounts for about 89.8% (6.25 × 10^8 m^3). While the water out flow consist; total evapotranspiration (ET) accounts for about 26.16% from the total out flow (1.82

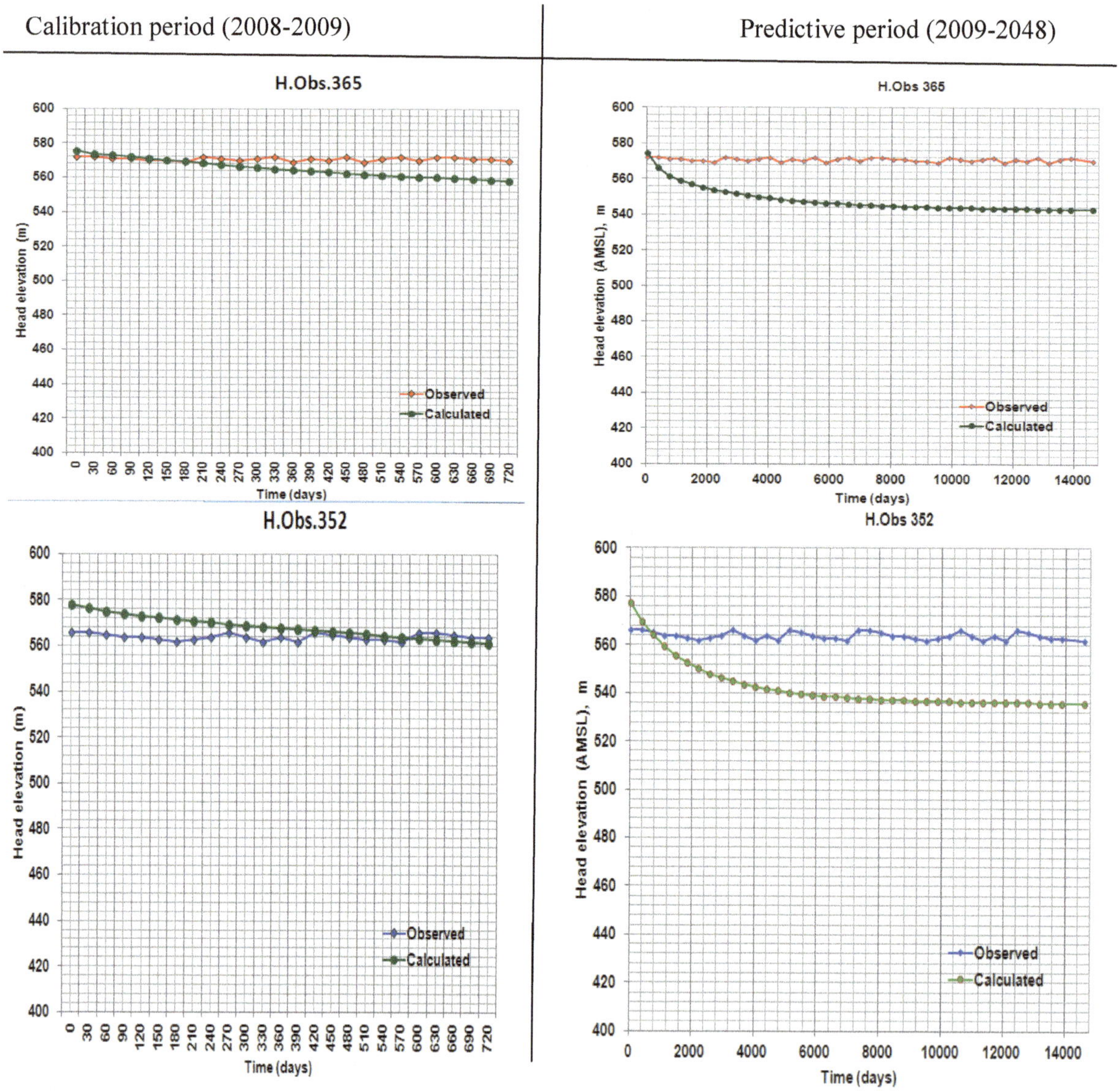

Figure 8b. Fit the head elevation during the calibration and predictive period for head observations .H.obs.365 and H.Obs.352.

$\times 10^8$ m^3), stream leakage out flow is accounts about 1.02% (0.071 × 10^8 m^3) , the pumping wells withdrawn is about 0.5% (0.035 ×$10^8$$m^3$), and groundwater out flow (storage out flow) is about 72.31% (5.03 × 10^8 m^3). The total inflow is about 6.96 × 10^8 m^3 and the total out flow is about 6.956 × 10^8 m^3. Hence, the total inflow minus total out flows about 0.004 × 10^8 m^3. The cumulative of volumes into and out of the system during (2008 to 2009), which were employed in V.MODFLOW indicated that there were no negative impact between the volume in flow and out flow. However, the difference between inflow and out flow is decreasing by increasing time; it is indicated that the water out from the basin is relatively large in amount than water into the basin. The mass balance error for simulation of about 0.01% is considered acceptable.

Conclusions

A groundwater model for Baoyang Irrigation Aquifer Region (BIAR) in the centre of Guanzhong Basin was developed, calibrated, and validated using observed groundwater levels for the period of 2006 to 2009. During the study we found that the maximum and minimum annual precipitation through (1981 to 2009) according to

Table 3. Illustrated the maximum, and average drawdown for 12 wells during the calibration and predictive periods (2008-2009, 2009-2048).

Well no.	Location		Calculated DDW (2008-2009)		Calculated DDW(2009-2048)	
	Longitude	Latitude	Max (m)	Average (m)	Max (m)	Average (m)
50	108° 19' 54"	34° 13' 29"	25.87	17.65	44.62	40.32
28-1	108° 42' 54"	34° 22' 11"	15.17	11.94	12.3	8.35
69-2	108° 29' 18"	34° 16' 7"	14.86	10.95	12.55	11.01
352	108° 21' 13"	34° 28' 05"	16.13	9.14	41.57	35.01
361	108° 09' 29"	34° 25' 42"	9.17	4.55	12.4	9.77
365	108° 07' 50"	34° 27' 50"	11.46	4.59	26.89	21.65
366	108° 06' 46"	32° 25' 56"	5.9	2.84	14.85	10.82
371	108° 10' 41"	34° 22' 40"	28.93	20.39	36.75	33.2
401	108° 22' 47"	34° 33' 0"	21.74	13.79	53.72	45.74
411	108° 33' 39"	34° 26' 50"	14.9	8.74	38.25	33.17
B201-1	108° 03' 54"	34° 32' 33"	36.28	31.64	34.19	31.05
B203	108° 03' 14"	34° 20' 52"	33.78	28.03	27.8	13.6

Table 4. Statistics report of calibration (2008-2009) and validation (2006-2009) processing.

Variable	Time (days)	Well No.	Calibration results				Validation results			
			SEE (m)	RMSE (m)	N.RME%	R^2	SEE (m)	RMSE (m)	N.RME%	R^2
Head	0	61	1.000	11.177	6.009	0.98	1.015	11.241	6.043	0.978
Head	30	61	1.082	11.477	6.171	0.98	1.026	10.917	5.87	0.978
Head	60	61	1.277	12.965	6.971	0.97	1.037	11.193	6.018	0.978
Head	90	61	1.418	14.161	7.655	0.96	1.047	10.498	5.675	0.978
Head	120	61	1.516	14.923	8.023	0.96	1.038	9.904	5.325	0.978
Head	150	61	1.596	15.46	8.312	0.95	0.991	9.203	4.948	0.98
Head	180	61	1.626	15.771	8.389	0.95	0.961	8.827	4.695	0.98
Head	360	61	1.908	16.749	9.103	0.94	0.974	8.238	4.477	0.98
Head	390	61	1.954	16.928	9.151	0.94	0.982	8.445	4.565	0.98
Head	480	61	2.08	17.799	9.621	0.94	1.06	8.687	4.696	0.976
Head	510	61	2.161	18.202	9.786	0.93	1.065	8.567	4.606	0.976
Head	540	61	2.164	18.386	9.938	0.93	1.052	8.598	4.647	0.976
Head	600	61	2.221	18.001	9.678	0.93	1.011	7.878	4.235	0.978
Head	660	61	2.281	18.612	10.006	0.93	1.029	8.165	4.39	0.978
Head	690	61	2.316	19.043	10.293	0.93	1.013	8.164	4.413	0.978
Head	720	61	2.365	19.506	10.487	0.93	1.011	8.693	4.674	0.978
Average			1.800	16.200	8.700	0.90	1.020	9.200	4.950	0.980

the available data are ranged between 895 mm in 1983, and 303.4 mm, in 1997, respectively. The statistic analysis through the time period showed that the average of rainfall such as maximum, minimum and mean is 603.4, 475.9, and 534.9 mm respectively. Also, the result found that the water table in the aquifer is slightly falling down due to continuous withdrawal of water, and the Groundwater in the model area discharges to the QiShihe River, and Wei He River. Calibration of the conceptual model was performed by adjusting the input parameters such as hydraulic conductivity, specific yield and recharge rate. Performance statistics indicated that simulated groundwater levels followed trends and magnitudes in the observed historical groundwater levels in the underlying Baoyang Aquifer. Overall, calibration results yielded average coefficients of determination of 0.9 and 0.98 and average RMSE 16.2 m and 9.2 m, N.RMSE% is 8.7 and 4.95, SEE is 1.8 and 1.02 for the

Table 5. Optimization Records (Initial conditions) and Optimization Results of Parameters.

Parameter name	Start time (day)	End time (day)	Initial values	Estimated values	Sensitivity	Rel. sensitivity
Kx_1	0	720	7.68	5.012	0.4099	0.2869
Ky_1	0	720	7.68	7.68	0.000	0.000
Kz_1	0	720	0.23	0.23	0.000	0.000
Ss_1	0	720	0.00001	0.00001	0.000	0.000
Sy-1	0	720	0.037	0.0246	0.3840	0.6177
Recharge	0	720	0.035-11.105	0.035-17.29	$9.9*10^{-5}$-$1.3*10^{-4}$	$2.1*10^{-6}$-$6.6*10^{-4}$
Average of recharge			4.62	7.89		

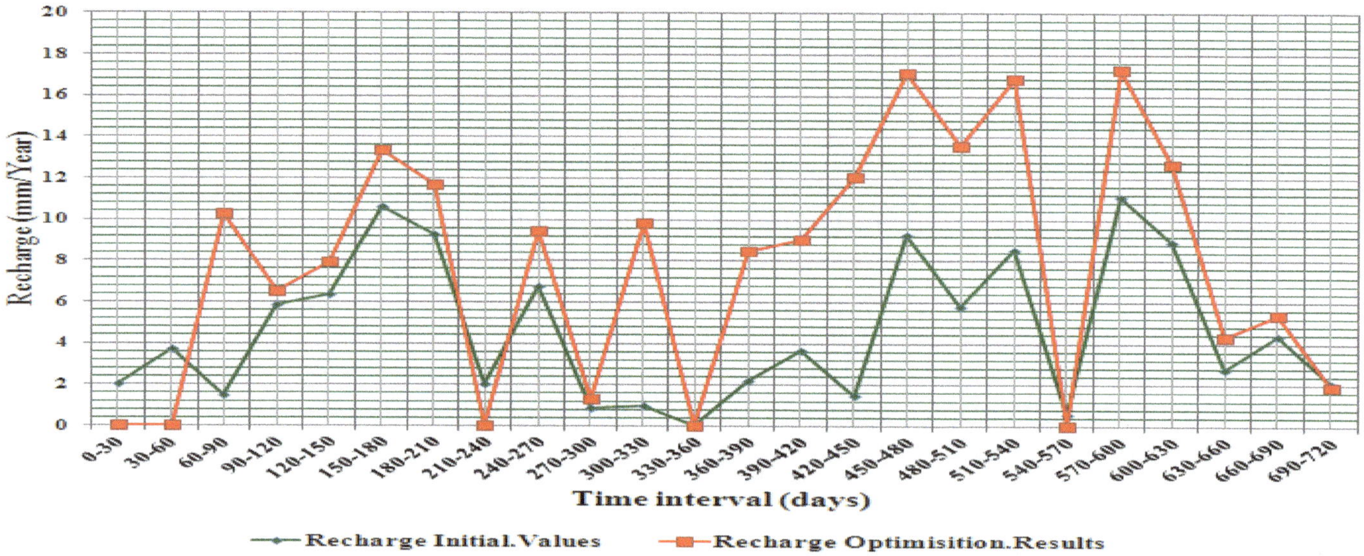

Figure 9. The initial values and optimization results of recharge (mm/year), 2008-2009.

calibration and validation processes, respectively. These results are indicative of a good agreement between predicted and observed groundwater levels. The water balance was checked for differences between the volume of water leaving and entering into the system. The cumulative of water volume into and out of the system during 2008 to 2009 is about 6.96×10^8 m^3 and the total out flow is about 6.956×10^8 m^3 respectively. The optimization results of parameters were calibrated automatically during the model processing using the Parameter Estimation Package "PEST". The initial values of all parameters, which are calibrated and comparison with estimated values were presented in Table 5. While the initial values of recharge and optimization results were visualized in Figure 7. The predictive was carried for 40 years (that is, 2009 to 2048) during this period; all parameters and well exploitation of 2009 were maintained constantlly. According to 2009 total water consumption about 1021.5 Mm^3, for irrigation, animals and industrial. 31% (314.67 Mm^3) of them from groundwater withdrawn per year and about 752.73 Mm^3 is withdrawn for irrigation purposes 27% (203.75 Mm^3) from groundwater. Likewise, according to 2010 about 725.46 Mm^3 is withdrawn for irrigation purpose; 27.7% (200.88 Mm^3) of them is from groundwater.

ACKNOWLEDGEMENTS

The author would like to thank the Chinese Government and Northwest A & F University for fellowship. For data sources, thanks to Northwest A & F University College of Water Resources and Architectural Engineering, and for everyone who gave fulfilled help. And also wish to express appreciation to Editor office, Board etc, and Anonymous journal reviewers for their valuable comments and suggestions concerning this manuscript.

REFERENCES

Allen RG, Pereira LS, Raes D, Smith M (1998). Crop evapotranspiration – Guidelines for Computing Crop water requirements. FAO Irrigation and Drainage Paper. Roma, Italy. N° 56.

Anderson MP , Woessner WW (1992). Applied groundwater modeling, simulation of flow and Advective-transport. Academic Press, San Diego, USA.

ASTM (1984). Standard practices for evaluating environmental fate models of chemicals. Annual Book of ASTM standards. American Society For Testing and Materials, Philadelphia, PA, pp. E978-984.

Ataie-Ashtiani B, Lockington DA, Volker RE (1999). Truncation errors in finite difference models for solute Transport equation with first order reaction. J. Contam. Hydrol. 34:409-428.

Baalousha H, Köngeter J (2006). Stochastic modeling and risk analysis of groundwater pollution using FORM coupled with automatic differentiation. Adv. Water Resour. 29(12):1815-1832.

Barnett B, Townley LR, Post V, Evans RE, Hunt RJ (2012).Australian groundwater modeling Guidelines. Sinclair Knight Merz and National Centre for Groundwater Research and Training. Waterlines Report Series, No. 82.

Bredehoeft JD, Pinder GF (1973). Mass Transport in Flowing Groundwater. Water Resour. Res. 9:194-210.

Chen HY (2010). Calculation of groundwater resources and mineable yield in Mamas Plain, Xinjian. Water Conserv. Sci. Technol. Econ. 16:905-907.

Chen Lin (2011). Groundwater Evaluation In Baoyang Irrigation. M.Sc dissertation Northwest A&F University.

Doherty J, Brebber L, Whyte P (1994). PEST - Model-independent parameter estimation. User's Manual Watermark Computing. Asrtalia.

Domenico PA (1972). Concepts and Models in Ground Water Hydrology. McGraw-Hill Co. New York pp. 405.

Freeze RA, Witherspoon PA (1966). Theoretical Analysis of Regional Groundwater Flow: 1. Analytical and Numerical Solutions to the Mathematical Model. Water Resour. Res. 2(4):641-655.

Freeze RA, Witherspoon PA (1967). Theoretical Analysis of Regional Groundwater Flow: 2. Effect of water table configuration and subsurface Permeability Variation. Water Resour. Res. 3(2):623-634.

Hao J, Junmin L (2011). The Improved Technology of GM (1, 1) Model in predicting groundwater of Xianyang City. China Academic Journal Electronic Publishing House.

McDonald MG, Harbaugh AW, Banta ER, Hill MC (2000). MODFLOW-2000, the U.S. Geological Survey modular ground-water model—User guide to modularization concepts and the Groundwater flow Press. US. Geological Survey Open-File Report. http://water.usgs.gov/nrp/gwsoftware/modflow2000/ofr00-92.pdf.

Hong W, Jian-en G, Chun-hong Z, Xiu-quan X, Yuan-xing Z, Hui S (2009). The Optimal Allocation of Water Resources in Baojixia up-tableland Irrigation District based on the Ecological Basic Flow of Weihe River. National Engineering Research Center for Water-saving Irrigation at Yangling,Northwest A&F University

LeVeque R (2004). Finite Difference Methods for Differential Equations. University of Washington. A. Math. Notes 585:6.

Li LI, Xiaomei WEI, Yezhen HAN (2010). Spatial Variability on Crop Water Requirement Based on Geostatistics. Res. Soil Water Conserv. Article ID 100523409. 0120239204, 17:1.

Liu J (1991). The New Aquifer of under groundwater in the Irrigated area on the Weibei Highland. Act. Univ. Agric. Boreali-occidentalis. 22:1

Liu J, Wei XM, Ma Y, Wang S, Song X (2001). Loess Taiyuan Irrigation District –Sustainability of groundwater exploitation. Northwest A&F University, College of Water Resources and Architectural Engineering.

Liu P, Yang Q (1989). Xian yang City practical hydrological Manual-Xianyang City. Shaanxi Province Water Conservancy Bureau Soil Water Conserv. 88:007.

Liu YW, Jin DL (2004).Evaluation of groundwater resources in Plain area. Express Water Resour. Res. Hydropower Inf. 12:5-8.

Luka F, König, Jonas L (2008). Groundwater: Modeling, management, and contamination. Weiss (editors), (p.cm. ISBN 978-1-60741-395-0 (E-Book), Published by Nova Science Publishers, Inc. New York. pp.163-196.

McDonald MG, Harbaugh AW (1988). A modular three-dimensional finite-difference groundwater flow model. U.S. Geological Survey Techniques of Water-Resources Investigations. Book 6, Chap. A1:586.

McDonald MG, Harbaugh AW (1984). A modular three dimensional finite difference groundwater flow model. US Geol. Surv. Open-File Rep. 83(875):528.

Moldrup P, Kruse CW, Yamaguchi T, Rolston DE (1996). Modelling diffusion and reaction in soils: I.A diffusion and reaction corrected finite difference calculation scheme. Soil Sci. 347(54):161.

Oreskes N, Shrader-Frechette K, Belitz K (1994). Verification, Validation, and Confirmation of Numerical Models in the Earth Sciences. Science 263(5147):641-646.

Poeter EP, Hill MC (1998).Documentation of UCODE, a computer code for universal inverse modeling. U.S. Geol. Surv. Water-Resour. Investig. Rep. 98:4080.

Remson I, Hornberger GM, Molz FJ (1971). Numerical Methods in Sub-surface Hydrology, with an introduction to the finite Element Method. Wiley (Interscience), New York p. 389.

Yacor T (2003). P.cm. principles and cases from developing/ includes bibliographic references index. ISBN: 1-891853-76-7 (hard cover: alk.paper). HD1702, p. 745.

Ye Z (2011). Groundwater Flow and Solute Transport modeling. Department of Geology and Geophysics, University of Wyoming.

Zhang L, Wei XM (2007). Study on Rational Disposition of Irrigation District Water Resources Based on water Supply and Water Demand. China: Northwest A&F University, pp. 10-11.

Zhang YP, Li H (2008). Economical Water-use Model Research of Baojixia irrigation district Based on WEB GIS". China: Northwest A&F University, pp. 24-36.

Zhao Y, WANG F, Wenquan L (2003). Climatic ecological environment, climate change and agr-icultural productivity in Loess Plateau. Agric. Res. Arid Areas 21(4):142-146.

Assessment of Silyaninov index for detecting climatic changes and droughts in the central Sudan

M. A. Shamseddin[1] and H.S.M. Hilmi[2]

[1]Water Management and Irrigation Institute, University of Gezira, P. O. Box 20, Wadmedani, Sudan.
[2]Faculty of Agriculture, Alzaiem Alazhari University, Khartoum North, P. O. Box 1432, Sudan.

Drought is a constraint upon development in Sudan. This paper attempts to understand drought and climate change in the central Sudan using the Silyaninov index (SI) because of its simplicity and its required datasets (monthly rainfall and temperature) are routinely collected in developing countries. Also, the ecoclimatological relationships for the natural vegetation cover were investigated using the normalized difference vegetation index (NDVI) and ancillary climatic data. The climate variability in the central Sudan is found to be highly generated by the variability in rainfall rather than temperature. Rainfall experienced a significant decreasing trend (≈ 3.5 mm per annum) coupled with a significant increasing trend in temperature (≈ 1.4°C per annum) during the period 1960 to 2010. Accordingly, the aridity increased significantly at 50% of the studied stations. SI is found to be more effective in detecting drought than using rainfall dataset alone. However, when the temperature dataset is anomalies-free it could explain effectively most of the historical meteorological droughts witnessed in central Sudan. Using SI, the majority of the drought events were observed in 1970 to 1990, with the exception of Damazine (1998 to 2002) and Ed Duim stations (2000 – 2005). The common wetted years outweighed the common drought years, revealing the localized behavior of the rainfall. The analysis of NDVI showed that the vegetation cover experiences a decreasing pattern under the semi dry (Ed Duim station) and semi humid (Damazine station) climatic zones during the period 2000 to 2010. The relationships of NDVI-SI and NDVI-rainfall were found better than the NDVI- temperature.

Key words: Central Sudan, drought, climate change, natural vegetation cover, Silyaninov index, normalized difference vegetation index (NDVI).

INTRODUCTION

According to the Intergovernmental Panel on Climate Change, IPCC (2007) climate change is defined as a change in the state of the climate that can be identified (e.g. using statistical tests) by changes in the mean and/or the variability of its properties and that persists for an extended period, typically decades or longer. Thus, it refers to any change in climate over time, whether due to natural variability or as a result of human activity. These changes will likely lead to changes in rainfall, atmospheric moisture and regional climate variability, especially in the tropics and sub-tropic areas (Gitay, 2002). Most of the climate change studies have been done in large scales. This is necessitated downscaling studies, as the impacts of climate change will be different from one place to another (IPCC, 2007). For instance, it is expected that Africa will likely be the most vulnerable continent (Challinor et al., 2007).

Drought is an ordinary phenomenon in the eastern African countries, including Sudan. It has affected 100% of the land in Somalia, Eritrea and Djibouti, and 61 to 87% of the land in Ethiopia, Kenya and Sudan (Mati, 2005). The periods 1970s and 1980s were the most decades that have witnessed catastrophic results due to droughts that is, mass lost of souls, destroyed vegetation covers, failure in biomass production, displacement and social stability (Osman and Shamseldin, 2002; Sivakumar et al., 2005; Tilahun, 2006). In Sudan, drought has been related to famine, civil unrest, ill-health and desertification and in turns as a constraint upon development (Hulme, 1987). Ayoub (1999) has reported that the repeated droughts of the late 1960s to mid 1970s have resulted in destroying the whole of *Acacia tortilis* belt of north Kordofan, and in increasing the number of dust storms per year during the period of 1970 to 1980 in El Fasher area, Darfur region.

Climate change and drought pose great threats to human life, economic sectors and environments. Therefore, their quantitative monitoring is very important from many perspectives of views: risk management, assessing adaptive capacity, planning and decision making process. The better drought preparedness depends on monitoring drought's onset, progress and real extent (Morid et al., 2006). However, drought is very difficult to define. For example, Logan et al. (2010) cited more than 150 definitions for drought. This is because drought is not a distinct event, it is as a result of complex factors and it has neither a distinct start nor a finish; it is only after a period of time it can be recognizable (Oladipo, 1985). One of the definitions of drought that had been formulated was that "a sustained period of significantly below normal precipitation". This definition encounters two problems. Firstly, there was no universal agreement on the definition of "a sustained period". Secondly, due to the high variability of rainfall the term norm precipitation is questionable in the African Sahel (Hulme, 2001). In Sudan, Ayoub (1999) has called that there is a southward shifting of the desertification boundaries, which can be considered as a climatic change. It goes without saying that aridity is universally distinguished from the drought as the former is a permanent climate feature of a given region resulting from low precipitation.

Generally, there are four wide types of drought; viz: meteorological, agricultural, hydrological and socioeconomic droughts. The widely natural and human factors used to designate drought are climate, soil water, precipitation, soil type, water storage distribution, number of population and land use (Oladipo, 1985). In order to monitor drought, many indices have been devised, such as the ratio of precipitation and potential evapotranspiration, the standardized precipitation index (SPI), the evaporation ratio, the Palmer drought severity index (PDSI) and the soil moisture index (Arora, 2002; Ntale and Gan, 2003; Sivakumar et al., 2005; Bates et al., 2008; Mpelasoka et al., 2008; Logan et al., 2010). Ntale and Gan (2003) found that SPI is the most appropriate method for detecting the regional east African droughts, agrees with Morid et al. (2006) results for Tehran Province conditions in Iran. However, the preciseness of the SPI depends on the rainfall distribution and its length of record (Wu et al., 2005).

Recently, the drought indices have been used for detecting climate change. Mavromatis (2010) found that the self-calibrated version of PDSI has the potential to be used for climate change impacts assessment studies. In Sudan, climate change impacts are poorly documented. Rainfall remains the most studied climatic variable (Hulme, 1986, 1987; Eldredge et al., 1988; Ayoub, 1999; Fud Elmoula, 2004). Temperature datasets are commonly available but not so studied. Thus, from cost effective and data availability point of view both temperature and rainfall could be effectively used in studying climate change and drought in Sudan.

The Silyaninov index is historically used in classifying the climatic zone of Sudan, seven climatic zones (Table 1). By such mean, detecting any prolonged changes in SI values may indicate a climatic change. There are discernable reasons behind using of SI to detect climate change and drought for the central Sudan: (1) Adam (2002) designated the critical values for every climatic zones in Sudan, these values can be used as baselines, given that the persistent changes in those values can be considered as a climate change; (2) SI is very easy to estimate since its estimation depends on the most two reasonably available climatic variables in Sudan, viz. rainfall and temperature.

Agriculture and livestock breeding are the dominant livelihood generations in central Sudan. Thus, drought and climate change are expected to have detrimental impacts on communities stability. It is therefore, studying the relationship between climate and vegetation cover is very important. Recently, remote sensing techniques were widely used in monitoring vegetation cover, agriculture, natural resources, weather conditions, water balance and droughts (Bastiaanssen et al., 2003; Omuto and Shrestha, 2007; Tan, 2007; Elhag and Walker, 2009; Mavromatis, 2010). Also, remote sensing has been used recently to define aridity as Lioubimtseva et al. (2005) reported that a period with a natural difference vegetation index (NDVI) value of less than 0.07 (a unit less) is classified as arid. The huge advance in the remote sensing techniques makes them capable for studying precisely the climate changes impacts on vegetation cover dynamics in central Sudan. As yet the define of drought years in the central Sudan is depended on memories of local people (Abusin, 1986; Telku et al.,

Table 1. Sudan's climatic zones, on the basis of Silyaninov index (SI).

Climate type	Value of SI
Desert	< 0.1
Sem-desert	0.1-0.3
Dry	0.3-0.5
Semi-dry	0.5-0.7
Sub-humid	0.7-1.3
Humid	1.3-2.0
Super-humid	> 2.0

Source: Adam (2002)

1991), the objectives of this study were to detect climate change and drought using the Silyaninov index (SI) because of its simplicity and required datasets (monthly rainfall and temperature) are generally available in developing countries in the central Sudan. Also, the study examines the dynamic of the natural vegetation cover using the NDVI.

Description of the study area

The study area was located between 11.00 to 14.38 °N and 24.8 to 35.4 °E, which roughly represents the central Sudan with a total area of 434 000 km^2 (Figure 1). The climate is generally hot and dry, ranging from dry to sub humid zones. Rainfall and length of dry season are the most significant climatic variables (Food and Agriculture Organization of the United Nations, FAO, 2006). Rainfall extends for a short period (June to October) with a low magnitude, ranging from 100 in the northern part to 600 mm in the southern part. This is coupled with a high annual evapotranspiration of 2200 mm. Thus, the region suffers shortage in the water supply for a long period of time (8 months).

Central Sudan is distinguished as the main region of both animals and agricultural productions. It includes most of the Sudanese irrigated and rainfed agricultural projects. There is however, a delicate balance between man needs and natural resources as the vast majority of the livelihood generations, basically agriculture and rearing livestock are climate-land-based, especially in rural areas. Both climate and land however, have been deteriorated and in turns disturbed the region's socio-economic stability (Hulme, 1986; Ayoub, 1998). Add to this, the region belongs to the African Sahel belt where drought is a repeatable problem. Moreover, the evidence of the 1984/1985 famine states that drought is the main cause of famine in Sudan (Teklu et al., 1991).

MATERIALS AND METHODS

The first step for detecting climate change and droughts in the central Sudan is to choose the representative stations. To ensure the statistical validity, the station with a minimum record length of 25 years was chosen (Table 2). This restricted the number to only eight stations. Many stations have been found out of service since 1970s and 1980s (Figure 1), especially at the western part, that is, Kordufan and Darfur, due to drought episodes that is leading to instability and insecure situations. Therefore, each station is expected to cover about 54 000 km^2 from the total area, indicating poor spatial meteorological measurements in the central Sudan. Therefore, stationary data series are used instead of the regional ones.

Testing data quality

In order to detect trends in the time series, the following tests were applied in advance.

Test for homogeneity

The nonparametric Thom test was used to examine the homogeneity of the time series, following the procedure mentioned in Rodrigo et al. (1999), Modares and Silva (2007) and Nasri and Modares (2009). The null hypothesis (the time series is homogeneous) is verified at the 0.99 confidence level if $|Z| < 2.58$. The time series of rainfall were found homogeneous. However, temperature time series of 50% of the stations (namely Damazine, El Obied, Wadmedani and Gedarif stations) were found to be inhomogeneous. Consequently, the cumulative residuals method was used to apply the appropriate corrections as described in Allen et al. (1998).

Test for randomness

The presence of a positive correlation will increase the possibility of rejecting the null hypothesis (no trend) while it is actually true and the vice versa is correct for the negative correlation (Yue and Wang, 2004). The detection of trend in a time series entails that the datasets is serially independent (Yue and wang, 2004). Therefore, the Durbin-Watson test was used. It is found that all the stations, with the exception of Gedarif and Damazine, have shown significant positive serial correlations ($P = 0.05$). To eliminate such serial correlation, the pre-whitening approach was adopted, following the approach described in Burn and Hag Elnour (2002).

Test for trend

The non-parametric Mann-Kendall test was used to detect trends in time series as described in Burn and Hag Elnur (2002), Yue et al. (2002), Smith et al. (2004), Yue and Wang (2004), Modarres and Silva (2007), Luo et al. (2008), Basistha et al. (2009), Nasri and

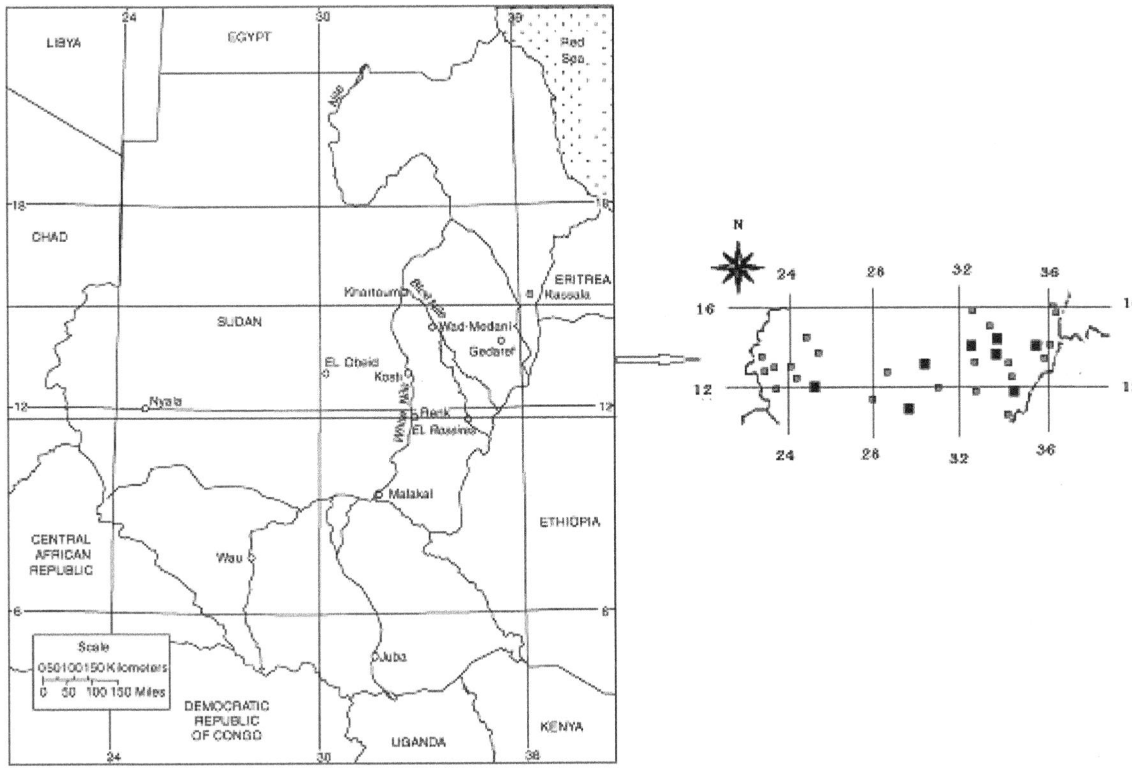

Figure 1. The studied stations (the black squares) in the central Sudan. The grey squares refer to the stations with problems in data availability

Table 2. The locations, altitudes and sample sizes of rainfall (R-N) and temperature (T-N) for the selected stations in the central Sudan.

No.	Station	Latitude (°N)	Longitude (°E)	Altitude (m)	R-N (year)	T-N (year)
1	Kadugli	11.00	29.72	500	47	41
2	Damazine	11.82	34.40	470	43	41
3	Nyala	12.00	24.80	675	36	39
4	El Obied	13.17	30.23	570	43	43
5	Sennar	13.55	33.56	420	36	40
6	Ed Duim	14.00	32.33	378	36	40
7	Gedarif	14.03	35.40	600	43	42
8	Wadmedani	14.38	33.48	405	47	41

Modarres (2009), and Xu et al. (2010). The magnitude of change, expressed as percentages of mean, is determined by the Theil and Sens's method (Burn and Hag Elnour, 2002; Basistha et al., 2009).

Detecting drought and climatic changes using Silyaninov index (SI)

Several methods have been used to classify Sudan's climate, among them Silyaninov index (SI) was found the best (Adam, 2002; Fud Elmoula, 2004; El Fadni, 1998). It's goodness is likely rose from the climatic variables used in its estimation, viz. rainfall and temperature, that is, moisture and thermal factor, which both play a vital role in the climate of the tropics (Fud Elmoula, 2004). For example, Tan (2007) points out that rainfall and temperature control most of the differences in the Earth's vegetation cover, affecting growth rate and plant reproduction. The SI is defined as a ratio between the annual rainfall and the cumulative annual temperature above 10°C as described below (Adam, 1995; Fud Elmoula 2004):

$$SI = \frac{R * 10}{C.T} \tag{1}$$

Where, R is the mean annual rainfall (mm) and $C.T$ is the cumulated annual temperature (°C). The stationary data lengths are shown in Table 2.

The preciseness of the SI in detecting drought is calibrated using the fragmentally documented historical drought events in Sudan, which have been well cited in Leku et al. (1996) and Abusin (1986). The recent drought events (1996 to 2010) were judged using the average crop yields of the period 1974-1982 as a baseline. Data of crop yields were obtained from the records of the ministry of agriculture and FAO reports. Type I errors i.e. the rate of rejecting the null hypothesis of no drought is computed by Clark and Schkade (1969)

$$\sigma p = \sqrt{\frac{\pi(1-\pi)}{n}} \quad (2)$$

$$z = \frac{p_c - \pi}{\sigma p} \quad (3)$$

Where, P_c is the critical value of the sample percent, π is the value of the universe percent designated in the null hypothesis and σ_p is the standard error of the percent, z is taken as 1.96 which is corresponding to the value of 0.025 in one tail of the normal distribution and n is the random sample size. In this case, type I errors (rejecting the null hypothesis while in fact it is true) can be determined by management. The critical values were taken as the means of the upper and lower values designated to each climatic zone in Table 1.

NDVI estimation

This section deals with expected impacts of climate change and drought on the vegetation cover in central Sudan.Two Acacia Senegal forests (Perennial trees) were chosen through a surveying trip using a Global Positioning System (GPS). The first forest is located (11.56 °N and 34.23 °E) in the Damazine region (Blue Nile state, southern-eastern part of the study area) where the climate is sub humid. The second forest (13.24 °N and 35.85 °E) locates in the Gedarif region (Eastern part of the studied area) under semi dry climate conditions. The soil of both forests is heavy clay with a water holding capacity of 230 mm/ m depth. Thus, these forests are anticipated to reflect the long term impacts of climate change on the vegetation cover, soil moisture content and groundwater. The sizes of both forests (roughly 22 000 ha) are large enough to be compatible with the satellite images of coarse resolutions such as of the Moderate Resolution Imaging Spectroradiometer (MODIS). The MODIS-derived NDVI values were downloaded with a spatial resolution of 250 x 250 m and a temporal resolution of 15 days for eleven consecutive years (2000-2010). The ecoclimatological single variable relationships of NDVI, rainfall, temperature and aridity were examined on the basis of Pearson correlation (r). In order to eliminate seasonality impacts the standardized values were used by subtracting the mean and then dividing by the standard deviation.

RESULTS AND DISCUSSION

Trends in time series

The descriptive statistics of annual rainfall and temperature (mean, coefficient of variation (CV), skewness (CS) and kurtosis coefficient (CK)) are presented in Tables 3 and 4. The mean value of annual rainfall and temperature are found to be 219 to 662 mm and 25.5 to 29.3°C, respectively. The coefficient of variation of rainfall (30%) is higher than that of the temperature (2%), suggesting that rainfall controls the climate variability in the central Sudan.

The obtained mean SI value is ranged between 0.33 and 1.01, revealing that there were three distinct climatic zones in the central Sudan. These are dry, semi dry and sub humid. However, these climates are very variable since SI is associated with a high coefficient of variation of 29%. The coefficient of variations of both SI and rainfall are found equal, restating that rainfall is the determinant factor in the region's climate. This finding has been further tested as the relationship between the peaked-nesses of SI (wettest situation) and rainfall is higher ($r \approx 0.89$) than that of the temperature ($r \approx 0.47$). Trends in rainfall, temperature and SI are shown in Tables 5 and 6. At 55% of the cases, the annual rainfall shows a statistically significant decreasing trend of 3.5 mm per annum. This is coupled with a statistically significant increasing trend of 1.4°C per annum in temperature at 100% of the cases. Accordingly, SI showed a decreasing trend, suggesting that in the climatic conditions of central Sudan is deteriorated. This deterioration is found significant at 50% of the stations, that is, Kadugli (South Kordufan), Damazine (Blue Nile), Nyala (South Darfur) and Wadmedani (Gezira), which are very vital areas for the agricultural production in Sudan. For instance, Wadmedani station represents the Gezira climate where the world largest singly managed irrigated scheme (Gezira scheme) exists. Consequently, the increase of aridity in this region would result in increasing the crop water requirements, considering the limited water resource, that is, Sennar dam with a storage capacity of 0.5 km^3). On the other hand, Kadugli, Damazine and Nyala stations represent the traditional rainfed agricultural areas where most of the Sudanese people favored food (sorghum) is produced. It is therefore the resulted increase inaridity is basically jeopardized food security and socioeconomic stability in central Sudan confirming the results obtained by Hamid et al. (1995) and Farah et al. (1997). Consequently, the Sudan's government became more willing to increase sorghum cultivated area under irrigated conditions at the expensive of other cash crops such as cotton (Guvele, 2002).

Detecting climate change by SI

The climatic zones in Sudan have been ascribed by with specific SI ranging values. It is important to restate that the higher the SI value is the better, that is, good climatic conditions (Table 1). Therefore, any persistent changes on those SI values for an extended period may be indicated as climatic changes. According to Table 6, all the stations experienced a decreasing trend in SI values, that is, their aridity is increased. The investigated period is 50 years, which is a persistent period, that is, IPCC in

Table 3. Mean rainfall (Rm), coefficient of variation (CV), coefficients of skewness (CS), coefficients of kurtosis (CK) and time-series period for the stations in the central Sudan.

Station	Rm (mm)	CV	CS	CK	Time period
Kadugli	662.5	0.22	0.42	0.01	1962-2009
Damazine	647.5	0.21	-0.82	0.63	1962-2005
Nyala	355.7	0.36	-0.03	-0.01	1962-1998
El Obied	315.0	0.36	0.54	0.39	1962-2005
Sennar	418.0	0.22	-0.82	0.30	1970-2006
Ed Dulm	219.0	0.43	0.70	2.35	1960-1996
Gedarif	589.3	0.22	-0.06	-0.36	1962-2005
Wadmedani	280.6	0.29	0.09	0.01	1960-2005

Table 4. Mean temperature (Tm), coefficient of variation (CV), coefficients of skewness (CS), coefficients of kurtosis (CK) and time-series period for the selected stations in the central Sudan

Station	Tm (°C)	CV	CS	CK	Time period
Kadugli	28.4	0.02	0.34	-0.63	1960-2010
Damazine	27.9	0.03	-1.15	2.02	1960-2010
Nyala	27.6	0.02	-0.36	-0.92	1960-2010
El Obied	26.7	0.03	0.57	-0.23	1960-2010
Sennar	28.3	0.03	-0.53	0.10	1960-2010
Ed Duim	29.3	0.02	-0.35	0.41	1960-2010
Gedarif	28.9	0.01	-0.03	-0.39	1960-2010
Wadmedani	28.8	0.03	0.14	-0.82	1960-2010

Table 5. The time series trend by Man-Kendall test for rainfall (R-MK), temperature (T-MK) and Silyaninov index (SI-MK).

Station	R-MK	T-MK	SI-MK
Kadugli	decreasing*	increasing*	decreasing**
Damazine	decreasing**	increasing*	decreasing**
Nyala	decreasing**	increasing**	decreasing**
El Obied	decreasing	increasing**	decreasing
Sennar	decreasing	increasing*	decreasing
Ed Duim	decreasing	increasing**	decreasing
Gedarif	decreasing	increasing*	decreasing
Wadmedani	decreasing*	increasing**	decreasing*

* Statistically significant at $P < 0.05$ and ** statistically significant at $P < 0.01$.

2007 suggests a decade. Thus, on the basis of the SI the central Sudan climatic zones experienced climatic changes. The significant decreasing trend in SI values were found to be 37% for Wadmedani (dry), 76% for Nyala (semi dry), 31% for Kadugli (sub humid) and 37% for Damazine (sub humid). These values suggest that aridity is increased in the central Sudan during the period of 1960 to 2010, especially in the semi dry climate, which may be sooner converted to dry conditions.

Detecting drought by SI

In the central Sudan, during the period of 1960 to 2010 the SI showed successive runs of dry years (Figure 2), which can be ascribed to the below normal rainfall. At this point it is difficult to designate those periods as drought cycles as yet there is no wide applicable definition for the term drought. For example, Abusin (1986) mentioned that in the central Sudan a drought year implies one of the

Table 6. Average changes in rainfall (mm/year), temperature (°C/year) and Silyaninov index (SI) by the Theil and Sens's method.

Station	Rainfall	Temperature	SI
Kadugli	4.8	1.1	1.00
Damazine	3.8	0.4	1.00
Nyala	3.8	1.6	0.62
El Obied	0.1	2.4	0.51
Sennar	1.8	1.2	0.64
Ed Duim	2.1	1.5	0.33
Gedarif	1.1	0.5	0.87
Wadmedani	2.1	2.1	0.45

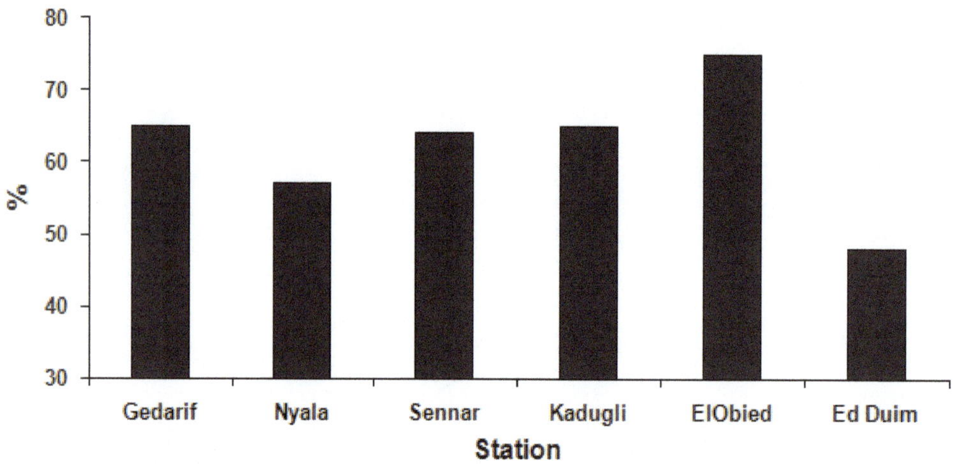

Figure 3. The performance of the Silyaninov index (SI) in detecting correctly the drought and wet years at selected stations in the central Sudan.

followings: lack of plant cover, crops failure and livestock losses, which in aggregate jeopardized the family's welfare. Thus, the drought definition will be different even between individuals of one society and from one society to another.

In this study, the power of SI in designation given year as drought or wet is evaluated using the confirmed historical events. This is firstly done without considering the drought's degree of intensity. The results suggest that generally SI is effective in detecting drought, with the exception of Gedarif and Ed Duim stations, as shown in Figure 3. The differences in detecting drought performance were largely attributed to the presence of anomalies in annual temperature. Hence, where the temperature data is anomalies-free the rainfall alone might explain much of the meteorological droughts in the central Sudan. This conclusion agrees with that of Ntale and Gan (2003) for the droughts in east African.

Hitherto, the comprehensive documentation of drought years in the central Sudan is poor. Consequently, SI is used to detect the stationary drought years. The obtained results were shown in Figure 4. It is obvious that Ed Duim and Damazine stations experienced the highest and lowest percentages of drought years of 59 and 38%, respectively. Based on the tested years, the region of central Sudan is dominated by wet conditions.

The common/regional wetted years were 1963, 1964, 1978 and 2007. The year 1984 is the only common drought year. This is revealed that rainfall of the central Sudan is characterized by its localized behavior.

According to the SPI approach, if the standardized precipitation index is continuously ≤ -1.0 there is drought event and vice versa. Therefore, each drought event has a well-defined duration (Ntale and Gan, 2003). Following the same procedure the stationary longest drought events in the central Sudan were determined (Table 7). It is clear that most of the drought events occurred during the periods of 1970 to 1990, with the exception of Damazine (1998 to 2002) and Ed Duim station (2000 to 2005). It is found also that the drought events with one year in length are dominant. Since that rainfall is found to be the dominant climatic variable in the SI estimation, the drought intensity classification scale of the SPI could be applicable to SI results. Consequently, the most detected stationary drought events in the central Sudan were classified (Table 7). It is clear that most of the drought

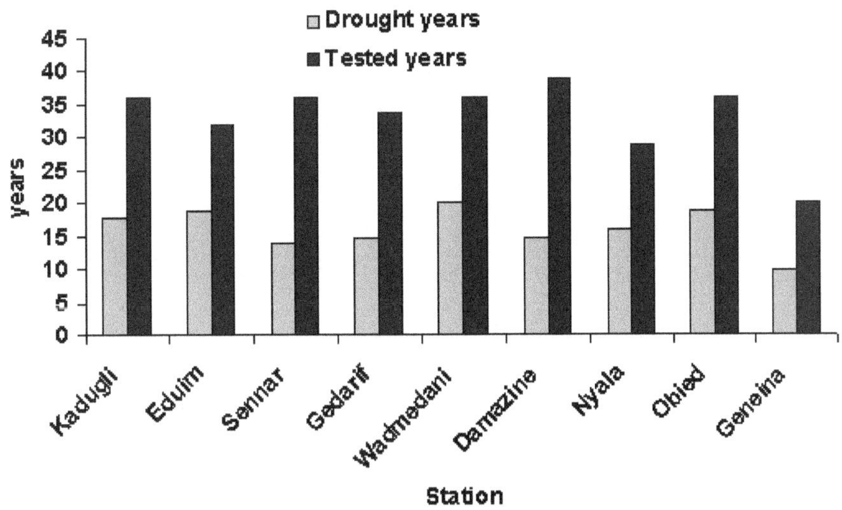

Figure 4. The drought years percentages at selected stations in the central Sudan. The trends of the standardized precipitation index, SPI (the smoothed line) and the Silyaninov index, SI (the dashed line) at selected stations in the central Sudan. The circles refer to years with different values of SPI and SI.

Table 7. The longest drought events, the driest years and their standardized Silyaninov index values (ZSI) and their drought scale at selected stations.

Station	Drought event time	Driest year	ZSI	Drought scale
Kadugli	1982-1987	2001	-1.61	Severe
Damazine	1998-2002	2004	-2.20	Extreme
Nyala	1982-1987	1984	-1.80	Severe
El Obied	1982-1985	1982	-1.95	Severe
Sennar	1965-1970	1981	-2.18	Extreme
Ed Duim	2000-2005	1984	-1.93	Severe
Gedarif	1970-1977	1984	-1.83	Severe
Wadmedani	1966-1970	2000	-1.50	Severe

cycles can be classified as severe ($-1.5 \geq SI \geq -1.99$), with the exception of Damazine and Sennar stations, which show extreme drought events ($SI \leq -2.0$). The use of regional mean in the central Sudan is found questionable and deceptive. For instance, on the basis of the regional rainfall average the central Sudan experienced a long-term drought cycle of 30 (1972 to 2002) as shown in Figure 5, which is unrealistic. Thus, the use of regional average in central sudan is not recommended.

Relationships of natural vegetation cover with rainfall, temperature and SI

Most of the obtained NDVI values were found to be below the general average for the period of 2000 – 2010 (Figures 6 and 7). This is attributed to the deterioration in the climatic conditions of the central Sudan, agreeing with the results of the Vicente et al. (2006) who found that aridity has caused a general decreasing of NDVI and increasing of coefficient of variation, in Spain. On the other hand, the result disagreed with results of Hermann et al. (2005) who indicated that there is a seasonal increasing trend in the greenness of the African Sahel region using the NDVI time series analysis (1982-2003).

Forestry system (90 million hectares, Mha) is the largest vegetation cover in Sudan (FAO, 2006). Its area however, is diminished by 3% during the period of 1976-1990 (Elbashir, 1994). Oldeman et al. (1991) stated that about 5 Mha of land in Africa as lost its original biotic functions due to degradation, to a level that their rehabilitation is economically been difficult. In Sudan, forestry system has an appreciated potential economic values that is, gum Arabic production, a sustainable timber industry, wildlife tourism, woods, etc (Glover, 2005). Thus, one of the steps that can be immediately taken regarding forest rehabilitation is the re-enforcement

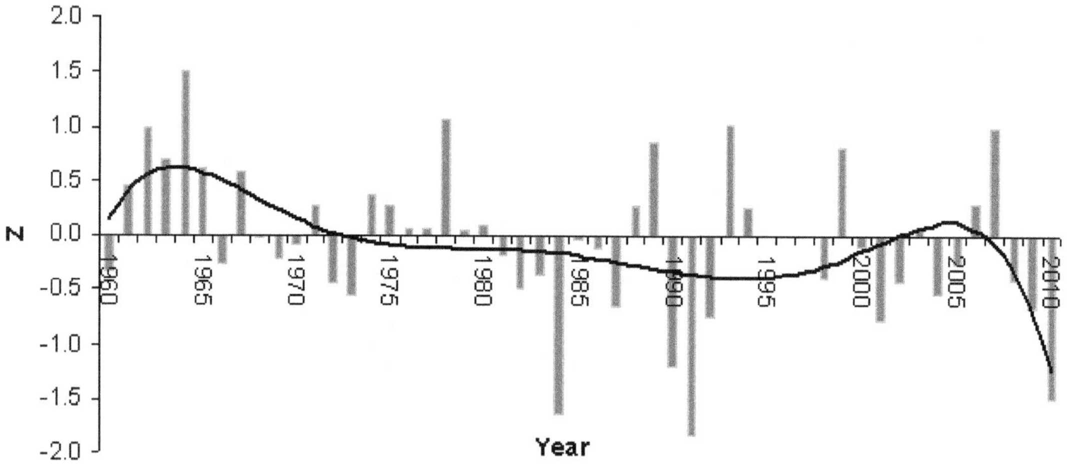

Figure 5. The standardized regional SI value.

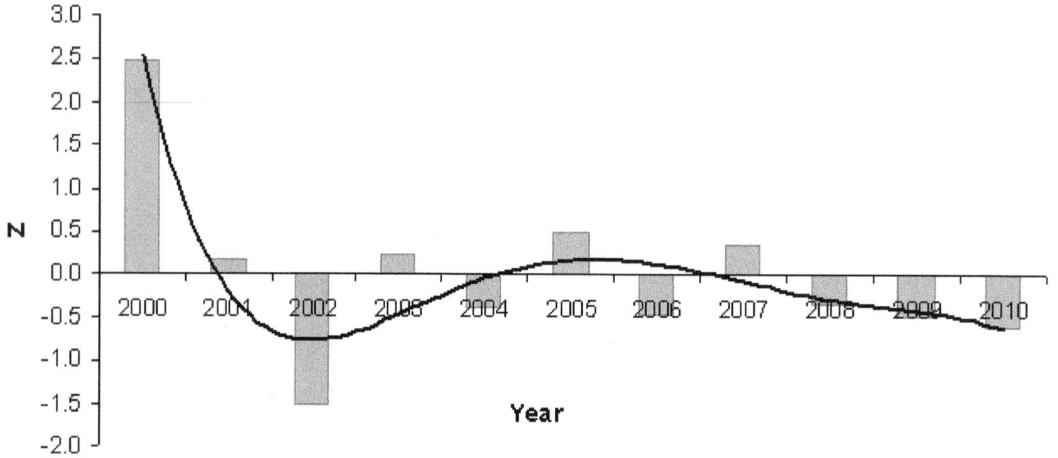

Figure 6. The standardized values of the NDVI for the forest, Damazine area. The smoothed line is the trend of NDVI

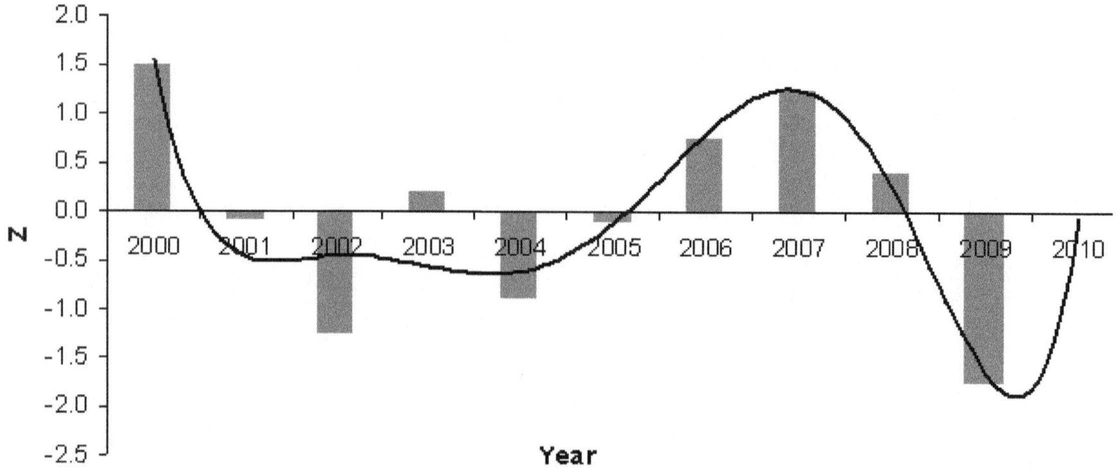

Figure 7. The standardized values of the NDVI for the forest, Gedarif area. The smoothed line is the trend of NDVI.

of the 10% law that is, set aside 10% of the farmer's land for forest production. Also, there is an urgent need for reforestation plans, which can be defined as the establishment of tree crops on deforested or cleared land, either on land previously degraded by agriculture, mining, or other activity, or on lands cleared of native vegetation specifically for the purpose of plantation establishment (Bowyer, 2001). However, care should be taken as some studies showed that improper forestation plans were lead to soil compaction, erosion and depletion of soil organic matter and thus degradation of physical and nutritional properties of soil (Bowyer, 2001). Thus, the good vegetation restoring plan needs the suitable vegetation to be re-stored in its right place. This required in-depth studies, e.g. studies on soil moisture variability are important since soil moisture is crucial to ecosystem restoration (Ma et al., 2004).

The detected vegetation-ecoclimatic relationships between NDVI-rainfall, NDVI-temperature and NDVI-SI were found nonlinear (exponential) under both the semi dry and sub humid climates. This is in conformity with the results of Vicente et al. (2006) who suggested a non-linear relationship between aridity and NDVI. Hermann et al. (2005) however reported a direct relationship between NDVI and rainfall at a threshold below 1000 mm per year, in the African Sahel zone.

Generally, the resulted Pearson correlations in the semi dry climate were observed higher than that of the sub humid. They were 0.88 for NDVI-temperature, 0.79 for NDVI-rainfall and 0.81 for NDVI-SI in the semi dry climate compared with 0.85, 0.62 and 0.61 for the semi dry climate, respectively. Because temperature is an indirect indicator for energy available for the plant development a strong relation between NDVI and temperature is anticipated (Tan, 2007). On the tested relations, NDVI-temperature was the highest; but it was statistically insignificant, which is attributed to the minimum variation in the annual temperature values (CV ≈ 3%). On the other hand, the relations NDVI- rainfall and NDVI- SI were found significant at 0.05 and 0.0005 levels of significance for sub humid and semi dry climates, respectively. It is worth mentioning that the slope of the relation NDVI-temperature is found negative and the opposite holds true for the relations NDVI-rainfall and NDVI-SI. This is indicated that the continued increasing in temperature may degrade the greenness of the forestry system in Sudan. Accordingly, rainfall is found to be the most significant climatic variable that affects the dynamics of the vegetation cover in the central Sudan.

Conclusion

Most of the livelihood generations are climate depended in the central Sudan. This study concerned on providing simple and cost effective approach for detecting climate change, drought and vegetation dynamics in the central Sudan using Silyaninov index (SI) and the normalized difference vegetation index (NDVI). Both indices showed good results. The most important result is that drought disturbed the climatic conditions of central Sudan, if the current trend of degradation is continued sooner the semi arid zone would experience arid conditions and the latter would change to desert conditions. Accordingly, the agricultural sector/food security will not sustain without irrigation, which is limited by water resources availability.

The accuracy of the SI results depends on two factors; the quality of inputs (rainfall and temperature) and availability of data (length of record). The former found to be poor since the homogeneity of the datasets were questionable. The length of record was disrupted mainly because of direct drought consequences (civil unrest) and fund issues as most of the meteorological stations in central Sudan were out of service since 1980s. This entails rehabilitation projects for the meteorological stations as well as capacity building programs.

REFERENCES

Abusin M (1986). Man's socio-economic response to drought in the White Nile. In: Rural Development in White Nile Province, Sudan (eds H. R. J. Davies). United Nations University Press, Tokyo, Japan.

Adam H (2002). Agroclimatology. University of Gezira Press. Wadmedani, Sudan.

Allen R, Pereira L, Raes D, Smith M (1998). Crop evapotranspiration. Guidelines for Computing Crop Water Requirements. FAO Irrigation and Drainage Paper 56. Rome, Italy.

Arora V (2002). The use of the aridity index to assess climate changeeffect on annual runoff. J. Hydr. 265:164-177. http://dx.doi.org/10.1016/S0022-1694(02)00101-4

Ayoub A (1998). Extent, severity and causative factors of land degradation in the Sudan. J. Arid Environ. 38:397-409. http://dx.doi.org/10.1006/jare.1997.0346

Ayoub A (1999). Land degradation, rainfall variability and food production in the Sahelian zone of the Sudan. J. Land Degrad. Dev. 10:489-500.http://dx.doi.org/10.1002/(SICI)1099-145X(199909/10)10:5<489::AID-LDR336>3.0.CO;2-U

Basistha A, Arya D, Goel N (2009). Analysis of historical changes in rainfall in the Indian Himalayas. Int. J. Clim. 29:555-572. http://dx.doi.org/10.1002/joc.1706

Bastiaanssen W, Ahmad M, Tahir Z (2003). Upscaling water productivity in irrigated agriculture using remote-sensing and GIS technologies. In: Water productivity in agriculture: Limits and Opportunities for Improvement (eds J.W. Kijne, R. Barker and D. Molden). CAB International 2003.

Bates B, Kundzewicz Z, Wu S, Palutikof J (2008). Climate Change and Water. Technical Paper of the Intergovernmental Panel on Climate Change, IPCC Secretariat, Geneva.

Bowyer J (2001). Environmental implications of wood production in intensively managed plantations. J. Wood Fiber Sci. 33(3):318-333

Burn D, Hag Elnur M (2002). Detection of hydrologic trends and variability. J. Hydrol. 255:107-122. http://dx.doi.org/10.1016/S0022-1694(01)00514-5

Burton I, Lim B (2005). Achieving adequate adaptation in agriculture. J. Clim. Change. 70:191-200. http://dx.doi.org/10.1007/s10584-005-5942-z

Challinor A, Wheeler T, Garforth C, Craufurd P, Kassam A (2007). Assessing the vulnerability of food crop systems in Africa to climate change. J. Clim. Change 83:381-399. http://dx.doi.org/10.1007/s10584-007-9249-0

Clark C, Schkade L (1969). Statistical methods for business decisions. South Western Publishing Co. New York.

Elbashir A (1994). Population and Environmental Degradation in the Sudanese agriculture and forestry. Proceedings of the Khartoum Nile

2002 Conference 29th January- 1st February 1994. PMid:7756939
Eldredge E, Khalil S, Nicholds N, Abdallah A, Rydjeski D (1988). Changing rainfall patterns in western Sudan. Int. J. Clim. 8:45-53. http://dx.doi.org/10.1002/joc.3370080105
El Fadni O (1998). Climatic zones in Sudan. M. Sc. thesis. Faculty of Agriculture Umdurman University, Sudan.
Elhag M, Walker S (2009) Impact of climate variability on vegetative cover in the Butana area of Sudan. Afr. J. Ecol. 47:11-16. http://dx.doi.org/10.1111/j.1365-2028.2008.01044.x
Farah S, Salih A, Taha A, Ali Z, Ali A (1997). Grain sorghum response to supplementary irrigations under post-rainy season conditions. J. Agric. Water Manage. 33:31-41. http://dx.doi.org/10.1016/S0378-3774(96)01283-8
FAO (2010). Crop production and food security assessment for the northern states of the Sudan. Assessment mission report 2009/2010. Rome, Italy.
FAO (2006). Fertilizer use by crop in the Sudan. Land and Plant nutrition and management service. Land and Water Development division, Rome, Italy.
Gitay H, Suarez A, Watson R, Dokken D (2002). Climate change and biodiversity. IPCC Technical Paper 5.
Guvele CA (2002). Gains from crop diversification in the Sudan Gezira scheme. J. Agri. Syst. 70:319-333. http://dx.doi.org/10.1016/S0308-521X(01)00030-0
Hamid F, Yousif G, Mohamed I (1995). Potentials of the Sudan's irrigated sector in cereals grain production: Analysis of various policy options. J. Agric. Syst. 48:457-483. http://dx.doi.org/10.1016/0308-521X(94)00026-H
Hulme M (1986). Secular changes in wet season structure in central Sudan. J. Arid Environ. Academic Press Inc. (London) Limited.
Hulme M, Viner D (1998). A climate change scenario for the tropics. Clim. Change. 39:145-176. http://dx.doi.org/10.1023/A:1005376007729
Hulme M (2001). Climatic perspectives on Sahelian desiccation: 1973-1998. Global. Environ. Change. 11:19-29. http://dx.doi.org/10.1016/S0959-3780(00)00042-X
IPCC (2007). Climate Change, Synthesis Report. The 4th Assessment Report of the Intergovernmental Panel on Climate Change, Valencia, Spain.
Lioubimtseva E, Cole R, Adams J, Kapustin G (2005). Impacts of climate and land-cover changes in arid lands of Central Asia. J. Arid Environ. 62:285-308. http://dx.doi.org/10.1016/j.jaridenv.2004.11.005
Logan K, Brunsell N, Jones A, Feddema J (2010). Assessing spatiotemporal variability of drought in the U.S. central plains. J. Arid Environ. 74:247-255. http://dx.doi.org/10.1016/j.jaridenv.2009.08.008
Luo Y, Liu S, Fu S, Liu J, Wang G, Zhou G (2008). Trends of precipitation in Beijiang river basin, Guangdong province, China. J. Hydr. Proc. 22:2377-2386 http://dx.doi.org/10.1002/hyp.6801
Ma K, Fu B, Liu S, Guan W, Liu G, Lu Y, Anand M (2004). Multiple-scale soil moisture distribution and its implications for ecosystem restoration in an arid river valley, China. J. Land Dev. 15:75-85.
Mati B (2005). An overview of water and soil nutrient management under small holder rain-fed agriculture in east Africa. Working Paper No. 105. Colombo, Sri Lanka: International Water Management Institute (IWMI).
Mavromatis T (2010). Use of drought indices in climate change impact assessment studies: an application to Greece. Int. J. Clim. 30:1336-1348
Modarres R, Silva V (2007). Rainfall trends in arid and semi arid regions of Iran. J. Arid Environ. 70:344-355 http://dx.doi.org/10.1016/j.jaridenv.2006.12.024
Morid S, Smakhtin V, Moghaddasi M (2006). Comparison of seven meteorological indices for drought monitoring in Iran. Int. J. Clim. 26:971–985 http://dx.doi.org/10.1002/joc.1264
Mpelasoka F, Hennessy K, Jones R, Bates B (2008). Comparison of suitable drought indices for climate change impacts assessment over Australia towards resource management. Int. J. Clim. 28:1283-1292. http://dx.doi.org/10.1002/joc.1649
Nasri M, Modarres R (2009). Dry spell analysis of Isfahan province, Iran. Int. J. Clim. 29:1430-1438. http://dx.doi.org/10.1002/joc.1805
Ntale H, Gan T (2003). Drought indices and their application to east Africa. Int. J. Clim. 23:1335-1357. http://dx.doi.org/10.1002/joc.931
Oldeman L, Hakkeling R, Sombroek W (1991). World Map of the Status of Human-induced Soil Degradation: An Explanatory Note. ISRIC-UNEP Report, Netherlands.
Oladipo E (1985). A comparative analysis of three meteorological drought indices. J. Clim. 5:655-664. http://dx.doi.org/10.1002/joc.3370050607
Omuto C, Shrestha D (2007). Remote sensing techniques for rapid detection of soil physical degradation. Int. J. Rem. Sens. 28(21):4785-4805 http://dx.doi.org/10.1080/01431160701260357
Osman Y, Shamseldin A (2002). Qualitative rainfall prediction models for central Sudan and southern Sudan using Elnino-Southern Oscillation and Indian Ocean sea surface temperature indices. Int. J. Clim. 22:1861-1878 http://dx.doi.org/10.1002/joc.860
Rodrigo F, Esteban P, Vazquez D, Diez Y (1999). A 500-year precipitation record in Southern Spain. Int. J. Clim. 19:1233-1253. http://dx.doi.org/10.1002/(SICI)1097-0088(199909)19:11<1233::AID-JOC413>3.0.CO;2-L
Sivakumar M, Das H, Brunini O (2005). Impacts of present and future climate variability and change on agriculture and forestry in the arid and semi arid tropics. Clim. Change. 70:31-72. http://dx.doi.org/10.1007/s10584-005-5937-9
Smith M, Koren V, Zhang Z, Reed S, Pan, Mored F (2004). Runoff response to spatial variability in precipitation: An analysis of observed data. J. Hydrol. 298:267-286. http://dx.doi.org/10.1016/j.jhydrol.2004.03.039
Tan S (2007). The influence of temperature and precipitation climate regimes on vegetation dynamics in the US Great Plains: A satellite bioclimatology case study'. Int. J. Rem. Sens. 28(22):4947-4966. http://dx.doi.org/10.1080/01431160701264268
Teklu T, Braun J, Zaki S (1991). Drought and Famine relationships in Sudan: Policy implications. Research Report No. 88. International Food Policy Research Institute.
Tilahun K (2006). Analysis of rainfall climate and evapo-transpiration in arid and semi-arid regions of Ethiopia using data over the last half a century. J. Arid Environ. 64:474-487
Vicente-Serrano S, Cuadrat-Prats J, Romod A (2006). Aridity influence on vegetation patterns in the middle Ebro Valley (Spain): Evaluation by means of AVHRR images and climate interpolation techniques. J. Arid Environ. 66:353-375. http://dx.doi.org/10.1016/j.jaridenv.2005.10.021
Wu H, Hayes M, Wilhite D, Svoboda M (2005). The effect of the length of record on the standardized precipitation index calculation. Int. J. Clim. 25:505-520. http://dx.doi.org/10.1002/joc.1142
Xu Z, Liu Z, Fu G, Chen Y (2010). Trends of major hydroclimatic variables in the Tarim River basin during the past 50 years. J. Arid Environ. 74:256-267. http://dx.doi.org/10.1016/j.jaridenv.2009.08.014
Yue S, Pilon P, Cavadias G (2002). Power of Mann-Kendall and Spearman's rho tests for detecting monotonic trends in hydrological series. J. Hydrol. 259:254-271. http://dx.doi.org/10.1016/S0022-1694(01)00594-7
Yue S, Wang C (2004). The Mann-Kendall Test Modified by Effective Sample Size to Detect Trend in Serially Correlated Hydrological Series. Water Res. Manage. 18:201-218. http://dx.doi.org/10.1023/B:WARM.0000043140.61082.60

Geo-electrical data analysis to demarcate groundwater pockets in Ado – Ekiti, southwest, Nigeria

Oyedele, E. A. A.[1] , Oyedele, T.[2] and Oyedele, K.[3]

[1]Physics Department, University of Ado - Ekiti, Nigeria.
[2]Geological Services, Ogun State Ministry of Commerce and Industry, Abeokuta, Nigeria.
[3]Omega – Inlands and Associates, Abeokuta, Nigeria.

Geo-Electrical Analysis of data obtained from Vertical Electrical Sounding (VES) measurements conducted in Ado-Ekiti has been carried out to demarcate groundwater pockets in the metropolis within the Basement Complex of Southwest, Nigeria. 49 'vertical electrical soundings' were carried out. The layer parameters of resistivity and thickness were determined and used to generate spatial distributions of relevant geo-electric parameters including overburden thicknesses, the bedrock relief and the weathered basement thickness/resistivity. The depth – to – bedrock varied from 3.8 to 56.2 m with a mean of 21.04 ± 14.62 m. The weathered basement has resistivity values ranging from 6.5O to 1214 Ω-m with weathered basement thickness ranging from 3 to 39.5 m. The basement rocks were observed to be variably fractured. Groundwater potential zones such as valley fills/basement depressions, weathered zones and fractured zones were delineated in the study area based on the geoelectrical data.

Key words: Groundwater, geoelectric, saprolite, bedrock, basement.

INTRODUCTION

The exploration for groundwater in hard rock terrains is a complex task. The crystalline basement rocks are generally of low porosity and permeability, and as such, have no water storage capacity in their fresh (unaltered) form; consequently, their groundwater resources are limited and commonly restricted to features produced by weathering and tectonic processes (Olayinka and Olorunfemi, 1992). The application of the Vertical Electrical Sounding (VES) lends itself as an efficient and effective method for mapping potential zones that are relevant to the development of groundwater resources. The electrical resistivity method is a non-invasive geophysical method. VES measurements are useful in groundwater studies as neither the structure nor the dynamics of the soil is disturbed (Adiat et al., 2009; Ariyo and Adeyemi, 2009). Ado-Ekiti, Southwest Nigeria, is underlain by the Precambrian basement Complex Rocks of Southwestern Nigeria with heavy dependence on rain water, surface water and groundwater for its water supplies (Rahaman, 1988). In tropical and equatorial regions, weathering process creates superficial layers, with varying degrees of porosity and permeability. The unconsolidated superficial materials often constitute reliable aquifer units if significantly thick and appropriately porous and permeable. The concealed basement rock may contain faulted areas, incipient joints and fracture systems derived from tectonic events earlier experienced. The detection and delineation of these hydrogeologic structures may facilitate the location of groundwater prospect zones in a typical basement setting (Omosuyi et al., 2003).

Increasing population with the attendant human activities geometrically leads to increasing demand for groundwater resources. The surface water, where available and accessible, cannot guarantee the required water quality status required for most domestic activities.

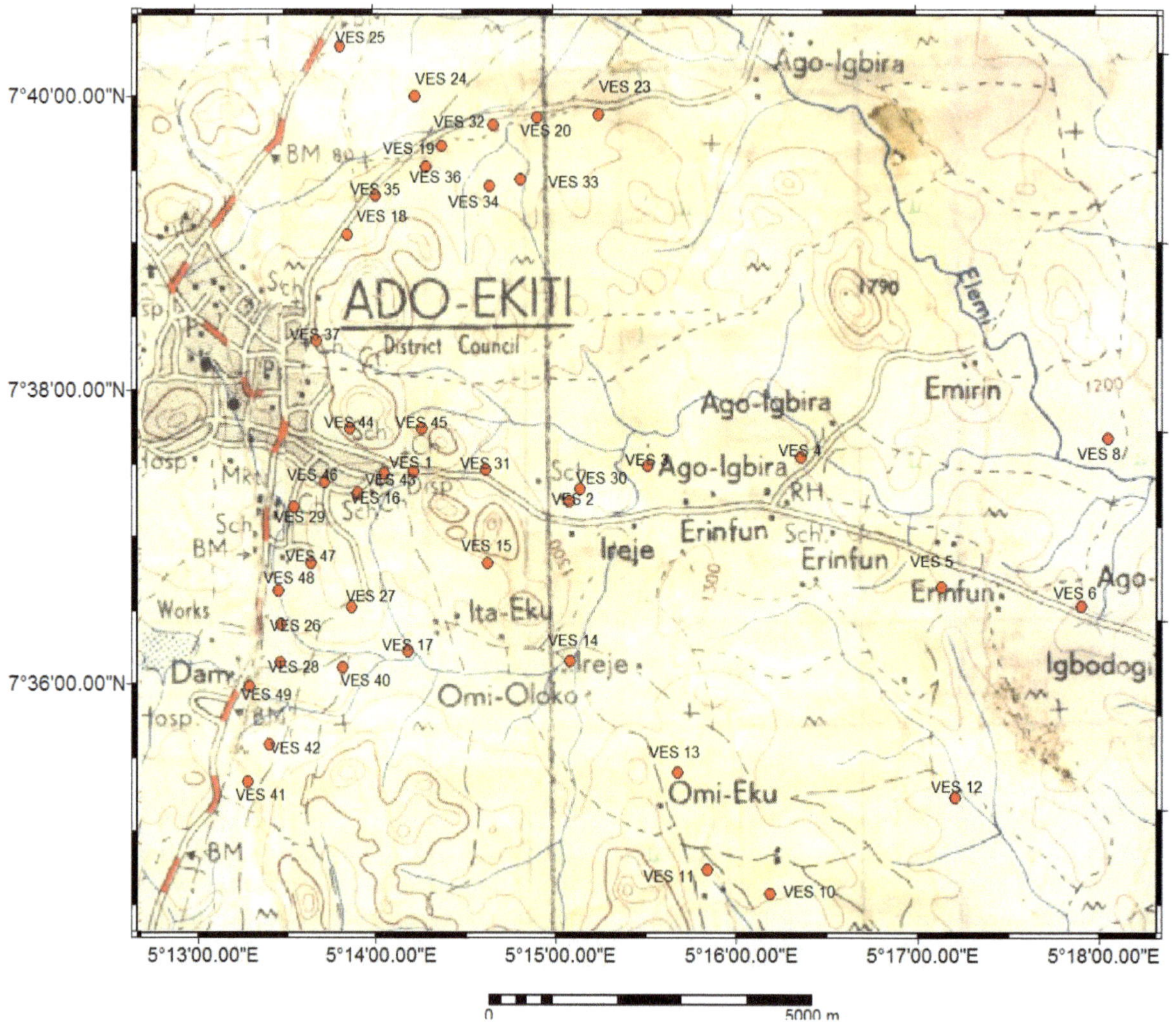

Figure 1. Map of the study area showing the vertical electrical sounding (VES) locations (Federal surveys, 1966).

This underlines the natural recourse to groundwater (Adeyeye and Abulude, 2004). This study was undertaken to delineate groundwater potential zones in Ado-Ekiti using geo-electrical analysis of VES data.

LOCATION AND GEOLOGY/HYDROGEOLOGY

Ado-Ekiti is located (within latitude 7° 33' and 7° 42' N and longitude 5° 11' and 5° 20' E) in South Western Nigeria on a low-land surrounded by several isolated hills and inselbergs. The highlands are separated sometimes by extensive lowlands. The centre portion of the metropolis is a high-rise thus contributing to the drainage pattern observed in the area. The topography (Figure 1) of the area is rugged due to the presence of crystalline basement rocks like charnockite and quartzite ridges which rise abruptly above the surrounding country rocks (Figure 2). The area is underlain by the Precambrian basement complex rocks of South-western Nigeria and drained by River Ireje, Elemi, Omisanjana and Awedele Stream. They flow into River Ose and Owena and empty into the Atlantic Ocean (Rahaman, 1988; Ebisemiju, 1993). The crystalline basement rocks generally lack primary porosity and permeability. As such, secondary porosity (joints, lineaments and the weathered zone) is the main source for groundwater occurrence, movement and transmission. Groundwater occurs under unconfined conditions in shallow, moderately weathered zones and in semi-confined conditions in joints, fissures, and fractures that extend beyond the weathered zones (Olayinka and Olorunfemi, 1992; Olorunfemi et al., 1999; Adelusi et al., 2004).

MATERIALS AND METHODS

The geoelectric data acquisition was carried out using vertical electrical sounding (VES) technique. The instrument used for the

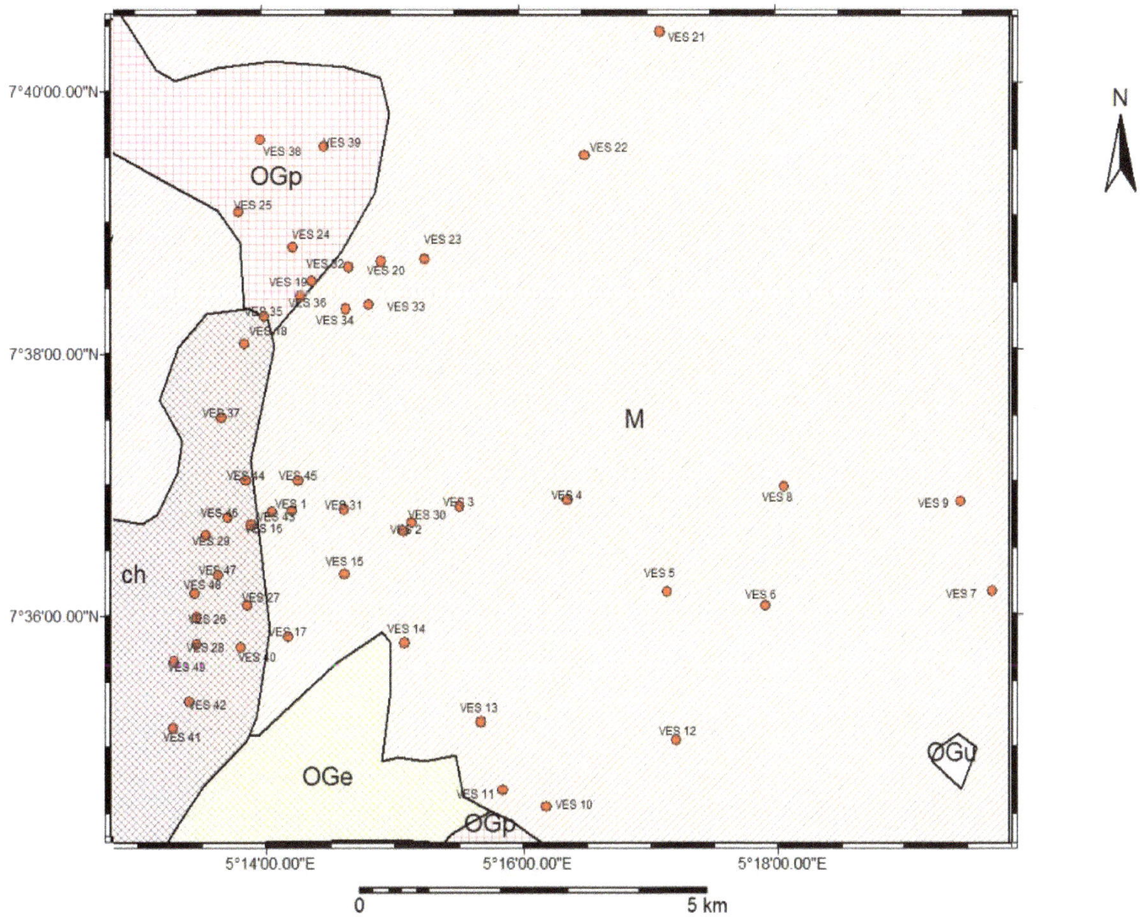

ch: Charnockitic meta-intrusives

OGp: Coarse porphyritic biotite & biotite-hornblende

OGe: Medium-to coarse-grained biotite-granite

OGu: Undifferentiated older granite mainly granitized gneiss with porphyroblastic gneiss & some migmatite

M: Viariably migmatized, undifferentiated, biottite & biotite-hornblende-gneiss with intercalatted amphibiolite

Figure 2. Geological map of Ado Ekiti and VES locations (geological map of Akure sheet 56).

resistance measurement was the ABEM Terrameter Signal Averaging System, SAS 300B with ABEM 2000 Booster. A total of 49VES points were fully occupied as shown in Figures 1 and 2. The VES data were acquired using the Schlumberger array with maximum electrode separation (AB/2) of 100 m. This has been decided based on well inventory data. A few soundings were however conducted up to (AB/2) of between 120 and 150 m. The apparent resistivity obtained from the field was plotted against half the current electrode separation (AB/2) on the bilogarithmic graph paper to obtain the initial parameters, which were used as inputs in a computer program, RESIST Version 1.0, to obtain the final parameters (Vander Velpen, 1988). The geoelectric parameters of layer thickness and resistivity obtained were used for the analysis (Olorunfemi and Okhue, 1992; Sheriff et al., 2002). Global positioning system (GPS) survey was run through the VES stations using a GERMIN CHANNEL 12 GPS Unit to obtain the vertical and horizontal positions (elevations and horizontal coordinates). All the processed geoelectric parameters and GPS data were used to generate the relevant maps using Surfer 8. Analysis of the maps would form the basis of the discussions.

RESULTS AND DISCUSSION

The results of the study are presented as coverage maps as shown in Figures 3 to 7. These maps are discussed in turn.

The isopach map of the overburden in the study area

The isopach map of overburden of the study area was produced as shown in Figure 3 with a contour interval of 5 m. The overburden encompasses all materials above the presumably fresh bedrock. It is defined as the total depth from the surface to the top of bedrock at each of the VES station. The map provides a general view of the geometry of the aquifers in the study area. The depth - to – bedrock varies from 3.8 to 56.2 m with a mean of

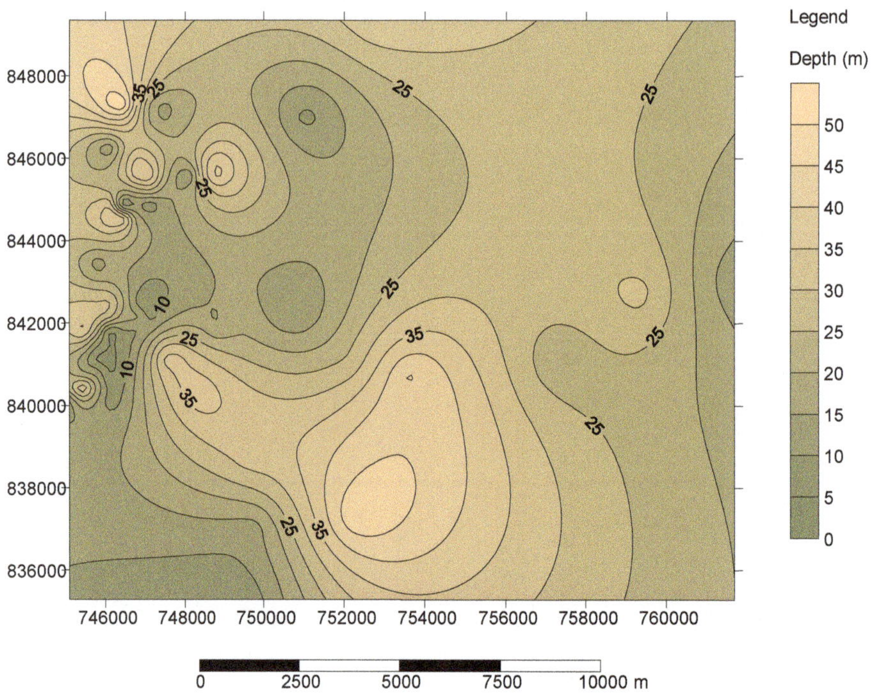

Figure 3. The isopach map of the overburden in the study area.

21.04 ± 14.62 m. The result agrees with Okhue and Olorunfemi (1991) which predicted a range of 4.0 to 79.2 m as possible overburden thickness in a similar geological terrain of the basement complex area of Ile - Ife, Southwest, Nigeria. It is evident from the map that the bedrock is not a smooth and continuous surface, rather, it exhibits undulating configuration. This depth is zero where the basement rocks outcropped (Mallam and Emenike, 2008).

A sizeable percentage of the study area has overburden cover ≥ 25 m. The isopach map of the overburden shows areas with thick overburden cover (H ≥ 30 m) in the North Western/Southern flanks of the study area. Pockets of such points also occur in the Western and Central parts of the area. These areas correspond to basement depression zones while zones with relatively thin overburden which are indicative of basement highs/ridges occur on the Western/Eastern and Southwestern flanks.

Generally, areas with thick overburden cover and less percentage of clay in which the intergranular flow has either dominant or important role are known to have high groundwater potential particularly in the basement complex area. These points are priority zones for groundwater development in the study area (Olorunfemi and Okhue, 1992; Mallam and Emenike, 2008).

The weathered basement thickness distribution map of the study area

The weathered basement thickness map (Figure 4) is a veritable tool to analyse and identify promising/potential aquifers. The map shows the variation of the weathered basement layer across the study area. The thickness of the layer ranges from 3 to 39.5 m with a mean of 15.40 ± 10.53 m. The weathered basement layer is seen to be thickest around the Southern/South Western portions with isolated portions along the Western/North Western portions of the study area. As the weathered basement layer is believed to be the major aquifer, the said regions offer strong appeal for groundwater development. In particular, average weathered basement layer thickness of about 34.04 m recorded at VES 12, 14, 15, 18, 19, 23 and 29 demarcates those zones of the study area as promising aquifers.

The weathered basement resistivity distribution map of the study area

The resistivity values of weathered basement at the various VES stations occupied in the area were contoured to give the 'weathered basement resistivity map' as shown in Figure 5. The resistivity values range from 6.5 to 1214 Ω-m. This indicates that the material composition is largely clay/sandy clay/clayey sand (Olayinka and Olorunfemi, 1992; Adelusi et al., 2004; Mallam, 2004). The basement rocks of the study area have varieties of texture of granites ranging from coarse to fine grained materials. The coarse grained granites weathered into water bearing residuals soil with a good water yield while the fine grained type weathered into clayey residual soil, which has poor water-yielding

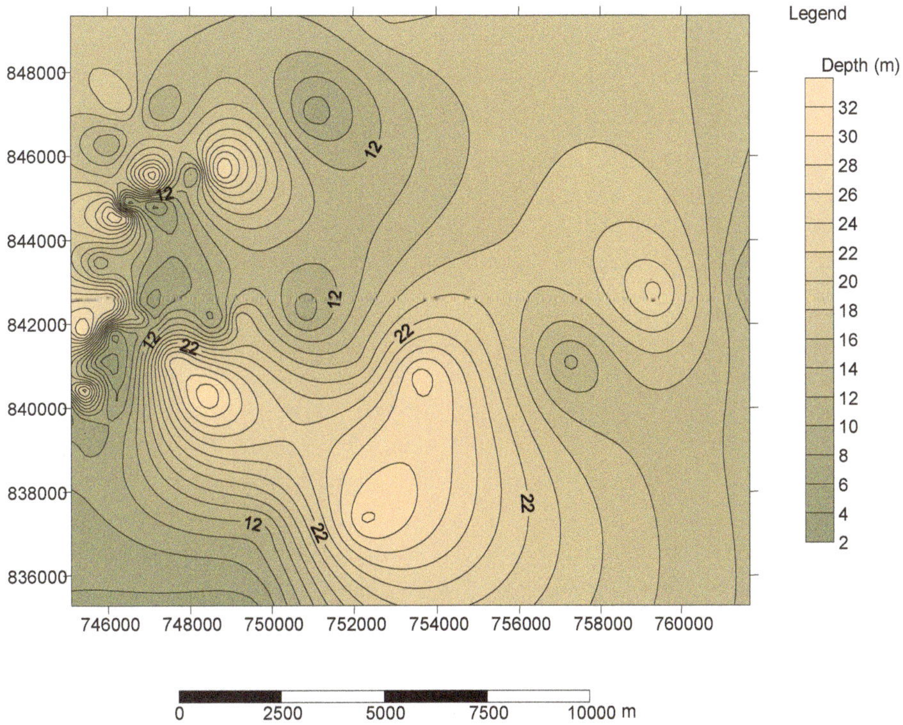

Figure 4. The weathered basement thickness distribution map of the study area.

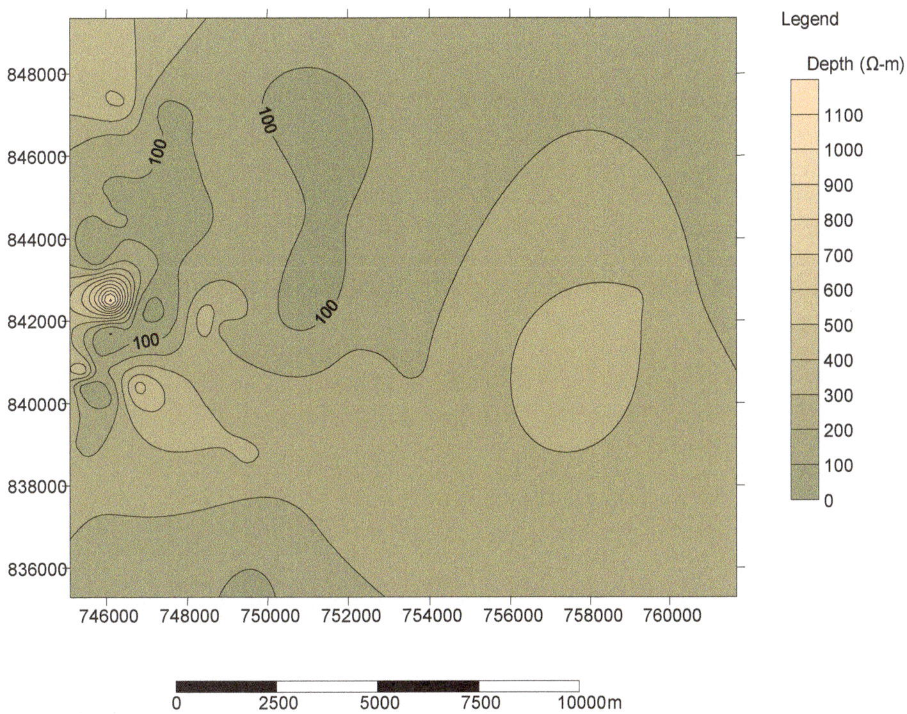

Figure 5. The weathered basement resistivity distribution map of the study area.

aquifer unit wherever if forms underlying basement rock. The map discriminated between high water bearing weathered basement zones and the low water bearing ones. It also indicated the variations in the degree of weathering/saturation across the study area. The low resistivity values characterizing the "low weathered

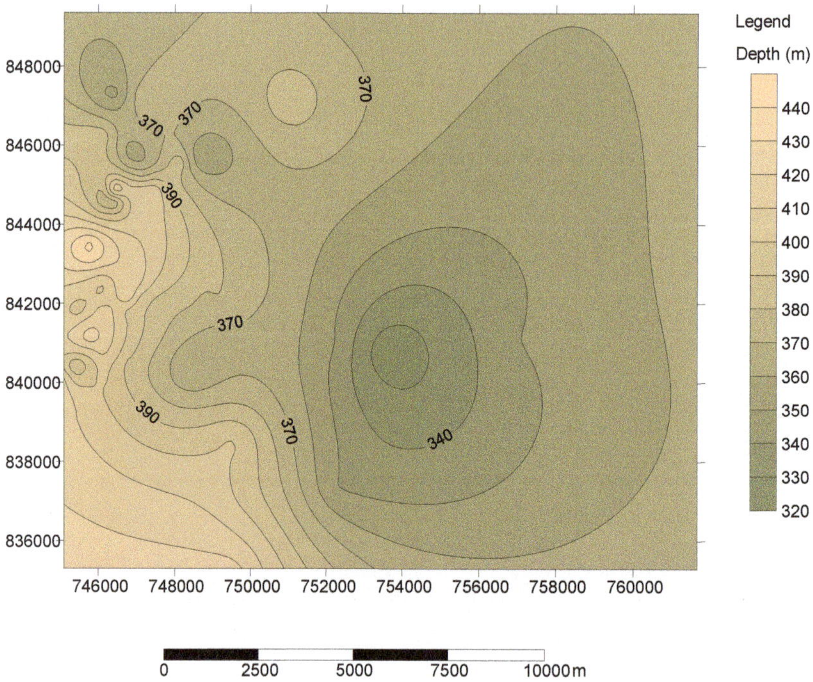

Figure 6. The bedrock relief map of the study area.

basement resistivity zones" are probably due to the highly weathered nature of the weathered basement layer at such points, tending towards clay and also due to the fact that the weathered basement layer is saturated with water at the zones. The very thin nature of the weathered basement layer across the zones and the presence of substantial amount of clay, render the potentials for groundwater low.

The zones underlain by thick overburden cover and less percentage of clay with appreciable degree of satuation as delineated on the map hold high potentials for groundwater exploitation (Okhue and Olorunfemi, 1991; Shemang, 1993).

The bedrock relief map of the study area

The bedrock relief map (Figure 6) is a contoured map of the bedrock elevation beneath all the VES stations. These elevations were obtained by removing the overburden thickness from the surface elevation at the VES stations. The bedrock relief map reflects the bedrock topography and its disposition. The map delineates a series of bedrock ridges and depressions within the surveyed area. The bedrock ridges show relatively thin overburden cover while the depressions are characterized by thick overburden. The depressions also exhibit low resistivity values. The bedrock depressions are diagnostic of groundwater collecting centres. They are thus demarcated as priority areas for groundwater development (Olorunfemi and Idornigie, 1992; Mallam and Emenike, 2008).

Bedrock resistivity distribution map of the study area

Over 50% of the area is underlain by fresh unfractured basement rocks with resistivities ≥ 5000 Ω-m (Figure 7). The basement rocks were observed to be variably fractured. The fractured nature of the rocks in other parts of the survey area is displayed by their low bedrock resistivity values. Aquiferous units in the basement complex terrain are mainly found in the thick and porous weathered overburden (saprolite zone) and the fractured part of the bedrock. The presence of these fractures further support the groundwater potentials of those zones (Shemang, 1993; Aboh and Osazuwa, 2000; Mallam, 2004).

Conclusion

Geo-electrical analysis of data obtained from vertical electrical sounding (VES) measurements has been carried out to demarcate groundwater pockets in Ado - Ekiti, Southwest, Nigeria. Feasible groundwater pockets like valley fills/basement depressions, weathered zones and fractured zones were delineated in the study area. The bedrock relief maps delineated a series of bedrock ridges and depressions within the surveyed area. The isopach map of the overburden showed relatively high overburden thicknesses (of up to 56.2 m) within the depressions. The bedrock depressions are observed to be variably fractured. The relatively low resistivity weathered basement constitutes the main aquifer unit. The highest groundwater yield is to be found in area

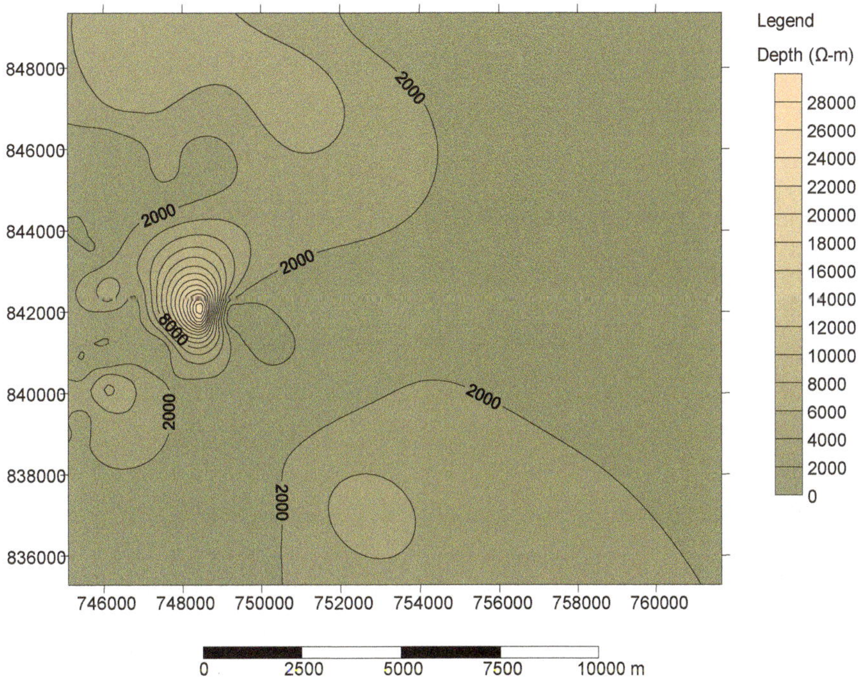

Figure 7. Bedrock resistivity distribution map of the study area.

where thick, weathered layer overburden (saprolite zone) with relatively low resistivity overlies the fractured basement. These zones were delineated in the study area.

ACKNOWLEDGMENTS

We appreciate the assistance of everyone who has contributed to this work. We acknowledge the useful comments and suggestions of the anonymous reviewers.

REFERENCES

Aboh HO, Osazuwa IB (2000). Lithological deductions from Regional Geoelectric Investigation in Kaduna, Kaduna State Nigeria. Nig. J. Phys. 12:1-7.

Adelusi AO, Ayuk MA, Oyedele AA (2004). Hydrogeophysical Analysis of Owode Area around Oba-Ile near Akure, SW Nigeria. Afr. J. Develop. Stud. 4(1&2):11-14.

Adeyeye EI, Abulude EO (2004). Analytical Assessment of some surface and groundwater resources in Ile–Ife, Nigeria. J. Chem. Soc. Nig. pp. 98-103.

Adiat KAN, Olayanju GM, Omosuyi GO, Ako BD (2009). Electromagnetic Profiling and Electrical Resistivity Soundings in Groundwater Investigation of a Typical Basement Complex – A Case Study of Oda Town Southwestern Nigeria. Ozean J. Appl. Sci. 2(4): 333-359.

Ariyo SO, Adeyemi GO (2009). Role of Electrical Resistivity Method for Groundwater Exploration in Hard Rock Areas: A Case Study from Fidiwo/Ajebo Areas of Southwestern Nigeria. Pac. J. Sci. Technol. 10(1):483-486.

Ebisemiju FS (1993). Ado-Ekiti Region - A Geographical Analysis and Master Plan. Alpha Prints, p. 15-30.

Mallam A (2004). Fresh Basement: Revealed from Resistivity Method. Zuma J. Pure Appl. Sci. 6(1): 6-9.

Mallam A, Emenike EA (2008). Preliminary Findings of Subsurface Characteristics From Direct Current Resistivity Survey of The Federal Capital Territory (FCT), Nigeria. Int. J. Pure Appl. Sci. 2(2):68-76.

Okhue ET, Olorunfemi MO (1991). Electrical Resistivity Investigation of a Typical Basement Complex Area -The Obafemi Awolowo University Campus Case Study. J. Min. Geol. 27(2):63-69.

Olayinka AI, Olorunfemi MO (1992). Determination of Geoelectric Characteristics in Okene Area and Implications for Borehole Siting. J. Min. Geol. 28(2):403-411.

Olorunfemi MO, Idornigie AI (1992). A Geoelectric Mapping of Basement Structures of South – Central Part of Bida Basin and its Hydrogeological Implication. J. Min. Geol. 28:94-98.

Olorunfemi MO, Ojo JS, Akintunde OM (1999). Hydro-geophysical Evaluation of the Groundwater Potentials of the Akure Metropolis, Southwestern Nig. J. Min. Geol. 35(2):207-228.

Olorunfemi MO, Okhue EJ (1992). Hyrogeologic and Geologic significance of a Geoelectric Survey at Ile–Ife, Nig. J. Min. Geol. 28:242-350.

Omosuyi GO, Ojo JS, Enikanselu PA (2003). Geophysical Investigation for Groundwater around Obanla – Obakekere in Akure Area within the Basement complex of South-Western Nigeria. J. Min. Geol. 39(2):109-116.

Rahaman MA (1988). Recent advances in the study of the basement complex of Nigeria. In Precambrian Geology of Nigeria. Geological Survey of Nigeria, Kaduna South, pp. 11 - 43.

Shemang EN Jnr (1993). Groundwater Potentials of Kubanni River Basin, Zaria, Nigeria from DC Resistivity Study. Water Resour. 4(1&2):36-42.

Sheriff MA, Malgwi DI, Hassan H (2002). Geoelectrical Depth Sounding Using Schlumberger Array in Yimirshika Area of Borno State, Nigeria. Zuma J. Pure Appl. Sci. 4(2):127-133.

Vander V (1988). RESIST Version 1.0. MSc Research Project. Delft, Netherlands.

11

Impact of soil organic matter on the sorption and phase distribution of 20% butanol-blended gasoline in the vadose zone

Ejikeme Ugwoha and John M. Andresen

Department of Chemical and Environmental Engineering, University of Nottingham, UK.

Batch microcosm experiments were conducted to examine the impact of soil organic matter (SOM) on the sorption and phase distribution of 20% butanol-blended gasoline (B20) compounds in the vadose zone. SOM was found to promote the sorption of B20 gasoline compounds in the vadose zone. It also affected the phase distribution of B20 gasoline compounds in the vadose zone by increasing adsorption on the soil solid and reducing concentrations in the soil air and soil water. These impacts increased with decreasing hydrophobicity of gasoline compounds. However, when compared with unblended gasoline, the SOM-induced increase in sorption coefficient for B20 gasoline compounds was reduced by a maximum of 66% for the cycloalkanes, 60% for the aromatics and 37% for the alkanes, implying a general reduction in the sorptive capability of SOM. Therefore, the use of high butanol volume such as 20% in gasoline to combat climate change is likely to reduce the retentive capability of SOM in the vadose zone and thus may increase groundwater contamination with gasoline compounds.

Key words: Butanol-blended gasoline, sorption, phase distribution, vadose zone, soil organic matter.

INTRODUCTION

Biobutanol is now gaining increasing interest as a gasoline oxygenate following butanol production from renewable and cost-effective cellulosic materials (Kumar and Gayen, 2011). Companies such as BP, Cobalt Biofuels, DuPont, Gevo, Green Biologics, METabolic Explorer and Tetravitae Bioscience plan to produce biobutanol on industrial scale to be used as a new gasoline extender (EBTP, 2009) where the first commercial-scale biobutanol facility are expected to be operational by 2013. With this near future commercialization of butanol-blended gasoline, potential large spills will be likely. Therefore, knowing the sorption and phase distribution of butanol-blended gasoline in the vadose zone will be useful both in making informed decision on its use and in assessing risks from spills.

As a fuel alcohol, butanol has several advantages over ethanol. These advantages include higher energy content, higher miscibility with gasoline, lower vapour pressure, lower water absorption and higher compatibility with existing gasoline pipelines (EBTP, 2009; USEPA, 2005). In a direct comparison of 20% blends in gasoline, Mariano et al. (2009) found that butanol may enhance the biodegradation of gasoline compounds in soil more than ethanol. As ethanol has been in the fuel market for decades, several studies on its impact on gasoline compounds transport in the vadose zone have been conducted (Adam et al., 2002; Chen et al., 2008; Corseuil et al., 2004; Dakhel et al., 2003; Lahvis, 2003; McDowell and Powers, 2003; Powers et al., 2001; Yu et al., 2009). These studies, especially with 10 and 20% ethanol blends, have shown that ethanol can reduce the interfacial and surface tensions of gasoline with respect to water, thereby increasing the rate of gasoline transport from the vadose zone to the saturated zone (McDowell and Powers, 2003; Yu et al., 2009). Also, ethanol enhances the mobility of dissolved gasoline compounds by exerting a cosolvent effect that decreases retention in the vadose zone (Alvarez et al., 2001; Powers et al., 2001; Yu et al., 2009). This leads to reduced retention capability of the soil (Adam et al., 2002). However, since butanol is still new in the fuel market, little is known about its potential impact on the transport of gasoline in the vadose zone. More so, to the best of our knowledge, the extent that soil organic matter (SOM) will impact the sorption and phase distribution of gasoline compounds in the vadose zone has not been addressed in the literature. Recent studies on butanol and butanol-blended gasoline have focused on the biodegradation of butanol and the effect of butanol on the biodegradation of gasoline compounds. These studies have revealed that butanol is readily biodegraded (Fairbanks et al., 1985; Mariano et al., 2009) and can affect the biodegradation of gasoline compounds in soil (Gomez and Alvarez, 2010; Mariano et al., 2009).

The SOM content of soils is the component that primarily regulates the sorption of organic contaminants by soils (Bohn et al., 2001; Celis et al., 2006; Chen et al., 2007; Li et al., 2009; Liu et al., 2008; Shi et al., 2010; Sparks, 1989). Consequently, the total removal of SOM from natural soils has been found to decrease sorption substantially (Shi et al., 2010). Besides the SOM content of soils, contaminant hydrophobicity is another important factor that can influence the sorption of contaminants in the vadose zone (Chen et al., 2007; Liu et al., 2008). Since butanol is a hydrophilic compound, its addition to gasoline could reduce the hydrophobicity of gasoline, thereby affecting the sorption of gasoline compounds by soils. Such impact could also alter the overall interactions between gasoline, soil water and soil particles.

Considering the near use of butanol-blended gasoline and the poor knowledge associated with its sorption by soil and phase distribution in the vadose zone, this study aimed to evaluate the impact of SOM on the sorption and phase distribution of butanol-blended gasoline in the vadose zone. The study also examined the effect of butanol on the SOM sorption potency. To achieve these aims, a number of laboratory batch microcosm experiments were performed. The soils used consisted of uncontaminated sand mixed with 0 to 5% SOM. Contamination involved a vapour phase injection of 20% butanol-blended gasoline as well as unblended gasoline into the headspace of microcosms, and allowing sorption to and distribution among the phases in the soils. The data obtained indicated that the addition of 20% butanol in gasoline could reduce the sorptive capability of SOM thereby increasing the risk of groundwater contamination with gasoline compounds.

MATERIALS AND METHODS

Butanol-gasoline blend composition

The butanol-blended gasoline used in this study was a synthetic gasoline blended with 20% butanol, referred to as B20. The butanol (99.8%) was purchased from Sigma-Aldrich chemical company, UK.The synthetic gasoline (referred to as B0) was prepared from six typical fuel compounds (Table 1). The six fuel compounds were all of high purity (>99.5%) and were purchased from Sigma-Aldrich chemical company, UK. The selected fuel compounds are typical constituents of petroleum fuel (Christophersen et al., 2005; Dakhel et al., 2003; Höhener et al., 2003; Pasteris et al., 2002; Speight, 2002) and represent the three major hydrocarbon groups in gasoline, namely alkanes, cycloalkanes (or naphthenes) and aromatics. The weight percentages of the fuel compounds have been chosen according to the typical fuel composition noted by Pasteris et al. (2002) and Höhener et al. (2003). The high blending of 20% butanol by volume was motivated by the consistent increase in the utilization of ethanol as gasoline oxygenate since the phase out of methyl tertiary butyl ether (MTBE) as gasoline oxygenate in the early 2000s.

Description of the soils

The soils used in this study comprised a mixture of SOM-free sand and peat as the source of SOM. The sand was oven dried and sieved through a 2 mm sieve to obtain uniform sand with a particle size distribution of coarse (20%), medium (53%) and fine (27%). The Peat was obtained from Westland Horticulture, UK and contained ≈96% SOM in its dry state, as determined by the "Loss on Ignition" (LOI) method (Sutherland, 1998). The peat was sieved through a 2 mm sieve to obtain particles generally regarded as SOM (Page, 1982). The sand and the peat were mixed on a dry basis to obtain soils consisting of 0 to 5% SOM fraction by weight (Table 2). These percentages of SOM were chosen because they are within the reported range of SOM fraction for typical soils (Bohn et al., 2001; Sparks, 2003).

Batch microcosm experiments

All microcosm experiments were performed with glass vials of 60 mL (H*Ø = 140*27.5 mm) sealed with a polytetrafluoroethylene screw cap for injection and extraction without leaks. The soils (1%f_{om}, 3%f_{om} and 5%f_{om}) were autoclaved by heating in an oven set at 160°C for 16 h and thereafter wetted with hot water (50°C) to a volumetric water content of 11%, similar to the water content used by Höhener et al. (2006). A 65 g mass of the wetted soils were packed into the glass vials and compacted alike by tapping vials on the worktable until stable heights were obtained. Heights in vials were 84 mm for 1%f_{om}, 92 mm for 3%f_{om} and 100 mm for 5%f_{om}, leaving headspaces of 56, 48, and 40 mm, respectively, in the glass vials for vapour phase sampling. The porosities of the soils were 0.46, 0.48 and 0.48 for 1%f_{om}, 3%f_{om} and 5%f_{om}, respectively. The control experiments were prepared with 0%f_{om}, similarly treated and packed. Height in vial was 80 mm leaving a headspace of 60 mm. The porosity of the 0%f_{om} was 0.45. All experiments were performed in triplicates. Before adding the B20 vapours, the vials were stored in a Thermostatic bath/circulator (L*W*H = 52*32*21.5 cm) set at 25°C for 24 h. Then 10 mL of air were extracted from the microcosms and 10 mL headspace of a glass vial containing the B20 at 25°C were injected using a stainless steel hypodermic needle (L*Ø = 50*0.63 mm) fitted to a 10 mL gas-tight syringe. The decreases of the gasoline compounds in the headspace of the microcosms were measured for up to 15 days as described in "Vapour phase analysis by gas chromatography" part of this work. The microcosms were maintained at 25°C in the Thermostatic bath/circulator throughout the duration of the experiment to enhance the unbiased extraction of both the highly and less volatile gasoline compounds. The increase in sorption resulting from SOM was obtained by comparing the a verage sorption in SOM-

Table 1. Synthetic gasoline composition.

Gasoline compound	Formula	Weight in mixture (%)	Volume (mL)	Vapour pressure at 20°C, Pa[a]	Density at 25°C, g/mL[a]	Henry's law constant[c]
Pentane	C_5H_{12}	9.6	15.3	57900	0.626	51.4
Octane	C_8H_{18}	25.8	36.7	1470	0.703	211
MCP	C_6H_{12}	19.5	26.0	17732[b]	0.75	14.7
MCH	C_7H_{14}	32.3	41.9	4930	0.77	17.5
Benzene	C_6H_6	3.2	3.7	9950	0.874	2.26E-01
Toluene	C_7H_8	3.2	11.1	2910	0.865	2.65E-01
			Gasoline additive			
Butanol	C_4H_9OH	-	0 &20%	500	0.81	4.99E-04

MCP – Methylcyclopentane; MCH – Methylcyclohexane; [a] values obtained from Sigma Aldrich Material Safety Data Sheet; [b] value obtained from Pasteris et al. (2002) Supporting Information; [c] obtained from Yaws (2008).

Table 2. Soil composition and characterisation.

Soils	Description	Properties					
		pH -	ρ_s g/mL	ρ_b g/mL	n -	SA m^2/g	TPV cm^3/g
$0\%f_{om}$	Sand + 0% SOM	9.10	2.5	1.22	0.51	0.82	4.04E-3
$1\%f_{om}$	Sand + 1% SOM	8.51	2.4	1.17	0.52	1.04	4.47E-3
$3\%f_{om}$	Sand + 3% SOM	7.34	2.3	1.07	0.53	1.47	5.33E-3
$5\%f_{om}$	Sand + 5% SOM	6.17	2.1	0.97	0.54	1.91	6.20E-3

ρ_s = Particle density; ρ_b = Dry bulk density; n = Porosity; SA = Surface area; TPV = Total pore volume.

containing porous media with that in $0\%f_{om}$. The impact of butanol on the sorptive capability of SOM for gasoline compound was obtained by establishing a baseline from experiments using unblended gasoline (B0).

Vapour phase analysis by gas chromatography

The vapour phase concentrations of gasoline compounds in the headspace of the microcosms were measured by extracting 40 µL of vapour and injecting into a HR-5300 mega series Gas Chromatography (Carlo Erba, UK) equipped with a Chrompack Poraploto column (27.5 m * 0.32 mm * 10 µm) and Flame Ionization Detector (FID). The injector was heated to 200°C. The column temperature was held at 200°C for 16 min (isothermal temperature program). The carrier gas was helium at a flow rate of 4 mL/min.

Estimation of mass distribution of gasoline compounds amongst phases

The mass of gasoline compounds that partitioned to the air and water phases of the vadose zone were estimated using Equations (1) and (2) (Kerfoot, 1991).

$$M_a = C_a * V_a \tag{1}$$

$$M_w = \frac{C_a}{H} * V_w \tag{2}$$

where M_a, and M_w are the masses of gasoline compound in the air and water phase, respectively, C_a is the concentration of gasoline compound measured at the headspace of microcosm, V_a and V_w are the volumes of air and water in the microcosm, respectively, and H is the dimensionless Henry's law constant.

By applying mass balance in a closed system, the mass adsorbed on the soil solids (M_s) was estimated using Equation (3).

$$M_s = M_t - M_a - M_w - M_L \tag{3}$$

where M_t is the total mass of gasoline compound injected into the microcosm and M_L is the mass of gasoline compound lost via sampling.

Estimation of sorption coefficient of gasoline compounds

Sorption coefficient (K_d), also called the soil-water distribution coefficient, was estimated as the ratio of the concentration of gasoline compound in the soil solid (C_s) to the concentration in the soil water (C_w) (Kerfoot, 1991):

$$K_d = \frac{C_s}{C_w} \tag{4}$$

Estimation of retardation factor of gasoline compounds

The degree of retardation, called retardation factor (R), of the migration of gasoline compounds in the vadose zone as a result of sorption on soil solid was estimated using Equation (5) (Site, 2001).

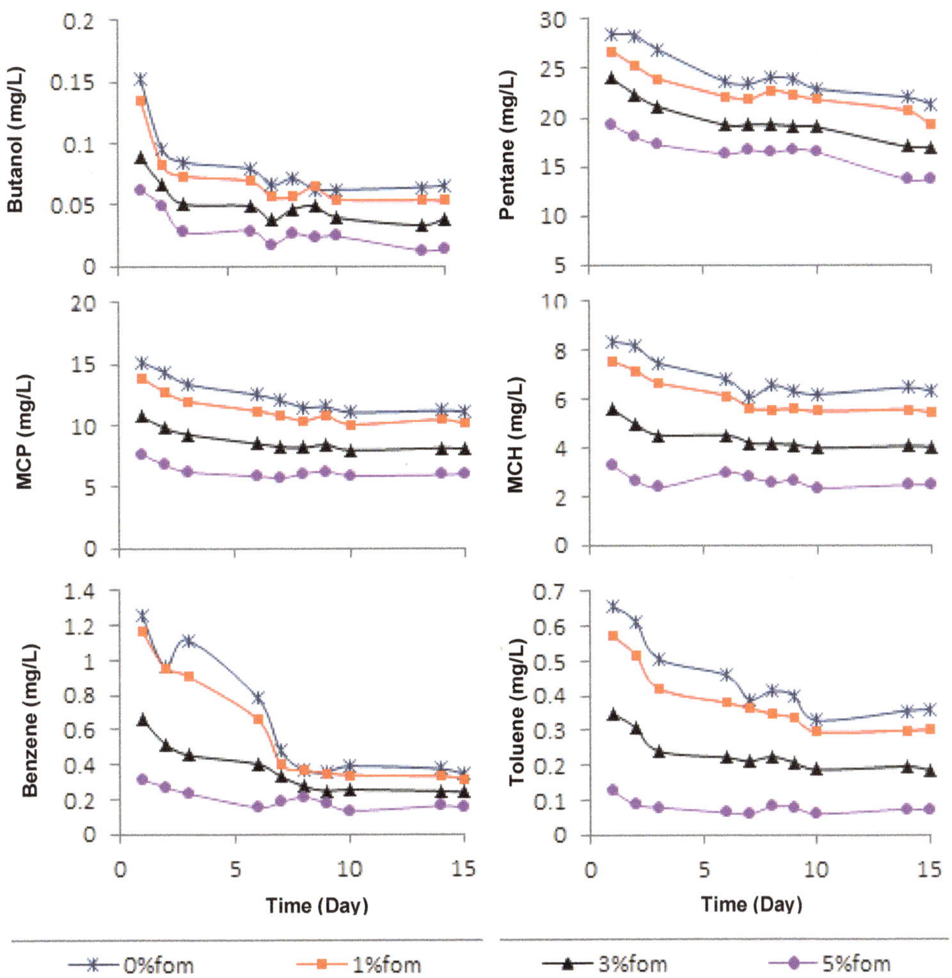

Figure 1. Headspace concentrations of B20 gasoline compounds with time as a function of SOM fraction of soils.

$$R = 1 + \left(\frac{\rho_b}{n}\right) K_d \quad (5)$$

where ρ_b and n are the bulk density and porosity of the vadose zone, respectively.

RESULTS AND DISCUSSION

Impact of SOM on the sorption and phase distribution of B20 gasoline compounds

Sorption of B20 gasoline compounds

Figure 1 presents the concentrations of B20 gasoline compounds in the headspace of microcosms as a function of time and SOM fraction of soils. Octane, the second alkane in the gasoline mixture, was poorly detected by the GC-FID used for analysis hence was not included in Figure 1. The headspace concentration of the different gasoline compounds decreased with specific trend. All gasoline compounds attained equilibrium within the 15 days duration of the experiment, except for pentane that still maintained an approximate linear concentration decrease with time. The decrease in the headspace concentration of gasoline compounds observed indicated increase in sorption to the soils.

The increase in the SOM fraction of soils resulted in greater sorption of all compounds. This enhanced sorption of compounds by SOM could be due to the increased porosity, surface area, and total pore volume of soils with increasing SOM fraction as listed in Table 2. Our ethanol-blended gasoline study indicated that SOM had no significant impact on the sorption of ethanol. Therefore, the impact of SOM on the sorption of butanol confirmed that butanol has lower water absorption than ethanol, as published by the United States Environmental protection Agency (USEPA, 2005), and higher octanol-water partition coefficient (K_{ow}) as reported by Yaws (2008). Therefore, butanol is likely to have less adverse effect, compared with ethanol, on the sorption of gasoline compounds by SOM. Among the gasoline compounds, SOM impacted the sorption of the aromatics to a greater extent (75 to 81%) than the cycloalkanes (50 to 61%) and the alkanes (32%). This impact was relatively consistent throughout the duration of the experiment except for benzene that dwindled from 75% 4 h after contamination to 55% on Day 15, suggesting that the sorption of benzene by SOM was via physisorption. Since physisorption involves relatively weak bonding forces between dipole moments of contaminant and soil molecules (Bhandari et al., 2007; Weber et al., 1991), it therefore implies that benzene retained by SOM in the vadose zone after a spill of B20 is likely to leach out with time.

Overall, this result indicates that SOM promoted the sorption of all B20 gasoline compounds. This effect was more on the aromatics than on the cycloalkanes and alkanes. Also, it indicates that SOM-compound bonding forces would be stronger for the more hydrophobic gasoline compounds than for the less hydrophobic gasoline compounds. Hence, less hydrophobic gasoline compounds retained by SOM are more likely to undergo leaching with time.

Soil-water interaction of B20 gasoline compounds in the vadose zone

The impact of SOM on the soil-water interaction of B20 gasoline compounds was examined by comparing the sorption coefficient (K_d) values estimated from 0, 1, 3 and 5%f_{om} soils as shown in Figure 2. K_d was estimated for each sampling day and the average K_d was used as the representative K_d. The standard deviation of all the K_d was calculated and used as the error bar. The K_d values obtained are comparable to those reported in the literature. For example, Joo et al (2008) reported K_d values of 0.2 and 0.3 L/kg for benzene and toluene, respectively, with aquifer sand of negligible SOM fraction. In contrast, we have K_d values of 0.6 and 0.7 L/kg for benzene and toluene, respectively, with 0%f_{om} soil which is comparable to aquifer sand in terms of SOM content. The slight difference in the K_d values can be attributed to the difference in the soil and contaminant mixture used (Allen-King et al., 2002). Joo et al. (2008) mixture consisted of only aromatics, while our mixture consisted of aromatics, cycloalkanes, alkanes and alcohol.

The K_d values for all gasoline compounds increased with increasing SOM fraction (f_{om}) of soils, suggesting increase in adsorption on soil solid and decrease in dissolution into the soil water. The increased surface area (SA) of the soils associated with increasing f_{om} could be the possible reason for the observed increase in K_d. As reported by Site (2001), the K_d values for benzene, toluene and xylenes were lower in kaolinite (SA = $3*10^4$ m^2kg^{-1}) than in illite (SA = 10^5 m^2kg^{-1}) and montmorillonite (SA = $8*10^6$ m^2kg^{-1}), indicating that increase in SA of soils could result in increase in K_d of gasoline compounds. The impact of SOM was greatest for the aromatics where the K_d increased 7 times, compared with the 4 times for the cycloalkanes and 2 times for the alkanes, as the f_{om} increased from 0 to 5%. This direct relationship between K_d and SOM have also been reported for a wide range of single organic contaminants and adsorbents, but it is to the authors knowledge shown for the first time for butanol-blended gasoline. Thus, the data indicate that SOM increased the adsorption of B20 gasoline compounds on the soil solid, but reduced their dissolution into the soil water. This effect is likely to be greatest for the aromatics compared with the alkanes and cycloalkanes.

Retardation of the migration of B20 gasoline compounds in the vadose zone

The degree of retardation of the migration of gasoline compounds in the vadose zone by SOM was investigated by observing the variations in the retardation factor (R) of gasoline compounds in 0, 1, 3 and 5%f_{om} soils. Figure 3 shows the variations in R with increasing SOM fraction (f_{om}) of soils for the B20 gasoline compounds. Generally, the R values for all gasoline compounds increased with increasing f_{om}, suggesting that partitioning into SOM is a major process causing retardation. This effect increased with increasing hydrophobicity, hence was greatest for the cycloalkanes (0 to 64) and least for the aromatics (0 to 9.4). The difference in the R values for gasoline compounds would result in different distributions for the gasoline compounds in soils (Johnson and Perrott, 1991).

A comparison of our R values for benzene and toluene in 0%f_{om} with those reported by Höhener et al. (2006) in sand shows a good agreement despite the slight difference in contaminant mixture. We obtained R values of 2.29 for benzene and 2.53 for toluene while Höhener and co-workers reported R values of 1.24 for benzene and 1.93 for toluene. Also, the R values of 1.61 for benzene and 1.91 for toluene obtained from Joo et al. (2008) K_d values of 0.2 L/kg for benzene and 0.3 L/kg for toluene were in good agreement with our R values. However, our R values were found to differ greatly with those reported for clay, 117.2 for benzene and 734 for toluene (Myrand et al., 1992), and for fine silty loam, 193.0 for benzene and 218.0 for toluene (Johnson and Perrott, 1991). The vast difference between the R values from different soils, as compared to the R values from different contaminant mixture, suggest that R of gasoline compounds could be highly site specific and that the properties of soils is likely to have a greater influence on the R of gasoline compounds than the composition of contaminants. Overall, our data indicate that SOM reduced the migration of B20 gasoline compounds in the vadose zone. This effect increased with increasing

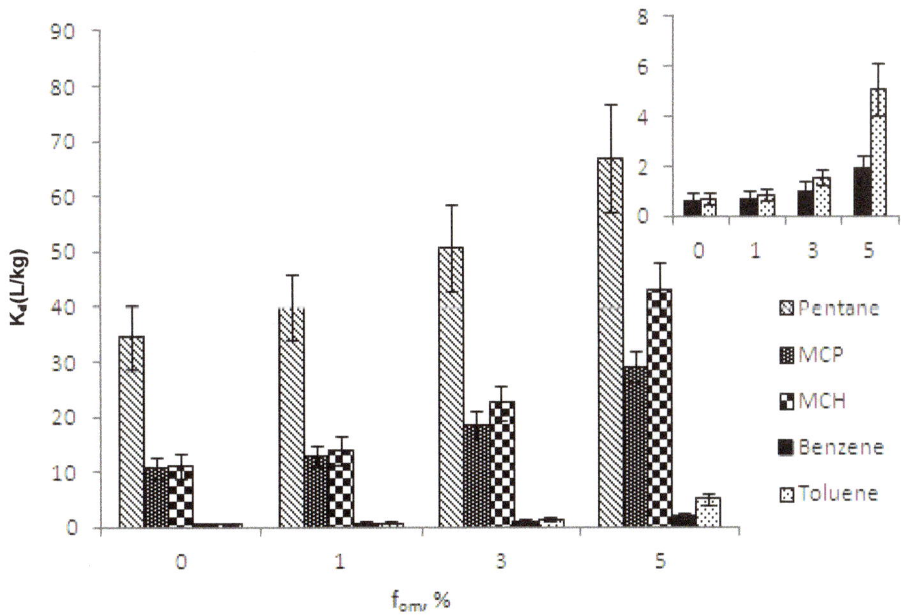

Figure 2. Sorption coefficient (K_d) of B20 gasoline compounds as a function of SOM fraction of soils.

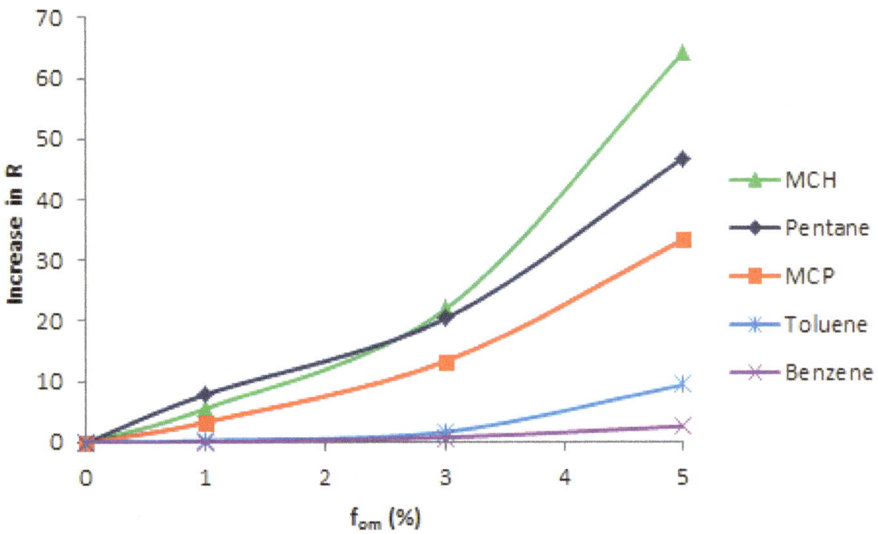

Figure 3. Impact of SOM on the retardation factor (R) of B20 gasoline compounds.

hydrophobicity of gasoline compounds.

Mass distribution of B20 gasoline compounds amongst the vadose zone phases

Table 3 lists the changing masses of B20 gasoline compounds in the soil air, soil solid and soil water with changing SOM fraction of soils. The increase in the SOM fraction of soils resulted in an increased mass distribution of gasoline compounds from the soil air to the soil solid, and in a reduced mass distribution to the soil water. Although this effect was observed throughout the duration of the experiment, it was generally greatest on Day 1 and affected the aromatics (29 to 49%) more than the cycloalkanes (28 to 35%) and the alkanes (20%). The reduction in the amount of available soil water in the microcosm with increasing SOM fraction of the soils could be the possible reason for the greater impact of SOM on the adsorption of the less hydrophobic gasoline compounds such as the aromatics. However, on Day 15, when equilibrium is assumed to have been reached, SOM

Table 3. Mass distribution of B20 gasoline compounds in the vadose zone as a function of SOM fraction of soils.

Day	Pentane			MCP			MCH			Benzene			Toluene		
	Air %	Solid %	Water %	Air %	Solid %	Water %	Air %	Solid %	Water %	Air %	Solid %	Water %	Air %	Solid %	Water %
								$0\%f_{om}$							
1	48.3	51.6	0.2	48.2	51.1	0.6	52.1	47.3	0.8	34.1	37.0	28.9	20.7	64.4	14.9
8	40.8	58.7	0.2	36.5	62.8	0.5	40.8	58.4	0.5	10.0	81.3	8.5	13.0	77.6	9.4
15	32.8	60.6	0.1	35.3	63.7	0.5	39.3	59.7	0.4	9.4	82.5	7.9	11.3	80.4	8.1
								$1\%f_{om}$							
1	43.9	56.0	0.2	42.7	56.8	0.6	45.5	54.0	0.5	30.7	42.5	26.8	17.4	69.6	13.0
8	37.3	62.2	0.1	31.9	67.3	0.4	33.5	65.8	0.4	9.7	81.6	8.5	10.5	81.5	7.9
15	31.5	67.9	0.1	31.1	68.0	0.4	32.8	66.4	0.4	8.3	84.2	7.3	9.2	83.8	6.8
								$3\%f_{om}$							
1	37.6	62.2	0.2	31.6	67.9	0.4	32.3	67.3	0.4	16.6	68.2	15.2	10.2	81.9	7.9
8	30.3	69.3	0.1	24.2	75.3	0.3	24.0	75.5	0.3	7.0	86.6	6.4	6.5	88.3	5.1
15	26.6	72.9	0.1	23.8	75.4	0.3	23.2	75.1	0.3	6.1	88.3	5.6	5.4	90.3	4.2
								$5\%f_{om}$							
1	28.0	71.9	0.1	20.7	79.0	0.3	17.4	82.4	0.2	7.3	85.5	7.2	3.4	93.7	2.9
8	23.8	75.9	0.1	16.5	83.0	0.3	13.7	86.0	0.2	5.0	90.0	4.9	2.2	95.9	1.9
15	19.9	79.6	0.1	16.4	83.1	0.3	13.2	86.5	0.2	3.6	92.7	3.6	1.9	96.4	1.6

promoted the adsorption of gasoline compounds based on hydrophobicity, with the cycloalkanes promoted to greater extent (19 to 26%) than the alkanes (19%) and the aromatics (10 to 16%). This implies that the bonding forces between the dipole moments of the SOM and the aromatics are weaker than those of the cycloalkanes and the alkanes. Hence, any aromatics retained by SOM in the vadose zone after a spill of B20 are more likely to leach out with time faster than would any cycloalkanes or alkanes. Overall, the data indicate that SOM could impact the phase distribution of B20 gasoline compounds in the vadose zone. Among the gasoline compounds, the impact is likely to vary with time. While the less hydrophobic gasoline compounds may be impacted to a greater extent few hours after spills, the more hydrophobic gasoline compounds are likely to be impacted to a greater extent afterwards.

Impact of butanol on SOM sorptive capability

The impact of butanol on SOM sorptive capability for gasoline compounds was investigated by comparing the K_d of B0 and B20 gasoline compounds in $0\%f_{om}$ and $5\%f_{om}$ soils. The difference in K_d increase with increasing SOM for B0 and B20 gasoline compounds is presented in Figure 4. The presence of butanol caused a substantial reduction in the K_d of all gasoline compounds in both 0 and $5\%f_{om}$, signifying reduction in the amount of gasoline compounds adsorbed on the soil solid. The presence of butanol also caused a significant reduction in the SOM-induced increase in K_d for all B20 gasoline compounds, implying reduction in the SOM sorptive capability for gasoline compounds. This effect increased with increasing molecular weight and thus was greatest for the cycloalkanes compared with the aromatics and alkanes. The SOM-induced increase in K_d for B20 gasoline compounds decreased by 55 to 66% for the cycloalkanes, 43 to 60% for the aromatics and 37% for the alkanes. Generally, the reduction in the K_d and in the SOM-induced increase in K_d of gasoline compounds would mean decrease in the amount of gasoline compounds retained in the vadose zone. Therefore, the data indicate that the full sorptive capability of SOM for gasoline compounds is unlikely to be realised for B20 gasoline compounds.

Conclusions

Soil organic matter (SOM) showed a great impact on the sorption and phase distribution of B20 gasoline compounds in the vadose zone. This impact increased with decreasing hydrophobicity of gasoline compounds. By increasing the SOM fraction of sand from 0 to 5%, the sorption coefficient (K_d) of B20 gasoline compounds increased 7 times for the aromatics, 4 times for the cycloalkanes and 2 times for the alkanes. However, when

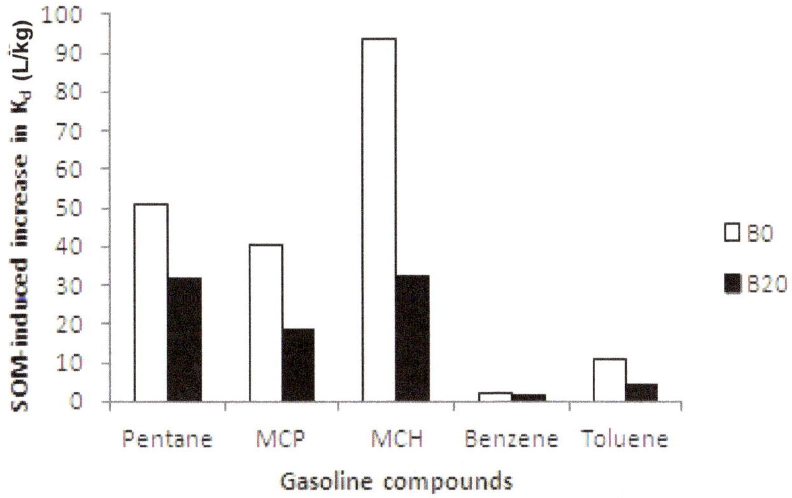

Figure 4. Impact of butanol on SOM capability to increase the K_d of gasoline compounds.

compared with B0 gasoline compounds, the SOM-induced increase in K_d for B20 gasoline compounds decreased by 55 to 66% for the cycloalkanes, 43 to 60% for the aromatics and 37% for the alkanes, implying a general reduction in the sorptive capability of SOM. Therefore, the use of high butanol volume such as 20% in gasoline to combat climate change is likely to reduce the retentive capability of SOM in the vadose zone. This may increase groundwater contamination with gasoline compounds.

ACKNOWLEDGEMENTS

The authors thank the Niger Delta Development Commission (NDDC) of Nigeria for sponsoring this research.

REFERENCES

Adam G, Gamoh K, Morris DG, Duncan H (2002). Effect of alcohol addition on the movement of petroleum hydrocarbon fuels in soil. Sci. Total Environ. 286:15-25.

Allen-King RM, Grathwohl P, Ball WP (2002). New modeling paradigms for the sorption of hydrophobic organic chemicals to heterogeneous carbonaceous matter in soils, sediments, and rocks. Adv. Water Resour. 25:985-1016.

Alvarez PPJJ, Ruiz-Aguilar GML, Fernandez-Sanchez JM, Kim D, Beller HR, Kane SR (2001). Effect of ethanol and MtBE on BTEX biodegradation in the saturated zone: kinetic studies in: Rice DW, Depue RT (Ed.), Environmental assessment of the use of ethanol as a fuel oxygenate: Subsurface fate and transport of gasoline containing ethanol.

Bhandari A, Surampalli RY, Champagne PP, Ong SK, Tyagi RD, Lo IMC (2007). Remediation technologies for soils and groundwater. American Society of Civil Engineers (ASCE), Reston.

Bohn HL, McNeal BL, O'Connor GA (2001). Soil chemistry, 3rd edition ed. John Wily & Sons, Inc., New York.

Celis R, De Jonge H, De Jonge LW, Real M, Hermosín MC, Cornejo J (2006). The role of mineral and organic components in phenanthrene and dibenzofuran sorption by soil. Eur. J. Soil Sci. 57:308-319.

Chen D, Xing B, Xie W (2007). Sorption of phenanthrene, naphthalene and o-xylene by soil organic matter fractions.Geoderma 139:329-335.

Chen H, Chen S, Quan X, Zhao H, Zhang Y (2008). Sorption of polar and nonpolar organic contaminants by oil-contaminated soil. Chemosphere 73:1832-1837.

Christophersen M, Broholm MM, Mosbæk H, Karapanagioti HK, Burganos VN, Kjeldsen P (2005). Transport of hydrocarbons from an emplaced fuel source experiment in the vadose zone at Airbase Værløse, Denmark. J. Contam. Hydrol. 81:1-33.

Corseuil HX, Kaipper BIA, Fernandes M (2004). Cosolvency effect in subsurface systems contaminated with petroleum hydrocarbons and ethanol. Water Res. 38:1449-1456.

Dakhel N, Pasteris G, Werner D, Höhener P (2003). Small-Volume Releases of Gasoline in the Vadose Zone: Impact of the Additives MTBE and Ethanol on Groundwater Quality. Environ. Sci. Technol. 37:2127-2133.

EBTP (2009). Biobutanol.European Biofuels Technology Platform.

Fairbanks BC, Schmidt NE, O'Connor GA (1985). Butanol degradation and volatilization in soils amended with spent acid or sulphuric acid. J. Environ. Qual. 14:83-86.

Gomez DE, Alvarez PJJ (2010). Comparing the effects of various fuel alcohols on the natural attenuation of Benzene Plumes using a general substrate interaction model. J. Contam. Hydrol. 113:66-76.

Höhener P, Dakhel N, Christophersen M, Broholm M, Kjeldsen P (2006). Biodegradation of hydrocarbons vapors: Comparison of laboratory studies and field investigations in the vadose zone at the emplaced fuel source experiment, Airbase Værløse, Denmark. J. Contam. Hydrol. 88:337-358.

Höhener P, Duwig C, Pasteris G, Kaufmann K, Dakhel N, Harms H (2003). Biodegradation of petroleum hydrocarbon vapors: Laboratory studies on rates and kinetics in unsaturated alluvial sand. J. Contam. Hydrol. 66:93-115.

Johnson RL, Perrott M (1991). Gasoline vapor transport through a high-water-content soil. J. Contam. Hydrol. 8:317-334.

Joo JC, Shackelford CD, Reardon KF (2008). Sorption of nonpolar neutral organic compounds to humic acid-coated sands: Contributions of organic and mineral components. Chemosphere 70:1290-1297.

Kerfoot HB (1991). Subsurface Partitioning of Volatile Organic Compounds: Effects of Temperature and Pore-Water Content. Ground Water 29:678-684.

Kumar M, Gayen K (2011). Developments in biobutanol production: New insights. Appl. Energy 88:1999-2012.

Lahvis MA (2003). Evaluation of small-volume releases of ethanol-blended gasoline at UST sites. API Soil Groundwater Res. Bull. No.19.
Li R, Wen B, Zhang S, Pei Z, Shan X (2009). Influence of organic amendments on the sorption of pentachlorophenol on soils. J. Environ. Sci. 21:474-480.
Liu P, Zhu D, Zhang H, Shi X, Sun H, Dang F (2008). Sorption of polar and nonpolar aromatic compounds to four surface soils of eastern China. Environ. Pollut. 156:1053-1060.
Mariano AP, Tomasella RC, Di Martino C, Filho RM, Seleghim MHR, Contiero J, de Angelis DF (2009). Aerobic biodegradation of butanol and gasoline blends. Biomass Bioenerg. 33:1175-1181.
McDowell CJ, Powers SE (2003). Mechanisms Affecting the Infiltration and Distribution of Ethanol-Blended Gasoline in the Vadose Zone. Environ. Sci. Technol. 37:1803-1810.
Myrand D, Gillham RW, Sudicky EA, O'Hannesin SF, Johnson RL (1992). Diffusion of volatile organic compounds in natural clay deposits: Laboratory tests. J. Contam. Hydrol. 10:159-177.
Page AL (1982). Method of soil analysis part 2: chemical and microbiological properties. . American Society of Agronomy, Inc. and Soil Science of America, Inc., Madison, Wisconsin USA.
Pasteris G, Werner D, Kaufmann K, Höhener P (2002). Vapor Phase Transport and Biodegradation of Volatile Fuel Compounds in the Unsaturated Zone: A Large Scale Lysimeter Experiment. Environ. Sci. Technol. 36:30-39.
Powers SE, Hunt CS, Heermann SE, Corseuil HX, Rice D, Alvarez PPJJ (2001b). The transport and fate of ethanol and BTEX in groundwater contaminated by gasohol. Crit. Rev. Environ. Sci. Technol. 31:79-123.
Shi X, Ji L, Zhu D (2010). Investigating roles of organic and inorganic soil components in sorption of polar and nonpolar aromatic compounds. Environ. Pollut. 158:319-324.
Site AD (2001). Factors affecting sorption of organic compounds in natural sorbent/water systems and sorption coefficients for selected pollutants: A review. J. Phys. Chem. Ref. Data 30:187-439.
Sparks DL (1989). Kinetics of soil chemical processes. Academic Press, Inc, San Diego, California.
Sparks DL (2003). Environmental soil chemistry, 2nd edition ed. Academic Press San Diego, California.
Speight JG (2002). Handbook of petroleum product analysis. Wiley-Interscience, New York.
Sutherland R (1998). Loss-on-ignition estimates of organic matter and relationships to organic carbon in fluvial bed sediments. Hydrobiologia 389:153-167.
USEPA (2005). Progress report: A novel fermentation process for butyric acid and butanol production from plant biomass
Weber Jr WJ, McGinley PM, Katz LE (1991). Sorption phenomena in subsurface systems: Concepts, models and effects on contaminant fate and transport. Water Res. 25:499-528.
Yaws CL (2008). Handbook of properties for environmental and green engineering. Gulf Publishing Co, Houston.
Yu S, Freitas JG, Unger AJA, Barker JF, Chatzis J (2009). Simulating the evolution of an ethanol and gasoline source zone within the capillary fringe. J. Contam. Hydrol. 105:1-17.

12

Determination of rainfall-recharge relationship in River Ona basin using soil moisture balance and water fluctuation methods

Oke, M. O., Martins, O. and Idowu, O. A.

[1]Department of Geography and Environmental Management, Tai Solarin University of Education, Ijagun, Nigeria.
[2]Department of Water Resources Management and Agricultural Meteorology, University of Agriculture, Abeokuta, Ogun State, Nigeria.

A quantitative evaluation of spatial and temporal distribution of groundwater recharge is a pre-requisite for the management of ground water resources system in an optimal manner. The amount of groundwater recharge depends upon the rate and duration of rainfall, as rainfall is the principal means for replenishment of moisture in the soil water system and recharge to ground water. This paper investigated the relationship between rainfalls and groundwater recharge within Ona River basin, southwest Nigeria, using soil moisture balance and water table fluctuation. Analysis of rainfall trends within the Ona River basin suggests that there is considerable high annual rainfall occurrence, with a mean of 1623.48. It must be noted that the mean annual lost due to evapotranspiration of 1361.68 mm is very high when compared to the rainfall (83.9%). The results obtained from the soil moisture balance when considering the three dominant soil types within the basin, that is, sandy loam, clay and find sand, having water capacity of root zone value of 70, 70, and 50 respectively, suggests that groundwater recharge follows a positive trend as the corresponding rainfalls. However, empirical relationships of: y = 0.540x – 606.2, with a coefficient of determination (r^2) value of 0.719, for sandy loam and clay; and y = 0.552x – 621, with a coefficient of determination (r^2) value of 0.726 for fine sand was established for the basin area. On the other hand, recharge ranging from 220.25 to 40.50 mm was computed from the water table fluctuation method.

Key words: Rainfall–recharge relationship, soil types, soil moisture balance, Ona River basin, water table fluctuation.

INTRODUCTION

The need to determine local rainfall recharge relationship through the use of locally determined empirical formula as a viable option for prediction of groundwater recharge is the main focus of this paper. Groundwater recharge is understood as the downward flow of water recharging the water table, forming an addition to the groundwater reservoir (Backundukize et al., 2011). But Kumar (1973) asserted that the quantification of the rate of natural groundwater recharge is a basic pre-requisite for efficient groundwater resource management. The case study happens to be in the humid region where the amount of recharge in wet season is usually high because such region receive large amount of rainfall and the relative proportion of these components fluctuate according to the climatic conditions, geology, and geomorphology. Although in the humid tropics region, the most important mechanisms of ground recharge are considered to be indirect recharge by infiltration from floods through the beds of ephemeral streams (Marechal et al., 2008), there are several methods for estimating groundwater recharge, which are being used until date. The use of methods depends on the temporal and spatial resolutions of required estimates. Estimation of groundwater recharge is normally with errors and uncertainties (Dages et al., 2009; Sophocleous, 2004; Fitzsimons and Misstear, 2006). The best way to minimize these uncertainties is to use a combination of several methods (Scanlon et al., 2002). However, the Food and Agricultural Organization of the United Nation (FAO) recommends the use of Penman-Monteith equation as the standard for estimating reference evapotranspiration because it approximates better with lysimeters (Jabloun and Sahli; 2008; Trajkovic, 2001). Thus, in this paper, reference evapotranspiration was calculated using Penman-Monteith equation. This is useful for estimation of recharge using the soil moisture balance. A water table fluctuation method was also used to estimate groundwater recharge in the Ona River basin.

The aim of this paper is to estimate groundwater recharge within Ona River basin using both soil moisture balance and the water table fluctuation methods; and to establish an empirical relationship between recharge and rainfall. In other to carry out the main objective, the following sub-objective is considered: estimation of monthly and annual evapotranspiration using Penman-Monteith equation; estimation of annual groundwater recharge using soil moisture balance, and water table function methods; and establishment of empirical relationship between the groundwater recharge and rainfall.

Justification

Two major cities in Nigeria are located within Ona River Basins with other smaller urban settlements. They are Ibadan and Ijebu Ode. These cities are with considerable high population. For example, the population of Ibadan city is hovering around 6 million people and Ijebu Ode around 1 million. Equally, major industries are located within the drainage basin in question. However, there exist challenges in water sourcing in these cities. Although the cities are blessed with various rivers, streams and rivulet, they are often polluted and the State Government is finding it difficult in satisfying the water provision using surface water. Also, lots of buildings, both surface and underground were observed in these cities; so also asphalt overlay and pavement construction capable of preventing recharge into the water table is eminent in the study area. Whereas the majority of the people rely on groundwater for domestic, industrial and irrigational purposes, the activities in the cities is preventing recharge into the groundwater. But natural recharge by downward flows of water through the unsaturated zone is generally the most important mode of groundwater recharge. It is therefore crucial to estimate groundwater recharge for safe and efficient management of groundwater resource (Fitzsmons and Misstera, 2006).

As a result of possible excess withdrawals from groundwater reservoirs against natural replenishment which lead to regular lowering of water table, detailed study of groundwater recharge within Ona basin is essential.

Study area

River Ona is located in Southwestern part of Nigeria, an area whose boundaries are approximately latitudes 6° 34' N and 7° 38' N, and longitudes 3° 26' E and 3° 59' E (Figure 1). The basin occupies an area of 6,800 km^2 with its greater part in Ogun State before it terminates in the Lagos lagoon. The Ona basin is located west of the Oshun River basin as well as the Owa, Ibu and Omi rivers.

The climate of the Ona basin is similar to what obtained in the southwestern Nigeria. It is influenced by the movement of the inter-tropical convergence zone (the ITCZ), a quasi-stationary boundary zone which separates the sub-tropical continental air mass over the Sahara and the equatorial maritime air mass over the Atlantic Ocean. The former air mass is characterized by the dry northeasterly winds known as Harmatan, the latter by the rain-bearing southwesterly winds from the gulf of Guinea.

The ITCZ moves northwards beyond the basin at the peak of the raining season in June and July, and southward to the coast in the middle of the dry season in December and January. The change from the raining season to the dry season is rather abrupt while the onset of the rains after the dry season is gradual. Data obtained from the basin shows that February and March are the hottest months of the year. During these months, temperatures are high over the entire area. The mean daily maximum temperature for February is 31.4°C in the south and as high as 34.6°C in the north. The lowest mean minimum temperature in the north are recorded in December (17°C), that is, during the harmatan; in the southern part, the lowest mean temperature of 22.8°C was recorded in July during the raining season. In general humidity decreases northward in the basin. The lowest mean monthly humidity at 12.00 GMT is 62% in the south and 50% in the north, while the mean annual humidity varies from 75% in the south to 55% in the north. The raining season begins early in the south, which occurs in March and continues until the end of October or early November giving at least seven month of rainfall. In late July and early August, the mean wet season rainfall varies from 1020 to 1520 mm in the south

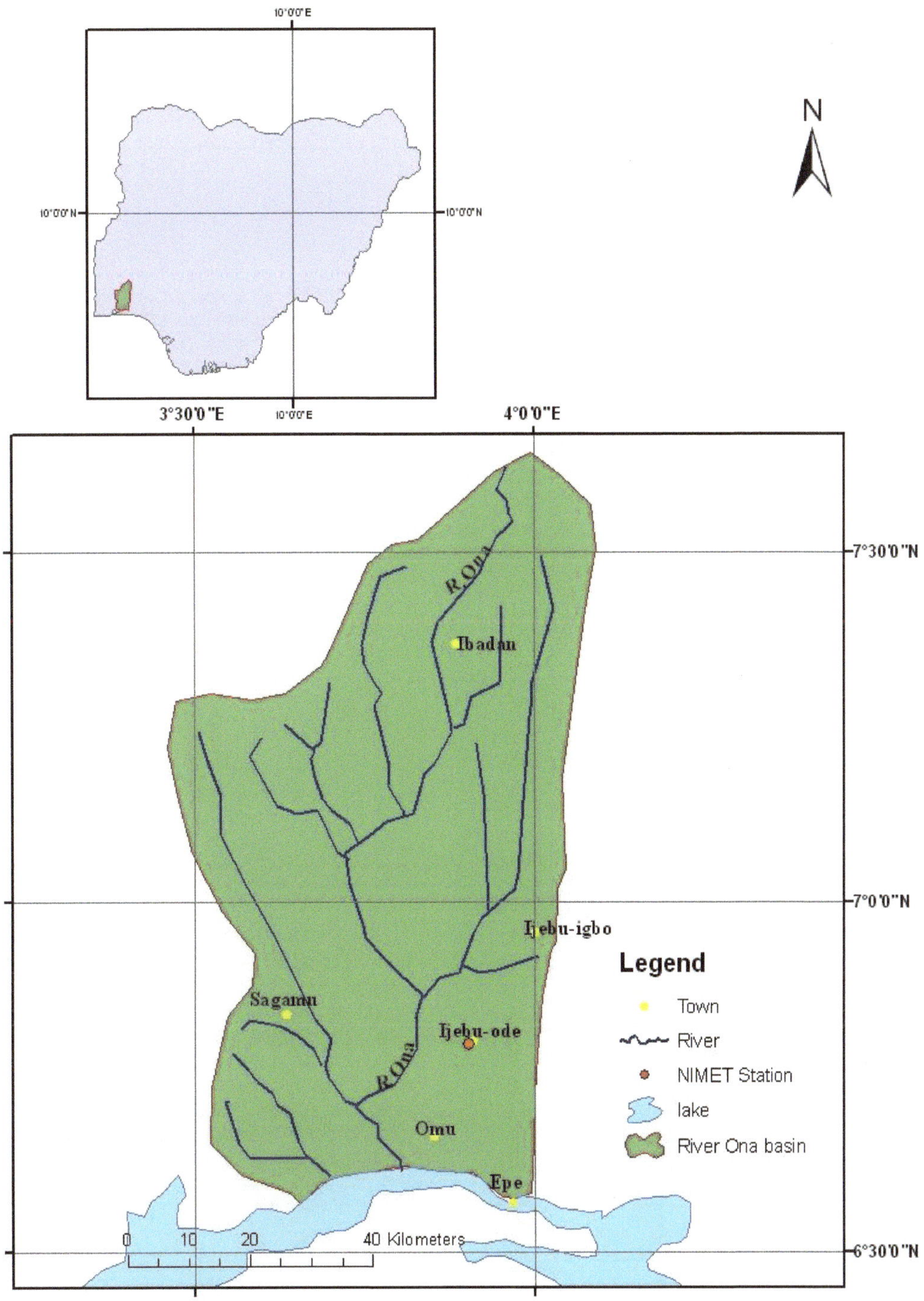

Figure 1. Map of Ona River basin.

of the basin and it is less than 1020 mm in the north. The mean dry season rainfall, on the other hand varies from 127 to 178 mm in the north and from 178 to 254 mm in the south.

Geologically, sedimentary rocks of cretaceous and latter deposit are found in the southern section of Ona River basin; the remaining section are composed of crystalline rock of the basement complex consisting folded gnesis, schist, and quartzite complexes which belong to the older intrusive series.

METHODOLOGY

The main data employed in this project are rainfall data and water

Table 1. Annual soil water budget calculation (Thornthwaite and Matter, 1957) after Backundukize et al. (2011).

	S_B = CAP	Wet season [SUR = (P-Ro) – PET > 0]		Dry season [SUR = (P-Ro) – PET < 0]
		S_B < CAP		
		(P-Ro) – PET ≤ CAP - S_B	(P-Ro) – PET > CAP - S_B	
S_B	CAP	S_B + (P-Ro) – PET	CAP	$CAP^*e^{-APWL/CAP}$
R_N	(P-Ro) - PET	0	(P-Ro) – PET – (CAP - S_B)	0
AET	PET	PET	PET	(P-Ro) + ΔS_B
DEF	0	0	0	PET - AET

P = Precipitation (mm); Ro = runoff (mm); PET = potential evapotranspiration (mm); APWL = accumulated potential water loss (mm) [PET - (P-Ro)] accumulated for subsequent dry months; AET = actual evapotranspiration (mm); S_B = water stored in soil: $S_B = CAP^*e^{-APWL/CAP}$; CAP = soil capacity (mm): maximum water content of soil, without gravitational water (= average rooting depth (mm) * water content at field capacity (in volume %); ΔS_B = Change in S_B; DEF = deficit (PET - AET) (mm); SUR = surplus [(P-Ro) – AET] (mm); R_N = natural groundwater recharge (SUR - ΔS_B) (mm).

levels data from wells within the river basins.

Rainfall data

Rainfall data were obtained from the Nigeria Meteorological Agency (NIMET) Ijebu-Ode, Ogun State station. The meteorological parameter obtained at this station include precipitation, maximum and minimum temperature, relative humidity, sunshine duration and cup counter anemometer. The time series of meteorological data covering 22 calendar years, that is, 1990-2012 was obtained. Time series of relative humidity, wind speed and solar radiation are only available for a period of 10 years (2000 to 2010). Missing data in time series were filled using arithmetic mean of adjacent month (Backundukize et al., 2011). Larger gaps were filled using arithmetic mean of the previous to the recent data for the period of 2000 to 2010. Time series data including solar radiation, precipitation, maximum and minimum temperature, and relative humidity are available and this enable us to compute the potential evapotranspiration using the standard Penman Monteith equation.

Water level from wells

Water level indicator was used to measure the water level daily between the hours of 4 pm for a year. The total number of wells monitored was twenty (20) scattered within the river basin.

Penman-Monteith equation

The standard Penman-Monteith method for estimating evapotranspiration can be mathematically expressed as follows (Allen et al., 1998):

$$E_0 = (\Delta/\gamma H + E_a)/\ \Delta/\gamma$$

Where Δ/γ is an empirical parameter depending on temperature. H is calculated as

$$H = (1-r)\ Rin - Ro$$

where Rin (incoming radiation) is given by:

$$(1-r)\ Rin = 0.95^*Ra\ (0.18 + 0.55\ n/N)$$

Where Ra is the solar radiation, Ro is the outgoing radiation, r is the albedo (0.05 for water), and n/N is the ration between actual sunshine hours and possible sunshine hours. The term n/N can also be estimated using the cloudiness, e.g., a cloudiness of 60% gives an *n/N* of 40% (=100 - 60). R_0 is calculated by:

$$R_0 = \sigma\ T_a 4\ (0.56 - 0.09\sqrt{ed})(0.10+0.90\ n/N)$$

Where ed is the actual vapor pressure and σT_a^4 is the theoretical black body radiation.

Ea is calculated by:

$$Ea = 0.35(0.5 + u^2/100)\ (ea - ed)$$

Where u^2 is the wind speed in m/s and ea is the saturation vapor pressure. Remember that the relative humidity RH = ed/ea.

Soil moisture balance method

Each parameter of soil moisture balance are computed separately in an excel sheet (Table 1). In these method the concepts of water balance of the unsaturated zone (Thornwaite and Matter, 1957) is applied. It consists of keeping track of the accumulated potential water loss (APWL) and the amount of water in the soil (Sb). Calculation to determine Sb and APWL are performed for each month using monthly precipitation (p) and potential evapotranspiration (PET) (Table 3).

The monthly climatic data available were first rearranged into hydrologic years, a hydrologic year in south-west Nigeria starts with April, which is the beginning of the rainy season, and terminates at the end of March, that is, the end of the dry season.

Actual evapotranspiration (AET): (1) In wet months, when there is enough rain, that is, when P-Ro > PET, the AET is at its maximum value, which is equal to the PET (PET = AET).

(2) In dry months, when there is not enough rain, that is, when P-Ro < PET,

AET = PET + P-Ro

Soil capacity (CAP): The soil water-holding capacity of the root zone is typically expressed in mm and can be obtained by multiplying the water content at field capacity by the effective depth of the root-zone. Hence assuming a uniform water-holding capacity of 30% over the entire the root-zone and a rooting depth of 0.25 m for shallow rooted crops, the water capacity of the root zone becomes 75 mm (Table 2).

Table 2. Showing groundwater recharges estimation using soil moisture balance method.

Year	Month	P	Ro	P-Ro	PET	(P-Ro) - PET	PET - (P-Ro)	APWL	S_B	ΔS_B	AET	DEF	SUR	R_N	Annual R_N
1996	April	183.1	43.28	139.82	115.95	23.87	-23.87	0	75	-75	115.95	0.00	23.87	-51.13	
	May	87	20.57	66.43	122.02	-55.59	55.59	55.59	35.74	39.259	105.69	16.33	-39.26	0.00	
	June	360.4	85.20	275.20	127.14	148.06	-148.06	0	75	-39.26	127.14	0.00	148.06	108.80	
	July	391	92.43	298.57	125.75	172.82	-172.82	0	75	0	125.75	0.00	172.82	172.82	
	August	204.2	48.27	155.93	132.92	23.01	-23.01	0	75	0	132.92	0.00	23.01	23.01	
	September	119.5	28.25	91.25	129.4	-38.15	38.15	38.15	45.1	29.903	121.15	8.25	-29.90	0.00	
	October	338	79.90	258.10	105.44	152.66	-152.66	0	75	-29.9	105.44	0.00	152.66	122.75	
	November	1.4	0.33	1.07	86.2	-85.13	85.13	85.13	24.11	50.895	51.96	34.24	-50.89	0.00	
	December	11.6	2.74	8.86	104.5	-95.64	95.64	180.77	6.734	17.371	26.23	78.27	-17.37	0.00	
1997	January	0	0.00	0.00	131.92	-131.92	131.92	227.56	3.609	3.1254	3.13	128.79	-3.13	0.00	
	February	0	0.00	0.00	76.16	-76.16	76.16	208.08	4.679	-1.07	-1.07	77.23	1.07	0.00	
	March	272.4	64.40	208.00	109.94	98.06	-98.06	0	75	-70.32	109.94	0.00	98.06	27.74	
Total		1968.6													403.99

Table 3. Annual rainfall of the 22 years (1990 to 2012).

Year	Annual rainfall
1990/1991	1767.40
1991/1992	1601.80
1992/1993	1686.60
1993/1994	1526.30
1994/1995	1414.90
1995/1996	1908.50
1996/1997	1968.60
1997/1998	1510.08
1998/1999	1337.38
1999/2000	1756.90
2000/2001	1774.55
2001/2002	1349.90
2002/2003	1449.84
2003/2004	1616.70
2004/2005	1865.60
2005/2006	1610.90
2006/2007	1585.25
2007/2008	1572.57
2008/2009	1576.69
2009/2010	1650.50
2010/2011	1599.18
2011/2012	1586.32

Runoff (R_O): The value of runoff is precipitation multiplied by runoff coefficient (R_c), where (R_c) is equal 0.2364 (Ayoade, 1975).

$R_O = P * R_c$

Change in soil moisture storage (ΔS_b): This is the difference between the current soil moisture and the previous one.

Deficit (D): The deficit is the difference between the actual evpotranspiration and potential evapotranspiration.

D = PET - AET

Surplus (S): Surplus (SUR) is computed as the difference between P-Ro and the actual evapotranspiration (AET).

Groundwater recharge (RN): Groundwater recharge occurs when there is a surplus and the soil moisture is at its field capacity. It is calculated as the remaining surplus after the soil moisture has been brought to field capacity.

$Rn = \Delta s + SUR$

Water-table fluctuation method

Recharge is calculated as:

$R(tj) = Sy*\Delta H(tj)$

Where R(tj) (cm) is recharge occurring between times t_0 and tj, Sy is specific yield (dimensionless), and ΔH(tj) is the peak water-table rise attributed to the recharge period (cm).

Key assumption and critical issues inherent in water table fluctuation that have greatly being on its successful application are: the observed well hydrograph depicts only natural water-table fluctuations caused by ground-water recharge and discharge; Sy is assumed in relation to the geologic properties of the area, and that is constant over the interval of the water-table fluctuation. For this study Sy was assumed to be 1.5%, that is for gneiss the pre-recharge water-level can be extrapolated to determine ΔH (tj) Rasmussen and Andreasen, 1959; while ΔH was computed with the graphical approach as the difference between the peak water level during a recharge even and the predicted level to which water levels would have declined to recharge event had not occurred. ΔH (tj) is estimated as the difference between the peak of a water-level rise and the value of the extrapolated antecedent recession curve at the time of the peak. This recession curve is the trace that the well hydrograph would have followed had there been any precipitation (Figure 2).

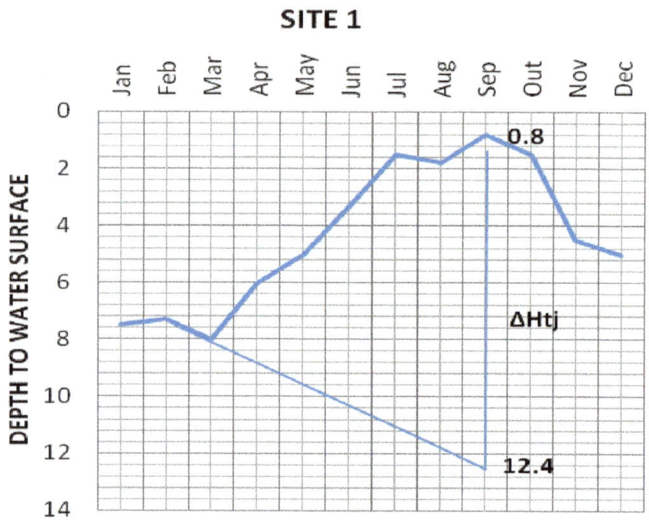

Figure 2. Depth to water surface chart.

Table 4. Descriptive statistics of the annual rainfall.

Years	Range	Minimum	Maximum	Mean	Standard deviation
22	631.22	1337.38	1968.6	1623.48	167.13

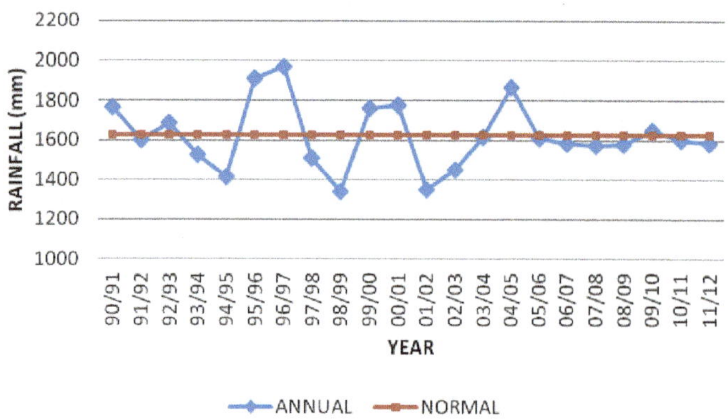

Figure 3. Trend of variation of annual rainfall for 22 years (90/91-11/12).

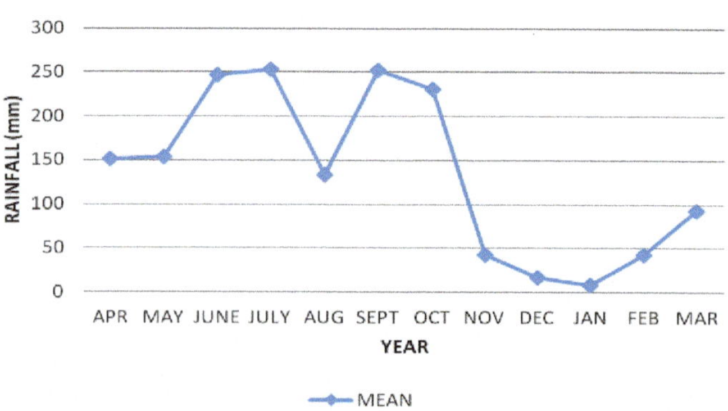

Figure 4. Trend of variation in mean monthly rainfall within the 22 years of study.

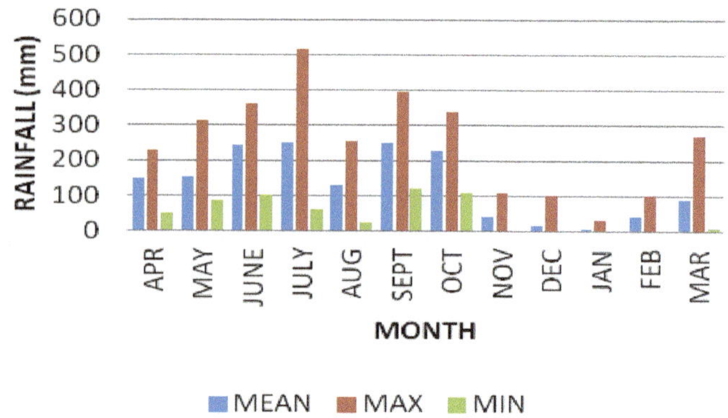

Figure 5. The mean, maximum and minimum monthly rainfall.

RESULTS AND DISCUSSION

This part encompasses the descriptive analysis of rainfall data obtained from Nigeria Meteorological Station (NIMET), the computed potential evapotranspiration and the analysis of the ground water recharge results obtained from the two methods, that is, soil moisture results balance for the duration of 22 years of the study and water table fluctuation analysis of 20 wells for the 2011.

Descriptive analysis of rainfall

Annual rainfall

The annual rainfall occurrence of the Ona River basin for the years of study (1990 to 2012) is presented in Tables 3 and 4. It can be observed that a maximum annual rainfall of 1968.6 mm was recorded for the hydrologic year 1996/1997, a minimum value of 1337.38 mm for 1998/1999. Also, a mean value of 1625.9 mm was observed for the 22 years of study according to Table 3 while a standard deviation of 167.12 mm shows that there is high variability in rainfall within the 22 years of study; this also indicate that the rainfall data deviate largely from their average value of 1625.9 mm.

The trend of variation of annual rainfall can also be observed from the line graph in Figure 3, however, there are 14 dry years (that is, years below normal rainfall of 1625.93 mm), while the remaining 8 were wet years.

Monthly rainfall

In Figure 4, the average monthly variation of rainfall was analyzed. The monthly rainfall for the Ona River basin can therefore be categorized into three seasons; which includes: two wet seasons, ranging from April to October; and a dry season from November to March; the mid break, that is, August break can also be observed.

However, it can be observed from the column chat in Figure 5 that the maximum rainfall ranges from 517.6 to

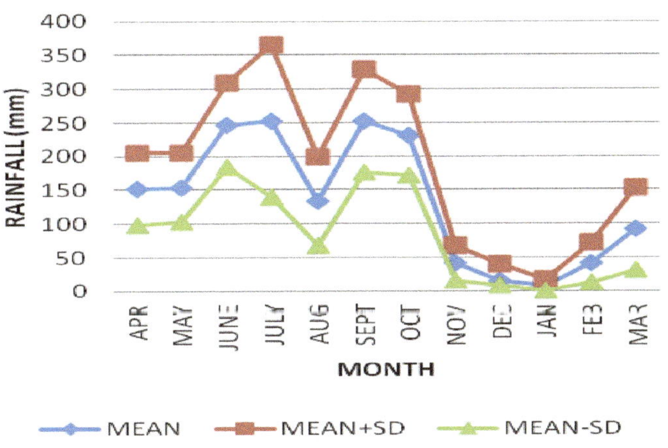

Figure 6. Variations in monthly rainfall pattern.

Table 5. Annual potential evapotranspiration computed from Penman equation.

Year	PET
1990/1991	1379.45
1991/1992	1390.44
1992/1993	1335.51
1993/1994	1353.5
1994/1995	1312.65
1995/1996	1393.01
1996/1997	1367.34
1997/1998	1376.76
1998/1999	1363.42
1999/2000	1332.19
2000/2001	1327.44
2001/2002	1334.61
2002/2003	1315.03
2003/2004	1353.09
2004/2005	1364.66
2005/2006	1434.12
2006/2007	1306.76
2007/2008	1331.8
2008/2009	1410.78
2009/2010	1415.68
2010/2011	1363.72
2011/2012	1395.04

Table 6. Descriptive statistics of annual PET.

Years	Range	Minimum	Maximum	Mean	Standard deviation
22	127.36	1306.76	1434.12	1361.68	35.31

31.8 mm, occurring in the month of July and January respectively. It can also be seen that the mean or normal monthly rainfall ranges from 252.61 mm in the month of July to 8.06 mm occurring in January. While the minimum

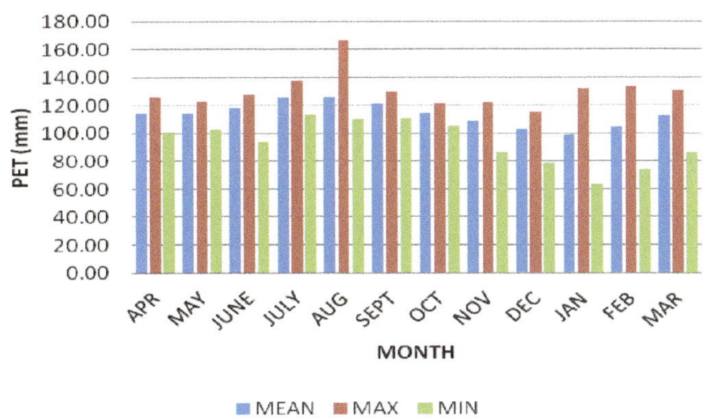

Figure 7. Monthly pattern of PET for the 22 years of study.

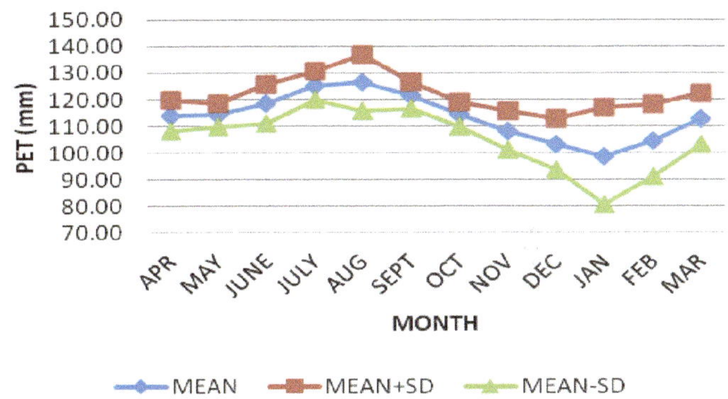

Figure 8. Seasonal PET patterns.

monthly rainfall range from 119.5 mm in September to 0 mm occurring from November to March within the 22 years of study.

It can be observed in the Figure 6 below that rainfall variation from normal occurring within each month for the 22 years of study reduces as we move into the dry season.

Potential evapotranspiration

It can be observed from Tables 5 and 6 that the maximum annual potential evapotranspiration falls in the hydrologic year 2005/2006 and with the value of 1434.12 mm, while minimum value of 1306.76 mm was observed in the hydrologic year 2006/2007; the standard deviation of 35.31 mm indicates a little deviation of annual PET from the mean value of 1361.68 mm per year. However, the mean of 1361.68 mm compared to the mean annual rainfall of 1623.48 mm shows that about 83.9% of the annual rainfall is lost to both evaporation and transpiration from plants.

It can be observed in the column chat in Figure 7 that August has the highest maximum PET with a value of 166.42 mm, while January records the lowest minimum PET with a value of 63.25 mm. In addition, it can be observed in the Figure 8 that there are high variations in

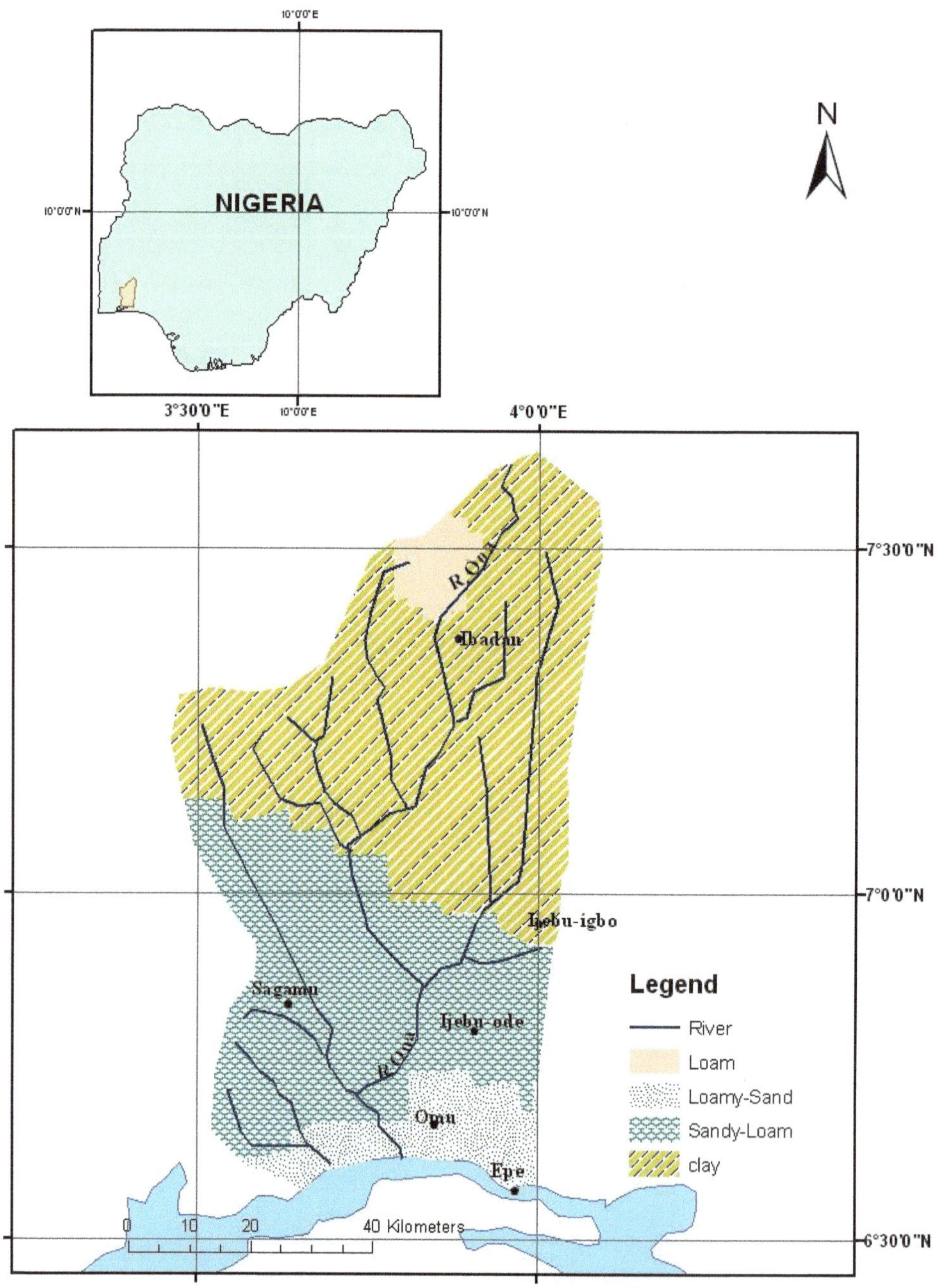

Figure 9. Soil map of the Ona River basin.

the dry months as compared to the wet months. The months with high deviation include August, along with November to March.

Analysis of groundwater recharge results obtained from soil moisture balance method

The soil dominant soil types within the Ona basin was considered for the estimation of groundwater recharge. The soils considered are clay, sandy load and fine sand (Figure 9).

From the Table 7, it can be seen that there are slight variation between recharge obtained for clay and sandy loam soils, and that obtained for fine sand.

For clay and sandy loam, the ground water recharge ranges from 509.01 to 140.82 mm, with a percentage recharge which also ranges from 24.33 to 9.20% of rainfall within the years of study.

On the other hand, the ground recharge obtained for fine sand ranges from 478.99 to 133.35 mm, with a percentage recharge, that ranges from 25.86 to 9.71% of rainfall.

From Table 8, the mean annual recharge of 273.75 mm

Table 7. Result obtained for different soils.

Year	Rainfall	Recharge			
		CAP=75		CAP=50	
		Clay and Sandy loam	%	Fine sand	%
1990/1991	1767.40	426.20	24.11	410.32	23.22
1991/1992	1601.80	229.42	14.32	229.86	14.35
1992/1993	1686.60	267.26	15.85	280.24	16.62
1993/1994	1526.30	195.22	12.79	207.70	13.61
1994/1995	1414.90	271.49	19.19	277.07	19.58
1995/1996	1908.50	338.28	17.72	341.75	17.91
1996/1997	1968.60	478.99	24.33	509.01	25.86
1997/1998	1510.08	213.53	14.14	213.53	14.14
1998/1999	1337.38	140.87	10.53	153.10	11.45
1999/2000	1756.90	422.81	24.07	426.75	24.29
2000/2001	1774.55	431.61	24.32	431.61	24.32
2001/2002	1349.90	156.84	11.62	156.84	11.62
2002/2003	1449.84	133.35	9.20	140.82	9.71
2003/2004	1616.70	266.43	16.48	280.33	17.34
2004/2005	1865.60	443.20	23.76	460.69	24.69
2005/2006	1610.90	245.13	15.22	252.06	15.65
2006/2007	1585.25	274.07	17.29	275.06	17.35
2007/2008	1572.57	213.53	13.58	213.53	13.58
2008/2009	1576.69	155.08	9.84	155.12	9.84
2009/2010	1650.50	227.83	13.80	230.48	13.96
2010/2011	1599.18	183.22	11.46	183.22	11.46
2011/2012	1586.32	243.41	15.34	245.50	15.48

CAP= (water capacity of root zone), % = (recharge/rainfall)*100.

Table 8. Descriptive statistics of the recharge results obtained.

Years	Soil type	SB	Maximum	Minimum	Mean	SD
22	Clay	75	478.99	133.35	273.75	111.78
22	Sandy loam	75	478.99	133.35	273.75	111.78
22	Fine sand	50	509.01	140.82	280.18	115.27

is computed for clay and sandy soil samples; while a mean of 280.18 mm is computed for fine sand soil sample, this shows a slight variation in results obtained (Figure 10). Consequently, the standard variation of 111.78 and 115.27 mm proves that the ground water recharge obtained for the 22 years of study deviate largely from their means.

Rainfalls-recharge relationships

Figure 11 show that the annual ground water recharge computed follows the same trend with the corresponding rainfall. It can be observed from Table 9 that the computed ground water recharge for the three types of soil has high positive correlations of 0.848 and 0.852 with the rainfall, the value 0.00 shows that these correlations

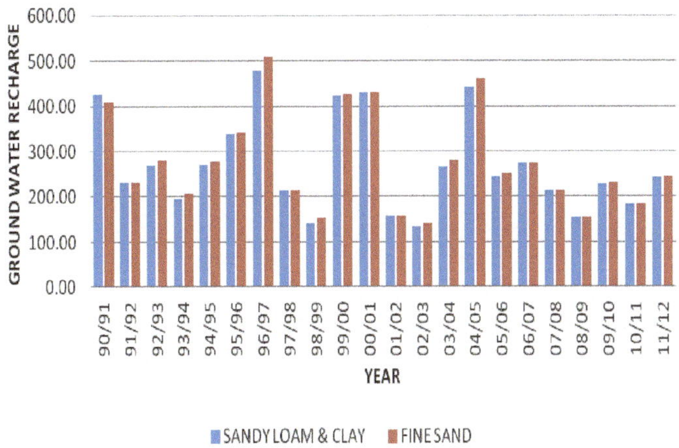

Figure 10. Column chat showing trend of variations in the computed recharge.

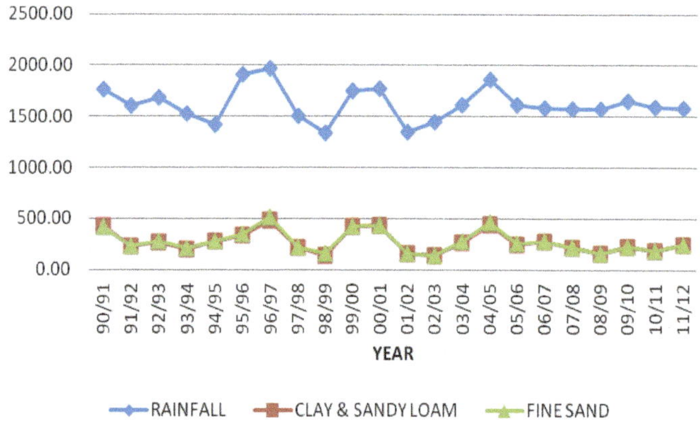

Figure 11. Trend of variation of annual rainfall and recharge.

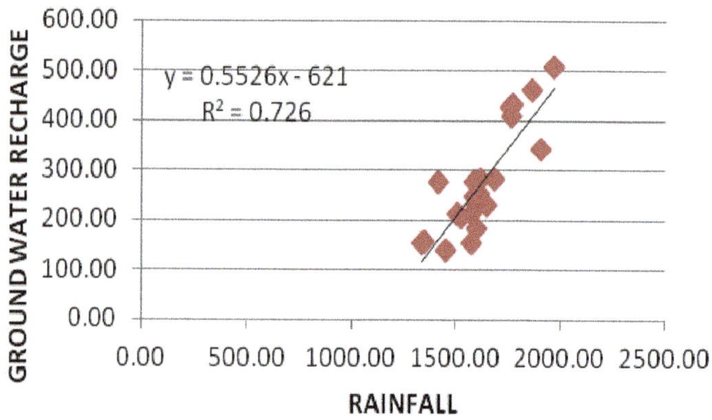

Figure 13. Regression analysis of recharge in sandy loam and rainfall.

Table 9. Correlations between recharge and rainfall.

Recharge correlation		Rainfall
Recharge clay and sandy loam	Pearson correlation	0.848**
	Sig. (2-tailed)	0.000
	N	22
Recharge fine sand	Pearson correlation	0.852**
	Sig. (2-tailed)	0.000
	N	22

**: Correlation is significant at the 0.01 level (2-tailed).

Table 10. Ground water recharge computed.

Rainfall (mm)	SY	Range of ΔH (M)	Rn (cm)	%
1579.1	0.015	3.0	4.5	2.8
	0.015	13.5	20.25	12.8

Sy= specific yield, ΔH=peak water table rise, Rn= recharge, %= (Rn/Rainfall)*100.

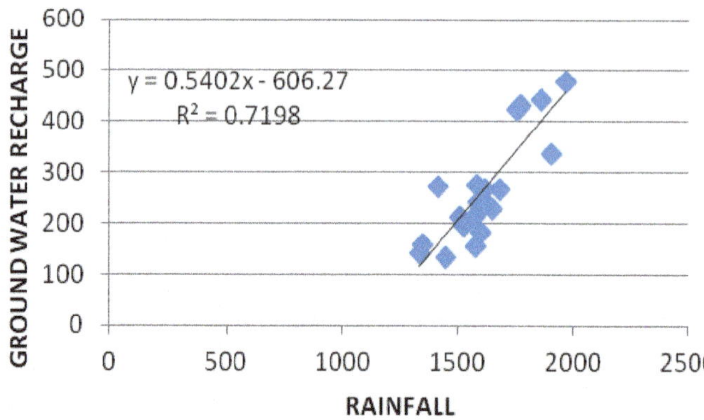

Figure 12. Regression analysis of recharge in clay and sandy loam and rainfall.

are significant at a level of p = 0.01; hence this relationship is not by chance.

The Figure 12 shows the regression analysis between ground water recharge in clay and sandy loam, and rainfall. The relationship shows a positive relationship as confirmed in the correlation analysis: y = 0.540, x – 606.2, and a coefficient of determination (r^2) value of 0.719. However, the regression analysis between recharge computed for fine sand soil and rainfall also shows a positive relationship, that is, y = 0.552 x – 621, and a coefficient of determination (r^2) value of 0.726.

The r^2 values of 0.719 and 0.726 shows that there is a high relationship between observed values of rainfall and predicted values of groundwater recharge when the two equations are used (Figure 13).

Analysis of groundwater recharge result obtained from water table functuation method

It was assumed that the geologic material for the area is gneiss with specific yield (sy) of 1.5%. By multiplying the specific yield with the change in peak water rise, a recharge ranging from 4.5 to 20.25 cm (Table 10) was computed. In addition, the recharge percentage of rainfall was found to be between 2.8 and 12.8%.

Comparison of the two results

The percentage of rainfall that eventually becomes groundwater recharge in the Ona River basin ranges from 24.44 to 9.20% for the soil moisture balance method, and 19.4 to 1.9% for the water table fluctuation method. This shows that it can be concluded that the total percentage of rainfall that becomes groundwater recharge in the Ona River basin ranges from 2.8 to 24.44% (Table 11).

Table 11. Annual groundwater recharge computed from the two methods.

Method	No. of years	Range RN (mm)	Range %Rn
SMBM	22	478.99 - 133.35	24.44 - 9.20
WTFM	1	220.25 - 40.5	12.8 - 2.8

SMBM = Soil moisture balance method, WTFM =Water table fluctuation method, Rn = Recharge.

Conclusion

A quantitative evaluation of spatial and temporal distribution of groundwater recharge is a pre-requisite for the management of ground water resources system in an optimal manner. The amount of groundwater recharge depends upon the rate and duration of rainfall, as rainfall is the principal means for replenishment of moisture in the soil water system and recharge to ground water. This paper has investigated the relationship between rainfalls and groundwater recharge within Ona River basin southwest Nigeria, using two methods. The soil moisture balance and the water table fluctuation.

Analysis of rainfall trends within the Ona River basin suggests that there is considerable high annual rainfall occurrence, with a mean of 1623.48. Only eight years of the 22 years of study can be considered wet years, as the rainfall values of the eight years exceed the mean. It must be noted that the mean annual loss due to evapotranspiration of 1361.68 mm is very high when compared to the rainfall (83.9%).

The results obtained from the soil moisture balance when considering the three dominant soil types within the basin, that is, sandy loam, clay and find sand, having water capacity of root zone value of 70, 70, and 50 respectively, suggests that groundwater recharge follows a positive trend as the corresponding rainfalls. However, empirical relationships of: $y = 0.540x - 606.2$, with a coefficient of determination (r^2) value of 0.719, for sandy loam and clay; and $y = 0.552x - 621$, with a coefficient of determination (r^2) value of 0.726 for fine sand was established for the basin area. On the other hand, recharge ranging from 220.25 to 40.5 mm was computed from the water table fluctuation method.

The general conclusions drawn from this study are groundwater recharge within the Ona River basin follows a positive trend as rainfall, and that this recharge ranges from 2.8 to 24.44% of rainfall.

RECOMMENDATIONS

In other to carry out this study successfully, data from NIMET synoptic stations were obtained. As a result, the study was limited to little data recorded at this station only; therefore, it is recommended that more synoptic stations should be installed across the basin area in other to have a consistent data which would result in a more accurate study of groundwater recharge.

River basin development authorities such as the Ogun-Oshun River Basin Development Authority (OORBDA) should carry out comprehensive studies concerning the groundwater resources of the Ona River basin, as well as provision of sophisticated equipments that will aid groundwater studies. However, to improve the reliability of ground water recharge estimates, the authority must monitor aquifer behavior on a continuous or periodic basis to ensure that adequate data are available.

A 1998 international recharge estimation workshop concluded that no single comprehensive estimation technique can yet be identified from the spectrum of method available; all are reported to give suspect result. Hence, groundwater recharge estimation must be treated as an iterative process that allows progressive collection of aquifer-response data and resources evaluation. This can only be achieved by professionals in the field of ground water hydrology; therefore, the employment of professionals must be addressed.

REFERENCES

Allen RG, Pereira LS, Races D, Smith M (1998). Crop evapotranspiration; guideliness for computing crop water requirement. FAO irrigation and drainage papers. Rome, Italy. 56:328.

Ayoade JO (1975). Water resources and their development in Nigeria', Hydrological Sciences-~Bulletin~des Sciences Hydrologiques, XX, 4 12/1975

Backundukize C, Van Camp M, Walraevens K (2011). Estimation of groundwater recharge in Bugesera region (Burundi) using soil moisture budget approach, GEOLOGICA BELGICA. 14(1-2):85-120.

Central Groundwater Board (2008). Groundwater Scenario, Bhilwara District Rajastan. Ministry of Water Resources Government of India

Dages C, Voltz M, Bsalbes A, Prevot L, Huttel O, Louchart X, Garnier F, Negros S (2009). Estimating the role of a ditch network in groundwater recharge in a Mediterranean Catchment using a water balance approach. J. Hydrol. 375:498-512.

Fitzsimons VP, Misstear BDR (2006). Estimating groundwater recharge through tills, a sensitivity analysis of soil moisture budgets and till properties in Ireland. Hydrol. J. 14:548-561.

Jabloun M, Sahlic A (2008). Evaluation of FAO-56 methodology for estimating reference evapotranspiration using limited climatic data; Application to Tunisia. Agric. Water Manage. 95:707-715.

Kumar CP (1973) Estimation of Ground Water Recharge. J. Hydraul. Eng. 3(1):61-74.

Marechal JC., Varma MRR., Riotte J, Vouillamoz JM., Kumar MSM, Ruiz L, Sekhar M., Braun JJ (2008). Indirect and direct recharges in a tropical forested watershed Mule Hole, India. J. Hydrol. 364:272–284.

Rasmussen WC, Andreasen GE (1959). Hydrologic budget of the Beaverdam Creek Basin, Maryland. US Geol. Surv. Water-Supply Pap. 1472:106

Scanlon BR, Healy RW, Cook PG (2002). Choosing appropriate techniques for quantifying groundwater recharge. Hydrol. J. 10:18-39.

Sophocleous M (2004). Ground water Recharge. In Sylveira L, Usunoff EJ (EDS). Groundwater in Encyclopedia of life support system (EOLSS), developed under the Auspices of the UNESCO. E class Publishers, Oxford, UK. http;//www.ecolss.net). p. 14.

Thornwaite CW, Matter JR (1957). Instructions and tables for computing potential evapotranspiration and the water balance. Laboratory of Climatology, Drexel Institute of Technology, Centerton, New jersey USA. Publication in Climatology. 10(3):183-311.

Trajkovic S, Kolakovic S (2001). Evaluation of reference evapotranspiration equation under humid conditions. Water Resour. Manage. 23:3057-3067.

13

Validating the Soil Conservation Service triangular unit hydrograph (SCS-TUH) model in estimating runoff peak discharge of a catchment in Masvingo, Zimbabwe

Nhamo Luxon , Manyamba Christopher and Chilonda Pius

International Water Management Institute (IWMI), Pretoria, South Africa.

The knowledge of runoff discharged by a catchment at its outlet is important for water accounting and water allocation to competing uses. Runoff generated by a catchment is important in determining the catchment water balance, estimating pollutant loads, and quantifying sediment yield and delivery ratio. The Soil Conservation Service triangular unit hydrograph (SCS-TUH) model was used in this study to simulate water discharged into Siya Dam from Rosva River Catchment in Masvingo Province, Zimbabwe. The SCS-TUH model is preferred because it takes into consideration the physical and hydrological conditions of a catchment, as well, as the volume and flow variations of the entire rainfall event. The model simulates the peak runoff rate of a catchment using daily rainfall data. An overview of the methodology and the different steps followed are given. The model results were validated by comparing the simulated values with measured values recorded from a gauging station at the catchment outlet. The plotted hydrographs of both simulated and measured values coincided very well in height as well as in shape with great precision. The results showed that the SCS-TUH model can be used for simulating runoff discharge in ungauged catchments in the region.

Key words: Peak runoff rate, water availability, water allocation, water management, water balance.

INTRODUCTION

Most catchments in developing countries do not have gauging stations at their outlets to measure peak runoff rates. Yet the knowledge of water discharged into reservoirs on a daily basis is important in determining water allocation to the downstream uses. Regular recording of runoff discharge helps decision makers regarding water distribution to the multiple water uses as they know the amount of water entering a reservoir on daily basis. Therefore, the knowledge of runoff discharged by a catchment is an important part of water resource management (Manley, 1978; Young et al., 2000). With the declining rainfall amounts in recent years, there is a growing need for developing countries to increase land under irrigation in order to ensure food security for a growing population. Because most catchments are not gauged for water flow observation, this study evaluates the suitability of the triangular unit hydrograph (TUH) model in simulating runoff discharge at a catchment outlet.

In the absence of measured records, runoff discharge is usually estimated by using physical and/or statistical relationships between flow, climate and catchment characteristics (Holmes et al., 2002). Water discharge models range from simple water balance models for mean discharge to complex physically based rainfall models designed to replicate the entire observed hydrograph (Manley, 1978; Institute of Hydrology, 1980; Moore, 1985; Gustard et al., 1992; Holmes, 2002).

This study, done in the Rosva River Catchment (RRC) in Masvingo Province of Zimbabwe, applies the Soil Conservation Service triangular unit hydrograph (SCS-TUH) model (Sherman, 1932, 1941; SCS, 1972; Fang et al., 2005) to simulate runoff discharge rate at the outlet of the catchment. The aim of the study is to evaluate the suitability of the SCS-TUH model in simulating runoff peak discharge of ungauged catchments. The simulated peak discharge rates of the studied catchment are compared with measured records in order to validate the model. Of the many unit hydrographs in literature, the SCS-TUH model is used in this study because it takes into consideration the physical and hydrological conditions of a catchment, as well, as the volume and flow variations of the entire rainfall event (Weaver, 2003; Durrans, 2003; Fang et al., 2005).

The maximum amount of flow discharged from a catchment at its outlet is related to the amount of time required for the entire catchment to be contributing to the flow (Durrans, 2003; Fang et al., 2007). A unit hydrograph (also called storm hydrograph) is a graph which shows the rate of water flow (river, canal, etc.) as a function of time at a determined section of a catchment (Ramirez, 2000). It is the integral expression describing the physiographic and climatic characteristics which control the relationships between precipitation and runoff in a catchment (Chow, 1994; Weaver, 2003).

MATERIALS AND METHODS

Study area

The Rosva River Catchment (RRC) is one of the catchments that supplies Siya Dam with water. Siya Dam, one of the major dams in Masvingo Province of Zimbabwe, is the main source of water for the various irrigation schemes and estates downstream, and also supplies water to the communities that live along the water canal that starts from the dam. Figure 1 is map showing the location, elevation and shape of RRC (31°24`-31°35` E, 20°08`-20°09` S). It is a sub-catchment of the Save Catchment in South East Zimbabwe. The average annual rainfall of the catchment is 650 mm. Rainfall is highly variable and occurs mainly during the summer, that is, from November to March. The catchment area is 198.9 km^2, with a broad and compact shape that make its tributaries to be few and long as shown in Figure 1.

It is mainly mountainous with isolated flat lands where agriculture is practised. Its average height is 1064 m. These geographic characteristics make the runoff to reach the catchment outlet quickly (Taffa, 2002). The soils are sandy loamy soils formed on granite rocks.

Data sources

Data that includes Landsat ETM satellite imagery, SOTER soil classification map (FAO, 2012), ASTER GDEM and weather data obtained from Siya Weather Station was used. The 30 m resolution ASTER GDEM was used to delineate the catchment boundaries and derive the river network using Geographic Information Systems (GIS). Landsat satellite imagery was used to generate the landuse/landcover map of the catchment, and the results were rectified using Google Earth. Measured daily runoff discharge recordings were obtained from the Zimbabwe National Water Authority (ZINWA), Station Number 114.

The geographic attributes (average height, length of longest river and catchment area, etc.) of the catchment, necessary for the calculation of the parameters important for estimating the peak runoff rates, were calculated using Geographic Information Systems (GIS).

The SCS-TUH model and method

The SCS-TUH (SCS, 1972; Sherman, 1932, 1941) analysis procedure is a widely used model among hydrology practitioners due to its predictability, and stability (Hann et al., 1982, 1996; Maidment, 1993; Chow et al., 1994; Viessman and Lewis, 2002; McCuen, 2005; Pongsai et al., 2010). The method takes into consideration the base time (T_b), that is, the time of the rise of the hydrograph (the time from the beginning of runoff to the time of peak runoff), and the area of the catchment and the runoff depth of a rainfall event (Ramirez, 2000).

The peak runoff rates (P_r) for 21 rainfall events that occurred within the catchment, and recorded within a period of 24 h were calculated using Equation 1, which is derived from the SCS-TUH analysis procedure (SCS, 1972; Huggins and Burney, 1992; Ponce and Hawkins, 1996; Hrissanthou, 2005). The peak runoff rate is calculated as:

$$P_r = \frac{0.278(AQ)}{T_b} \quad (1)$$

where, P_r is the peak runoff rate (m^3/s), A is the area of the basin (km^2), Q is the runoff depth (mm), and T_b is the base time, (h).

The Q for each rainfall event was calculated using the SCS Curve Number method (SCS, 1972; Ponce and Hawkins, 1996; Liu, 2008; Hawkins et al., 2009), expressed as;

$$Q = \frac{(P-0.2S)^2}{P+0.8S} \quad (2)$$

where, Q is the runoff depth (mm), P is the daily rainfall (mm), S is the maximum retention capacity of soil (mm).

Equation 2 is only valid when $P >$ the initial abstractions (I_a) (that is, $P > 0.2S$). I_a is calculated as $I_a = 0.2S$. $P = 0$ when $P \leq 0.2S$ (Ponce and Hawkins, 1996; Durrans, 2003). The I_a is the total abstractions that occur before any runoff takes place. Selected days from 1997 to 2000 that had rainfall of more than the calculated I_a of 6.9 mm, were used to estimate the Q of each of the selected daily rainfall events. The I_a means that when rainfall is less than 6.9 mm, there is no runoff produced, thus it is used to determine rainfall events that produced runoff in that particular catchment.

The use of Equation 2 requires an estimate of the maximum retention capacity (S) which is also used to calculate the I_a. The S is a function of the curve number (CN) and is calculated using Equation 3 (SCS, 1972; Williams and LaSeur, 1976; Ponce and Hawkins, 1996):

$$S = 254\left(\frac{100}{CN} - 1\right), \text{ in mm} \quad (3)$$

The composite CN for the catchment is calculated as (Hawkins, 1993, 2009; Williams and LaSeur, 1976; Liu and Li, 2008; D'Asaro and Grillone, 2010):

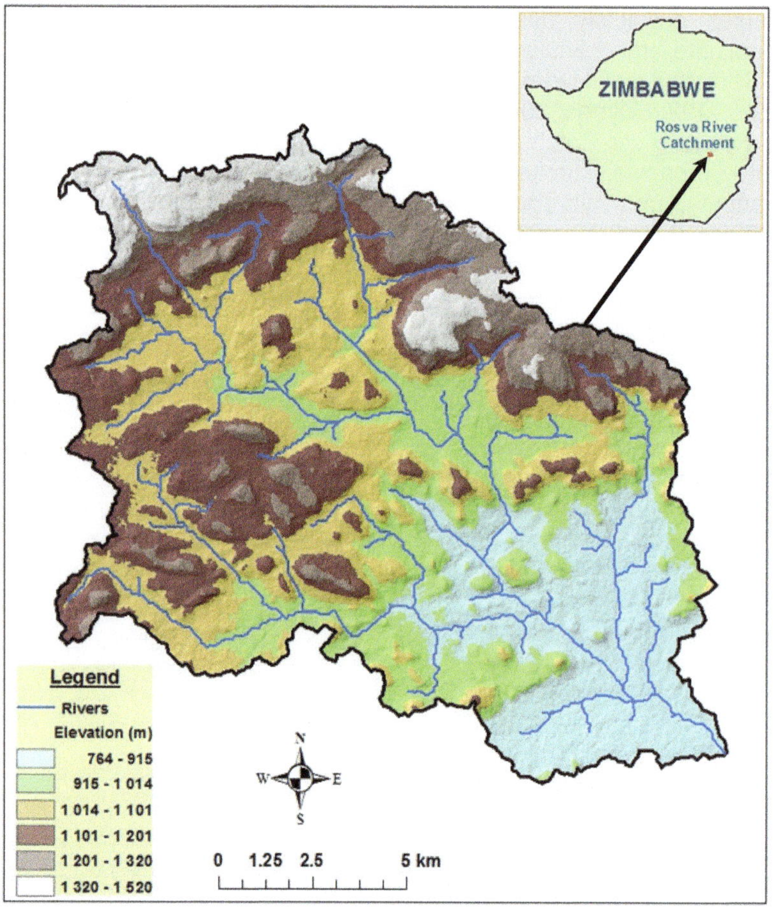

Figure 1. Location, elevation and shape of Rosva River Catchment.

$$CN = \frac{\sum_{i=1}^{n} CN_i A_i}{A}, \quad (4)$$

where, CN is the composite curve number of the catchment (dimensionless), A is the basin area (ha), CN_i is the curve number of each uniform plot of land within the catchment, and A_i is the area of each uniform plot within the catchment (ha). Using Equation 5, the composite CN for RRC was found to be 88.

The T_b for Equation 1 was calculated as a function of the time of concentration of the catchment. The Kirpich formula (Kirpich, 1940; Huggins and Burney, 1992; Xing et al., 2008) was used to calculate the T_b as a function of the time of concentration, which is expressed as:

$$T_b = 2.67\left(\frac{T_r}{2} + 0.6T_c\right) \quad (5)$$

where, T_b is the base time (h), T_r is the duration of the rainfall event, (h), which is 24 h in this case, T_c, is the concentration time (h), which is calculated as follows (Durrans, 2003; Xing et al., 2008):

$$T_c = \left(\frac{0.87L^3}{H}\right)^{0.385} \quad (6)$$

where, T_c is the concentration time (h), L is the length of the main river channel of the basin (km), and H is the difference in elevation (m). The method assumes that the maximum runoff rate of a catchment is reached when all parts of the catchment are contributing to the outflow. This happens when the T_c is reached (Dawod et al., 2011). The Kirpich method was originally developed for small catchments of less than 0.45 km^2; however, it has been applied to larger basins (Chow et al., 1994; Ponce and Hawkins, 1996; Watts and Hawke, 2003; Garen and Moore, 2005; Fang et al., 2007; Granato, 2012).

The flow of water originating from different locations in a catchment will have different travel times to the catchment outlet as some points are closer to the outlet than others (McGlynn et al., 2004). The two common measures of runoff travel times are time of concentration (T_c) and lag time (T_L). T_c is the time required for a drop of effective precipitation falling at the most remote point of a catchment to reach the catchment outlet (Ramirez, 2000; Durrans, 2003), whereas T_L is time from the centroid of rainfall excess to the time of peak runoff for a catchment (Green and Nelson, 2002; Wanielista et al., 1997). T_L and T_c are variables used to compute surface runoff using unit hydrograph methods, and there is no distinct advantage of one method over the other. In this study we used the T_c (Durrans, 2003; Ramirez, 2000; Granato 2012).

RESULTS AND DISCUSSION

Determination of model parameters

The input parameters for the model were derived from data collected from selected 21 rainfall events that

Table 1. Peak runoff rate input parameters for Rosva River Catchment.

Initial abstraction (I_a) (mm)	Maximum retention (S) (mm)	Curve number	Difference in elevation (m)	River length (Km)	Area (Km^2)	T_c (h)	T_p (h)	T_r (h)	T_b (h)
6.9	34.6	88	1064.17	27.62	198.88	3	14.2	24	38.3

Table 2. Comparison between measured and simulated discharge values, Rosva River Catchment.

Day	Rainfall (mm)	Evaporation (mm)	Effective rainfall (mm)	Runoff depth (mm)	Discharge (m^3/s)		Relative error (%)	Absolute error	Standard deviation
					Simulated	Measured			
18 Dec 1997	8.0	0.4	7.6	0.013	0.02	0.17	-88.87	0.15	0.11
19 Dec 1997	9.0	1.2	7.8	0.022	0.03	0.40	-92.12	0.37	0.26
19 Mar 1998	12.5	1.5	11.0	0.430	0.62	0.71	-12.50	0.09	0.06
08 Oct 1998	8.5	2.4	6.1	0.020	0.10	0.07	42.29	-0.03	0.02
01 Jan 1999	17.0	3.2	13.8	1.141	1.65	1.26	30.74	-0.39	0.27
02 Jan 1999	15.5	5.1	10.4	0.318	0.46	0.96	-52.18	0.50	0.35
03 Jan 1999	11.0	2.6	8.4	0.061	0.09	1.63	-94.62	1.54	1.09
21 Jan 1999	11.5	2.2	9.3	0.153	0.22	0.93	-76.22	0.71	0.50
22 Jan 1999	12.0	2.6	9.4	0.166	0.24	1.09	-78.03	0.85	0.60
23 Feb 1999	8.5	3.3	5.2	0.090	0.13	1.18	-88.99	1.05	0.74
04 Mar 1999	15.0	2.6	12.4	0.749	1.08	1.90	-43.07	0.82	0.58
06 Mar 1999	16.0	1.9	14.1	1.234	1.78	1.71	4.16	-0.07	0.05
09 Mar 1999	12.0	3.3	8.7	0.087	0.13	1.42	-91.15	1.29	0.92
24 Nov 1999	7.5	4.5	3.0	0.501	0.72	0.77	-6.10	0.05	0.03
09 Dec 1999	14.5	4.3	10.2	0.284	0.41	0.33	24.24	-0.08	0.06
01 Jan 2000	18.0	6.4	11.6	0.558	0.80	1.16	-30.61	0.36	0.25
11 Jan 2000	15.9	4.4	11.5	0.535	0.77	0.66	17.10	-0.11	0.08
23 Feb 2000	14.0	2.0	12.0	0.650	0.94	1.11	-15.42	0.17	0.12
24 Feb 2000	15.0	2.8	12.2	0.699	1.01	1.07	-5.69	0.06	0.04
17 Mar 2000	24.0	5.3	18.7	2.992	4.32	6.16	-29.88	1.84	1.30
10 May 2000	12.0	0.8	11.2	0.471	0.68	0.68	0.02	0.00	0.00

occurred between 1997 and 2000 as shown in Table 2. The calculation of the P_r using the SCS-TUH model relied mainly on the runoff depth (Q) of the rainfall event. The Q is a function of the amount of rainfall and the maximum retention capacity of the soil as given by Equation 2. The Q reflects the P_r potential.

According to the curve number (CN) and the maximum retention capacity (S) of the RRC, rainfall events produce runoff only if they are above the initial abstraction (I_a) of 6.9 mm as shown in Table 1. This means that the I_a is dependent on the CN of the catchment.

Table 1 gives the input parameters calculated for RRC and used to run the SCS-TUH model in the catchment. Table 1 also gives geographic attributes and the time variables estimated to calculate the runoff depth (Q) and the base time (T_b) according to Equations 2, 5 and 6 respectively.

The 21 rainfall events shown in Table 2 were recorded on a daily basis and at equal intervals, meaning that they had durations of 24 h each. The time of concentration (T_c) of 3 h was calculated using Equation 6. Using the calculated T_c and the duration of the rainfall event (T_r), the base time (T_b) was calculated as 38.3 h. The simulated P_r results shown in Table 2 were calculated from effective rainfall obtained from subtracting evaporation from total rainfall. Effective rainfall is the amount of rainfall that will eventually result in runoff. The simulated and measured peak discharge rates for the catchment range between 0.02 and 4.32 m^3/s, and between 0.07 and 6.16 m^3/s respectively. The measured rates tend to be higher than the simulated ones but the differences are minimal.

Model validation

In order to validate the SCS-TUH model output, the daily rainfall data of the 21 events that occurred in RRC were used to estimate the daily peak runoff rates (P_r). The validity and feasibility of the model was then verified by comparing the simulated P_r with measured values recorded from a gauging station located at the outlet of

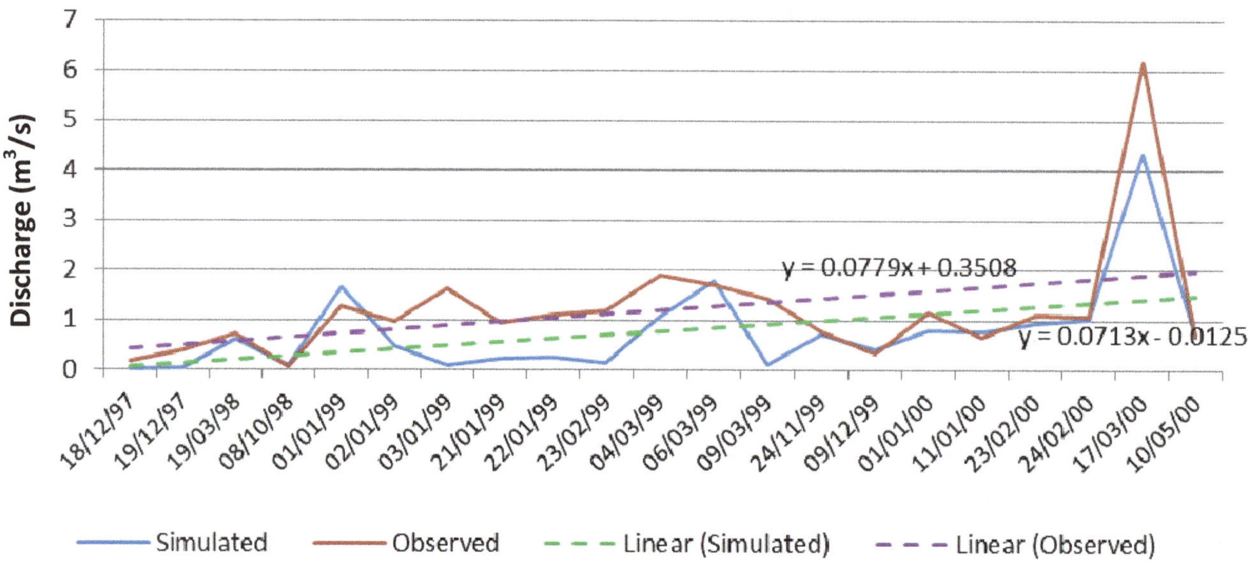

Figure 2. Comparison of measured and simulated discharge by the triangular unit hydrograph model.

the catchment. Table 2 shows both the simulated and measured runoff discharge rates as well as the statistical comparison between them.

The absolute errors between the simulated and measured discharge values range from 0 to 1.3 mm and the relative errors from -94.62 to 42.29% which are within the permissible limit. The standard deviation of the sample ranges from 0 to 1.3 and the group standard deviation averages 0.38, a low standard deviation indicating that the data points tend to be very close to the mean of the measured and simulated values. The correlation coefficient between the simulated and the measured discharge values is 0.88 (88%), implying a positive association between simulated and measured trends as the coefficient is close to the best fit of 1.

Figure 2 shows the hydrographs of both the simulated and measured data for the 21 rainfall events, and the linear estimate trend lines for the two sets of results. The hydrographs of both the simulated and measured discharge processes coincided very well in height as well as shape except for a single event (03/01/99). The linear estimate trend lines are close to each other and the equations are almost the same showing the closeness of both simulated and measured data. Generally, the results indicate that there is a minimal difference between the exact values and the approximation, a scenario that validates the SCS-TUH in estimating runoff discharge rates at the outlet of a catchment.

The analysis derived from both the hydrographs shown in Figure 2 and the data presented in Table 2 demonstrate that the simulated results of the SCS-TUH are consistent with the actual situation. Thus the SCS-TUH model could be applied to estimate runoff discharge for ungauged catchments as it has shown to be precise and practical.

Conclusions

This study used the SCS triangular unit hydrograph to estimate runoff discharge (peak runoff rate) at the outlet of Rosva River Catchment. The simulated results were compared with measured discharge data recorded from a gauging station located at the catchment outlet. The results of the analysis have shown that the simulated discharge results are in good agreement with measured data, and the simulated accuracy is over 88%, showing that the triangular unit hydrograph provides a powerful tool for runoff rate simulation for ungauged catchments.

The SCS Curve Number model and the Kirpich formula were used to estimate the runoff depth and the time of concentration respectively. These are very important parameters in the estimation of runoff rate using the triangular unit hydrograph. Therefore, the results of the analysis may also validate both the SCS Curve Number method and the Kirpich formula.

ACKNOWLEDGEMENTS

The authors would like to thank the Swiss Agency for Development and Cooperation (SDC) for funding the project on the rehabilitation of small scale irrigation schemes in Masvingo Province, Zimbabwe from where this research came. Special thanks to the International Water Management Institute (IWMI), who coordinated the project.

REFERENCES

Chow VT, Maidment RD, Mays LW (1994). Applied Hydrology. McGraw Hill, New York, USA p. 232.

D'Asaro F, Grillone G (2010). Runoff Curve Number Method in Sicily: CN Determination and Analysis of the Initial Abstraction Ratio. Paper presented at the Proceedings of the 2nd Joint Federal Interagency Conference, Las Vegas, NV.
Dawod MG, Koshak AN (2011). Developing GIS-Based Unit Hydrographs for Flood Management in Makkah Metropolitan Area, Saudi Arabia. J. Geogr. Inf. Syst. 3:153-159.
Durrans SR (2003). Stormwater Conveyance Modeling and Design. Haestad Press, Waterbury, CT.
Fang X, Prakash K, Cleveland TG, Thompson DB, Pradhan P (2005). Revisit of NRCS Unit hydrograph procedures. Proceeding of the ASCE, Texas, Section Spring Meeting, Austin, USA.
Fang X, Thompson D, Cleveland T, Pradhan P (2007). Variations of Time of Concentration Estimates Using NRCS Velocity Method. J. Irrig. Drain Eng. 133(4):314-322.
FAO/IIASA/ISRIC/ISSCAS/JRC (2012). Harmonized World Soil Database (version 1.2). FAO, Rome, Italy and IIASA, Luxemburg, Austria.
Garen DC, Moore DS (2005). Curve number hydrology in water quality modeling: uses, abuses, and future directions. J. Am. Water Resour. As. 41(2):377-388.
Granato GE (2012). Estimating basin lag-time and hydrograph-timing indexes used to characterize storm flows for runoff-quality analysis. U.S. Geological Survey, Scientific Investigations Report 2012–5110, 47, Reston, Virginia, USA.
Green JI, Nelson EJ (2002). Calculation of time of concentration for hydrologic design and analysis using geographic information system vector objects. J. Hydroinform. 4(2):75-81.
Gustard A. Bullock A, Dixon JM (1992). Low flow estimation in the United Kingdom. Report 108. Institute of Hydrology, Wallingford, UK.
Hann CT, Barfield BJ, Hayes JC (1996). Design hydrology and sedimentology for small catchments. Academic Press, San Diego, CA.
Hann CT, Johnson HP, Brakensiek DL (1982). Hydrologic modelling of small watersheds. ASAE Monograph No. 5. St. Joseph, MI.
Hawkins RH (1993). Asymptotic Determinations of Runoff Curve Numbers from Rainfall-Runoff Data. J. Irrig. Drain. E-ASCE. 119(2):334-345.
Hawkins RH, Ward TJ, Woodward DE, van Mullem JA (2009). Curve Number Hydrology: State of practice. Am. Soc. Civil Eng., Reston, Virginia (USA). ISBN 978-0-7844-1044-2.
Holmes MGR, Young AR, Grew R, Gustard A (2002). A Region of Influence approach to predicting Flow Duration Curves within ungauged catchments. Hydrol. Earth Syst. Sci. 6:721-731.
Hrissanthou V (2005). Estimate of sediment yield in a basin without sediment data. Catena 64:333-347.
Huggins LF, Burney JR (1992). Surface runoff, storage and routing. In: Haan, CT (Editors.). Hydrologic Modeling of Small Watersheds, ASAE Monograph 5:209-214.
Institute of Hydrology (IH) (1980). Low Flow Studies Report. Institute of Hydrology, Wallingford, UK.
Kirpich ZP (1940). Time of concentration of small agricultural watersheds. Civ. Eng. 10(6):362.
Liu X, LI J (2008). Application of SCS Model in Estimation of Runoff from Small Watershed in Loess Plateau of China. China Geogr. Sci. 8(3):235-241.
Maidment DR (1993). Handbook of Hydrology. McGraw-Hill, Inc., New York.
Manley RE (1978). The use of a hydrological model in water resources planning. Proc. Inst. Civil Eng. 2(65):223-235.
McCuen RH (2005). Hydrologic Analysis and Design. Third Edition. Prentice Hall, Pearson Education, Inc., Upper Saddle River, NJ 07458.
McGlynn BL, McDonnell JJ, Seibert J, Kendall C (2004). Scale effects on headwater catchment runoff timing, flow sources, and groundwater-streamflow relations. Water Resour. Res. 40:W07504.
Moore RJ (1985). The probability-distributed principle and runoff production at point and basin scales. Hydrolog. Sci. J. 30:273-297.
Ponce VM, Hawkins RH (1996). Runoff Curve Number: Has It Reached Maturity? J. Hydrol. Eng. 1(1):11-19.
Pongsai S, Vogt DS, Shrestha RP, Clemente RS, Eiumnoh A (2010). Calibration and validation of the Modified Universal Soil Loss Equation for estimating sediment yield on sloping plots: A case study in Khun Satan catchment of northern Thailand. Can. J. Soil Sci. 90:585-596.
Ramirez JA (2000). Prediction and Modelling of Flood Hydrology and Hydraulics. Inland Flood Hazards: Human, Riparian and Aquatic Communities, Chapter 11. Eds. Ellen Wohl, Cambridge University Press.
Sherman LK (1932). Stream Flow from Rainfall by the Unit Graph Method. Eng. News Rec. 108: 501-505.
Sherman LK (1941). The unit hydrograph and its application. Bull. Assoc. State Eng. Soc. 17:4-22.
Soil Conservation Service (SCS) (1972). Hydrology. In: National Engineering Handbook. Section 4. Washington DC: US Govt. Printing Offices.
Taffa T (2002). Soil and Water Conservation for Sustainable Agriculture. Mega Publishing Enterprise. CTA, Postbus 380, 6700 AJ Wageningen, The Netherlands.
Viessman WJ, Lewis GL (2002). Introduction to hydrology. Fifth edition, Prentice Hall, Pearson Education, Inc., Upper Saddle River, NJ 07458.
Wanielista MP, Kersten R, Eaglin R (1997). Hydrology: Water Quantity and Quality Control. Wiley, New York.
Watts LF, Hawke RM (2003). The effects of urbanisation on hydrologic response: a study of tow coastal catchments. J. Hydrol. 42(2):125-143
Weaver JC (2003). Methods for Estimating Peak Discharges and Unit Hydrographs for Streams in the City of Charlotte and Mecklenburg County, North Carolina. U.S. Geological Survey, Water-Resources Investigations Report 03–4108, Raleigh, North Carolina, USA.
Williams JR, LaSeur WV (1976). Water yield model using SCS curve numbers. J. Hydraul. Div., Proc. Am. Soc. Civ. Engrs. 102(HY9):1241-1253.
Xing F, Thompson DB, Cleveland TG, Pradhan DE, Malla R (2008). Time of concentration estimated using watershed parameters by automated and manual methods. J. Irrig. Drain. E.-ASCE 134(2).
Young AR, Gustard A, Bullock A, Sekulin AE, Croker KM (2000). A river network based hydrological model for predicting natural and influenced flow statistics at ungauged sites. Sci. Total Environ. 251/252:293-304.

14

Assessment of discharge and sediment transport from different forest cover types in lower Himalaya using Soil and Water Assessment Tool (SWAT)

J. V. Tyagi[1], S. P. Rai[1], Nuzhat Qazi[2] and M. P. Singh[2]

[1]National Institute of Hydrology, Jal Vigyan Bhawan, Roorkee - 247 667, Uttarakhand, India.
[2]Forest Research Institute, Dehradun - 248006, Uttarakhand, India.

The present study was carried out to examine the applicability of Soil and Water Assessment Tool (SWAT) in estimating daily discharge and sediment delivery from mountainous forested watersheds and to assess the impact of forest cover types on stream discharge pattern and sediment load. The study watersheds namely Arnigad and Bansigad, comprising of dense Oak forest (80%) and degraded Oak forest (83%) respectively, are located in lower Himalaya (India). Apart from hill topography, deforestation in the watersheds results in huge loss of productive soil and water as runoff. Daily discharge, sediment concentration and other hydro-meteorological data were monitored at the outlet of each watershed. SWAT was calibrated and validated for daily discharge and sediment concentration using the observed data. The performance of the model was evaluated using the statistical measures of coefficient of determination (R^2) and Nash-Sutcliffe efficiency (E_{NS}). The statistical analysis of calibration results for Arnigad watershed showed very good agreement between observed and simulated daily values, with an R^2 value of 0.91, and an E_{NS} of 84.48% in discharge simulation; and an R^2 value of 0.89, and an E_{NS} of 83.11% in sediment simulation. The model also exhibited high performance on Bansigad watershed with an R^2 value of 0.91, and an E_{NS} of 89.74% in discharge simulation; and an R^2 value of 0.86, and an E_{NS} of 82.07% in sediment simulation. The model performed equally well on validation data and estimated the discharge and sediment yield very close to the observed data. The simulated mean annual water yield and sediment yield were also comparable to observed values in both the watersheds. The mean annual surface runoff and water yield over the entire study period were simulated as 6 and 59.4% respectively of the mean annual rainfall in Arnigad watershed; and 6.9 and 63.7% respectively in Bansigad watershed. The results of the study indicated that SWAT is capable of estimating the discharge and sediment yield from Himalayan forested watersheds and can be a useful tool for assessing hydrology and sediment yield response of the watersheds in the region.

Key words: Oak forest, soil and water assessment tool (SWAT), discharge, sediment concentration, calibration, water yield.

INTRODUCTION

In mountainous watersheds, especially in Himalayan region, the spatial and temporal variability in terms of soil, land use/land cover, topography, rainfall and biotic forest cover, as well as young geologic materials have interventions is large. The steep slopes along with depleted been major factors in soil erosion and sedimentation in river reaches (Jain et al., 2004). Runoff and sediment yield data is scarcely available for Himalayan watersheds which are often required for operation and management of irrigation and hydropower projects in the region. Reliable measurement of various hydrological parameters including runoff and sediment yield is also a difficult task in remote and inaccessible areas. The use of simulation models can partially solve the problem of hydrologic evaluation of watersheds in conditions with limited and unavailable data of discharge and sediment yield. A suite of physically based, spatially distributed hydrological models are now available. The USDA-Agricultural Research Service (ARS) developed CREAMS model (Knisel, 1980) to simulate the long-term impact of land management on water leaving the edge of a field. Several other distributed models for hydrologic and pollutants transport modelling include ANSWERS (Beasley et al., 1980), GLEAMS (Leonard et al., 1987), EPIC (Williams et al., 1983), OPUS (Smith, 1992), AGNPS (Young et al., 1989) and SWRRB (Williams et al., 1985). These models were all developed for specific problems and have limitations for modelling watersheds with hundreds or thousands of sub-watersheds.

The soil and water assessment tool (SWAT) (Arnold et al., 1998), a physically based, spatially distributed model overcomes these limitations and is being increasingly used to assess the hydrological behaviour of large and complex watersheds. Rapid parameterization of hydrologic models can be derived using remote sensing (RS) and geographic information systems (GIS) as remotely sensed data provides valuable and up-to-date spatial information on natural resources and physical terrain parameters. Numerous studies have described the potential benefits and use of RS and GIS in hydrologic modelling (Hession and Shanholtz, 1988; Maidment, 1993; Srinivasan and Engel, 1991; Bhaskar et al., 1992; Pandey et al., 2005, 2009). Among others, the SWAT model has proven to be an effective tool for assessing water resource and nonpoint-source pollution problems for a wide range of environmental conditions. The model has been widely used in various regions and climatic conditions on daily, monthly and annual basis (Arnold et al., 1998; Mulungu and Munishi, 2007; Muttiah and Wurbs, 2002; Srinivasan et al., 2005; Tolson and Shoemaker, 2007) and for the watershed of various sizes and scales (Kannan et al., 2008, 2007). Rosenthal et al. (1995) tested SWAT predictions of stream flow volume for the Lower Colorado River basin (8927 km^2) in Texas. A GIS-hydrologic model link was used to aid in forming input files. Stream flow was simulated for nine years for four stream gauge locations with 60 sub-watersheds. With no calibration, the model closely simulated monthly stream flow with a regression coefficient (R^2) of 0.75. Bingner (1996) evaluated the SWAT model in the Goodwin Creek Watershed (21.31 km^2) located in northern Mississippi over a 10-year period. The land use of the watershed was primarily pasture and cultivated field. The Nash-Sutcliffe coefficients (E_{NS}) and R^2 values computed with observed monthly flow were all around 0.80. Srinivasan et al. (1997) used the SWAT model to simulate hydrology from 1960 to 1989 in the Rio Grande/Rio Bravo river basin (598,538 km^2) located in parts of the United States and Mexico. The simulated average annual flow rates were compared against USGS stream gauge records. Visual time-series plots and statistical techniques were used to evaluate the model performance.

In one of the few applications to study daily streamflow, Peterson and Hamlett (1998) used the SWAT model to simulate discharge in the Ariel Creek watershed (39.5 km^2) of north eastern Pennsylvania. Model evaluation of daily flow prior to calibration revealed a deviation of runoff volume of 68.3% and a R^2 of -0.03. Spruill et al. (2000) evaluated the SWAT model and parameter sensitivities were determined while modelling daily stream flow in a small central Kentucky watershed comprising an area of 5.5 km^2 over a two year period. Stream flow data of 1996 were used for calibration and of 1995 were used for evaluation of the model. The E_{NS} for monthly total flow was 0.58 for 1995 and 0.89 for 1996, whereas for daily flows it was observed to be 0.04 and 0.19. Oeurng et al. (2011) used SWAT to simulate discharge and sediment transport at daily time steps within the intensively farmed Save catchment in south-west France (1,110 km^2) and concluded that simulated daily values matched the observed values satisfactorily. Ayana et al. (2012) applied SWAT model to Fincha watershed (3,251 km^2), located in Western Oromiya Regional State, Ethiopia and estimated monthly sediment yield with R^2 of 0.82 and E_{NS} of 0.80 during calibration and R^2 of 0.80 and ENS of 0.78 during the validation period. SWAT has also been successfully used for simulating runoff, sediment yield and water quality of small watersheds for Indian catchments (Pandey et al., 2009, 2005; Tripathi et al., 1999).

Tripathi et al. (2003) applied the SWAT model for Nagwan watershed (92.46 km^2) with the objective of identifying and prioritizing of critical sub-watersheds to develop an effective management plan. The model was verified for monsoon season on daily basis for the year 1997 and on monthly basis for the years 1992 to 1998 for both surface runoff and sediment yield. Jain et al. (2010) calibrated and validated SWAT for estimating runoff and sediment yield from part of the Satluj river basin lying between Suni and Kasol in Western Himalaya. The R^2 values in estimating daily runoff and sediment yield were 0.33 and 0.26 respectively, while for monthly runoff and sediment yield these were computed as 0.62 and 0.47 respectively. They considered these values reasonably satisfactory for estimating runoff and sediment yield from remote watershed with limited data. The review of literature, in general, indicated that SWAT is capable of simulating hydrological processes with reasonable accuracy. However, studies related to applicability of SWAT to the watersheds located in Himalayan region of India are rarely available in literature. The present study

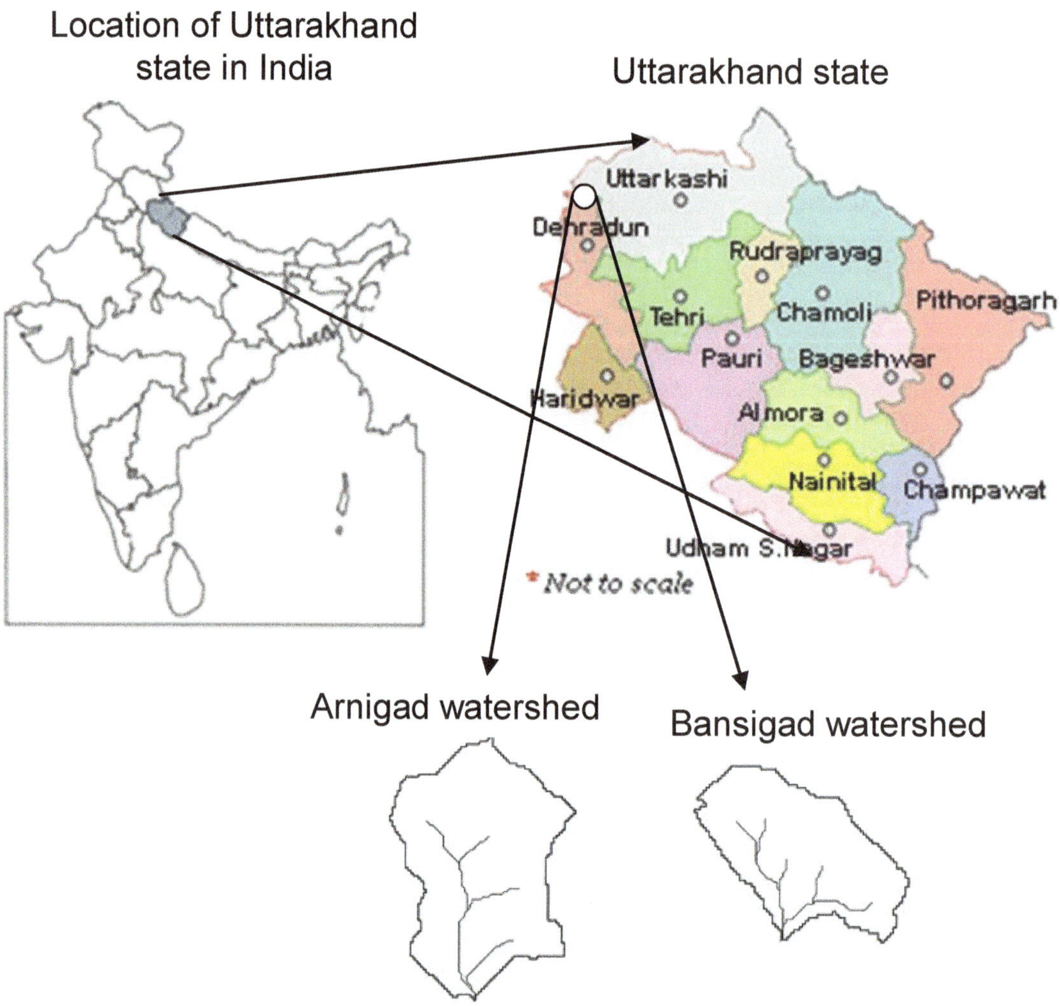

Figure 1. Location of study watersheds.

aims at examining the applicability of SWAT in simulating the hydrological and sediment response of dense and degraded Oak forest covers that predominantly occupy Arnigad and Bansigad watersheds respectively in lower Himalaya (India).

The Specific objectives of the study were to: i) measure rainfall, runoff and sediment concentration and other hydro-meteorological parameters in the watersheds, ii) calibrate and validate SWAT for the study watersheds using the measured data and assess its applicability in simulating daily discharge and sediment transport from forested mountainous watersheds, and iii) assess the impact of different forest cover types on stream discharge pattern and sediment yield.

STUDIED WATERSHEDS

Two small watersheds namely, Arnigad (30° 26' 13.9" N, 78° 05' 37.4" E) and Bansigad (30° 27' 9.1" N, 78° 02' 45.9" E), located 36 km North of Dehradun near Mussoorie (situated on the first mountain ridge beyond Dehradun) in Uttarakhand state of India were selected for the present study (Figure 1). The Arnigad (304.4 ha) and Bansigad (209.8 ha) watersheds are predominantly covered with moderately dense Oak forest and moderately degraded mixed Oak forest, respectively. The landform of both the watersheds consists of rugged, mountainous terrain with steep slopes. The elevation in Arnigad and Bansigad ranges between 2,220 to 1,640 m above msl and 2,160 to 1,620 m above msl, respectively. The mean orientation of both the watersheds is south. The drainage pattern of both the micro-watersheds is of dendritic type. The annual rainfall in Mussoorie is about 2005 mm of which 60 to 85% is received during monsoon season (June to September). In Mussoorie, the mean annual air temperature is 13.7°C. The hottest month is June with an average (1961 to 1995) air temperature of 19.8°C, and the coldest month is January with an average air temperature of 6°C.

The Mussoorie range, constituting the Proterozoic to lower Cambrian rocks of the lesser Himalaya is separated from the Cainozoic Siwalik Group and the Dun gravels by the MBT (Thakur and Pandey, 2004), that is a north–northeast dipping thrust along which the lesser Himalayan rocks are thrust over the Siwaliks (Rautela et al., 2010). The main parent material in this area consists of quartzite, schist, slates, phyllite, hard sandstones, limestone and dolomite (Bartarya, 1995).

BRIEF DESCRIPTION OF SWAT MODEL

SWAT 2005 with ArcSWAT interface was used in the present study. SWAT is a continuous, physically based distributed model that operates on a daily time step at watershed scale for long-term simulation of hydrology, sediment and agricultural chemical movement (Arnold et al., 1998). SWAT can analyse small or large catchments by discretising into sub-basins, which are then further subdivided into hydrological response units (HRUs) with homogeneous land use, soil type and slope. The SWAT system embedded within ARCGIS can integrate various spatial environmental data including soil, land cover, climate and topographical features. SWAT estimates daily volume of overland rainfall excess over each HRU by solving the water budget components of precipitation, runoff, evapotranspiration, percolation and return flow from subsurface and groundwater flow (Arnold et al., 1998). The model uses the Green-Ampt method or the modification of the SCS curve number method (USDA Soil Conservation Service, 1972) to compute surface runoff volume. Peak runoff rate is estimated using a modification of the 'rational method' (Chow et al., 1998). The measured daily potential evapotranspiration can be loaded directly for the watershed or determined using the Penman–Monteith method, the Priestley–Taylor method or the Hargreaves method (Arnold et al., 1998). Lateral subsurface flow is simulated using kinematic storage model, whereas empirical approaches are adopted for groundwater (Arnold et al., 1998; Borah and Bera, 2003; Neitsch et al., 2005).

In SWAT, Manning's equation is used to estimate flow rate and velocity through channels. Flow routing is based on either the variable storage or the Muskingum routing method (Neitsch et al., 2005). In the present study, SCS curve number and Muskingum routing methods, along with daily climate data, were used for surface runoff and streamflow computations. The Penman method was used to estimate potential evapotranspiration. SWAT uses the modified universal soil loss equation (MUSLE) (Williams, 1975) for computing the soil loss for each HRU. The sediment concentration is obtained from the sediment yield, which corresponds to flow volume within the channel on a given day. The transport of sediment in the channel is controlled by simultaneous operation of two processes: deposition and degradation. Whether channel deposition or channel degradation occurs depends on the sediment loads from the upland areas and the transport capacity of the channel network.

MODEL INPUT DATA

The basic spatial input datasets used by the model include the digital elevation model (DEM), land use/cover data, soil data and climatic data. The brief methodology for preparation of the data is described as follows:

Digital elevation model

DEM is one of the main inputs of the SWAT model to define topography of the study area. Elevation contours at 20 m interval were digitized from Survey of India toposheet (no. 53 J/3) at 1:50,000 scale using ARCGIS software. The digitized contours were used to generate DEM (Figure 2a and 2b) with a grid cell resolution of 30 m. The DEM was used to delineate the boundary of the watershed and analyze the drainage patterns of the land surface terrain. Terrain parameters such as slope gradient and slope length, and stream network characteristics such as channel slope, length and width were derived from the DEM.

Land use/cover data

Land use is one of the most important factors that affect runoff, soil erosion and evapotranspiration in a watershed during simulation (Neitsch et al., 2005). As per the Survey of India toposheet, major land use in Arnigad and Bansigad watersheds consists of oak forest with small areas under habitation and barren lands. The extent of various land use classes shown in the Survey of India toposheet (1:50,000 scale) was verified in the field and minor modifications were made in the boundaries of land use classes as per actual extent. For preparation of land use map, the field surveyed land use classes were digitized and converted to raster format with grid cell size of 30 m. The generated land use maps of Arnigad and Bansigad watersheds are shown in Figure 3(a and b). The various land use categories and their coverage in both the study watersheds are presented in Table 1.

Soil data

The soil textural and physicochemical properties required by SWAT model include soil texture, available water content, hydraulic conductivity, bulk density and organic carbon content for each soil type. Undisturbed soil samples were collected from the depths of 0 to 15 and 15 to 30 cm using core samplers from five locations in each watershed. The collected soil samples were analysed in a standard soil laboratory for particle size distribution, bulk density, soil organic carbon and hydraulic conductivity. The available water content was calculated by subtracting the moisture content at wilting point from that at field capacity. Texturally, the soils in both the study watersheds were sandy loam soils. The average values of soil properties for Arnigad and Bansigad watersheds are presented in Table 2.

Weather data

Meteorological observatories were established within each watershed to monitor daily rainfall, temperature, humidity and wind velocity. Rainfall was measured using tipping bucket rain gauge linked with a data-logger system, and also with ordinary rain gauge for cross check. Maximum and minimum temperature, relative humidity and wind velocity were measured with the help of maximum to minimum thermometers, dry-wet bulb thermometers and anemometer respectively. The meteorological data was collected from March 2008 to February 2011.

Hydrological and sediment yield data

The daily discharge and suspended sediment concentration for the period of March 2008 to February 2011 were measured at the outlets of each study watershed. A sharp-crested weir with apex angle of 120° was constructed at the medial line of flow and a digital stage level recorder was used to measure stream stage. Daily discharge was calculated using appropriate weir formula. The water samples were collected using Punjab bottle samplers and analysed in the laboratory for sediment concentration. Sediment

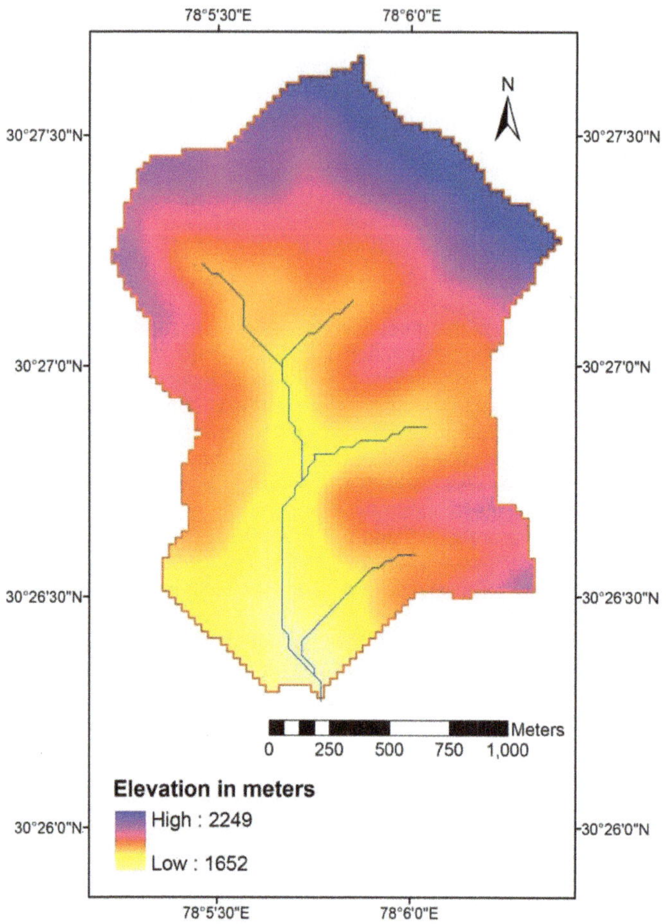

Figure 2a. DEM of Arnigad watershed.

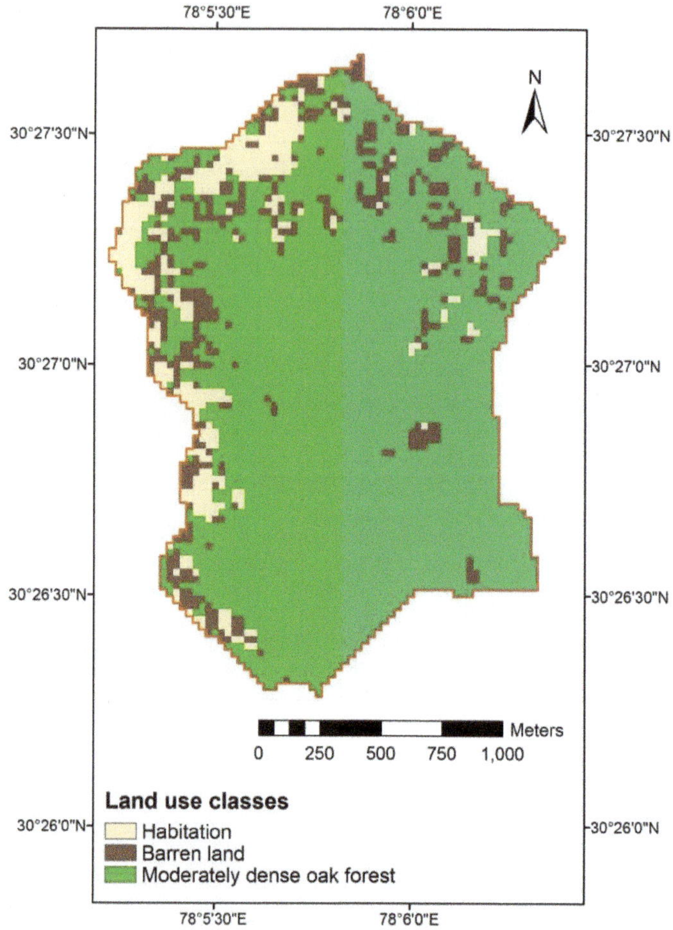

Figure 3a. Land use map of Arnigad watershed.

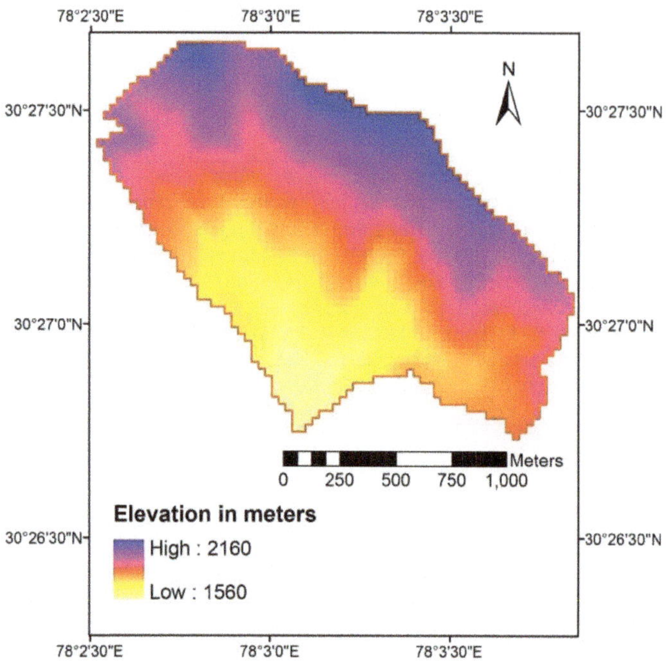

Figure 2b. DEM of Bansigad watershed.

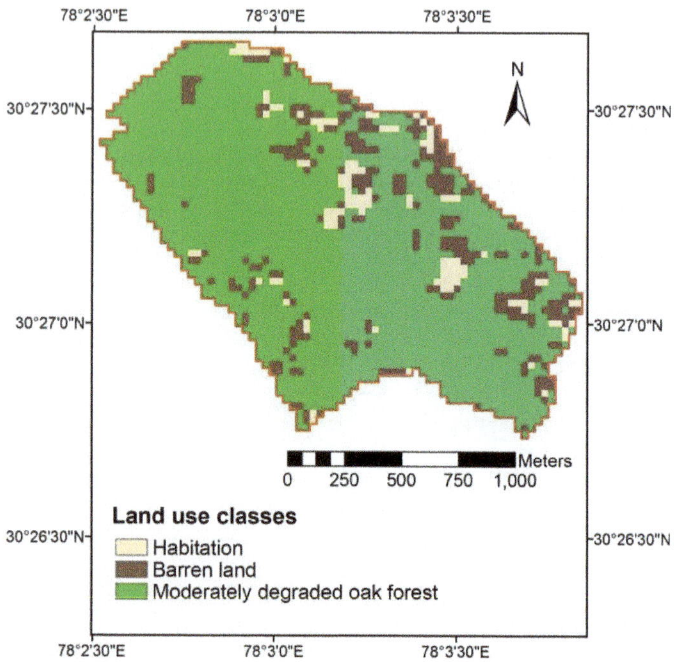

Figure 3b. Land use map of Bansigad watershed.

Table 1. Major land use classes in Arnigad and Bansigad watersheds

Land use	Arnigad watershed		Bansigad watershed	
	Area (ha)	% total	Area (ha)	% total
Dense Oak forest	244.2	80.22	0	0
Degraded Oak forest	0	0	174.2	83.03
Barren	33.8	11.10	25.7	12.25
Habitation	26.4	8.68	9.9	4.72
Total	304.4	100	209.8	100

concentration was measured by filtering samples through Whatman filter paper no. 42. The flow and suspended sediment concentration were measured over a range of hydrological conditions and daily values were calculated from the mean of instantaneous values for a given day.

APPLICATION OF SWAT

Model set-up

The ArcSWAT interface was used for the setup and parameterization of the model. A digital elevation model (DEM) was imported into the SWAT model. A masking polygon (in grid format) was loaded into the model in order to extract the area of interest, delineate the boundary of the watershed and digitize the stream network in the study area. The minimum threshold area for generation of streams was taken as 20 ha that divided Arnigad and Bansigad watersheds into seven and nine sub-watersheds (Figure 4a and 4b) respectively. The land use/cover and soil maps of the study watersheds (in grid format) were also imported into the model and overlaid to obtain a unique combination of land use, soil and slope. Multiple HRUs with 10% land use and 10% slope thresholds were set to eliminate minor land uses and slope classes in each sub-watershed as recommended in the SWAT user manual (Neitsch et al., 2002). A total of 15 and 21 HRUs were delineated in Arnigad and Bansigad watersheds respectively. The daily data of rainfall, minimum and maximum temperature, relative humidity, wind speed and solar radiation were prepared in the appropriate file format and imported into the model.

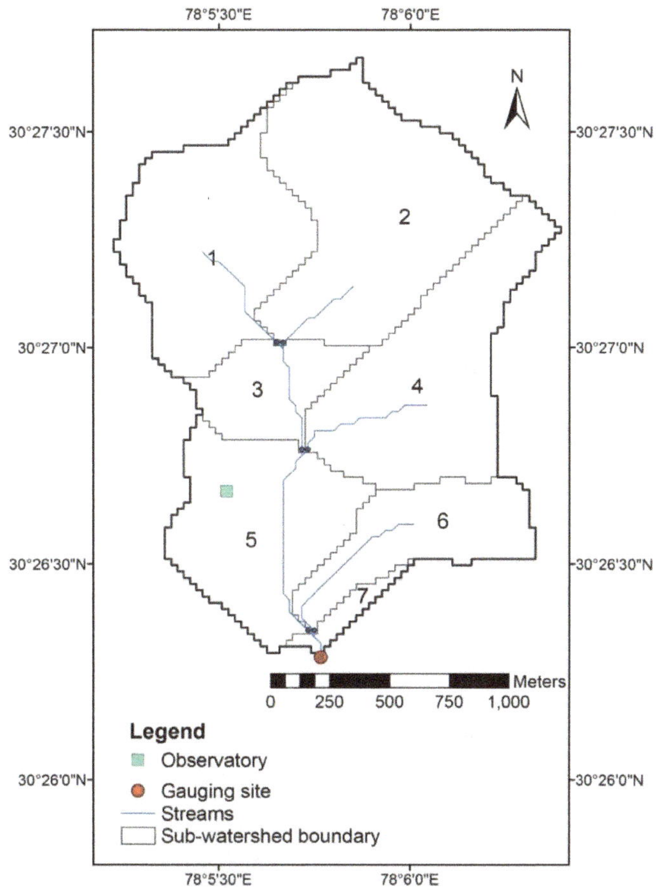

Figure 4a. Sub-watershed delineation in Arnigad watershed.

Model calibration and validation

The calibration and validation were carried out at daily time steps using flow and suspended sediment concentration data. The calibration was performed using the data from June 2008 to May 2010. The data for the period of March 2008 to May 2008 were utilized for warming up and initialization of the model variables. The warm up period was not used for evaluation of the model predictions. The SWAT model includes a large number of parameters that describe different hydrological conditions and characteristics across the watershed. These parameters need to be calibrated to adequately simulate streamflow and sedimentation processes in the study watersheds. Parameters can either be calibrated manually or automatically. In this study, the calibration was done manually based on physical catchment understanding and sensitive parameters from published literature (Bärlund et al., 2007; Xu et al., 2009) and calibration techniques from the SWAT user manual. The hydrological component and the erosion component of the model were calibrated sequentially until the average simulated and measured values were in close agreement. Results of many studies have indicated that SCS curve number (CN_2), a function of soil permeability,

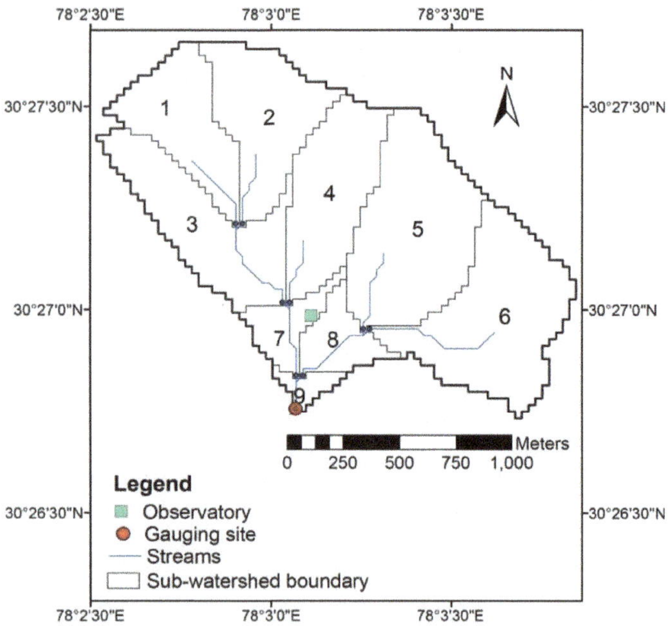

Figure 4b. Sub-watershed delineation in Bansigad watershed.

landuse and antecedent soil water conditions, is an important parameter for surface runoff (Oeurng et al., 2011; Das et al., 2007; Parajuli et al., 2007; Arabi et al., 2008; Wang et al., 2008). Since the base flow forms a significant part of the total flow in the study watersheds, the baseflow recession coefficient (ALPHA_BF) was calibrated for simulation of base flow.

The other important parameters that were calibrated for prediction of flow included 'soil evaporation compensation factor' (ESCO), 'plant water uptake compensation factor' (EPCO), 'surface runoff lag time' (SURLAG), 'groundwater delay' (GW_DELAY), 'deep aquifer percolation factor' (RCHRG_DP), 'Manning's "n" value for tributary channels' (CH_N1), 'Manning's "n" value for main channel' (CH_N2) and 'Maining's "N" for overland flow' (OV_N). SWAT uses MUSLE (Williams, 1975) for prediction of sediment concentration. Therefore, the MUSLE "crop cover and management factor' (C) and the channel sediment routing variables, namely, a linear parameter for calculating the maximum amount of sediment that can be entrained during channel sediment routing (SPCON), an exponential parameter for calculating the channel sediment routing (SPEXP) were adjusted during the calibration. In the validation process, the model was run with calibrated input parameters and the model predictions were compared with an independent set of observed data of the period of June 2010 to February 2011.

Criteria for model evaluation

Several statistical measures are available for evaluating the performance of a model. In the present study, the performance of the model in simulating discharge and sediment was evaluated graphically and by Nash–Sutcliffe efficiency (E_{NS}) and coefficient of determination (R^2). The Nash and Sutcliffe (1970) efficiency is one of the most frequently used criteria and is expressed in percentage form as:

$$E_{NS} = \left\{1 - \frac{\sum_{i=1}^{n}(O_i - S_i)^2}{\sum_{i=1}^{n}(O_i - \bar{O})^2}\right\} \times 100$$

Where O_i and S_i are the observed and simulated values, n is the total number of paired values and $\bar{O}$ is the mean observed value.

The efficiency varies from 0 to 100 with 100 denoting perfect fit. Generally, E_{NS} is very good when E_{NS} is greater than 75%, satisfactory when E_{NS} is between 36 and 75%, and unsatisfactory when E_{NS} is lower than 36% (Nash and Sutcliffe, 1970; Krause et al., 2005). However, a shortcoming of the Nash–Sutcliffe statistic is that it does not perform well in periods of low flow, as the denominator of the equation tends to zero and E_{NS} approaches negative infinity with only minor simulation errors in the model (Oeurng et al., 2011). This statistic works well when the coefficient of variation for the data set is large.

The coefficient of determination (R^2) is the proportion of variation explained by fitting a regression line and is viewed as a measure of the strength of a linear relationship between observed and simulated data. It is computed as:

$$R^2 = \left\{\frac{\sum_{i=1}^{n}(O_i - \bar{O})(S_i - \bar{S})}{[\sum_{i=1}^{n}(O_i - \bar{O})^2]^{0.5}[\sum_{i=1}^{n}(S_i - \bar{S})^2]^{0.2}}\right\}$$

Where $\bar{S}$ is the mean of simulated values, R^2 ranges between 0 and 1. The value of 1 implies that the computed values are in perfect agreement with the observed data.

RESULTS AND DISCUSSION

Assessment of calibration results

The observed and simulated daily runoff and sediment concentration during calibration period of June 2008 to May 2010 are graphically presented in Figure 5(a, b) for Arnigad and Figure 6(a and b) for Bansigad watershed. It can be observed that the simulated discharge generally followed the trend to observed discharge in both the watersheds. A critical comparison of the runoff hydrographs of Arnigad watershed (Figure 5a) shows that the flow peaks are simulated slightly higher than the observed peaks during monsoon seasons both in 2008 and 2009. However, the low flows simulated by the model

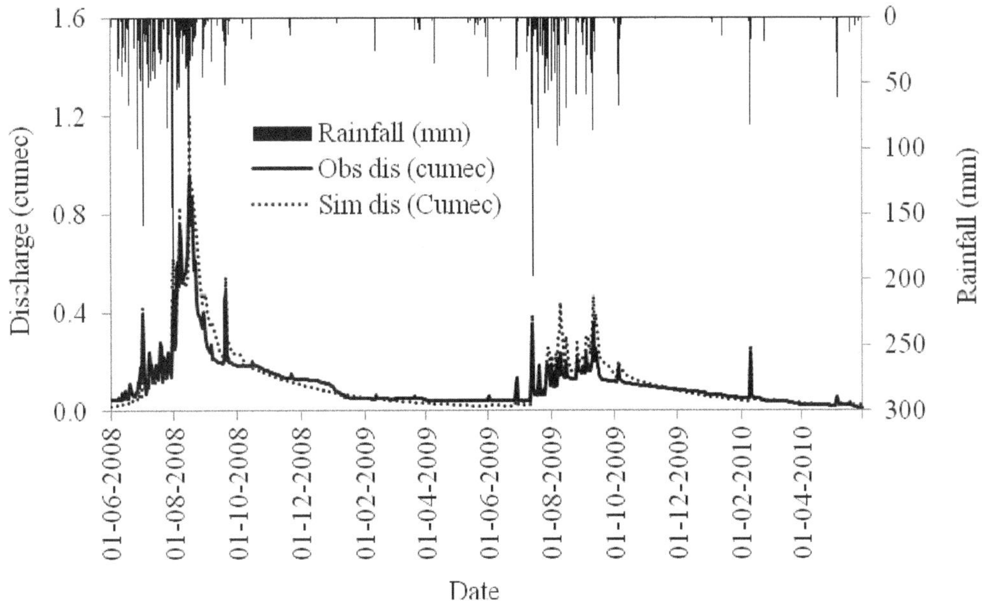

Figure 5a. Observed and simulated daily discharge during calibration in Arnigad watershed.

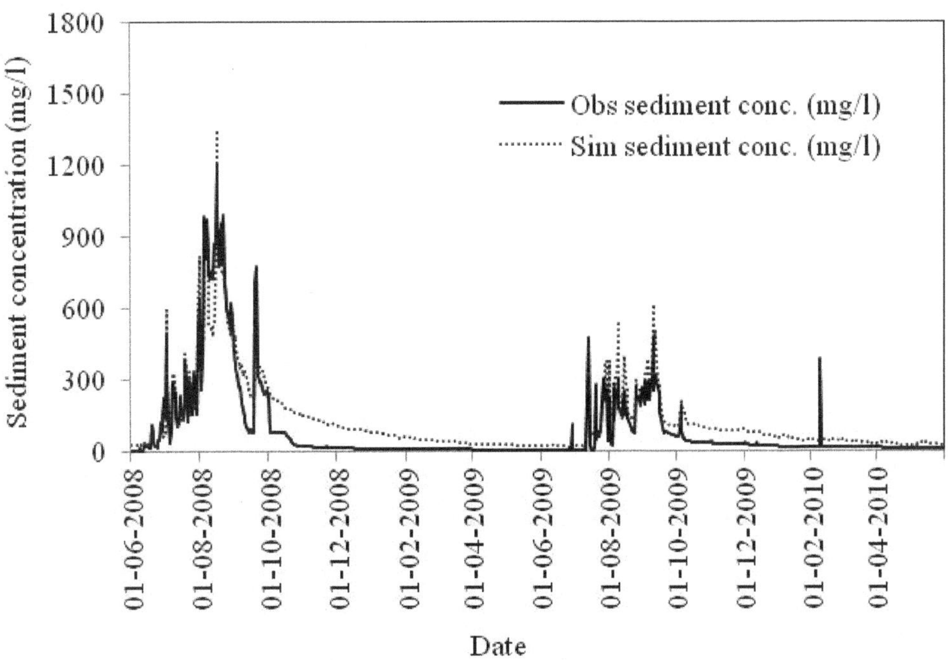

Figure 5b. Observed and simulated daily sediment concentration during calibration in Arnigad watershed.

generally match well with the observed values. In Bansigad watershed, a mixed trend is observed in simulating flows during 2008 and 2009; while, high and low flows are simulated reasonably well during 2008; the high flows appear to be underestimated and low flows overestimated during 2009. A comparison of observed and simulated suspended sediment concentration (Figures 5b and 6b) shows that simulated sediment concentration also generally followed the observed trend in both the watersheds. Although, model predicted peak values were found both higher and lower than the observed values at different times in both watersheds, the difference was within reasonable limits. The difference in simulated and observed values could occur due to the fact that in

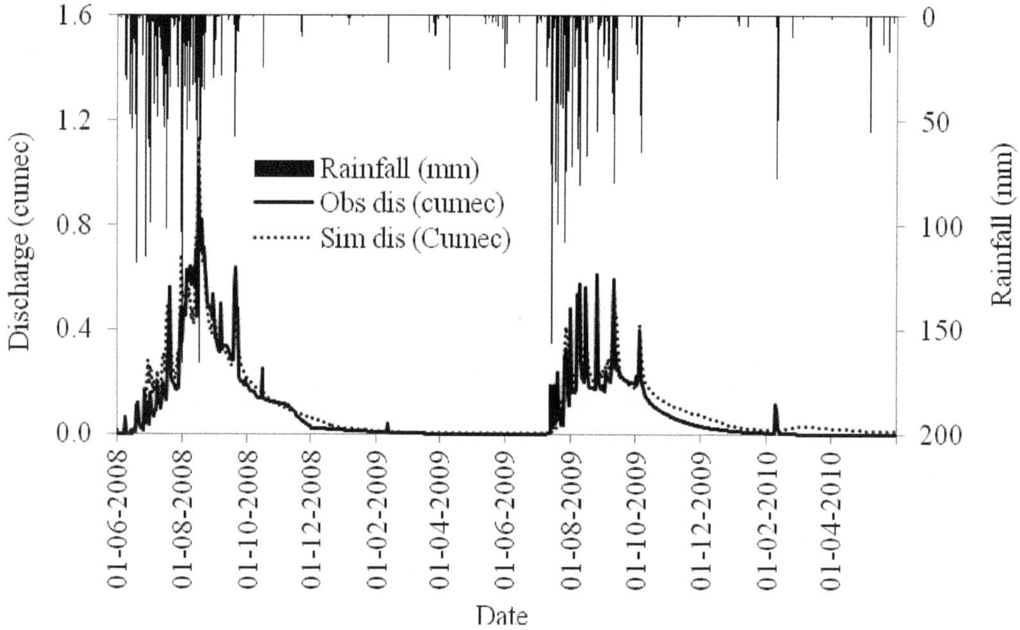

Figure 6a. Observed and simulated daily discharge during calibration in Bansigad watershed.

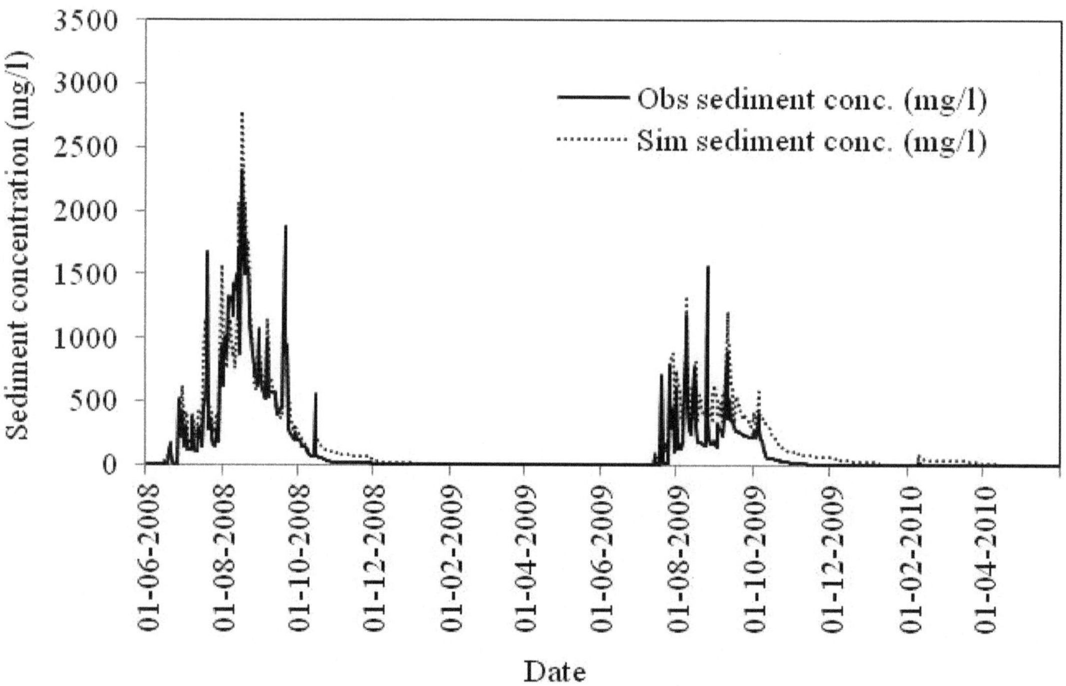

Figure 6b. Observed and simulated daily sediment concentration during calibration in Bansigad watershed.

practice, high-intensity and even short duration rainfall can generate more sediment than simulated by the model on the basis of daily rainfall (Xu et al., 2009). The simulated sediment concentration during non monsoon seasons was higher than the observed values in Arnigad watershed.

The obvious reason for higher sediment simulation is that the sediment response follows the simulated runoff rate as the sediment generation is largely determined by the runoff quantity. In Bansigad watershed, the simulation of sediment concentration during non monsoon seasons was reasonably good. The observed daily flows and sediment concentration were plotted against simulated daily flows and sediment concentration along with 1:1 line

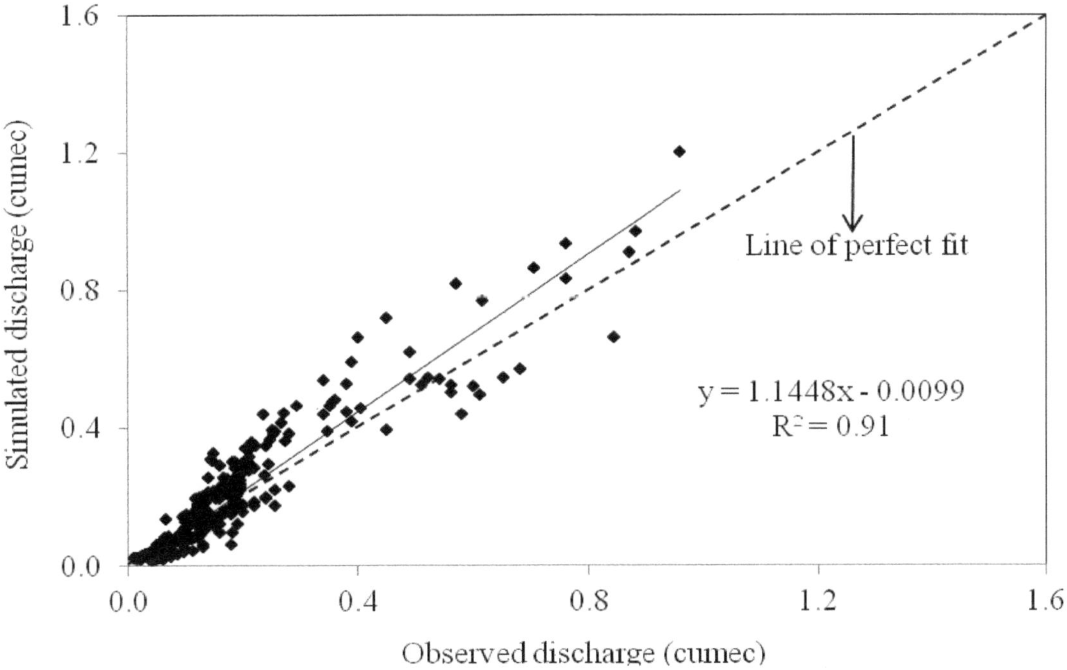

Figure 7a. Scatter plot of observed and simulated daily discharge during calibration in Arnigad watershed.

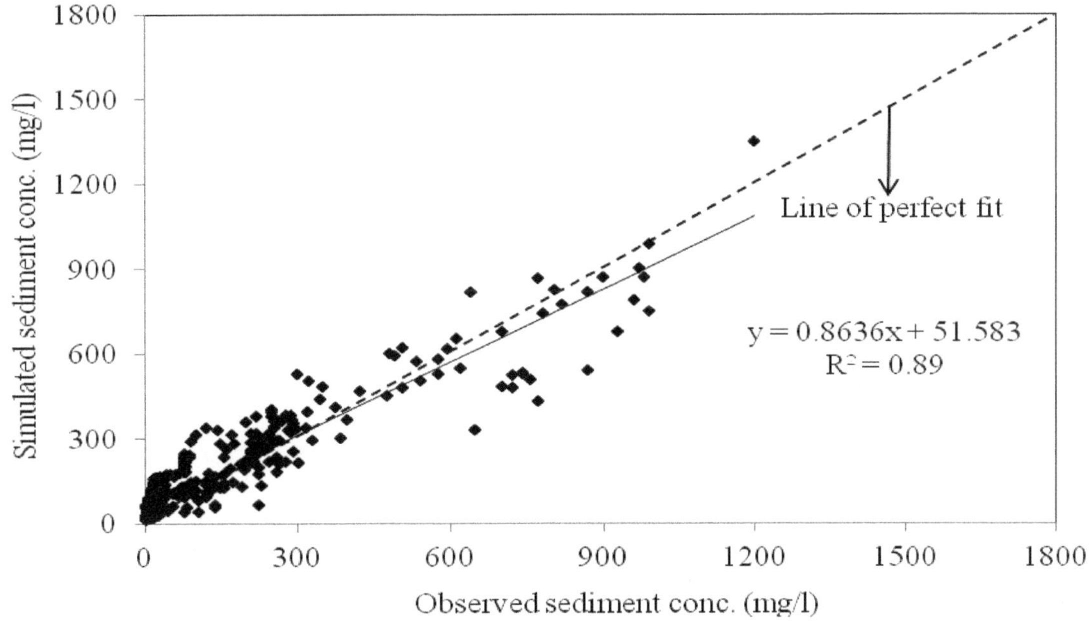

Figure 7b. Scatter plot of observed and simulated daily sediment concentration during calibration in Arnigad watershed.

(line of perfect fit) as shown in Figure 7(a, b) for Arnigad and Figure 8(a and b) for Bansigad watershed. It is observed from Figure 7a that the simulated runoff values are distributed uniformly about the 1:1 line for low values of observed runoff. For high values of observed runoff, majority of the simulated values are slightly above the line of perfect fit, indicating that the model over-predicts the high values of runoff. A close observation of Figure 8a shows two clusters of data scatter, one cluster for low flows where most data lie above line of perfect fit indicating overestimation, and another cluster for high flows where most data lie below line of perfect fit that

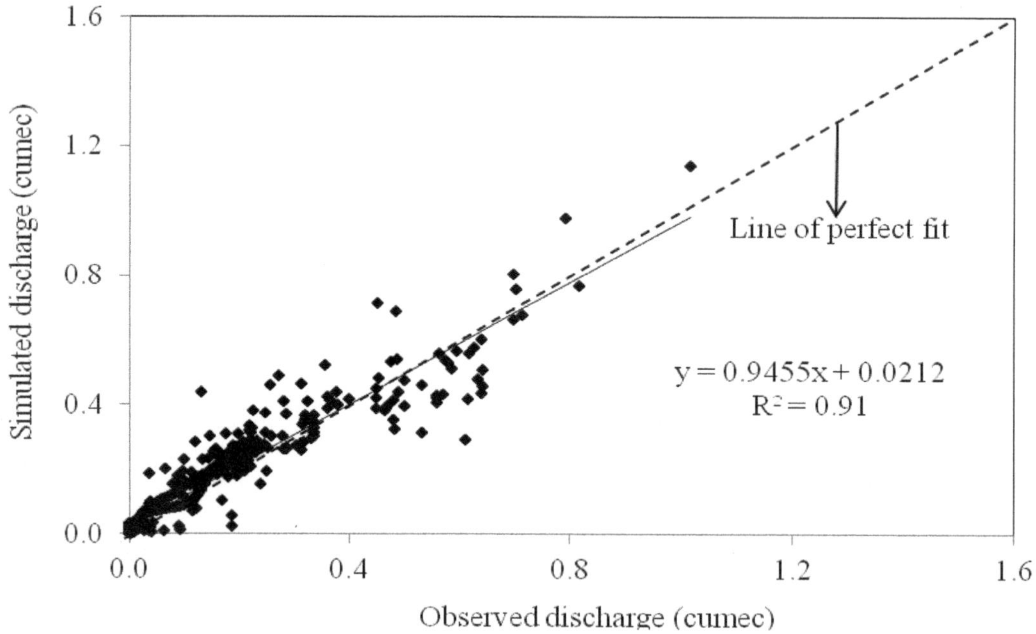

Figure 8a. Scatter plot of observed and simulated daily discharge during calibration in Bansigad watershed.

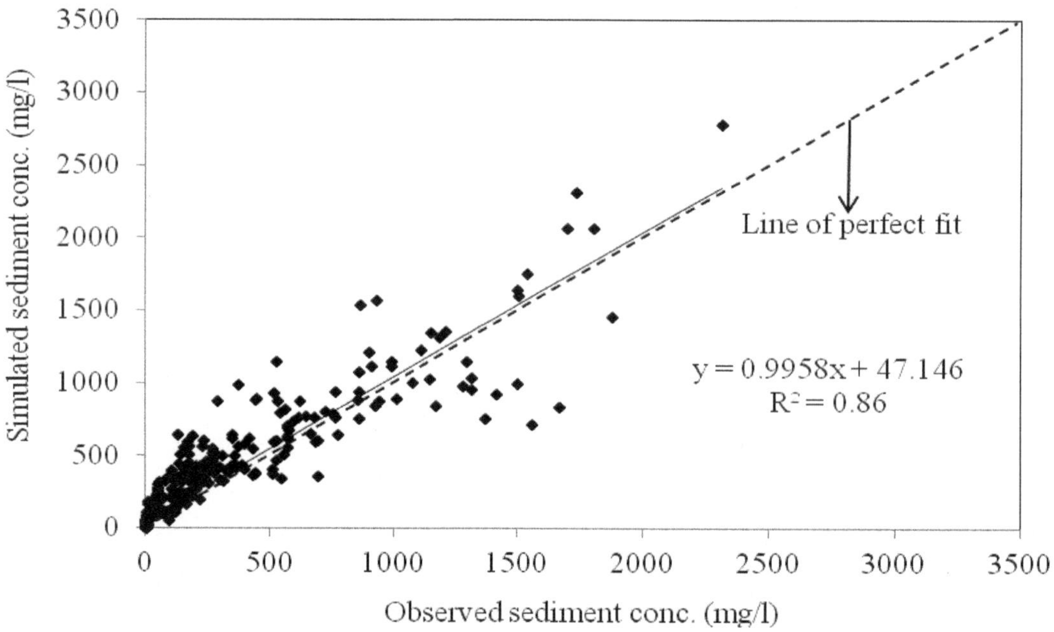

Figure 8b. Scatter plot of observed and simulated daily sediment concentration during calibration in Bansigad watershed

indicates under-prediction of the flows. Similar interpretations can also be made from the scatter plots of sediment concentration (Figures 7b and 8b).

The statistical indices of model performance are presented in Tables 3 and 4 for Arnigad and Bansigad watersheds, respectively. The analysis for Arnigad watershed showed very good agreement between observed and simulated daily values, with an R^2 value of 0.91, and an E_{NS} of 84.48% in discharge simulation; and an R^2 of 0.88, and an E_{NS} of 83.11% in sediment simulation. In Bansigad watershed too, the model exhibited a very good performance in simulating the discharge (R^2 = 0.91, and E_{NS} = 89.74%) and the sediment concentration (R^2 = 0.86 and E_{NS} = 82.07%).

Table 2. Soil properties of study watersheds.

Depth (cm)	Soil organic carbon (%)	Available water content (mm water mm soil^{-1})	B. D. (gm cm^{-3})	Hydraulic conductivity (mm h^{-1})	Sand (%)	Silt (%)	Clay (%)	Soil texture
Arnigad watershed (dense oak forest)								
0 - 15	3.31	0.142	1.01	71.42	67.75	14.00	18.51	Sandy loam
15 - 30	2.30	0.136	1.05	62.35	66.07	13.33	20.51	Sandy loam
Bansigad watershed (degraded oak forest)								
0 - 15	2.37	0.131	1.06	65.39	74.27	11.33	14.40	Sandy loam
15 - 30	1.91	0.124	1.09	58.36	72.93	11.33	15.83	Sandy loam

Table 3. Goodness of fit statistics for Arnigad watershed for calibration period.

Description	E_{NS} (%)	R^2	Mean		Standard deviation	
			Observed	Simulated	Observed	Simulated
Discharge (m^3 s^{-1})	84.48	0.91	0.111	0.118	0.120	0.143
Sediment concentration (mg l^{-1})	83.11	0.89	82.485	122.820	172.944	158.658

Table 4. Goodness of fit statistics for Bansigad watershed for calibration period.

Description	E_{NS} (%)	R^2	Mean		Standard deviation	
			Observed	Simulated	Observed	Simulated
Discharge (m^3 s^{-1})	89.74	0.91	0.092	0.108	0.153	0.152
Sediment concentration (mg l^{-1})	82.07	0.86	132.128	178.717	309.983	332.161

Assessment of validation results

For validation, the observed daily discharge and sediment concentration data of the period of June 2010 to February 2011 were utilized and compared with the model simulated values. A visual comparison of the observed and simulated daily discharge and sediment concentration is presented in Figure 9(a, b) for Arnigad and Figure 10(a and b) for Bansigad watershed. These results show a good general agreement between observed and simulated trends of discharge and sediment concentration in both the watersheds. Further, it is observed that similar to calibration results, the peak flows are slightly overestimated in Arnigad watershed and underestimated in Bansigad watershed. The flows during non monsoon season are simulated reasonably accurate in both the watersheds. In estimating suspended sediment concentration (Figures 9b and 10b), the peak values are observed to be overestimated in both the watersheds. The model also overestimated the sediment concentration during non monsoon period which is possibly due to the reason that the quantity of sediment generation also depends on the simulated discharge rate as mentioned earlier.

The scatter plots of observed and simulated daily discharge and sediment concentration are shown in Figure 11(a, b) for Arnigad and Figure 12(a and b) for Bansigad watershed. Although, the data points lying above and below the line of perfect fit show some overestimation and underestimation respectively, the closeness of the data points to the line of perfect fit indicates a very good performance of the model in estimating both discharge and sediment concentration for the study watersheds. It can be seen from Tables 5 and 6 that the R^2 of 0.94 and E_{NS} of 82.78% in discharge estimation, and R^2 of 0.88 and E_{NS} of 83.28% in sediment estimation are computed for Arnigad watershed; and R^2 of 0.92 and E_{NS} of 92.5% in discharge estimation, and R^2 of 0.94 and E_{NS} of 80.67% in sediment estimation are computed for Bansigad watershed. These R^2 and E_{NS} values are of the same order as obtained during calibration which explain that model has performed equally well on the data set used for validation purpose. The model performance with these high values of statistical indices can be rated as more than satisfactory in simulating discharge and sediment concentration from the study watersheds. The aforementioned results can be viewed in the light of the fact that the runoff and soil erosion process in hilly and mountainous forested catchments are highly complex phenomena and affected by interaction among rainfall, runoff, soil texture and structure, land use, land slope and conservation measures. Therefore, magnitude of randomness in daily simulated values may be large in mountainous catchments

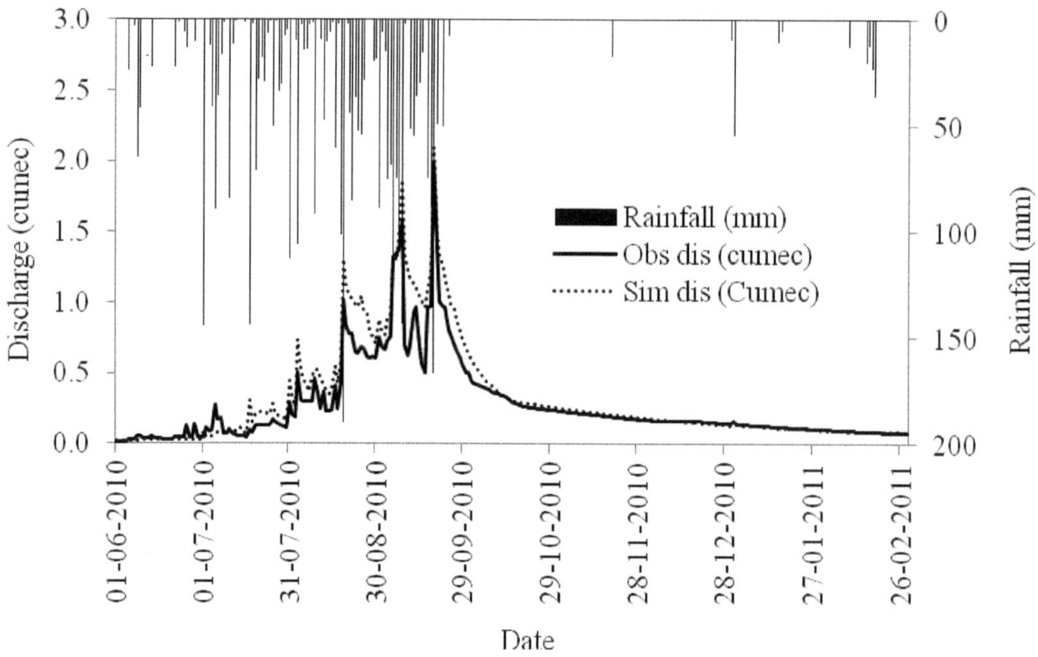

Figure 9a. Observed and simulated daily discharge during validation in Arnigad watershed.

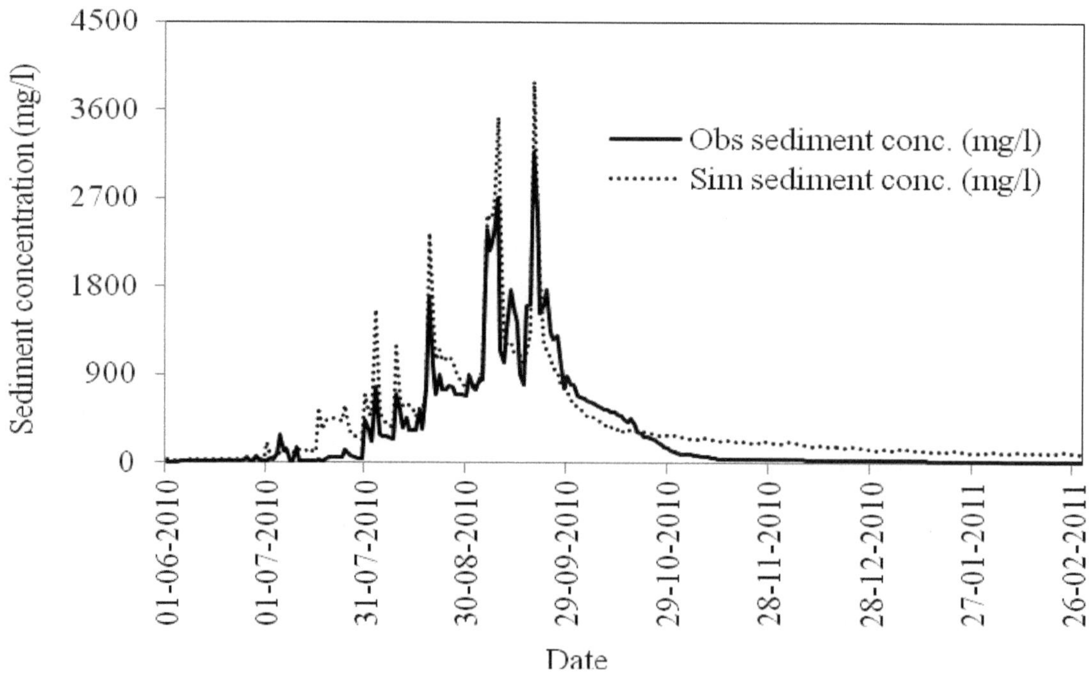

Figure 9b. Observed and simulated daily sediment concentration during validation in Arnigad watershed.

(Singh et al., 2011). Generally, poor correlation among daily values in SWAT simulation has been reported in literature (Peterson and Hamlett, 1998; Varanou et al., 2002; Spruill et al., 2000).

The results of the present study, however, indicate that SWAT can be used for estimation of daily discharge and sediment from forested watersheds in lesser Himalayas.

Assessment of the impact of forest cover types on stream discharge pattern and sediment yield

The assessment of runoff and sediment yield was made based on the total simulation period of three years. The model predicted that mean annual rainfall of 2925 mm over Arnigad watershed was mainly partitioned among

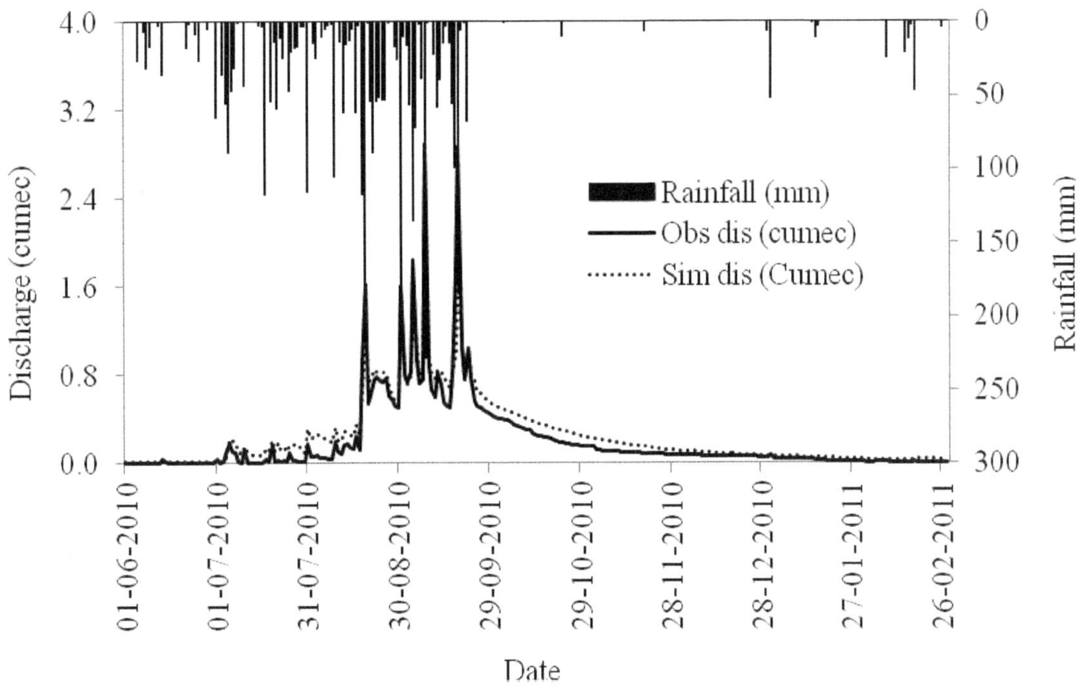

Figure 10a. Observed and simulated daily discharge during validation in Bansigad watershed.

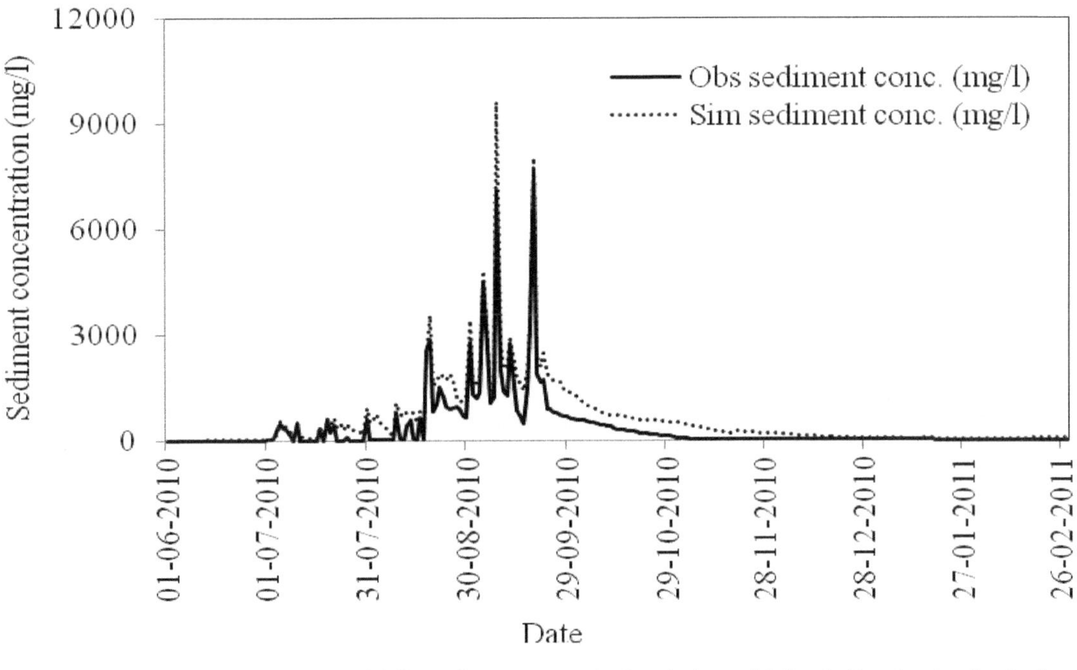

Figure 10b. Observed and simulated daily sediment concentration during validation in Bansigad watershed.

evapotranspiration, ET (27.3%), percolation/ groundwater recharge (62.2%), transmission loss/abstraction (4.5%) and surface runoff (6%). The simulated mean annual water yield amounted to 1738.5 mm (59.4%) against the observed water yield of 1622.4 mm (55.5%). In Bansigad watershed, the mean annual rainfall of 2926.5 mm was partitioned among ET (22.1%), percolation/groundwater recharge (65.8%), transmission loss/abstraction (5.2%), and surface runoff (6.9%) and the mean annual water yield was simulated as 2030.1 mm (69.3%) against the observed value of 1863.6 mm (63.7%). These values indicate that the water balance components in both catchments are almost identical. These results can be supported by the fact that the distribution of land use and

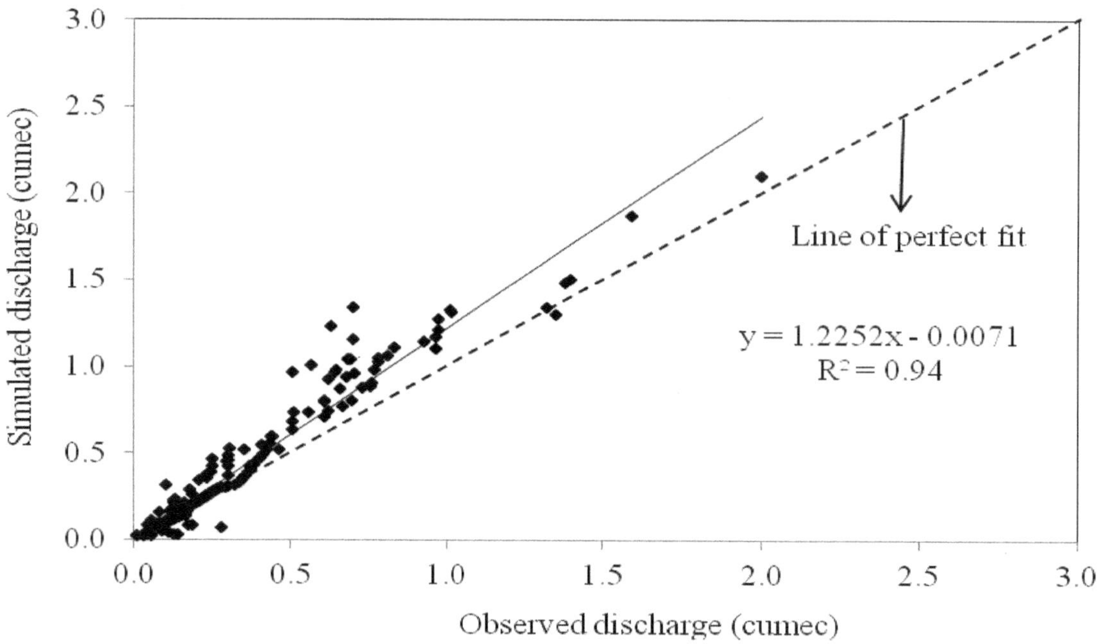

Figure 11a. Scatter plot of observed and simulated daily discharge during validation in Arnigad watershed.

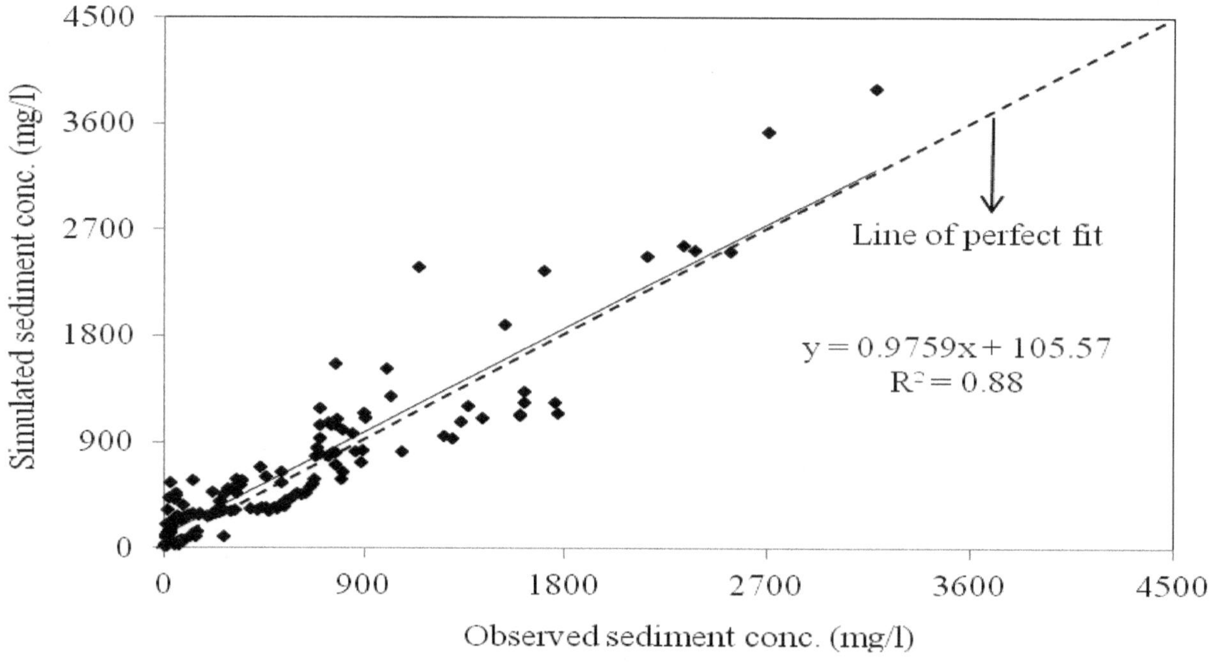

Figure 11b. Scatter plot of observed and simulated daily sediment concentration during validation in Arnigad watershed.

soil types in both the watersheds is almost similar. Higher ET and lower surface runoff in Arnigad than the Bansigad watershed is obvious due to the difference in forest cover types in Arnigad (dense Oak forest) and Bansigad watershed (degraded Oak forest). Although, higher water yield is obtained in Bansigad than the Arnigad watershed, the river flow in Bansigad ceases in the month of February or during early March, while Arnigad sustains the river flow throughout the year. Sharda and Ojaswi (2006) reported that root system of an oak tree is very extensive and soil-root complex system of each mature oak tree has a capacity to store several hundred litres of water, which is released as base flow during the lean season.

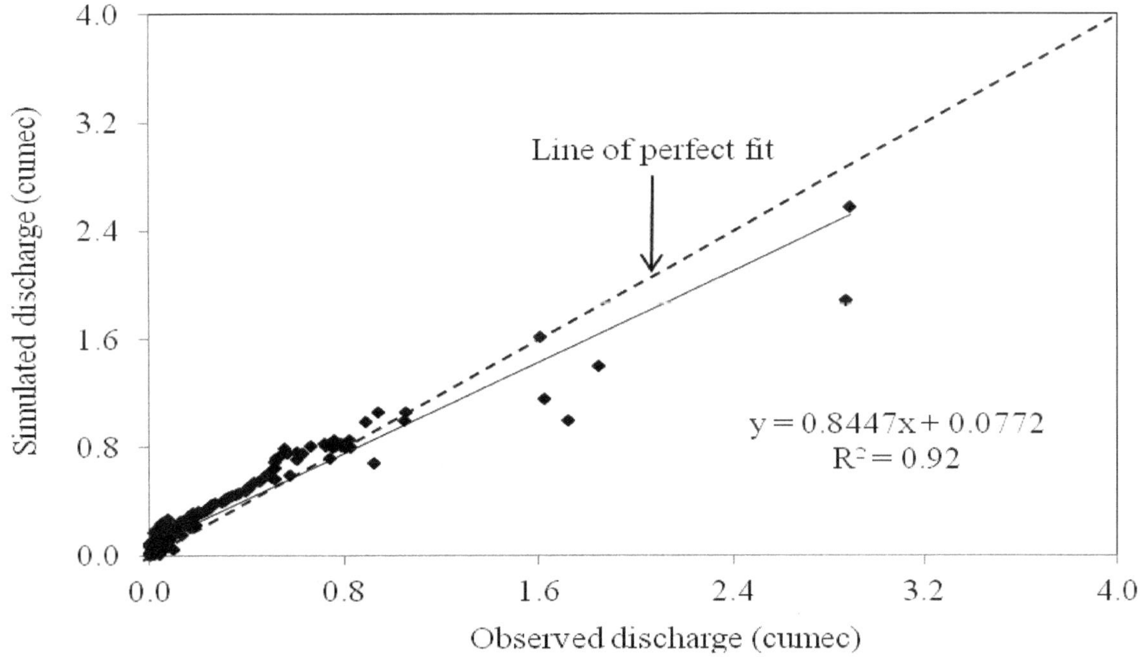

Figure 12a. Scatter plot of observed and simulated daily discharge during validation in Bansigad watershed.

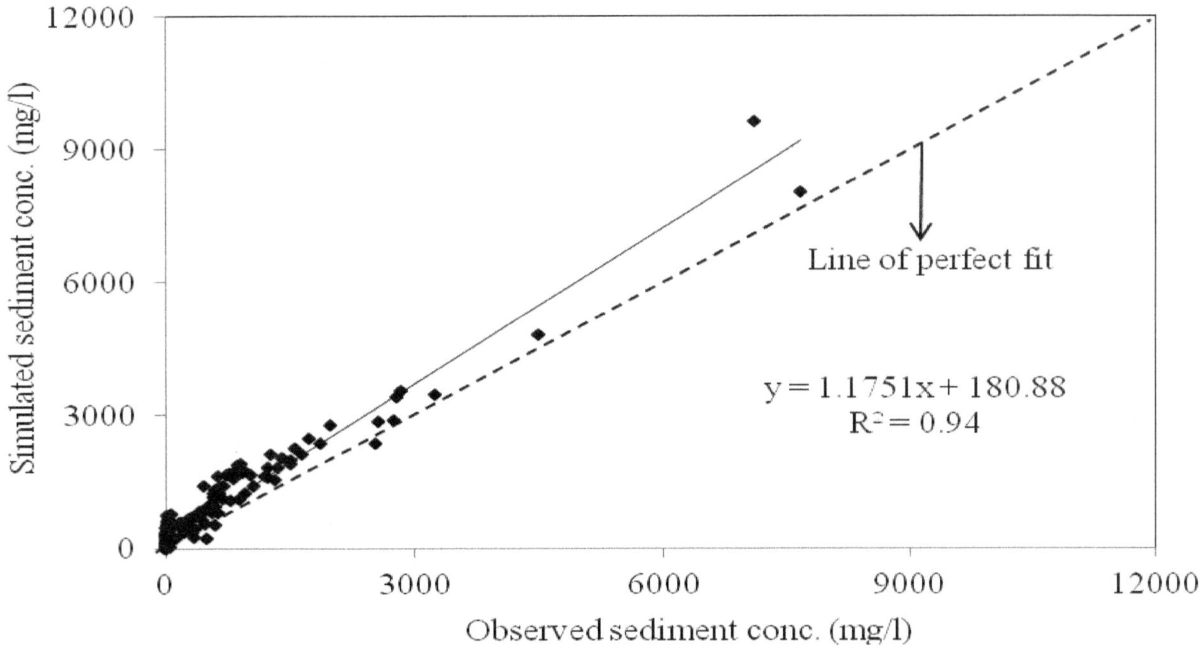

Figure 12b. Scatter plot of observed and simulated daily sediment concentration during validation in Bansigad watershed.

Table 5. Goodness of fit statistics for Arnigad watershed for validation period.

Description	E_{NS} (%)	R^2	Mean		Standard deviation	
			Observed	Simulated	Observed	Simulated
Discharge ($m^3 s^{-1}$)	82.78	0.94	0.273	0.327	0.293	0.369
Sediment concentration ($mg l^{-1}$)	83.28	0.88	298.989	397.358	520.740	541.964

Table 6. Goodness of fit statistics for Bansigad watershed for validation period.

Description	E_{NS} (%)	R^2	Mean		Standard deviation	
			Observed	Simulated	Observed	Simulated
Discharge ($m^3 s^{-1}$)	90.50	0.92	0.205	0.250	0.378	0.331
Sediment concentration ($mg l^{-1}$)	80.67	0.94	337.922	577.961	853.365	1032.541

The mean annual sediment loading from the Arnigad and Bansigad watershed was simulated as 8.45 and 21.97 t ha^{-1} respectively against the mean observed sediment yield of 10.70 and 24.46 t ha^{-1} in respective watersheds. The simulated sediment yield is comparable to observed values in both the watersheds. The high sediment yield in Bansigad watershed can be attributed to degraded forest cover and other anthropogenic activities in the watershed.

Conclusions

The present study was carried out to evaluate the applicability of physically based, distributed parameter SWAT model in estimating discharge and sediment yield from two forested watersheds in lower Himalaya (India) and to assess the impact of forest cover types on stream discharge and sediment yield. The following were drawn based on the results of the study:

1) The model simulated daily discharge and suspended sediment concentration followed the trend of observed values in both the watersheds.
2) R^2 values of 0.91 and above and E_{NS} values of 82.8% and above both in calibration and validation exhibited high performance of SWAT in simulating the discharge from the study watersheds.
3) Similarly, the model also performed more than satisfactory on both the study watersheds in simulating the sediment concentration with R^2 values of 0.86 and above and E_{NS} above 80%.
4) The model also simulated the mean annual water yield and sediment yield close to the observed values in both the watersheds. The mean annual surface runoff and water yield over the entire study period were simulated as 6 and 59.4% respectively of the mean annual rainfall in Arnigad watershed; and 6.9 and 63.7% respectively in Bansigad watershed. The mean annual sediment yield from the respective watersheds was simulated as 8.45 and 21.97 t ha^{-1} respectively.
5) The higher ET, lower mean annual surface runoff, lower water yield and lower sediment yield from dense oak forest than that from the degraded oak forest clearly indicated the effect of forest cover types on these hydrological variables.
6) The water balance components simulated by the model provided a useful insight for examining the hydrological behaviour of study watersheds, especially the ET needs and water delivery from the watersheds which are dominated by two different forest cover types.
7) The results indicated that SWAT is capable of estimating the discharge and sediment yield from Himalayan forested watersheds, the estimates of which are often required for operation and management of irrigation and hydropower projects in the region.

REFERENCES

Arabi M, Frakenberger J, Engel BA, Arnold JG (2008). Representation of agricultural management practices with SWAT. Hydrol. Proc. 22(16):3042-3055.

Arnold JG, Srinivasan R, Muttiah RS, Williams JR (1998). Large area hydrologic modeling and assessment Part I: Model development. J. Am. Water Resour. Assoc. 34(1):73-89.

Ayana AB, Edossa DC, Kositsakulchai E (2012). Simulation of sediment yield using SWAT model in Fincha Watershed, Ethiopia. Kasetsart J. Nat. Sci. 46:283-297.

Bärlund I, Kirkkala T, Malve O, Kämäri J (2007). Assessing SWAT model performance in the evaluation of management actions for the implementation of the Water Framework Directive in a Finnish catchment. Environ. Model. Softw. 22(5):719-724.

Bartarya SK (1995). Hydrogeology and water resources of intermountain, Doon valley. Himalayan Geol. 6(2):17-28.

Beasley DB, Huggins LF, Monke EJ (1980). ANSWERS User's Manual, Purdue University, West Lafayatte.

Bhaskar NR, James WP, Devulapalli RS (1992). Hydrologic parameter estimation using geographic information system. J. Water Resour. Plan. Manage. 118(5):492-512.

Bingner RL (1996). Runoff simulated from Goodwin Creek watershed using SWAT. Trans. ASAE 39(1):85-90.

Borah DK, Bera M (2003). Watershed-scale hydrologic and nonpoint-source pollution models: Review of mathematical bases. Trans. ASAE 46(6):1553-1566.

Chow VT, Maidment DR, Mays LW (Eds.) (1998). Applied Hydrology. McGraw-Hill Inc., New York.

Das S, Ruda RP, Gharabaghi B, Goel PK, Singh A, Ahmed I (2007). Comparing the performance of SWAT and AnnAGNPS model in a watershed in Ontario. ASABE publishing paper: 701P0207, ASABE, St. Joseph, MI, USA.

Hession WC, Shanholtz VO (1998). A geographic information system for targeting non-point source agricultural pollution. J. Soil Water Conserv. 43(3):264-266.

Jain SK, Kumar S, Varghese J (2004). Estimation of soil erosion for a Himalayan watershed using GIS technique. Water Res. Manage. 15(1):41-54.

Jain SK, Tyagi J, Singh V (2010). Simulation of runoff and sediment yield for a Himalayan watershed Using SWAT Model. J. Water Resour. Prot. 2:267-281.

Kannan N, White SM, Worrall F, Whelan MJ (2007). Hydrological modelling of a small catchment using SWAT 2000 - Ensuring correct flow partitioning for contaminant modelling. J. Hydrol. 334(1-2):64-72.

Kannan N, Santhi C, Arnold JG (2008). Development of an automated procedure for estimation of the spatial variation of runoff in large river basins. J. Hydrol. 359(1-2):1-15.

Knisel WG (1980). CREAMS: A Field-scale Model for chemical, runoff, and erosion from Agricultural Management Systems. USDA Conservation Research Report No. 26, Washington, D.C.

Krause P, Boyle DP, Bäse F (2005). Comparison of different efficiency criteria for hydrological model assessment. Adv. Geosci. 5:83-87.

Leonard RA, Knisel WG, Still DA (1987). GLEAMS: Groundwater loading effects of agricultural management systems. Trans. ASAE 30(5):1403-1418.

Maidment DR (1993). GIS and hydrologic modeling. In: Environmental Modeling with GIS, eds. MF Goodchild, BO Parks, L Steyaert. New York, Oxford University Press.

Mulungu DMM, Munishi SE (2007). Simiyu River catchment parameterization using SWAT model. Phys. Chem. Earth 32:1032-1039.

Muttiah RS, Wurbs RA (2002). Modeling the impacts of climate change on water supply reliabilities. Water Int. Int. Water Resour. Assoc. 27(3):407-419.

Nash JE, Sutcliffe JV (1970). River flow forecasting through conceptual models: Part I. A discussion of principles. J. Hydrol. 10(3):282-290.

Neitsch SL, Arnold JG, Kiniry JR, Srinivasan R, Williams JR (2002). Soil and Water Assessment Tool (SWAT) User's Manual, Version 2000. Blackland Research Center, Grassland, Soil and Water Research Laboratory, Agricultural Research Service: Temple, TX, USA.

Neitsch SL, Arnold JG, Kiniry JR, Williams JR (2005). Soil and Water Assessment Tool: Theoretical Documentation. Blackland Research Center, Grassland, Soil and Water Research Laboratory, Agricultural Research Service: Temple, TX, USA.

Oeurng C, Sauvage S, Sánchez-Pérez JM (2011). Assessment of hydrology, sediment and particulate organic carbon yield in a large agricultural catchment using the SWAT model. J. Hydrol. 401:145-153.

Pandey VK, Panda SN, Sudhakar S (2005). Modelling of an agricultural watershed using remote sensing and a geographic information system. Biosyst. Eng. 90(3):331-347.

Pandey VK, Panda SN, Pandey A, Sudhakar S (2009). Evaluation of effective management plan for an agricultural watershed using AVSWAT model. Rem. Sens. GIS. Environ. Geol. 56:993-1008.

Parajuli PB, Mankin KR, Barnes PL (2007). New Methods in Modeling Source Specific Bacteria Scale using SWAT. ASABE publication No. 701P0207. ASABE: St. Joseph, MI, USA.

Peterson JR, Hamlett JM (1998). Hydrologic calibration of the SWAT model in a watershed containing fragipan soils. J. Am. Water Res. Assoc. 34(3):531-544.

Rautela P, Joshi GC, Bhaisora B (2010). Seismic vulnerability and risk in the Himalayan township of Mussoorie, Uttarakhand, India. Curr. Sci. 99(4):521-526.

Rosenthal WD, Srinivasan R, Arnold JG (1995). Alternative river management using a linked GIS-hydrology model. Trans. ASAE 38(3):783-790.

Sharda VN, Ojaswi PR (2006). Hydrological evaluation of forest tree covers in different agro-ecological regions of India. Paper presented in Int. Workshop on Impacts of Reforestation of Degraded Lands on Landscape hydrology in the Asian region, 6 -10 March 2006, Roorkee, India.

Singh RK, Panda RK, Satapathy KK, Ngachan SV (2011). Simulation of runoff and sediment yield from a hilly watershed in the eastern Himalaya, India using the WEPP model. J. Hydrol. 405(3-4):261-276.

Smith R (1992). OPUS, An integrated simulation model for transport of nonpoint sources pollutants at the field scale, Volume I: Documentation. ARS-95. Washington, D.C.: USDA Agricultural Research Service.

Spruill CA, Workman SR, Taraba JL (2000). Simulation of daily and monthly stream discharge from small watersheds using the SWAT model. Trans. ASAE 43(6):1431-1439.

Srinivasan R, Engel BA (1991). A Knowledge Based Approach to Extract Input Data From GIS. ASAE Paper No. 91-7045.

Srinivasan R, Ramanarayanan TS, Jayakrishnan R,Wang H (1997). Hydrologic modeling of Rio Grande/Rio Bravo basin. ASAE Paper No. 97-2236. St. Joseph, Mich.: ASAE.

Srinivasan M S, Gerald-Marchant P, Veith TL, Gburek WJ, Steenhuis TS (2005). Watershed-scale modeling of critical source areas of runoff generation and phosphorus transport. J. Am. Water Resour. Assoc. 41(2):361-375.

Thakur VC, Pandey AK (2004). Late Quaternary tectonic evolution of Dun in fault bend/ propagated fold system, Garhwal Sub-Himalaya. Curr. Sci. 87(11):1567-1576.

Tolson BA, Shoemaker CA (2007). Cannonsville reservoir watershed SWAT2000 model development, calibration, and validation. J. Hydrol. 337(1-2):68-86.

Tripathi MP, Panda RK, Raghuwanshi NS (1999). Runoff estimation from a small watershed using SWAT model. Proceeding of International Conference on Water, Environment, Ecology, Socio-Economics and Health Engineering Seoul, Korea 143-152.

Tripathi MP, Panda RK, Raghuwanshi NS (2003). Identification and prioritisation of critical sub-watersheds for soil conservation management using the SWAT model. Biosyst. Eng. 85(3):365-379.

USDA Soil Conservation Service (1972). National Engineering Handbook. Hydrology Section 4 (Chapters 4–10).

Varanou E, Gkouvatsou E, Baltas E, Mimikou M (2002). Quantity and quality integrated catchment modeling under climate change with use of soil and water assessment tool model. J. Hydrol. Eng. 7(3):228-244.

Wang S, Kang S, Zhang L, Li F (2008). Modeling hydrological response to different land use and climate change scenarios in the Zamu River basin of northwest China. Hydrol. Proc. 22:2502-2510.

Williams JR (1975). Sediment yield prediction with universal equation using runoff energy factor. In: Present and Prospective Technology for Predicting Sediment Yield and Sources, Proceeding of the Sediment Yield Workshop. USDA Sedimentation Lab., Oxford, MS, USA. 28–30 November, 1972, p. 244-252.

Williams JR, Renard KG, Dyke PT (1983). EPIC: A new method for assessing erosion's effect on soil productivity. J. Soil Water Conserv. 38(5):381-383.

Williams JR, Nicks AD, Arnold JG (1985). SWRRB, a simulator for water resources in rural basins. ASCE Hydrol. J. 111(6):970-986.

Xu ZX, Pang JP, Liu CM, Li JY (2009). Assessment of runoff and sediment yield in the Miyun Reservoir catchment by using SWAT model. Hydrol. Proc. 23: 3619-3630.

Young RA, Onstad DD, Anderson WP (1989). AGNPS: A nonpoint source pollution model for evaluating agricultural watersheds. Soil and Water Cons. 44(2): 168-173.

Determination of appropriate hydraulic conditions to decrease the negative impacts of Dez dam flushing operation on the downstream

H. Samadi-Boroujeni[1]*, A. H. Haghiabi[2] and E. Ardalan[3]

[1]Water Engineering Department, Shahrekord University and Water Recources Research Center, Iran.
[2]Water Engineering Department, Lorestan University, Iran
[3]Water Engineering Department, Shahrekord University, Iran, Msc Water Engineering Graduated from Shahrekord University, Iran

The Dez dam is located in the Southwest Iran and was constructed in 1963, with a height of 203 m and an underground powerhouse of 520 MW, because of the high rate reservoir sedimentation, in recent years, therefore, performing flushing operations through the three irrigation gates of the dam has received great attention. Due to the release of high concentrations of sediments, on the other hand, performing such operations may cause considerable impacts downstream. The downstream limitations should, therefore, be taken into consideration during the flushing operations. In this paper, it has been tried to identify and introduce the downstream limitations of performing flushing operations for Dez dam, and to determine the most appropriate flushing hydraulic conditions in terms of discharge, concentration, and duration of such operations using model MIKE 11, so that the dam utilization system can perform these operations with the least damage to the downstream. Results show that if we take into consideration every factor involving the decision- making, then the flushing operations along with two stages of opening the irrigation gates (that is the evacuation of sediments) for 6 hours and the discharge of 30 m^3/s will be the best option. The interval between the two flushing operations should be at least 6 hours. Based on the results of this paper, appropriate conditions for hydraulic flushing can be determined. These conditions will be manipulated by the responsible administration as these operations would have the least damage and cost.

Key words: Dez dam, environmental impacts, sediment flushing, Mike 11 model.

INTRODUCTION

The Dez dam Project was constructed in the period from January 1960 to November 1962. The project is located in the Zagros Mountain range in the Southwest Iran and approximately 25 km north of Dezful city. The dominant foreseen elements of the project are the 203 m high double curvature arch dam and the 60 km long reservoir with the original volume of 3315 million m^3. The minimum and maximum control level is 300 and 352 m, respectively. An underground powerhouse contains eight 65 MW units for a total installed capacity of 520 MW has been generated an average of 2400 GWh/year energy production during last 46 years. Also Dezful re-regulating dam was constructed in 31 km downstream Dez dam to regulate the released water from Dez HEPP. A map of Dez project is shown in Figure 1.

Although the project has been well maintained, the project is now over 46 years old and the total amount of sediment accumulated on the dam reservoir amounts to 700 million cubic meters so it is reaching its mid-life period. Sedimentation in the Dez dam reservoir is one of the worth issues, which have assumed special importance in recent years, and the increased level of sediment behind the Dez dam body during the past years with a rate of 2 m/year has caused the sediment level to rise near the intake of Dez power plant. Another component of the project is Dezful re-regulating dam. The re-regulating dam with a height of 20 m is located in the vicinity of Dezful city and 31 km downstream of Dez dam.

*Corresponding author. E-mail: samadi153@yahoo.com.

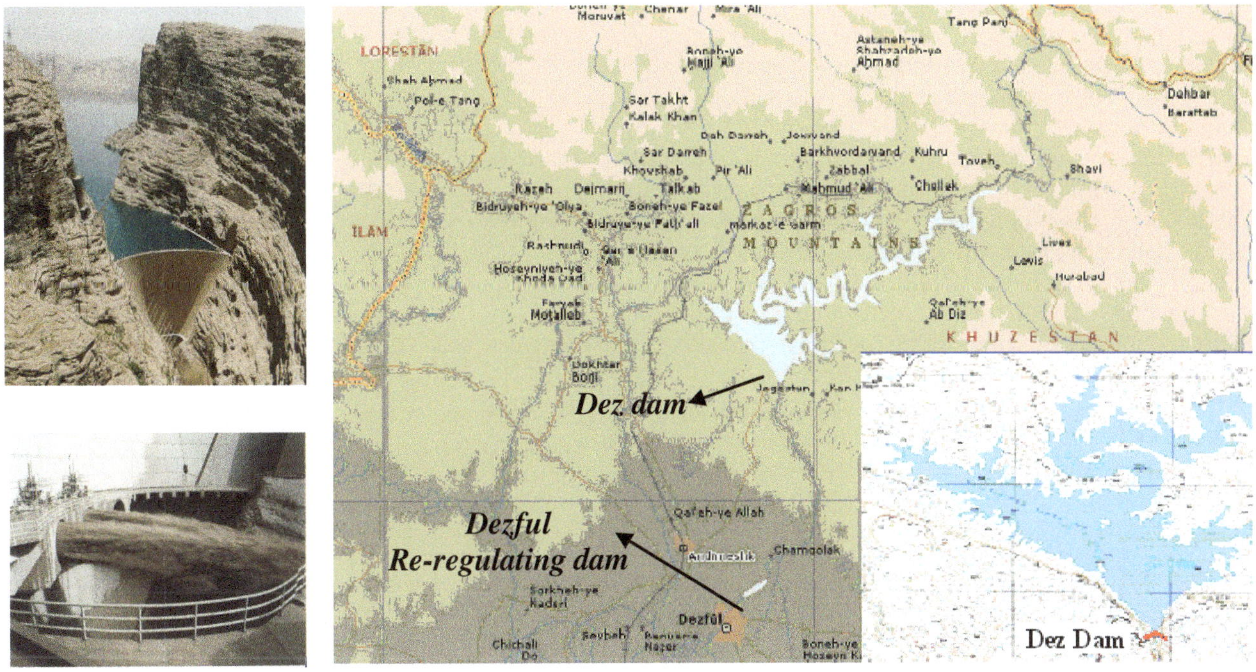

Figure 1. Map of Dez dam project and flushing br irrigation outlets.

It has been utilized since 1972 and the previous surveys show that 40% of the Dezful re-regulating dam reservoir capacity is already has been filled with sediment (Sadeghi, 2002).

Generally with particular regard to the elements affected by sedimentation, include:

- The reservoir, its remaining storage life and its ability to provide the degree of regulation originally envisaged for the project;
- elements already affected by the passage of water and sediment, including the irrigation outlets, and which may in the future be affected, including turbine equipment, particularly runners and wicket gates.
- Dezful re-regulating dam sedimentation. Since the main function of the re-regulating dam is to make regular release of water from Dez dam hydropower plant in downstream river, the useful storage volume is therefore, very important for this dam.

An overview of the reservoir sedimentation problem, its effects on the elements noted above, an indication of the impacts on the project, if remedial work is not implemented, and in this regards, one of the proposed measures to address the sediment management is the cone flushing of sediments through the dam irrigation outlets. Flushing of sediments has been conducted over the last few decades to augment irrigation flows as well as to remove the sediments that has accumulated immediately upstream of the irrigation outlets. This is surely has prompted the Khuzestan Water and Power Authority (KWPA) to consider performing the Dez dam flushing operations on the agenda. Predictions indicate that 1 - 1.5 million cubic meters of fine-grained sediments in the Dez dam reservoir will be discharged annually into the river downstream, in the next few years, as a result of performing flushing operations through the Dez dam irrigation outlets (Acres and Dezab, 2004). It should be noted that there are various important constraints during sediment flushing operation via the irrigation outlets such as environmental hazards downstream, loss of reservoir water storage, possibility of irrigation outlets clogging by large boulders, the effect of sediment spray around of the power house, and decreasing of power generation. Since the sedimentation in the re-regulating dam reservoir play an important role in Dez dam downstream flushing effects, it is essential that the sedimentation processes in the re-regulating dam reservoir to be simulated for surveying the negative impacts of Dez dam flushing operation on the downstream. It should be noted that sediments released from Dez irrigation outlets are mostly cohesive sediments therefore this matter should be considered for modeling of sediment transport processes in river downstream.

Since the hydraulic of flow in rivers is very complex and man has not managed to give a real explanation of this phenomenon yet, it is necessary, therefore, to carry out a lot of experiments and measurements to ascertain the conditions of flow. Accordingly, researchers and scientists have tried to apply mathematical methods and computer-aided facilities in order to develop mathematical models, so that they could be used to facilitate the measurements and experiments as far as possible and in order to give a simulation for the flow under various conditions. Recently, these models were developed to solve more difficult problems, such as the simulation of sediment movements. Generally, the basics for the development of hydro-mathematical models are the famous equations of Navier-Stokes and the mass transport equations. For open channel flows (canals and rivers), these equations have been replaced with Sant Venant equations and the sediment transport equation. Although these equations do not have analytical solutions for real problems, one can solve them with an acceptable precision through the application of numerical techniques (Scarlatos and Lin Li, 1997). In this respect, many mathematical models have been developed, and one of the most famous is MIKE11 Model, which was developed by Denmark Hydraulic Institute in 1992 which simulates the flow in an unsteady and one-dimensional state (DHI, 2000). The MIKE 11 computational modules with particular relevance to sediment transport studies besides the main module are AD (advection-dispersion module), ACS (advanced cohesive sediment module), ST (sediment transport module), GST (graded sediment transport module), XZ (sediment transport in stratified flow). The MIKE 11 Sediment transport modules are used for simulation of one-dimensional free surface flow and sediment transport in rivers where stratification can be neglected. The AD module for transport modeling (advection and dispersion) is based on the one-dimensional equation of conservation of mass of a dissolved or suspended material (e.g., sult or fine sediments). The module requires output from the hydrodynamic module (HD module), that is discharge and water level, cross-sectional area and hydraulic radius. This model can numerically solve the general form of the mass transport equations as follows:

$$\frac{\partial c}{\partial t}+\frac{\partial}{\partial x}\left[u(x,t).C-D(x,t)\frac{\partial c}{\partial x}\right]+\sum_{i=1}^{i=n}\left[w_i(1-\frac{\tau_b}{\tau_{di}})\frac{1}{h}\right]C_i=C_e.e^{-\beta.t} \quad (1)$$

Where u(x, t) is the water velocity (m/sec), C_e is the coefficient of erodibility, β is the non- dimensional exponent of erodibility, w is the settling velocity of particle (m/sec), τ_d and τ_b are the shear stress for the beginning of sedimentation and the shear stress of bed respectively (N/m^2), h is the depth of water (m), C_i is the concentration of suspended sediments related to its group (Kg/m^3), C is the concentration of suspended sediments (Kg/m^3), D(x, t) is the coefficient of diffusion (m^2/sec), x is the distance (m), and t is time (sec).

MATERIALS AND METHODS

Data collection

In this study, the following information and data were collected: Statistics and information concerning the Dez River cross–sections, the discharge-stage relationships of the project's hydrometery stations, the data of discharge and sediments in the project's hydrometery stations (including the stations of Dez dam, Dast-Mashan and Dezful), and the information concerning the Dez dam flushing operations (Samadi Boroujeni and Galay, 2005). The cross-sections used in this study were carried out in the area between the Dez dam and the Dezful re-regulating dam in 1991 and 2001 by Khuzestan Water and Power Authority (Samadi-Boroujeni, 2004).

For the formulation of scenarios, the information concerning the utilization of the Dez dam and the Dezful re-regulating dam, and the information concerning the environment of river were used.

Based on the collected information, the input file of MIKE11 was provided for 31 km of Dez River, in the area occurring between Dez dam and Dezful re-regulating dam.

Information concerning the Dez dam reservoir flushing

Following the over-accumulation of sediments behind the Dez dam body, flushing operations were performed for the first time in 1994, through opening of the dam irrigation outlets. However, these operations were subsequently repeated annually. As the flushing for the Dez dam is carried out without the discharge of reservoir, it is called flushing under pressure, where this type of flushing is repeated annually during the following years. As the flushing for the Dez dam is carried out without the discharge of the reservoir, called 'flushing under pressure'; in this type of flushing, parts of the reservoir sediment are discharged by opening the depth outlets for a relatively short time (less than 1 or 2 days); a hole in the shape of the cone is created behind the dam body, so after the formation of this hole, the outflow from depth discharge becomes clear (Morris and Fan, 1998). The previous observations of the Dez dam flushing operations have shown that these operations had a considerable contribution for the decreased level of sediment in front of the power plant intake. Accordingly, Khuzestan Water and Power Authority began the rehabilitation operations for the dam's irrigation outlets in 2004 (as they are the only outlets for sediments discharge) in order to perform these operations periodically. Accordingly, a schedule was provided for the regular flushing operations, 3 - 5 flushing operations for each year until the outlets are completely rehabilitated. It is predicted, therefore, that 1 - 1.5 million cubic meters of fine-grained sediments will be debouched into the downstream river annually (Acres and Dezab, 2004). It is worth mentioning that the sediments behind the dam body consist of 40% clay and 60% silt (Samadi Boroujeni et al, 2009). These sediments are cohesive, due to their properties, and they flocculate and deposit when they reach the reservoir of Dezful re-regulating dam during any flushing operation. The rate of sedimentation in the reservoir of re-regulating dam highly depends on the way a flushing operation is performed (i.e. the opening of irrigation outlets) and on the concentration of sediment released from the dam. The concentration of sediment release from the dam at the downstream cross-section of the dam (C_{River}) is obtained from the following relation, given the outflow discharge from the dam (Q):

$$C_{River}=\frac{\sum Q_i.C_i}{\sum Q_i}=\frac{Q_{Irr}.C_{Gate}+Q_{Tur}.C_{Tur}+Q_{Spill}.C_{Spill}}{Q_{Irr}+Q_{Tur}+Q_{Spill}}=\frac{Q_{Irr}.C_{Gate}}{Q_{Total}} \quad (2)$$

Where Spill, Tur, Irr and Gate indices belong to the spillway, the turbine outflow, and the irrigation outlets, respectively. The concentration of outflow from the irrigation outlets is more than 500 gm/l at the beginning of outlet opening (flushing) and then, finally, within one day, it becomes clear. The clear outflow of turbines and spillways, however, causes the concentration of sediment to

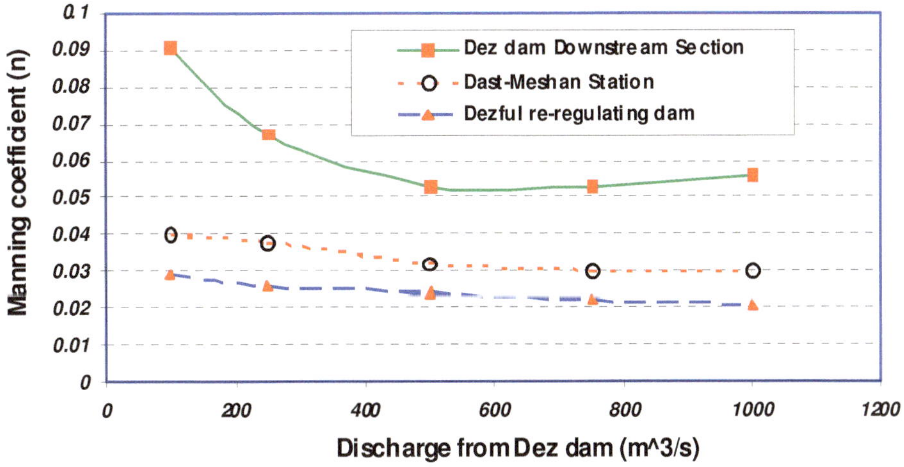

Figure 2. Calibrated Manning coefficient values via discharge in Dez River.

decrease at the downstream of the dam.

Calibration of MIKE 11 model

The model calibration is the most important part of the model running. In this study, two parts; the hydrodynamic and cohesive sediments transport in the model are used. the calibration of the model's hydrodynamic part is achieved through the selection of the coefficient of resistance (Manning coefficient), and the calibration of cohesive sediment transport model is achieved through the selection of the model's diffusion coefficient and the sedimentation and erosion extent rates.

Determination of the calibrated manning coefficient was achieved on the basis of the discharge–stage relationship of the downstream stations (i.e. stations of Dezful, Dast-Mashan, and Dez dam tail water). The results of model calibration are shown in Figure 2 for parts of hydrodynamic (Samadi-Boroujeni, 2004). These results show that amount of Manning coefficient "n" has low sensitivity to discharge changes. Accordingly, average amounts of manning coefficient, i.e. 0.064, 0.034 and 0.024 were used respectively for the Dez dam downstream section, Dast-Mashan station and Dezful station, in this paper. High variability of the n values refers to the river bed material properties.

The calibration of cohesive sediment transport part (TD model) was achieved by validation of measurement results obtained during the Dez dam flushing operation in 2000. The sediment flushing operation in 2000 took time of about one day and during it, maximum concentration at Dez dam downstream section, inflow to re-regulating dam reservoir and re-regulating dam outflow was measured to 51, 33 and 4.5 gm/l, respectively. Based on these data, MIKE 11 model was calibrated and diffusion coefficient and erodability coefficient of the model were calibrated as 30 m^2/s and 0.02 kg/m^2/s, respectively. Other parameters of the model were determined based on the performed measurements, such as settling velocity of the sediments, grain size and fluid viscosity (Fathi Moghadam et al., 2009).

Scenarios for running the model

In the study, different scenarios have been proposed for performing the flushing operations in the Dez dam, taking into consideration every important factor associated with these operations (Table 1). According to these options, maximum concentration of sediment at the Dez dam downstream section was considered to be 40 gm/l, given the environmental limitations. Considering the following relation one can determine the sediment concentration at the Dez dam downstream:

$$Q_{t,s} = C_G(\frac{Q_G}{C_0}) - Q_G \qquad (3)$$

$$Q_o = Q_{t,s} + Q_G \qquad (4)$$

Where; $Q_{t,s}$ is the total outflow discharge from turbines and spillways, C_G is the concentration of outflow from the irrigation outlets, C_o is the concentration of outflow from the dam at the downstream, and Q_G is the discharge from the irrigation outlets and Q_o is total outflow discharge from the dam. In this study, it has been estimated that the concentration of outflow from the irrigation outlets is on the average of 500 gm/l, and the discharge of outflow from the turbines is 250 m^3/s during the flushing operation. For safety precautions, it was assumed that two outlets at most, out of three irrigation outlets in the Dez dam, should be on manever simultaneously, where the maximum discharge for each outlet was considered to be 45 m^3/s.

For the environmental considerations of the river, the flushing duration is considered to be 12 h at maximum, and, in some options, the flushing is to be performed intermittently, where the duration for flushing is 4 h each time. In such case the interval between the two flushings will be at least 6 h finally, MIKE 11 model was run for each scenario. To run the model, the hydrograph has been envisaged at the upstream border (the Dez dam downstream) on the basis of the following rules:

- Up to half an hour before the beginning of flushing, the flow will begin from the spillway in amounts specified in Table 2.
- Up to half an hour after the flushing, the outflow from the spillway will continue in order to maintain the river's environment (Table 2).
- When the irrigation outlets are closed, the outflow discharge from the dam has been assumed to be equal to the outflow discharge from the turbines. The total hours of flushing, in all alternatives, has been assumed to be 30 h.

For example, the hydrograph of flow and sediments obtained from by using MIKE 11 model for option G2-40 -30 is given in Figure 3.

Table 1. Different scenarios of the Dez dam flushing to run MIKE 11 model.

Row	Scenario Code	Q_G (m^3/s)	Released sediment concentration (gm/l)	Total Flushing time (h)	Times of Flushing for each operation	Duration of flushing each time (h)	$Q_{t,s}$ (m^3/s)	Q_o^* (m^3/s)
1	G1-20-30	30	20	12	1	12	720	750
2	G1-30-30	30	30	12	1	12	470	500
3	G1-40-30	30	40	12	1	12	335	365
4	G2-20-30	30	20	12	2	6	720	750
5	G2-30-30	30	30	12	2	6	470	500
6	G2-40-30	30	40	12	2	6	335	365
7	G3-20-30	30	20	12	3	4	720	750
8	G3-30-30	30	30	12	3	4	470	500
9	G3-40-30	30	40	12	3	4	335	365
10	G1-20-45	45	20	8	1	8	1105	1150
11	G1-30-45	45	30	8	1	8	720	765
12	G1-40-45	45	40	8	1	8	523	568
13	G2-20-45	45	20	8	2	4	1105	1150
14	G2-30-45	45	30	8	2	4	720	765
15	G2-40-45	45	40	8	2	4	523	568
16	G1-20-90	90	20	4	1	4	2230	2320
17	G1-30-90	90	30	4	1	4	1470	1560
18	G1-40-90	90	40	4	1	4	1085	1175

*Total discharge of release from the dam. It should be noted that the safe discharge at downstream river is about 1100 m3/s (Samadi Boroujeni, 2004)

Table 2. Results obtained from MIKE 11 model simulation.

Scenario Code	$C_{o\text{-}max}$ (gm/l)	$C_{1\text{-}max}$ (gm/l)	$C_{2\text{-}max}$ (gm/l)	Amount of the sediments discharged (ton)			Te (%)
				Dez dam outflow	re-regulation dam reservoir inflow	Re-regulation dam outflow	
G1-20-30	20	17.6	6.9	674768	606492	229448	61
G1-30-30	30	18.6	8.6	674514	418776	188872	54
G1-40-30	40	23.6	9.8	656233	388387	154886	58
G2-20-30	20	13.2	6.9	701147	463671	239266	47
G2-30-30	30	18.6	8.8	700850	435576	195009	53
G2-40-30	40	23.6	9.8	681581	403434	159077	59
G3-20-30	20	13.2	7.1	727460	474934	236131	46
G3-30-30	30	21.6	8.8	772934	540766	201247	59
G3-40-30	40	24.3	9.8	706595	415347	159151	60
G1-20-45	20	13.9	8.2	703465	492375	285784	42
G1-30-45	30	19.8	10.4	701555	463239	240831	47
G1-40-45	40	25.2	12.2	694075	438806	206861	52
G2-20-45	20	14.2	8.3	743908	515924	289738	42
G2-30-45	30	20.2	10.6	727578	473178	234535	48
G2-40-45	40	25.9	12.4	733202	457981	206375	52
G1-20-90	20	15.3	9.9	750890	570698	366002	36
G1-30-90	30	21.6	13.5	756729	545640	331491	38
G1-40-90	40	27.8	16.6	759327	528881	303500	40

$C_{o\text{-}max}$: maximum concentration of sediment at Dez dam outflow.
$C_{1\text{-}max}$: maximum concentration of sediment at the re-regulation dam reservoir inflow.
$C_{2\text{-}max}$: maximum concentration of sediment at the re-regulation dam outflow.

$$T_e = \frac{(C_{1-\max} - C_{2-\max})}{C_{1-\max}} \times 100$$

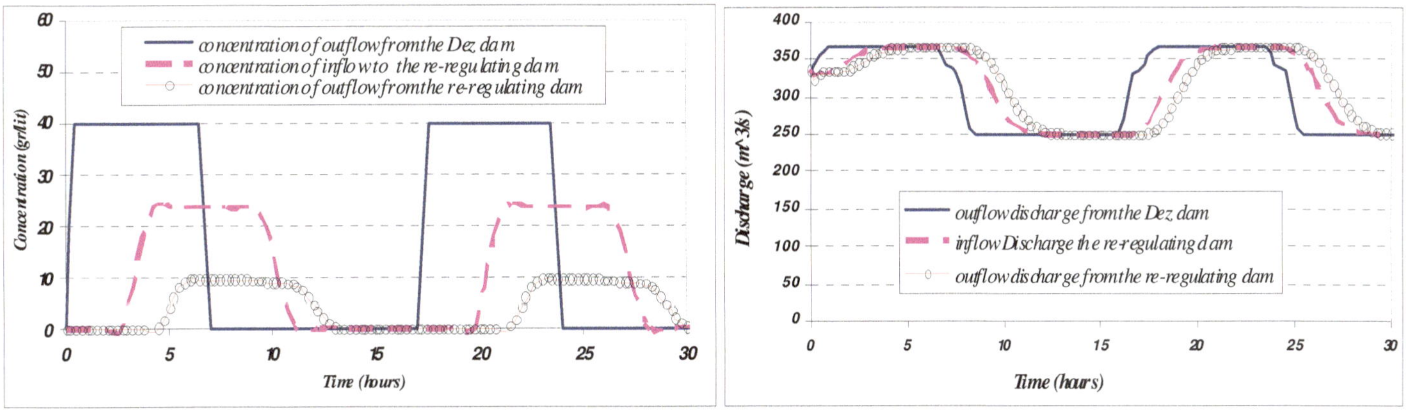

Figure 3. Hydrograph of flow and sediment concentration of the Dez River for the option G2-40 30 based on the model results.

Table 3. Introducing and ranking of criteria for the flushing operations in terms of scores.

Criterion	Degree of impact				
	Very high (0 points)	**High (2.5 points)**	**Moderate (5 points)**	**Low (7.5 points)**	**Very low (10 points)**
Ratio of water loss	W > 45	35< W<45	25< W<35	15< W<25	W<15
Threat of flooding downstream	Q_m2000<	<2000 Q_m1500<	<1500 Q_m1200<	<1200 Q_m1000<	Q_m<1000
Concentration at downstream (environmental aspect)	C_o > 12	12< C_o <10	10< C_o <8	8< C_o <6	C_o<6
re-regulating dam reservoir Sedimentation	T_e > 60	50< T_e <60	40< T_e <50	30< T_e <40	30 >T_e
Opening risk of the outlets	4 times opening	3 times opening	2 times opening	1 time opening	No opening

W: Ratio of Dez dam water outflow to the amount of sediment discharged during the flushing (million cubic meters of water /ton of sediment),
Q_m: Maximum discharge of outflow from the dam (m^3/s),
C_o: Maximum concentration of sediment at re-regulating dam outflow (gm/l),
Te: Percentage of trap efficiency of Dezful re-regulating dam reservoir.

RESULTS AND DISCUSSION

MIKE 11 model was run for different options, according to the scenarios given in Table 1. To meet the conditions of model stability, time-step calculations at the hydrodynamic part equal to one minute and at the cohesive sediment transport section as equal to 0.1 min were selected. Table 2 shows the important results obtained from using the model. Results show that travel time of flow between the Dez dam and the Dezful re-regulating dam is, on the average, 4 h. In the studied scenarios, on the average, 70% of the sediments released from the dam find their ways into the reservoir of Dezful re-regulating dam. The same results show that for the options under study, the trap efficiency (T_e) of the sediments is, on the average, 50% in the reservoir of Dezful re-regulating dam. It can be seen, sedimentation in the reservoir of Dezful re-regulating dam is one of the important consequences of flushing in the Dez dam. Although this is undesirable in terms of decreased useful capacity of the re-regulating dam reservoir, one may consider it as useful in terms of environmental considerations since the concentration of suspended sediments in the downstream river is decreased up to an allowable concentration level (which is 8 gm/l in the Dez River, given the natural regime of the river sediments). Obviously, the consequence of this is to dredge the reservoir of re-regulating dam.

It should be noted that the Dez dam flushing causes various negative impacts on the downstream, so it is necessary to select the best option in terms of every important associated factor and consequence. Accordingly, one needs criteria to rank options in accordance with priorities. Surveys show that until now, unfortunately, quantitative criteria for flushing operations in the Dez River have not been set. Therefore, given the results of previous studies and the experiences of the present authors, and after various consultations, the important criteria for the Dez dam flushing have been determined quantitatively and along with points or marks in Table 3. This table has been worked out tentatively by the authors. in other words, it is a preliminary determination. it is necessary to consider the degree of importance for each factor in terms of the conditions for the location. Given the great importance of environment, for example, it is necessary to multiply the mark of environment by an incremental coefficient. During the years of abundant water, the mark related to the loss of water should be

Table 4. The criteria scores for each flushing option.

Scenario Code	Score of ratio of water loss	Score of threat of downstream flooding	score of environmental aspects	score of re-regulating dam reservoir sedimentation	score of opening risk of the outlets	Sum of scores for each option
G1-20-30	7.5	10	7.5	0	7.5	32.5
G1-30-30	7.5	10	5	2.5	7.5	32.5
G1-40-30	10	10	5	0	7.5	32.5
G2-20-30	2.5	10	7.5	5	5	30
G2-30-30	7.5	10	5	2.5	5	30
G2-40-30	10	10	5	0	5	30
G3-20-30	2.5	10	7.5	2.5	2.5	25
G3-30-30	7.5	10	5	0	2.5	25
G3-40-30	10	10	5	0	2.5	27.5
G1-20-45	0	7.5	5	5	7.5	25
G1-30-45	5	10	2.5	5	7.5	30
G1-40-45	7.5	10	0	2.5	7.5	27.5
G2-20-45	0	7.5	5	5	5	22.5
G2-30-45	5	10	2.5	2.5	5	25
G2-40-45	7.5	10	0	2.5	5	25
G1-20-90	0	0	5	7.5	7.5	20
G1-30-90	5	2.5	0	7.5	7.5	22.5
G1-40-90	7.5	7.5	0	5	7.5	27.5

multiplied by a minimal coefficient, and in case the dam's irrigation outlets have difficulties, the mark related to the maneuver risk of outlets should be multiplied by an incremental coefficient. In this study, as shown in Table 4, it was tried to calculate the marks or 'points' of each option without considering the incremental or minimal coefficients in order to achieve a preliminary result. Option with minimum scores can be introduced as the best option for performing flushing. Results showed that the options G1-20-30, G1-30-30, and G1-40-30 had the highest scores, and the option G1-20-9 had the lowest score. At the same time, option G3-30-30, which had the lowest trapped sediments in the re-regulating dam reservoir, has got 25 as the total score, while the option G2-20-30, which had the lowest environmental damages, has gained 30 points. it can be seen, therefore, that it is necessary to take the incremental or minimal coefficient into consideration in each criterion. Accordingly, to decide which option the best is flushing, the beneficiary administration should calculate the total score for each option, while paying attention to the dominant conditions and taking the relevant incremental as well as minimal coefficients into consideration.

Conclusion

Based on the present study, the important results are as follows:

- Results obtained from the use of MIKE 11 model, in this study, showed that this model, which is able to solve the equation of mass transport, can be used for the simulation of cohesive sediment transport in the reservoir.

- Results obtained from MIKE 11 model showed that manner of carrying out the flushing for the Dez dam sediments influences the trapping efficiency of re-regulating dam reservoir and one can, therefore, reduce the sedimentation in the re-regulating dam reservoir through proper management at the Dez dam flushing. The same results indicate that lesser the duration of flushing operations at the Dez dam and more the discharge, lesser the trapping efficiency in the re-regulating dam reservoir. This, however, causes the concentration of outflow from the re-regulating dam to be more than the allowable level of concentration at the downstream river.

- If all factors are involved in decision-making, such as the loss of Dez reservoir water storage, threat of flooding downstream, sedimentation in the reservoir of re-regulating dam, downstream environment, and risk of opening the Dez dam outlets are considered to have equal degrees of impact, the option G2-20-30 will gain the lowest impacts and therefore, it will be the most appropriate option. In this option, the irrigation outlets (sediments discharge) are opened for two times with a total discharge of 30 m^3/s, the duration of performing the flushing operation being 6 h each time at a minimum 6-hour interval between two flushing operations. In the years of excess flooding, the option G2-20-45 may create better conditions. In this option, the flushing is performed two times, at a minimum of 6-hour interval between the two flushing operations, with a duration of each flushing (the opening of irrigation outlets) being 4 h, and the discharge of sediment outlets being 45 m^3/s in this option.

- To decide which option is the best flushing alternative, the beneficiary administration needs to pay attention to the relevant conditions and to consider the incremental as well as the minimal coefficients for each criterion, and to calculate the total scores of each option.

ACKNOWLEDGEMENTS

The authors wish to acknowledge the Khuzestan Water and Power Authority for their assistance and guidance provided in the preparation of this paper.

REFERENCES

Acres Dezab consulting company (2004). Dez Dam Rehabilitation Project- Stage2: Irrigation Outlet Study, Report to Khuzestan Water and Power Authority, Ahwaz, Iran. 79p.

DHI (2000). Reference Manual MIKE11 Model.

Fathi Moghadam M, Arman A, Samadi-Boroujeni H, Emamgholizadeh S (2009). Fall Velocity of Cohesive Sediments in Dez Dam Reservoir, Research J. Environ. Sci. 3(1): 71-79.

Morris G.L, Fan J (1998). Reservoir Sedimentation Handbook, McGraw-Hill, New York, USA, p.848.

Sadeghi MA (2002). Surveying the Trend of Sedimentation in the Dez Re-regulating Dam Reservoir Using the BRI–STARS Mathematical Model, Faculty of Agriculture, university of shahid-chamran, Ahwaz, Iran, Dissertation (M.S. Thesis in Farsi) p.155.

Samadi-Boroujeni H (2004). prediction of Sediment Concentration Resulting from the Dez Dam Hydraulic Flushing Operations at the Downstream River Along with a Survey of its Environmental Impacts, Khuzestan Water and Power Authority, Iran (Research Report in Farsi). p.149.

Samadi-Boroujeni H, Galay VJ (2005). Turbidity Current Measurements Within the Dez Reservoir, Iran, Proceeding, 17th Canadian Hydrotechnical Conference pp.127-127.

Samadi-Boroujeni H, Fathi oghaddam M, Shafaie-Bajestan M (2009). Investigation on bulk density of deposited sediments in Dez reservoir, J. Trend Appl. Sci. Res. 4(3): 148-157.

Samadi-Boroujeni H, Fathi-oghaddam M, Shafaie-Bajestan M, Mohammad.Vali.Samani H (2005) Modelling of Sedimentation and Self-Weight Consolidation of Cohesive Sediments, Sediment and Ecohydraulics Intercoh (2005.1stEdn), Elsevier B.V. Oxford, UK, pp165-191

Scarlatos PD, Lin L (1997). Analysis of Fine-grained Sediment Movement in Small Canals, J. Hydrol. Eng. 123(3): 200-207.

Advancing stormwater management practice in Iran using water sensitive urban design approach

Mohammadhossein Kashefizadeh[1] , Zulkifli bin Yusop[2] and Ali Hekmat[3]

[1]Construction Management, Civil Engineering Faculty, University Teknologi Malaysia, Skudai, Johor, Malaysia.
[2]Water Research Alliance, University Teknologi Malaysia, Skudai, Johor, Malaysia.
[3]Water Resource Management, Civil Engineering Faculty, University Teknologi Malaysia, Skudai, Johor, Malaysia.

More than 350 cities of Iran are subject to floods every year and flood phenomena cause annual 1,500,000,000,000 Rials loss to Iran. The reason for floods lie with lack of appropriate water management systems in terms of stormwater management. Current status of Iran water management calls for imitating the successful experience of other countries. The aim of this paper is to review better management frameworks for urban water management in regard to stormwater management in outstanding countries. In this paper, the concept of Water Sensitive Urban Design (WSUD) is initially reviewed, then the proposed frameworks of stormwater management in the world are reviewed and the initiatives associated with stormwater management are offered. Finally, an implemented case of Malaysia is studied to verify the use of initiatives in real similar scales. Among several variables in regard with stormwater management, three types of objectives are highlighted and two evaluation criteria are propsed. Analysis of the results is conducted based on proposed evaluation criteria. Result of the Analysis of the Putrajaya Stormwater Management System shows a remarkable contribution of WASUD initiatives in mitigating the stormwater in this city. Due to geometrical similarities of this city with Iranian ones, results show a great potential of employing WSUD initiatives in Iran.

Key words: Resource water management, Water Sensitive Urban Design (WSUD), sustainable management, Putrajaya stormwater management, water management criteria.

INTRODUCTION

In every year, more than 350 cities of Iran are subject to floods. The flood phenomena cause 1,500,000,000,000 Rials loss to Iran every year and in the recent years 70% of the annual credit of mitigation of natural disasters is devoted to combat the losses of floods (Organization of Geology and Mining Explorations, 2011; Iran Ministry of Energy, 2011). Likewise, Malaysia suffered from a loss of 300 Billion Dollars annually (Eighth Malaysian Plan, 2008). Many reasons account for the causes of floods, such as area's geographical location, old, poor, inadequate or insufficient drainage systems, and in particular surface drainages. Following the widespread realisation of the significance of climate change, urban communities are increasingly seeking to ensure resilience to future uncertainties in urban water supplies (Rutger, 2009; Iran Ministry of energy. 2011; Eighth Malaysian Plan, 2008). Both Iranian and Malaysian governments suffer from similar flood causes, despite insurmountable differences between their geographical aspects. Despite the efforts of Iran government to improve the reliability of water supply systems, enhancement is still unsatisfactory which arises from lack of a comprehensive guidelines nor practical samples of other countries for this purpose. Studying the course of action that Malaysia took in an effort to meet this problem is a good help to Iran floods problem.

WATER SENSITIVE URBAN DESIGN (WSUD)

Tony (2007) declares that the words 'water sensitive' define a new paradigm in integrated urban water cycle management that combined the various disciplines of engineering and environmental sciences associated with the provision of water services, including the protection of aquatic environments in urban areas. Water Sensitive Urban Design (WSUD) integrates the social and physical sciences. The inter-government agreement on a National Water Initiative defines WSUD as "the integration of urban planning with the management, protection and conservation of the urban water cycle that ensures that urban water management is sensitive to natural hydrological and ecological processes". Andrew and Grace (2000) asserted that sustainable water management leads to WSUD and it encompasses the integration of a number of levels in many different measures that have a range of suitable applications, resulting in many different realisations of this concept, depending on the site-specific conditions of the location. Mohamed et al. (2010) determined the environmental impacts of traditional urban water systems which are shown in Table 1. Table 2 indicates the differences between elements of conventional and sustainable water management and WSUD. Effective management of stormwater is being the principal focus of WSUD (Wong, 2006). WSUD espouses the need to integrate stormwater management into the planning and design of urban areas and applies across the entire spatial scale from a catchment-wide regional level to the precinct level to the local, building level (Wong, 2006; Thomas et al., 1997; Ecological Engineering, 2003; Australian Runoff Quality, Australia). Benefits of implementation of stormwater management initiatives are shown in Table 3.

PROPOSED FRAMEWORKS AND INITIATIVES OF STORMWATER MANAGEMENT

Andrew and Grace (2000) in their research, proposed new concept of stormwater management throughout a new water management approach. This concept, which is known as integrated urban water cycle, proposes the treatment of the development water in three steps of blackwater, greywater and stormwater before releasing the wastewater into the sewage system. In their method, balckwater, greywater and stormwater are stored in a pond and will receive treatment in order to recycle and reusing purpose. The remaining of each stage is released into the next stage and eventually sewage. This proposed model is very rudimentary one and therefore suffers from lack of detailed techniques and variety of initiatives. In 2002, Australia announced its guideline of water sensitive urban design and through the guideline; they proposed the scheme of WSUD to be implemented in the city of Australia. The guideline proposed to integrate three types urban water cycle including potable water, stormwater and wastewater and conservating, managing and minimising them respectively throughout recycling, reusing and recovering techniques. Despite sound solutions, the implementation required more advances in terms of administrative measures rather than only technical means (Australian Environmental Water Management, 2012). Gardner (2003) proposed healthy home initiative. In this framework, the focus is to manage water at consumer point as far as possible. The precipitation entered to the home will receive several treatments from the very first point. Rainwater is initially collected from the roofs and is sent to a tank located at the basement of the home and after undergoing disinfection by radiation of UV, it will be sent to home for use. The return water of this water is then stored in the greywater treatment tank and subsequently, this greywater is used for irrigation purposes and finally it is discharged to the sewer. In this method, focus is using the rainwater as the main source of water use and the urban water is assumed to be a supplementary for special circumstances. Despite its high efficiency in controlling stormwater management, this framework has several drawbacks. It takes a long time for implementation and also it is a costly method which is unjustified in small scales. Moreover, it can be implemented in those areas with high rate of precipitation and other areas with less rain amount cannot use this method expecting high efficiency. Brown et al. (2009) proposed a framework for stormwater management towards achieving water sensitive city. Their proposed model is based on providing water supply security, public health and flood protection, social amenity, considering natural resources and climate changes by by separating sewerage schemes, providing drainage channelisation, point and diffuse source pollution management, waterway protection and fit-for-purpose source scheme, multi-functional infrastructure and public awareness. Although their framework was comprehensive and covered the shortcomings of the previous proposed frameworks, it did not vividly point out the appropriate measure for every single step of water management. Singapore has proved itself as a premier country in sustainable stormwater management recently. Singapore's framework, which is called ABS water framework, consists of three major categories: regulatory and administrative framework, technology development and industry capacity. According to this framework, a water management initiative is not implemented unless the requirements of these three categories are met consecutively. The first category comprises policies in regard with using reservoir, catchments precinct and regional stormwater management and compliance validation. The second category involves R&E activities and at the third category

Table 1. Environmental impacts of conventional urban water systems.

Impact	Description	Urban water relation
Carcinogens	Damaging human health	Heavy metals in effluents
Climate change	Temperature variation, sea level changes	Waste water emissions (Greenhouse Gas)
Eco toxicity	Eco toxic substances damage air, water and soil	Eco toxics exist in wastewater
Minerals and fossil fuels	Gradual depletion of minerals	Water treatment requires consumption of minerals
Respiratory organics	Pollutants cause respiratory disease	Water treatment substances include respiratory organics
Eutrophication and acidification	Deposition of inorganic substances will damage environment	Substances in water or waste water treatment are subjected to acidification
Ozone layer	Ozone layer depletion	Water and waste water treatment substances affect ozone
Land use	Land cover changes affects ecosystem	The urban water infrastructures change landuse

Table 2. Differences between elements of conventional and sustainable WSUD water management (Peter et al., 1996).

Comparison of conventional and sustainable water management practices in different categories		WSUD
Water supply		
Conventional	Scheme water in standard fashion and groundwater bores for POS	The integrated management of the three urban water streams of potable water, wastewater and stormwater.
Sustainable	Supplemented by community groundwater schemes as a secondary supply for all outdoor use in households and for POS	
Water conservation		
Conventional	Neither interior nor exterior water conservation devices	The integration of the scale of urban water management from individual allotments and buildings, to precincts and regions.
Sustainable	Application of low water use applications and technologies	
POS		
Conventional	Design dominated by high water use plants	The integration of sustainable urban water management into the built form, incorporating building architecture, landscape architecture and public art.
Sustainable	Design based on WSUD techniques	
Sewerage		
Conventional	Standard reticulation systems	
Sustainable	Full recovery of water used in POS rather than direct linkage to main	The integration of structural and non-structural sustainable urban water management initiatives.
Stormwater		
Conventional	Large sums and piped systems	
Sustainable	Application of WSUD and BMP concepts	

technical requirements as well as training requirements are encompassed. Framework of stormwater management in city of Oslo in Norway encompasses a series of conditions and procedures in which treat water from the source to sewage. In this framework, water treatment is initially checked in order to comply with the elements of boundary conditions (Economic, social and environmental considerations) and control the performance within the boundary condition. In simpler terms, all management initiatives must be economic, social and environmental friendly so that they can yield a number of positive results. At the first stage of this framework, water is treated through the water supply and source point. Then water pipes and distributors are monitored and improved. Then wastewater of the consumer is managed and subsequently drainage systems are managed for reusing or recycling purposes in following stages. This framework however, is based on implementation of technologies in every stage for water management (Venkatesh and Helge, 2011).

EVALUATION CRITERIA

Basically, among several variables in regard with stormwater management, three types of objectives are highlighted: common good interests, sectoral interests and operational interests. Table 4 shows these interests and the associated objectives. As indicated in this table, objectives are of varying importance and each variable has its own weight. The objectives are ranked in order of importance. The highest achievable weight is absolute.

Table 3. Benefits of implementation of WSUD and its initiatives in outstanding countries.

Benefit	Amount (Or Rate)	Reference	Benefit	Amount (or Rate)	Reference
Wastewater amount reduction	50%	Andrew and Grace (2000)	Urban water management	40% Optimization	Seong-Rin et al. (2010)
Stormwater management	Twice faster				
Water consumption reduction	25% Reduction per capita	Gardner (2003)	Rainfall effects and mitigation	80% Mitigation	Burns et al. (2011)
Total suspended solids	80% Reduction	City of Melbourne WSUD guideline (2008)	Source control practices	More attractive than traditional water supply and stormwater infrastructure	Coombes et al. (2000)
Total nitrogen	45% Reduction				
Phosphorous amount	45% Reduction				
Litters of urban water	75% Reduction				
Green-house emission	2 To 4 times less than conventional ones				
Life cycle environmental impacts	40% Reduction	Fagan et al. (2010)			
Ecological footprint	Negligible impact				

Therefore the traditional water management system will convert from an ordinary system to a resilient system which is interpreted as Tony (2007):

i) The amount of disturbance the system can absorb and still remain within the same state;
ii) The degree to which the system is capable of self-organisation (versus lack of organisation, or organisation forced by external factors); and,
iii) The degree to which the system can build and increase the capacity for learning and adaptation.

Resilient system leads to a water sensitive city with following characters: (Tony and Rebekah, 2000):

(i) access to a diversity of water sources underpinned by a diversity of centralised and decentralised infrastructure;
(ii) provision of ecosystem services for the built and natural environment; and
(iii) A socio-political capital for sustainability.

The use of weights allows determining the PI on the basis of satisfying the objectives set. Equation (1) presents the overall structure for determining the PI during dry weather and storm events.

$$\mathrm{PI}_c = \sum_n O w_o, \quad \sum_n w_o = 1, \tag{1}$$

in which, PI: dimensionless performance index; c: system loading condition (dry weather or storm event); n: number of objectives; O: objective; W_o: weight of objective O. Satisfying objectives can be determined on the basis of meeting requirements, expressed in terms of water-system variables. Again, weighting is possible at this level. Subsequently, meeting the objectives O can be determined, using Equation (2):

$$O = \sum_m I w_I, \quad \sum_m w_I = 1, \tag{2}$$

in which, O: ratio of meeting objective; m: number of performance indicators; I: performance indicator; WI: weight of indicator, I.

CASE STUDY

City of Putrajaya is located at about 25 km south of Kuala Lumpur, the capital of Malaysia, and is anticipated to accommodate a population of 330,000. The centre-piece of the city is an artificial lake surrounded by 20 planning precincts. The primary goal of the stormwater management plan is to facilitate coordinated management of stormwater within each development precinct of Putrajaya to maximise

Table 4. interests of objectives and associated priority.

Interest	Priority	Description	Weight
Flood prevention (Common good)	1	Secure health and safety of people	Absolute
	2	Reduction of flooding material loss	Absolute
Ecology/nature/water recreation (Sectoral)	3	Reduction of pollution of surface water, canals, ponds and main rivers pollution, discharge of WWTPs	10 to 30
Water management (Operational	4	Reduction of operation cost	1
	5	Reduction of sedimentation	
	6	Reduction of odour problems	

ecological sustainability, mitigate development impacts, minimise the potential for future impacts, enhance water sensitive urban design principles, maximise social and economic benefits using sound stormwater, Enhance aesthetic and ecological values into stormwater management infrastructure and provide opportunity for the community to gain an appreciation of water as essential element of the urban environment (Angkasa, 1996). Principal features of Stormwater Management System for City of Putrajaya is based on Integrated strategy, meaning that, the stormwater management strategy is based on the philosophy of Avoiding pollution whenever possible through source control measures, Controlling and minimizing pollution by means of in-transit and end-of-pipe control methods where pollutant generation cannot be feasibly avoided, Managing the impacts of stormwater pollution by managing receiving waters and their appropriate utilization as a last resort (Angkasa, 1999). Structural measure include four major steps: 58 km of trunk main for drainage conveyance system, planting of macrophytes in wetlands and riparian parks for aesthetic perspective, 300 gross pollutant traps for water quality control system, and integration of 400 ha lake and 7 detention basins for stormwater detention system. Non-structural measure also comprise of Source controls, In-transit controls, and end-of-pipe controls. Seven measures taken into account of source control are: Community awareness, Landuse planning and regulation, construction sites management, Isolation of high pollutant source areas, Permissible discharge, Street cleaning and Sewerage management. In the second level (In-transit controls) measure consist of: detention basins, swale drains and GPTs. Two types of GPT initiatives are used including Closed GPTs: enclosed units used for underground drains and Open GPTs: open systems used for open channels. End of pipe measure include Reparian parks, lake and wetlands, submerge outlets (Angkasa, 1996; Angkasa, 1999). The result of implementation of the initiatives are evaluated by the method porposed by Johannes and Arnold (2002). Their proposed method was used to evaluate the urban water drainage system of Rotterdam city, Holand (Johannes and Arnold, 2002). The results for City of Putrajaya show a great conformity with expected results of a stormwater management system, with negligible discrepancies. Other countries enjoyed similar results of implementation of stormwater management initiatives as malaysia enjoyed, as shown in Table 5. Results of the case study of Malaysia have shown a great potential of desirable effect on the water management system (Putrajaya Wetlands Book, 1999).

RESULTS AND DISCUSSION

In this study the environmental impacts of conventional urban water systems were pinpointed which necessitates the need for new water practices. Followingly, the concept of WSUD was discussed and the differences between the WSUD and conventional systems were highlighted. Proposed Frameworks and initiatives of WSUD in regard to stormwater initiatives, which were implemented in different countries shows a wide range of benefits in terms of water management practice. The case study of city of Putrajaya in Malaysia, using the most important criteria in this study, showed a remarkable contribution in mitigation of stormwater and towards a sustainable water resource management practice by more than 60% runoff and pollutant mitigation as well as flood occurrence likelihood. However, the country of Iran has several similarities with Malaysia in terms of flood threats. Although there are major differences in location and topography features of these two countries, imitating general features of Malaysian management scheme will yield similar positive results in Iran as well. Critical issues in regard with water management practices and also critical issues on selecting the practices are similar in both countries and therefore using the exerience of Malaysia in the same field is a sound and wise decision. However, it must be considered that the essential prerequisites of WSUD, such as Regulatory Framework; Assessment and Costing; Technology and Design; and Community Acceptance and Governance must be taken into account first. Meanwhile, implementation of WSUD initiatives require integration with construction practices where many constructed projects have faced problems caused by a combination of poor engineering water design and

Table 5. Result of the analysis of the Putrajaya Stormwater Management System.

Objective	Description	Priority	Achievement		
Flood prevention	Secure health and safety of people	1	Absolute	Runoff water detention	60% of runoff is intercepted by Wetlands
	Secure safety of asset	2		Flood occurrence likelihood	70% enhance
Ecology and water recreation	Surface water pollution reduction	3	Satisfactory	Acceptable quality of water runoff entered to lake	Enhancement from ordinary <7% to 40% of total runoff
	Canals and ponds			Total area drainage	51.9 Km^2
	Rivers			Pollutants controls including: rubbish, debris, suspended sediment	
Water management	Reduction of cost of operation	4	Satisfactory	Nitrogen	60% reduce in the pollutants
	Sedimentation reduction in sewers	5		Oil, grease and oil derivatives	

construction practices.

REFERENCES

Andrew S, Grace M (2000). Integrated Urban Water Cycle, National Conference on Water Sensitive Urban Design–Sustainable Drainage Systems for Urban Areas, Melbourne 2000.

Australian Environmental Water Management (2012). National Water Commission, Common Wealth of Australia. May.

Brown R, Keath N, Wong T (2009). Urban water management in cities: Historical, current and future Regimes. Water Sci. Technol. 59(5):847–855.

Eighth Malaysian Plan (2008). Ministry of Natural Resources, Putrajaya, Malaysia.

Gardner EA (2003). Some examples of water recycling in Australian urban environments: A step towards environmental sustainability, Water Science and Technology. Water Suppl. 3(4):21–31© IWA Publishing.

Iran Ministry of Energy (2011). http://www.sabainfo.ir/fa/news/63308.

Johannes MU, Geerse A, Lobbrecht H (2002). Assessing the performance of urban drainage systems: 'General approach' applied to the city of Rotterdam. Urban Water 4:199-209.

Newman P, Mouritz M (1996). Principles and planning opportunities for community scale systems of water and waste management. Desalination IO, pp. 339-354.

Organization of Geology and Mining Explorations, Iran (2011). Geosci. J. No. 82.

Putrajaya Wetlands Book (1999). Perbadanan Putrajaya and Putrajaya Holdings Sdn. Bhd. July.

Tony HFW (2007). Water Sensitive Urban Design – The Journey Thus Far. J. BDEP Environ. Des. Guide.

Venkatesh G, Bratteb H (2011). Energy consumption, costs and environmental impacts for urban water cycle services: Case study of Oslo (Norway). Energy 36:792e800.

Wong THF (2006). An Overview of Water Sensitive Urban Design Practices in Australia. Water Pract. Technol. 1(1) © IWA Publishing 2006 doi: 10.2166/WPT.2006018.

17

Groundwater conditions and hydrogeochemistry of the shallow Benin Formation aquifer in the vicinity of Abraka, Nigeria

Irwin Anthony Akpoborie[1] **and Oghenevwede Efobo**[2]

[1]Department of Geology, Delta State University, Abraka.
[2]Centre for Research in Water and Environment, Abraka.

Shallow borehole and dug well data are used in describing groundwater conditions in the vicinity of Abraka in the Nigerian coastal plain. Drill cuttings from ten boreholes show that the Abraka area is underlain by reddish brown unconsolidated sands, followed by a succession of grey- off-white medium grained sands of the Benin Formation. Hydraulic conductivity estimated from grain size analysis of cuttings obtained from typical borehole screened horizons range from 0.12 to 0.19 msec^{-1}. Regional groundwater flow is from north east to southwest with local distortions on this regional trend resulting from ground water abstraction in densely populated areas. Maximum TDS in ground water was recorded at 28 and 85 mg/l from dug wells. The trilinear plots of major ions in water indicate a mixing of mainly sodium chloride and calcium chloride water types. The stiff diagrams are also suggestive of possible stratification of water chemistry with depth. Borehole water quality is well within WHO and Nigerian drinking water quality standards while that from dug wells contains minimal levels of fecal coliform. Ground water and surface water are determined to be suitable for irrigation. It is also shown that the quality of water in the River Ethiope, TDS 6.6 to 8.09 mg/l, *Escherichia coli* occurrence at less than an average of 50 cfu/100 ml from selected recreation sites meets WHO standards for body contact recreation.

Key words: Benin Formation, River Ethiope, Abraka, Niger Delta, groundwater, water quality.

INTRODUCTION

The Abraka area is located in the western Niger Delta and underlain by the Benin Formation that bestrides the River Niger and stretches from west of the Lagos area to the Calabar Flank in eastern Nigeria. The formation which contains prolific aquifer horizons is heterogeneous and has been studied in some detail in the Port Harcourt area by Amajor (1991) who characterized the water bearing sandy layers of its upper (300 m) horizon and described rapid horizontal and vertical variations in lithology and hydraulic characteristics. Ibe and Njemanze (1998) also identify at least three aquifer horizons that 200 m of the formation in the vicinity of Owerri, also in the are separated from each other by clay layers in the upper eastern sector. West of the River Niger, Oteze (2011) and Akujieze and Oteze (2006) have described the water bearing upper horizons of the formation and identify lateral variations in hydraulic properties in the vicinity of Benin City. Longe (2010) also reports the presence of three aquifer horizons that are separated by thin clay layers in the upper 100 m of the formation in the Shomolu area of Lagos State. There is a dearth of similar studies in the Abraka area and existing ground water conditions in the underlying Benin Formation are not well understood. *Laissez faire* exploitation of ground water

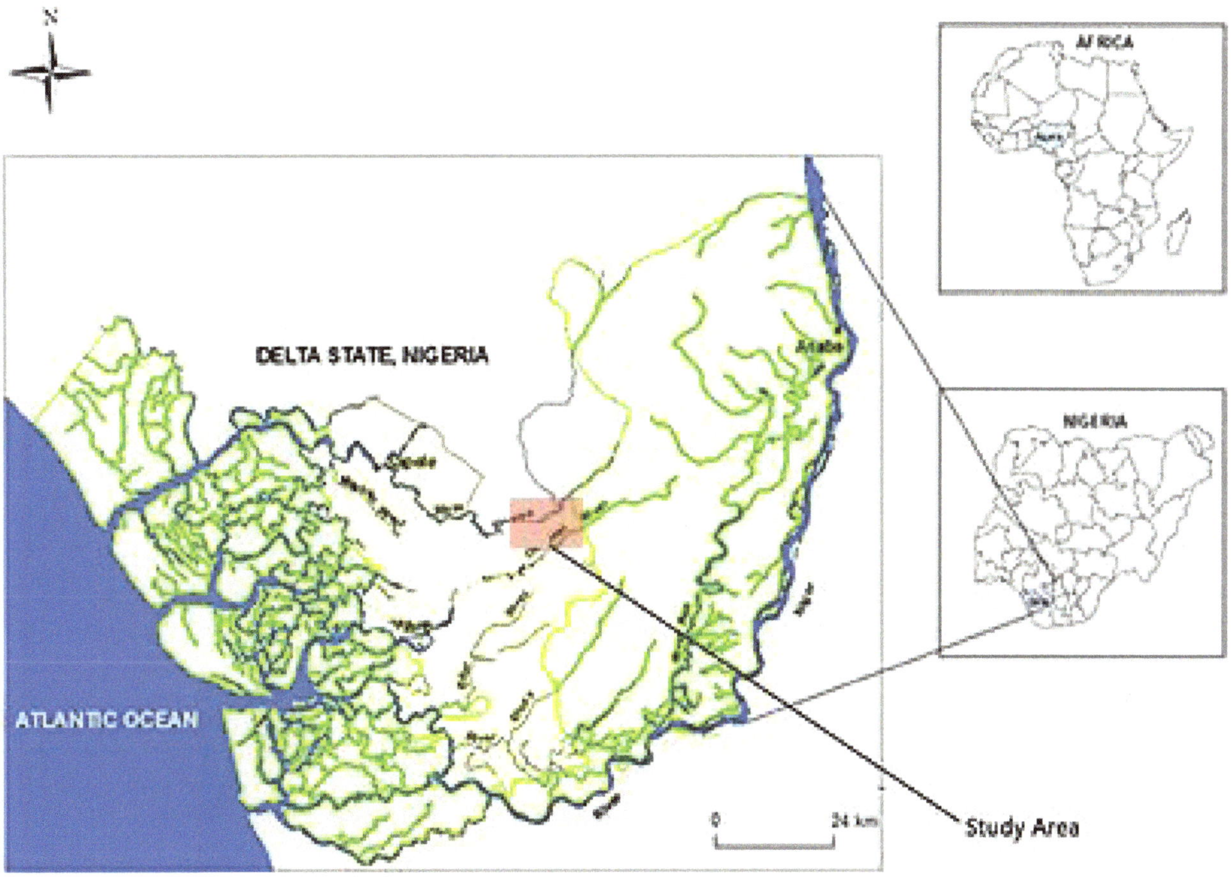

Figure 1. Map of Delta State Nigeria showing location of study area.

especially in this rapidly expanding university community that is witnessing rapid growth could potentially create problems for the ground water system in the long term. This is because in the absence of public water supply facilities to serve an estimated population of 90,000 (Oje and Origbo, 2012), all water supplies for domestic, commercial and industrial uses are obtained from privately owned boreholes that tap the shallow Benin Formation.

An appreciation of existing ground water conditions is not only desirable but a requirement for urban expansion and regional planning. The objective of this investigation is thus to gain an understanding of ground water conditions in the area, associated hydrogeochemistry, as well as determine the suitability of surface water and ground water for domestic use, irrigation, water contact recreation and industry. Furthermore, the results will add to and contribute to the expansion of current regional knowledge and understanding of this prolific, extensive and important aquifer.

STUDY AREA

The Abraka area, situated between latitude 5° 45' and 5° 50' N and longitude 6° and 6° 15' E is located on the south bank of the River Ethiope and is an agglomeration of several communities that are aligned linearly along the New and Old Sapele– Agbor highway (Figures 1 and 3). These communities include from the west, Oria, Ajalomi, Abraka PO, Ekrejeta, Urhuoka and Umeghe. Ajalomi, Abraka PO, Ekrejata and Urhuoka have grown in size and are now conjoined to form what is loosely referred to as Abraka Urban, the seat of all three sites of the Abraka Campus of the Delta State University. Southwards, the rural communities of Ugono, Aragba, Abraka Inland and Ughere - Uragbesa are as well, bona fide Abraka communities (Figure 3).

Physiography and climate

The Abraka area is a typical coastal plain terrain, monotonously lowland and flat with a gentle slope towards the Ethiope River. The climate is equatorial, hot (23 to 37°C) and humid (relative humidity, 50 to 70%). There is a dry season from about November to February, and a wet season that begins in March, peaks in July and October. Six -years (2000 to 2005) annual mean rainfall measured at the Delta State University weather station is

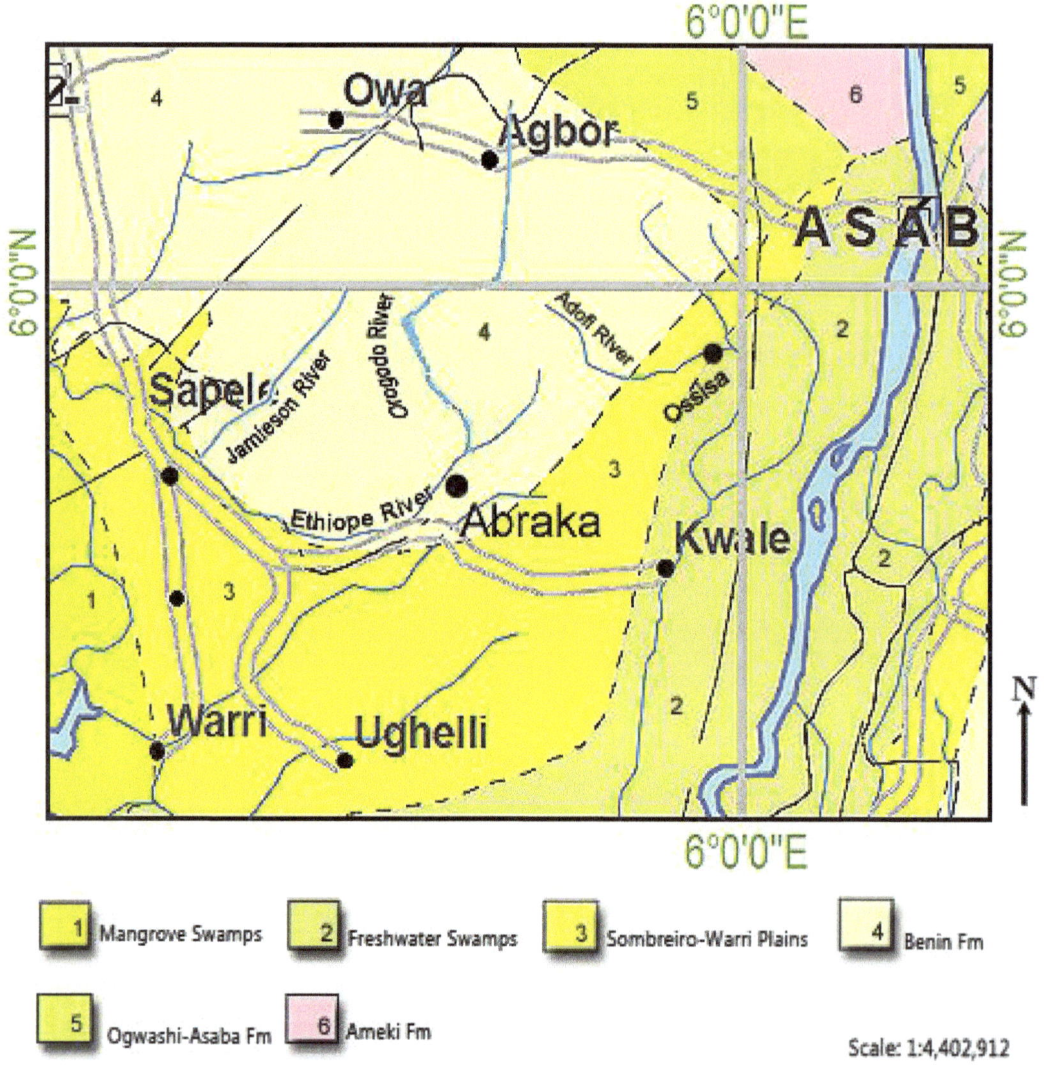

Figure 2. Regional geology of part of the Western Niger Delta showing the Abraka area (NGSA, 2006).

3317.8 mm. Vegetation is rainforest, most of which has been decimated and replaced with farmlands and secondary forest. However, lush, dense and swamp primary forest flanks the river banks.

Geology

The area is located in the Niger Delta Basin, a much studied and important petroleum province. Briefly, there is general agreement that the basin was formed as a result of an aulacogen type development that was triggered by the separation of the African and South American plates in the Jurassic. The resulting trough has been filled by a series of marine transgressions, regressions and deltas, the present Niger delta being the most recent. The basin fill has been described by Short and Stauble (1967), Reijers et al. (1996), Nwajide (2006), among many others and consists of three formations, namely, from the oldest to the youngest, the Akata Formation, Agbada Formation both of Eocene to Recent and the Miocene to Recent Benin Formation. However, west and just south of Abraka the Benin Formation is masked by the younger Holocene deposits of the Sombreiro-Warri Deltaic Plain, the Mangrove Swamp and Freshwater Swamp wetlands. These deposits which have not been assigned formal geological names because they are universally considered to be recent expressions of and a continuation of the Benin Formation are only identified by the physiographic terrains in which they occur. The aerial distribution of these delta top deposits coincides somewhat with the associated physiographic subdivisions shown in Figure 2. The inferred boundary between the Sombreiro–Warri Deltaic Plain and the Benin Formation outcrop as can be seen from Figure 2 passes through the Abraka area such that the southern

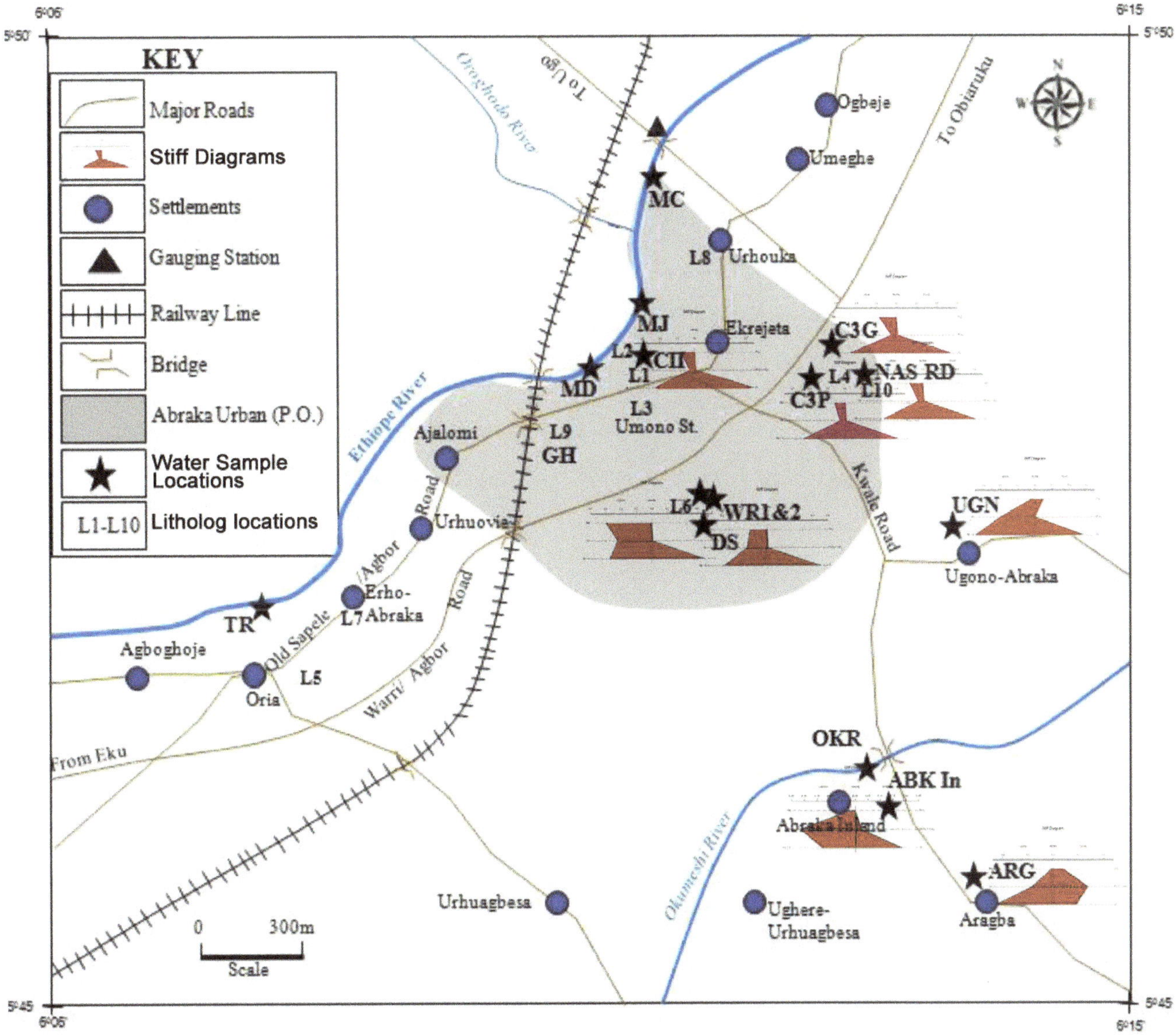

Figure 3. Borehole drill - cutting sample locations, water sample locations and associated stiff diagrams. L1(CII), Geology Lab; L2, Abraka Hall; L3, Omono Street; L4, Campus 3 New Administrative Block; L5, Oria; L7, Erho; L8, Urhuoka; L9, General Hospital; L10 (Nas Rd), Nassarawa Road; ARG, Aragba; ABK In, Abraka Inland; UGN, Ugono; CII, Geology Lab; C3G, Campus 3 Gate; C3P, Campus 3 Pool site; WR1, Winner's Road 1; WR2, Winner's Road 2; DS, Dump site; MC, McCarthy Beach; MJ, Majoroh Beach; MD, Mudi Beach; TR, The Turf; okr, Okemesi River; L6, Winner's Road (not WR1or WR2).

part of the area is underlain by Sombreiro-Warri Deltaic Plain deposits while the north rests on Benin Formation outcrop.

The Ethiope River, the most important physiographic feature in the area flows almost exclusively on the outcrop of the Benin Formation. The Benin Formation in the Abraka area is overlain everywhere by reddish brown colored ferruginized regolith that is usually no less than 2 m thick. This can be observed clearly at two locations on the Abraka – Ugo road (Figure 3). The first site is a large borrow pit located just before the bridge on the River Ethiope and from where material is being excavated for building and road construction and the second at the rail road crossing near the overpass on the Abraka – Ugo Road. At both locations, the formation is massive and deeply weathered.

METHODOLOGY

Drill cuttings were collected and described from ten sites (Figure 3) where shallow water supply boreholes were drilled for private home owners. Sieve analysis of drill cutting samples obtained from the screened and usually the lowermost horizon penetrated was undertaken at the laboratories of the Department of Geology, Delta

State University. The sampling procedure for groundwater involved the collection of replicates from six randomly located boreholes that could be accessed and three dug wells. Borehole water samples were collected after boreholes were pumped continuously for about 1 h. Surface water samples were also collected from selected recreational points along the River Ethiope, from the Okemeshi River as well as from a water filled quarry that is being used as a dump site. Sampling sites are shown in Figure 3. The set of samples designated for heavy metal analysis were stabilized *in situ* with nitric acid. Electrical conductivity and total dissolved solids were also measured *in situ* using the HACH conductivity/TDS meters, respectively. The pH was determined by means of a Schott Gerate model pH meter and temperature was determined using mercury–in-glass thermometer calibrated in 0.2°C units from 0 to 100°C. The Pye Unicam Atomic Absorption Spectrophotometer SP 2900 was employed in the determination of the heavy metals while the HACH Spectrophotometer was employed in determining the NO_3 ion using the cadmium reduction method. Na and K ion concentrations were obtained with a Jenway Clinical flame photometer. Sulphate content was determined by turbidimetry and Ca, Mg, HCO_3 and Cl with appropriate titrimetric methods (APHA, 1992). In order to determine the presence of total and fecal coliform, 100 ml of each water sample were passed through a membrane filter consisting of uniform pore diameter of 0.45 nm following which the membrane filter was placed in a petri-dish containing Mac-Conkey Agar and Eosin-Methylene blue Agar, in duplicate with the grid side up and incubated at 35 and 45°C for 18 to 24 h, respectively. Bacteria colonies if present were counted and expressed as numbers of coliform per 100 ml of water.

Four dug wells and six boreholes (Table 1) were selected on the basis of spread for depth to water level measurements. Depth to water level in each of the boreholes and dug wells was measured with an electronic well sounder. An Ertec model GPS instrument was used to determine wellhead coordinates. Because existing maps of the area are devoid of contours, averaged elevation readings from three GPS instruments at each site were used with the sparsely distributed benchmarks to approximate the elevation of each well location. Surfer 8 (Golden Software Inc., 2002) was employed in generating water table contour lines of equal head.

RESULTS AND DISCUSSION

Lithology

Cuttings recovered from boreholes show a succession of fine grained reddish brown sands underlain by typically clean, grey - off white and fine - medium - coarse grained, angular to sub- rounded well sorted quartz sands. In addition to remarkable lithological similarity, a layering and cyclical fining upward sequence is apparent at some locations (Figure 4). The consistent absence of clays gives an indication of the potential high vulnerability of the shallow aquifer to contamination from surface sources.

Groundwater conditions

Water table conditions prevail in the Abraka area. Depth to water varies from about 4 m at Abraka Inland to 26 m at Oria. Water table head above mean sea level determined at each location is shown in Table 1 and associated water table contour lines of equal head in meters above sea level are plotted in Figure 5. Sustained groundwater abstraction in the more densely populated Abraka Urban has resulted in a depression of the water table in the town centre that stretches from Oria through Winner's Road to Ugono and towards which groundwater is flowing from virtually all directions. This has thus caused a local distortion on regional south westward groundwater flow (Akpoborie, 2011; Aweto and Akpoborie, 2011) that mirrors the seaward slope of the coastal plain. There also appears to be a mound at the Abraka Inland community that extends to Umuebu from which water is moving in all directions.

Hydraulic characteristics

Sieve analysis curves for cuttings retrieved from boreholes at Geology Laboratory, Oria, Erho, behind the General Hospital and Nassarawa Road, locations L1, L5, L7, L9 and L10 respectively in Figure 3, were plotted as shown in Figure 6. Cuttings from all ten boreholes sampled in this investigation were all remarkably similar upon visual examination in terms of textural attributes and which similarity is reflected in the curves for the five representative samples shown in Figure 6. Following Odong (2007) and Fetter (2004), hydraulic conductivity was estimated with the Hazen (1892) approximation:

$$K = C\,(D_{10})^2$$

Where K is hydraulic conductivity in cm/s, D_{10} is the effective grain size in cm, C is a coefficient that is based on the aquifer matrix.

Estimated values of K obtained with C = 6 (Uma et al., 1989) range from 0.12 to 0.17 cm/s with uniformity coefficient ranging from 1 to 1.4. It should be stressed that all sampled boreholes are located on the outcrop of the Benin Formation and that maximum depth penetrated and sampled is a shallow 18 m which makes the results of limited application especially with respect to the reputed average 2000 m thickness of the Benin Formation. However, they are interesting to the extent that in the absence of formal aquifer tests, this is the first indication of hydraulic properties of this horizon of the Benin Formation in this area.

Water chemistry

Characteristics of surface and ground water

The physical and chemical characteristics of surface water from different recreational sites located on the south bank of the Ethiope River are shown in Table 2. One sample collected from the bridge crossing on the Okumesi (Warri) River, near Abraka Inland, and another

Table 1. Depth to water level measurements.

Eastings	Northings	Location	Well type	DWL	Head
6.1369	5.731333	Aragba	Dug Well	5.1	9.01
6.125633	5.746067	Abraka Inl	Dug Well	3.9	13.22
6.137133	5.782383	Ugono	Dug Well	4.5	5.7
6.097833	5.792	Geology Lab.	Borehole	3.69	16.3
6.103472	5.778833	Winners Rd	Borehole	14.63	4.27
6.117789	5.789828	Delsu, Site 3 Pool	Borehole	16.82	12.77
6.063889	5.75025	Oria	Borehole	16.03	3.87
6.148333	5.783875	General Hospital	Borehole	13.41	10.36
6.159883	5.752183	Umuebu	Dug well	4.2	11.42
6.148333	5.846183	Morka Primary School, Obiaruku	Borehole	31.2	3.10

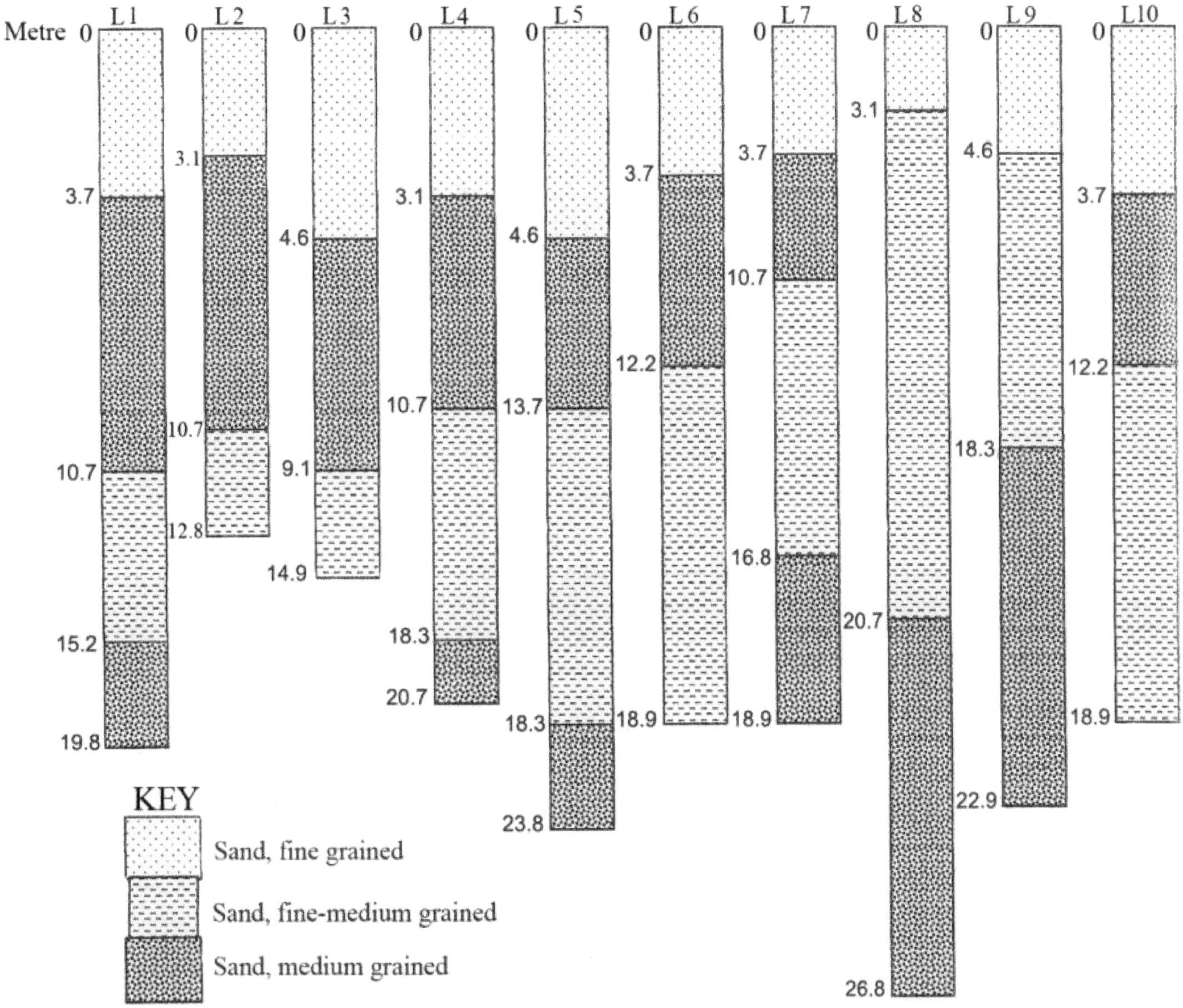

Figure 4. Lithologs from water supply wells in Abraka. L1, Geology Lab; L2, Abraka Hall; L3, Omono Street; L4, Campus 3 New Administrative Block; L5, Oria; L6, Winner's Road; L7, Erho; L8, Urhuoka; L9, General Hospital; L10, Nasarrawa Road.

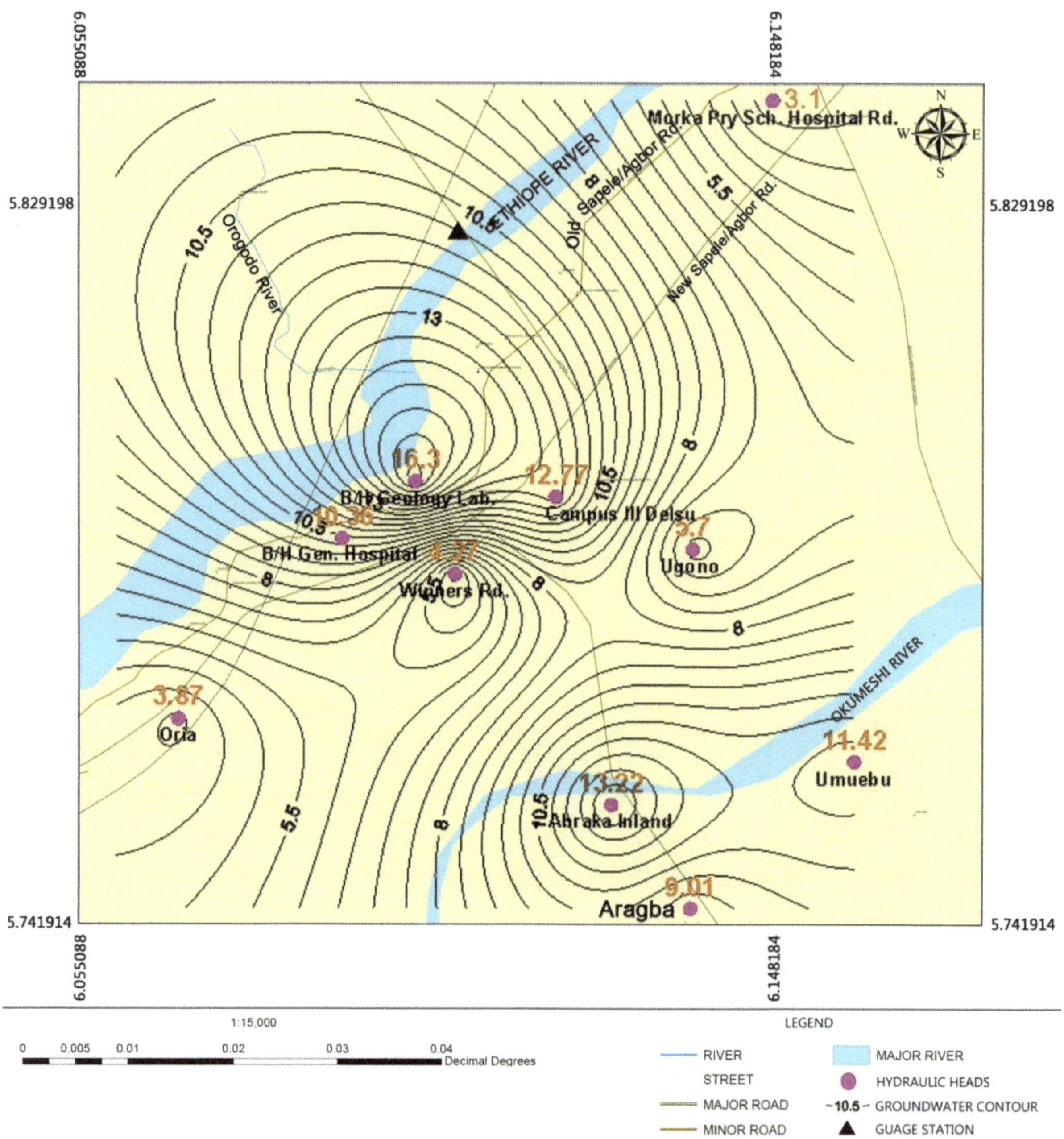

Figure 5. Water table contour lines of equal head in meters above sea level in the vicinity of Abraka.

from the Jamieson River are also included for comparison. Sampling sites except for the Jamieson River are shown in Figure 3. TDS at all sites is very low and all ions show very low levels of occurrence. However, TDS from Ethiope River is much lower at mean 7.18 mg/l than Okumeshi River water which returned a TDS value of 16.5 mg/l. Indeed, Kaizer and Osakwe (2010) in a regional comparative study of water quality from five rivers in Delta State show that the Ethiope River contains the lowest total dissolved solids. Water in both the Ethiope and Okumeshi Rivers is also slightly acidic, which as reported from previous studies (Efe, 2005; Olobaniyi and Efe, 2007; Kaizer and Osakwe, 2010) is reflective of associated gas flaring in the Niger Delta petroleum province. The physical and chemical characteristics of ground water collected from boreholes and dug wells in the Abraka area are shown in Table 3. TDS is low at mean 26.67 mg/l (standard deviation 21.3) and as expected, higher than TDS in surface water. The highest TDS at 85 mg/l was recorded from the dug well at Aragba, but water from this dug well is not represenative as it is uncovered and in the wet season regularly receives runoff from surrounding areas. Ground water is also weakly acidic at an average pH of 6.64. The acidity

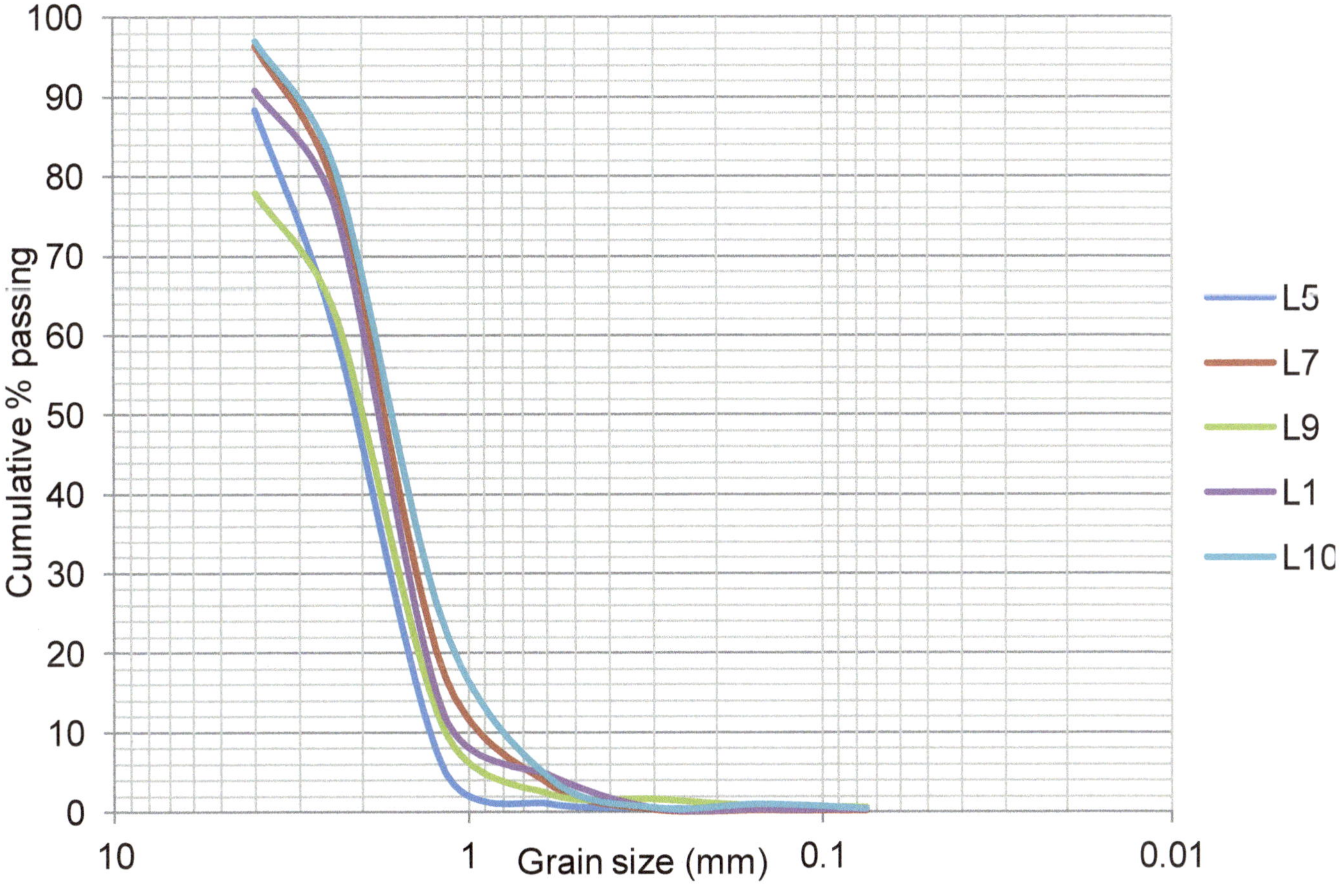

Figure 6. Typical grain size curves from screened horizons in selected boreholes in Abraka.

Table 2. Physical, chemical and microbiological characteristics of surface water from the Ethiope River and the Okumeshi River (mg/l, except where indicated), bacteria counts in (cfu/100 ml).

Parameters	Sampling points					
	MC	MJ	MD	TR	OKR	JMR
Temperature (°C)	27.8	28.2	27.5	27.5	28.1	-
TDS	6.62	8.09	7.21	6.83	16.5	8.91
EC (μ/cm)	12.74	15.51	13.87	13.14	33	17.85
pH	5.27	5.14	5.18	4.92	6.33	8.0
Ca	0.96	3.45	1.84	1.23	3.99	0.41
Mg	0.32	2.72	4.67	2.72	0.20	0.88
Na	1.91	2.33	2.08	1.97	0.86	0.42
K	0.51	0.62	0.55	0.53	1.46	0.10
HCO_3	1.01	3.05	0.61	0.16	2.20	0.65
SO_4	1.24	0.25	0.12	0.12	6.60	1.01
Cl	4.1	4.02	4.3	4.07	3.33	2.26
NO_3	0.00	0.65	0.26	0.18	0.20	1.25
Total bacteria	3.3×10^2	1.7×10^3	4.2×10^3	3.0×10^3	Nd	
Total coliform	11	35	45	80	Nd	0.0
Total *E. coli*	4.0×10	1.5×10	2.8×10	4.0×10	Nd	0.0
Total Salmonella	0	0	0	0	Nd	

MC = McCarthy Beach, MJ = Majoroh Beach, MD = Mudi Beach, OKR = Okumeshi River, TR = The Turf, JMR = Jamieson River, Nd = Not determined.

Table 3. Physical and chemical characteristics of groundwater (mg/l except where indicated).

Parameters	Sampling locations										
	CII	C3G	C3P	NAS Rd	WR2	WR1	UGN	ABK In	ARG	DS	SON
Temperature (°C)	28	27	28	27	28	27	27	27	28	28	
H	7.1	7.4	6.9	7.13	5.7	5.6	6.25	6.39	7.29	8.9	6.5-8.5
TDS	28	25	27	26.67	41.2	20.9	20	13	85	424	500
EC (μ/cm)	58.2	52.5	47.8	52.83	82.8	41.08	40	26	170	849	
Turbidity (NTU)	0	0.4	0.1	0.17	0.32	0.00	0.25	0.19	0.20	6.95	5
Ca	1.0	1.01	1.2	1.07	9.0	2.8	4.05	4.32	8.46	65.3	
Mg	1.8	2.1	1.5	1.8	7.3	1.4	0.26	0.30	0.41	54.6	0.2
K	3.5	3.6	2.0	3.03	4.32	2.43	4.09	0.84	9.95	74.4	
Na	7.2	8.5	9.8	8.5	10.11	5.21	5.58	2.49	20.34	109.1	
Cl	24.0	19.0	22.0	21.67	24.84	12.54	9.99	6.67	16.66	254.7	
HCO_3	1.3	1.0	1.4	1.23	5.01	3.22	<1.0	3.17	47.82	87.09	
SO_4	0	0	0.1	0.03	0.04	0.00	2.39	<1.0	10.07	9.7	200
NO_3	0	0	0	0	0.29	0.15	0.26	0.19	0.29	5.12	10
Pb	0	0.01	0	0	0	0	ND	ND	ND	0.04	0.01
Zn	0.1	0.2	0.1	0.13	0.1	0.13	ND	ND	ND	ND	3
Cd	0.01	0	0.02	0.01	0.001	0.00	ND	ND	ND	0.006	0.003
Fe	0.04	0.06	0.03	0.043	0.18	0.09	ND	ND	ND	7.65	0.3
Cr^{6+}	0	0	0	0	0.00	0.00	ND	ND	ND	0.008	0.05
Total coliform (cfu/100 ml)	0	0	0	0	6	0	8	8	8	54	
Faecal coliform (cfu/100 ml)	0	0	0	0	4	0	4	6	4	42	0

ARG, Aragba; ABK In, Abraka Inland; UGN, Ugono; CII, Geology Lab; C3G, Campus III Gate; C3P, Campus 3 Pool site; Nas Rd, Nassarawa Rd; WR1, Winner's Road 1; WR2, Winner's Road 2; DS, Dump Site; SON, Nigerian Drinking Water Standards.

is retained from that of recharging rainfall which as explained has been attributed to gas flares.

The unusually low TDS in the Ethiope River water could possibly be explained by the fact that its catchment is exclusively situated on the outcrop of the Benin Formation which is directly recharged perenially by rainfall. The Jamieson River its only tributary (Figure 2) was also sampled at its headwaters near Ugo, Edo State as part of this investigation and returned similar low TDS of 8.91 mg/l (Table 2). Reported regional variation in groundwater TDS (Olobaniyi and Owoyemi, 2004; Emeshili, 2008; Sarner PFM, 2011; SPDC, 2004, 2008) indicate that shallow groundwater from Benin Formation outcrop areas appears to be relatively lower in mineral content than that from areas where the formation is masked by younger layers.Thus, following from Domenico (1972), it does appear that while recharging, rainwater may have over time leached the upper layers of the Benin Formation of all soluble mineral matter with the resulting low total dissolved solids in shallow groundwater, it continues to dissolve mineral matter from overlying younger deposits where they occur. Because regional groundwater flow is from northeast to southwest, that is from Benin Formation outcrop areas southwestwards and contributes more than 80% to total flow of the River Ethiope (Akpoborie, 2011), it follows that river water would also be low in dissolved solids. Furthermore, this high contribution of low TDS ground water to stream flow could be partly responsible for the remarkable and unique clarity of water in the Ethiope River as well as the Jamieson River.

With respect to the microbial content, water from the boreholes in both campuses of the university at Abraka CII, C3G, C3P and Nas Rd is free of coliform bacteria. However, one borehole at Winners Road, WR2 is possibly receiving water from the heavily contaminated water in the abandoned quarry (DS, Table 3) as indicated by the relatively higher coliform count. That this contamination is limited for the time being to the immediate vicinity of the quarry is borne out by the fact that borehole WR1 which is also near the quarry but further away from it than WR2 is coliform free. Also, water in all the dug wells in rural Abraka at Ugono, Abraka Inland and Aragba contains faecal coliform which may be explained by the fact that the wells are shallow, uncovered and are often left in an unsanitary state.

Major ion geochemistry

Major ion content is used in interpretive diagrams (Hem, 1991; Piper, 1944; Keheew, 2001) to describe, classify and determine geochemical trends in ground water and has subsequently been used in several ground water

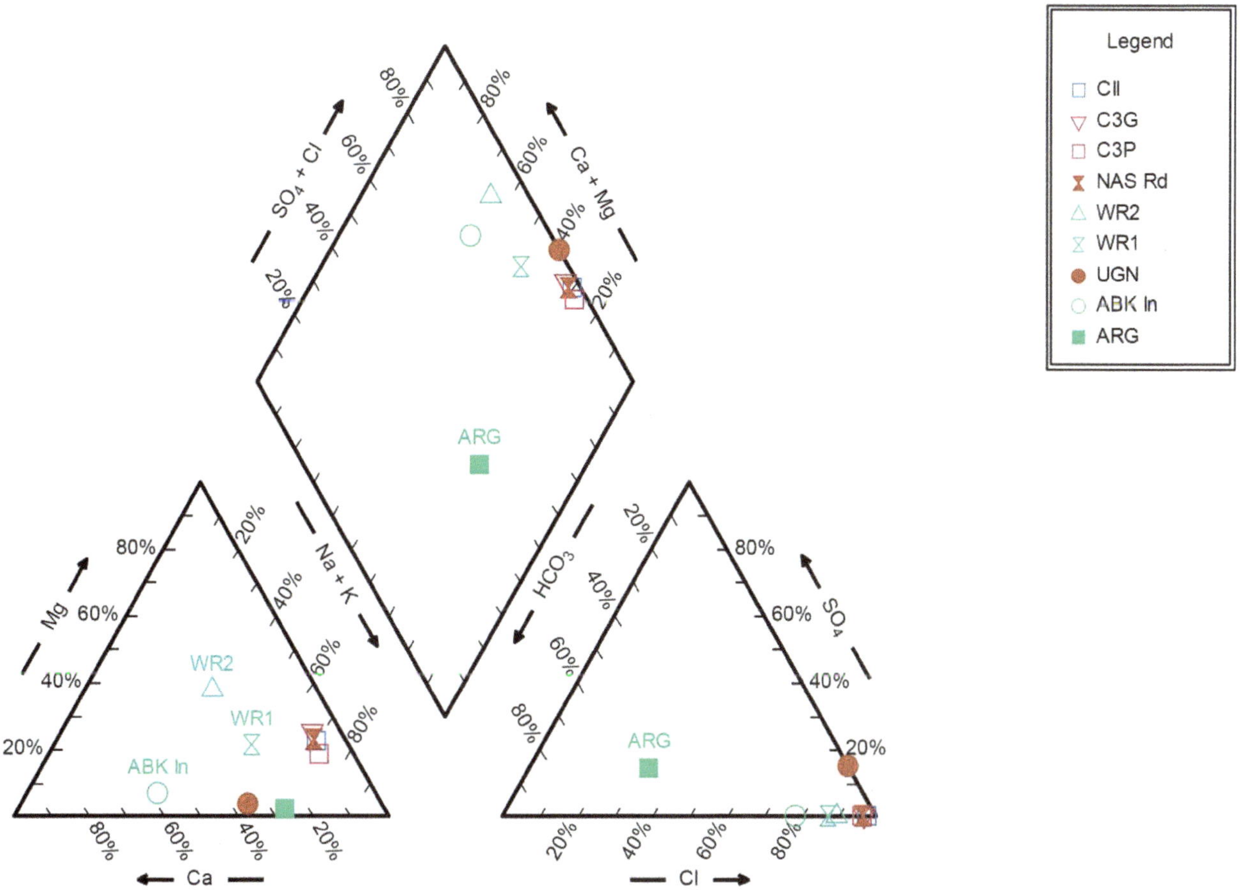

Figure 7. Piper diagram of groundwater analyses from Abraka area. ARG, Aragba; ABK In, Abraka Inland; UGN, Ugono; CII, Geology Lab; C3G, Campus 3 Gate; C3P, Campus 3 Pool site; WR1, Winner's Road 1; WR2, Winner's Road 2; Nas Rd,

studies (Arabi et al., 2010; Naseem et al., 2010; Smith and Wahl, 2003; Rafael et al., 1998). The Piper diagram plots of the chemical constituents in ground water presented in Table 3 are shown in Figure 7. All samples except Aragba (ARG) plot in the right hand quadrant of the diamond that stretches from about 20 to 60% calcium + magnesium, while sulphate + chloride are between 70 to 80% indicating a mixing trend as well as direct recharge by rainfall (Petalas and Diamantis, 1999). Ground water from both campuses of the University overlap at the 25% point on the Ca + Mg axis while Ugono plots slightly higher at 40%. Keheew (2001) classifies hydrochemical facies using the Piper plot and on the basis of which ground water from the boreholes at the university and at Ugono, UGN, may be described as the sodium chloride type while at Abraka Inland (ABK In), ground water is the calcium chloride facie. At Aragba, the water is dominated by sodium and potasium which constitute about 70% of the cations while the bicarbonate and chloride ions occur respectively at 60 and 35%; hence, the water is uniquely for this area, the sodium bicarbonate type. As has been noted however, this sample is from an unringed dug well that is always left uncovered and may not be considered representative of groundwater in the area.

The Stiff diagrams (Hem, 1989; Kehew, 2001) plotted for each sampling location and superimposed on Figure 3 confirm these facie classifications as well as further highlight the differences in shallow dug well water and water obtained from deeper in the aquifer from drilled wells. Ground water from boreholes located at the university for example presents a distinctive inverted funnel shape that may be considered representative of native ground water in Benin Formation outcrop area. At WR1, the funnel shape is mildly distorted due to relative enrichment in calcium occasioned by recharge from the nearby waste dump. At WR2, the borehole location closest to the dumpsite, ground water is already an admixture of native ground water and dumpsite water, and the funnel shape is completely distorted. In comparison to groundwater obtained from deeper in the aquifer, dug well water from shallower horizon and outside Benin Formation outcrop area at Aragba and Ugono present distintively different shapes than the university campus wells which is suggestive of possible hydrogeochemical stratification.

Suitability for domestic water supply

The depth at which water would be encountered in a well has always traditionally influenced the choice of domestic water supply source in the Niger Delta. The shallow water table at Umuebu, Aragba and Abraka Inland (Table 1), explains the preponderance of manually dug wells in these areas. Boreholes predominate at Abraka Urban and other areas located on the outcrop of the Benin Formation where the water table is deeper. The chemical parameters for both water supply sources in Table 3 are below the Nigerian guidelines for drinking water (SON, 2007). However, the presence of cadmium detected at three locations CII, C3P and NAS Rd where it occurs above the guideline value of 0.003 mg/l is worrisome and needs to be monitored because of its toxicity (Ifeagu and Ayankora, 2012). The source of cadmium enrichment in the area is unknown but its presence in stream sediments at elevated levels has been reported from nearby Ndokwa area by Emeshili (2008). Akpoborie (2011) and Aweto and Akpoborie (2011) have drawn attention to the potential problems that could be associated with continous ingestion of low TDS water in the long term especially when combined with very low levels of calcium and magnesium in drinking water supplies as reflected in the values for both ions in Tables 2 and 3.

Calcium and magnesium are essential micronutrients and their deficiency in drinking water supplies especially when dietary habits exclude foods rich in calcium and magnesium has been associated with increased risks of osteoporosis, nephrolithiasis (kidney stones), colorectal cancer, hypertension and stroke, among others (Ong, 2006; Kozisek, 2006). Enrichment of these micronutrients is therefore recommended if the water in this area is to be used as a source of raw water for public water supply schemes. Finally, gradual impairment of ground water with coliform bacteria (WR2, Table 3) by water from the abandoned quarry at Winner's Road confirms the potentially high vulnerability of the shallow and upper layers of the Benin Formation in this area to contamination as suggested by the predominantly sandy lithology.

Suitability for water contact recreation

The River Ethiope is the most prominent natural feature in Abraka area. The waters of the river are crystal clear, and the white river sands are of clean quartz grains with the result that the river is like a giant natural aquarium where fish and other organisms may be observed. This natural beauty is the main attraction for the rapid development of the hospitality industry along the river's south bank. Indeed, the river is an international tourist destination site and a gallery of spectacular pictures from some of the recreational beaches has been posted on the World Wide Web by Emiel Jegen (Jegen, 2004). In order to establish the suitability of river water for full body contact recreation, water samples from the more popular sites (Figure 3) were screened for the presence of coliform bacteria over a 30 day period in March 2011 in the middle of the dry season when the sites enjoy maximum patronage. The results which are shown in Table 2 indicate that occurrence of *Escherichia coli* is less than 50 cfu/100 ml at all sites. This level of occurrence is well below the guideline value of 126 cfu/100 ml for *E. coli* specified by the U.S. Environmental Protection Agency (USEPA, 2003) and affirmed by Wade et al. (2003) who reviewed the issue of using *E. coli* as the choice indicator of fecal contamination of bathing waters. Omo - Irabor and Olobaniyi (2007) report a mean *E. coli* occurrence of 145 MPN/100 ml in River Ethiope water with a range of 39 to 502 MPN/100 ml but do not provide results from individual sample sites thus precluding direct comparison with results shown in Table 2.

With respect to chemical quality, the World Health Organization (WHO, 2003) and the Australian National Health and Medical Research Council (NHMRC, 2008) are in agreement that health risks associated with chemical contamination during swimming are usually insignificant. The results in Table 2 indicate that the quality of water in the River Ethiope is suitable in this regard for swimming purposes.

Suitability for irrigation

The standard indices that are universally utilized to determine suitability for irrigation purposes have been determined for all samples and the results shown in Table 4 are summarized as follows:

Salinity hazard: High TDS in irrigation water affects soil efficiency, plant growth and yields. Following from Wilcox (1955) water with TDS that is less than 200 mg/l is excellent for irrigation and all water sampled in this study falls in this category.

The sodium adsorption ratio (SAR), the tendency of water to replace adsorbed calcium and magnesium with sodium was calculated as follows (Hem, 1991):

$$SAR = Na^{+} / \{[Ca^{2+} + Mg^{2+}] /2\}^{0.5}$$

Where the concentrations are expressed as milliequivalents per liter. All water sampled in this study falls within the low (< 10) SAR category (Richards,1954) and is thus suitable for irrigation.

Magnesium hazard was estimated with the Szabols and Darab (1964) relationship:

$$MH = (Mg^{2+} \times 100)/ (Ca^{2+} + Mg^{2+})$$

Table 4. Parameters for determining suitability for irrigation.

Sample site[a]	Parameter[b]				
	Facie	SH	SAR	MH	RSC
ARG	Na-HCO_3	Low	1.85	7.4	0.24
ABK INL	Ca-Cl	Low	0.312	10.3	0.0
UGN	Na-Cl	Low	0.725	9.57	N/A
CII	Na-Cl	Low	0.994	74.8	0.0
C3G	Na-Cl	Low	1.11	77.4	0.0
C3P	Na-Cl	Low	1.41	67.3	0.0
NAS Rd	Na-Cl	Low	1.16	73.5	0.0
WR1	Mg-Cl	Low	0.606	57.2	0.0
MC	Na-Cl	Low	0.476	35.5	0.0
MJ	Mg-Cl	Low	0.186	56.5	0.0
MD	Mg-Cl	Low	0.207	80.7	0.0
TR	Mg-Cl	Low	0.239	78.5	0.0
OKR	Ca-SO_4	Low	0.114	7.63	0.0

[a] ARG, Aragba; ABK In, Abraka Inland; UGN, Ugono; CII, Geology Lab; C3G, Campus 3 Gate; C3P, Campus 3 Pool site; WR1, Winner's Road 1; WR2, Winner's Road 2; DS, Dump site; MC, McCarthy Beach; MJ, Majoroh Beach; MD, Mudi Beach; TR, The Turf; okr, Okemesi River; Nas Rd, Nassarawa Rd. [b] SH, Salinity Hazard; SAR, sodium absorption ratio; MH, magnesium hazard; RSC, residual sodium carbonate.

Waters that have MH ratios > 50 are considered harmful to soils and may not be used for irrigation without appropriate treatment. The results in Table 4 indicate that water in the Abraka area has a high magnesium hazard and needs to be carefully evaluated in this regard before being utilized for irrigation.

Residual sodium carbonate (RSC) was obtained from the folowing relationship:

$$RSC = (CO_3^{+} + HCO_3^{2-}) - (Ca^{2+} + Mg^{2+})$$

Where all ionic concentrations are expressed in epm. Following the Eaton (1950) classification, RSC values from Abraka are all less than the + 1.25 threshold and may thus be safely used for irrigation.

Conclusions

This study has described for the first time aspects of the hydrogeology of the upper 20 m horizon of the Benin Formation aquifer in this part of the coastal plain and the Niger Delta. This horizon consists of a succession of fine grained reddish brown sands underlain by typically clean, grey – off white and fine – medium - coarse grained, angular to sub- rounded well sorted quartz sands. Depth to water level in the water table ranges from about 4 to 16 m. Sustained groundwater abstractions around the more densely populated Abraka Urban have resulted in a depression of the water table in the area towards which groundwater is flowing from virtually all directions in a distortion superimposed on regional southwestward groundwater flow. Hydraulic conductivity of the medium grained sands that are usually screened at an average 20 m depth in production wells is estimated from grain size analysis to range from 0.12 to 0.19 $msec^{-1}$. The major ion geochemistry indicates that groundwater is an admixture of NaCl, CaCl and $NaHCO_3$ facies and there is evidence of possible hydrogeochemical stratification. All water has very low TDS and this suggest that the perennial leaching of soluble minerals by recharging rainfall from the near surface horizon of the Benin Formation is probably responsible for the presence of minimal amounts of mineral content in water. It has also been determined that surface water and ground water are suitable for domestic use, irrigation and industrial purposes without prior treatment and that the water in this stretch of the Ethiope River is suitable for full body contact recreation.

ACKNOWLEDGEMENT

This investigation was undertaken with the aid of partial funding and resources provided by the Center for Research in Water and Environment (CREWE), Abraka and for which the authors are grateful.

REFERENCES

Akpoborie IA (2011). Aspects of the hydrology of the western Niger Delta wetlands: groundwater conditions in the Neogene (Recent) deposits of the Ndokwa area. Afr. Geosci. Rev. 18(3):25-36.

Akujieze CN, Oteze GE (2006). Groundwater Quality of Benin City Urban aquifer of the Pleistocene – Oligocene Benin Formation, Nigeria. Afr. Sci. 7(2):69-85.

Amajor LC (1991). Aquifers in the Benin Formation (Miocene Recent), Eastern Niger Delta, Nigeria: Lithostratigraphy, Hydraulics, and Water Quality. Environ. Geol. Water Sci. 17(2):85-101.E.

APHA (1992). Standard methods for the examination of water and waste water, 18th Edition, American Public Health Association, Washington, D.C.

Arabi SA, Funtua II, Alagbe SA, Zabosrki P, Dewu BBM (2010). Investigation of Groundwater quality for Domestic and Irrigation purposes around Gubrunde and Environs, northeastern Nigeria. J. Am. Sci. 6(12):664-672.

Aweto K, Akpoborie IA (2011). Geo-Electric and Hydrogeochemical Mapping of Quaternary Deposits at Orerokpe in the Western Niger Delta. J. Appl. Sci. Environ. Manage. 15(2):351-359.

Domenico PA (1972). Concepts and Models in Groundwater Hydrology: McGraw Hill, 405 pp.

Eaton FM (1950). Significance of carbonates in irrigated waters. Soil Sci. 69:127-128.

Efe SI (2005). Urban Effects on precipitation amount, distribution and rainwater quality in Warri Metropolis, Ph. D. Thesis, Delta State University, Abraka, Nigeria.

Emeshili ME (2008). Quality assessment of the potability of borehole and surface waters in the Ndokwa area of Delta State, Nigeria, Unpublished Ph.D. Thesis, Delta State University, Abraka, Nigeria, 254 pp.

Fetter CW (1994). Applied Hydrogeology. Prentice-Hall, New Jersey, p.691

Golden Software Inc. (2002). Surfer 8, Golden Software Inc. Co. USA.

Hazen A (1892). Some physical properties of sands and gravels, with special reference to their use in filtration, 24th Annual Report, Massachusetts State Board of Health, Pub.Doc. 34:539-556

Hem JD (1991). Study and Interpretation of the chemical characteristics of natural waters, 3RD Edition. US Geol. Survey, Water Supply Paper 2254.

Ibe KM, Njemanze GN (1998). The Impact of Urbanization and Protection of water Resources, Owerri, Nigeria. J. Environ. Hydrol. 6:1-9.

Jegen E (2004). "Jump" available via: www.pbase.com/emieljegen/abraka&gcmd.

Kaizer AN, Osakwe, SA (2010). Physicochemical Characteristics and Heavy Metal Levels in Water Samples from Five River Systems in Delta State, Nigeria. J. Appl. Sci. Environ. Manage. 14(1)83-87.

Kehew AE (2001). Applied chemical hydrogeology: Prentice-Hall, New Jersey, p. 368.

Kozisek F (2006). Health risks from drinking demineralized water, In: Nutrients in Drinking Water. WHO, Geneva, pp. 148-163.

Longe EO (2010). Groundwater Resources Potential in the Coastal Plain Sands Aquifers, Lagos, Nigeria. Res. J. Environ. Earth Sci. Res. 3(1):1-7.

Naseem S, Hamza S, Bashir E (2010). Groundwater geochemistry of Winder Agricultural Farms, Balochistan, Pakistan and Assessment for Irrigation Water Quality. Eur. Water 31:21-32.

National Health and Medical Research Council, (2008). Guidelines for Managing Risks in Recreational Water Available via: http://www.nhmrc.gov.au

NGSA (2006). Geological Map of Nigeria, Nigeria Geological Survey Agency, Abuja.

Nwajide CS (2006). A guide for geological field trips to Anambra and related sedimentary basins in southeastern Nigeria. PTDF Chair, University of Nigeria, Nsukka.

Odong J (2007). Evaluation of empirical formulae for determination of hydraulic conductivity based on grain-size analysis. J. Am. Sci. 3(3):54-60.

Ojeh VN, Origho T (2012). Socioeconomic Development of Rural Areas in Nigeria Using the Growth Pole Approach: A Case Study of Delta State University in Abraka. Global Adv. Res. J. Geogr. Regional Plan. 1(1):007-015.

Olobaniyi SB, Owoyemi FB (2004). Quality of groundwater in the Deltaic Plain Sands aquifer of Warri and Environs, Delta State, Nigeria. Water Res. 15:39-45

Olobaniyi SB, Efe SI (2007). Comparative assessment of rainwater and groundwater quality in an oil producing area of Nigeria: Environmental and health implications. J. Environ. Health Res. 6(2):111-118.

Omo-Irabor OO, Olobaniyi SB (2007). Investigation of the Hydrological Quality of Ethiope River Watershed, Southern Nigeria. J. Appl. Sci. Environ. Manage. 11(2):13-19.

Ong CN (2006). Minerals from Drinking Water: Bioavailability For Various World Populations And Health Implications, In: Nutrients in Drinking Water, WHO, Geneva, pp. 61-74.

Oteze GE (2011). Water supply, groundwater and flood control in Benin City, Presented at the NMGS, Benin Chapter sensitization workshop on water resources management and its implications in Benin City,22nd March.

Piper AM (1944). A graphic procedure in geochemical interpretation of water analyses. Trans. Am. Geophys. Union 25:914-923

Rafael HA, Teodoro MG, Rafael MR (1998). Patterns of groundwater hydrochemistry in Apan-Tochac sub-basin, Mexico. Hydrol. Sci. J. 43(5):669-685.

Reijers TJA, Petters SW, Nwajide CS (Eds.) (1996). The Niger Delta Basin, SPDC Corporate Reprographic Services, Warri.

Richards LA (1954). Diagnosis and improvement of saline and alkali soils, Agriculture handbook, Volume 60, US Department of Agriculture, Washington DC.

Sarner PFM (2011). Environmental Impact Assessment of Oleri Water Theme Park, Sarner Pfm, Warri

Short KC, AJ Stauble (1967). Outline of geology of Niger delta. Bull. Am. Assoc. Petr. Geol. 54(5):761-779.

Smith SJ, Wahl KL (2003). Changes in Streamflow and Summary of Major-Ion Chemistry and Loads in the North Fork Red River Basin Upstream from Lake Altus, Northwestern Texas and Western Oklahoma, 1945–1999, USGS Water-Resources Investigations Report 03–4086.

SPDC (2004). Environmental Impact Assessment Of Jones Creek Field Development Plan Project, http://www.shell.com.ng/environment-society/environment-impact-assessments.html (Accessed on 12 Sept. 2013)

SPDC (2008). Environmental Impact Assessment (EIA) of Oben Gas Development Project http://www.shell.com.ng/environment-society/environment-impact-assessments.html (Accessed on 12 Sept. 2013).

Standards Organization of Nigeria (SON) (2007). Nigerian Standard for Drinking Water Quality, NIS 554:2007, Abuja, 30pp

Szabolcs I, Darab C (1964). The influence of irrigation water of high sodium carbonate content of soils, In: Proceedings of 8th international congress of ISSS, Transaction II, pp 803-810

Uma KO, Egboka, BCE, Onuoha, KM (1989). New Statistical Grain-Size Method for Evaluating the Hydraulic Conductivity of Sandy Aquifers. J. Hydrol. 108:367-386

USEPA (2003). Bacterial Water Quality Standards for Recreational Waters (Freshwater and Marine Waters) Status Report, EPA-823-R-03-008

Wade TJ, Pai N, Eisenberg, JNS, Colford, Jr. JM (2003). Do U.S. Environmental Protection Agency Water Quality Guidelines for Recreational Waters Prevent Gastrointestinal Illness? A systematic review and meta-analysis. Environ. Health Perspect. 111:1102-1109

Wilcox LV (1955). Classification and use of irrigation water, U.S.D.A. Circ., 969, Washington, DC,

World Health Organization (WHO) (2003). Guidelines for safe recreational water environments Volume 1: Coastal and Fresh Waters WHO, 20 Avenue Appia, 1211 Geneva 27, Switzerland.

18

Estimate stage-discharge relation for rivers using artificial neural networks- Case study: Dost Bayglu hydrometry station over Qara Su River

Khosro Nezamkhiavy and Sohrab Nezamkhiavy

Young Researchers and Elite Club, Meshkinshahr Branch, Islamic Azad University, Meshkinshahr, Iran.

The prediction of discharge and its variability in a river is an essential component of surface-water planning. For that purpose, a functional relationship between stage and discharge is established with the help of field measurement and the relationship is expressed as a rating curve. Whereas direct discharge measurement in rivers is time consuming and expensive, and sometimes it is impossible, for this purpose the relation between discharge and stage as rating curve is estimated by using of measured data. The Qara Su River in Ardabil province, Iran, makes damages with its floods every year. Informing of flood discharge's value in upstream for management activity in downstream is necessary. In this research for estimate stage-discharge relation in Dost Bayglu hydrometry station over Qara Su River, using MATLAB software, Artificial Neural Network with different state and regression methods studied and finally compared together. As a result of Artificial Neural Network with a 1-2-3-2-1 structure, compared with a simple regression, has a better answer.

Key words: Artificial neural network, stage-discharge relation, Qara Su River, Dost Bayglu hydrometry station, regression.

INTRODUCTION

In hydrometry stations, stage value was usually read while measuring the discharge value of river. At present times engineers read the stage value by advanced devices and send it to the reference centers by using modern methods. If we measure the discharge and the stage value several times, we would find a mathematical relation between them, which was called the stage-discharge relation. Measuring the river discharge constantly, is very difficult and expensive task, even in normal conditions. However, it will be more difficult and sometimes impossible in flood situations or in unfavorable climate conditions. In some rivers, when there is a reasonable mathematical relation between discharge and stage, we can use stage value, to read and send it to the center, to obtain the discharge value. The stage-discharge relation refers to all possible concepts explaining how to calculate discharge based on measuring the flow parameters. Stage-discharge relation can be defined by determining discharge measuring points and stage parallel points on x and y axis, respectively. Forming of the stage-discharge curve is a function of control station geometry. At the mathematical

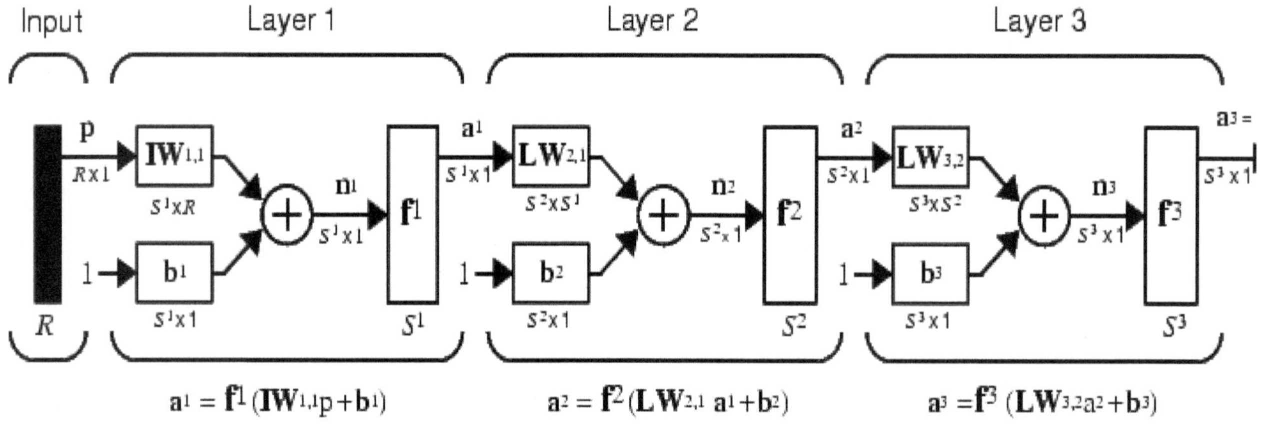

Figure 1. A schema of multiple-layer neural network.

point of view, most of stage- discharge curves haveparabolic form. But, at the regressional point of view, the stage-discharge curves could have various forms such as; polynomial, power, linear and etc.

Goel in his study discussed at the application of soft computing techniques such as back propagation feed forward neural network-based algorithm for modelling stage-discharge relation. The outcome of his study suggests that the back propagation feed forward ANN works quite well for the data sets and produced promising results in comparison to the linear regression technique (Goel, 2011). Bhattacharya and Solomatine (2000) studied the water level and discharge by using 9 years obtained datum from hydrometry station in Bagirathy River, India, by using artificial neural network. The results showed that the power model is the best, as its determination coefficient is about 0.988. Validation test showed that the error in this model is insignificant.

Jain and Chalisgaonkar (2000) studied stage-discharge relation by using artificial neural network in India. In this study, they used three informational layers in artificial neural network model for making stage-discharge curve. The results showed that there is high correlation coefficient in stage-discharge relation by this method.

ANNs are widely used in various areas of water-related research - rainfall-runoff modeling (Minns and Hal,l 1996; Dawson and Wilby, 1998), replicating behavior of hydrodynamic modeling systems (Solomatine and Avila Torres, 1996; Dibike et al., 1999), water level control (Lobbrecht and Solomatine, 1999). Muttiah et al. (1997) addressed the problem of discharge prediction, Thirumalaiah and Deo (1998) modeled the stage behavior (without considering discharge). Clair and Ehrman (1996) used ANN to model the relationship between variation of discharge and ecological parameters and climate change. However, extensive search of sources covering the problem of modeling the stage-discharge relationship delivered no result.

METHODOLOGY

In order to perform the research, we collected information and statistics related to discharge-stage on Dost Bayglu hydrometry station, by referring to regional water organization in Ardebil. Then, we analyzed the information and data. By employing simple regression, we obtained the relation between stage and discharge. To do this, in start by dividing data to four parts, we used 3/4 of data as a calibrated data and 1/4 as a validating data.

Artificial neural network (ANN)

Artificial neural network is a simplified model of human brain. The network has mathematical structure able to indicate non-linear, desired, process and compounds, and to find the relation between input and outputs of system. These networks will be used in the future to predict in the future through accomplishing the learning and teaching process.

Neural network consist of neural cells-called neuron, and some communicational units called axon. Neurons are simplified from of biological neurons. Although, the neuron made artificial neural network have higher speed than biological neuron, they have less capability when compared to them. A schema of multiple-layer neural network has been shown in Figure 1. As is shown in Figure 1, each artificial neural network consist of three layers, include: input layer, output layer and hidden layer.

There are some neurons as processor units on each layers, connected to each other by weighted connectors. In this system, errors reach to its least number through conducting some process. To transmit outputs of one layer to another transfer functions were used. Sigmoid, purelin and tan-sigmoid functions are examples of transfer functions. In general, artificial neural network divided to two groups:

(1) Feed forward networks
(2) Feed backward networks

Levenberg-Marquardt algorithm (LM)

The research, provide LM algorithm to identify the best method with the highest efficiency for learning network. This algorithm considers one approximation in varying the weights, for Heyzen matrix, like Newton's methods.

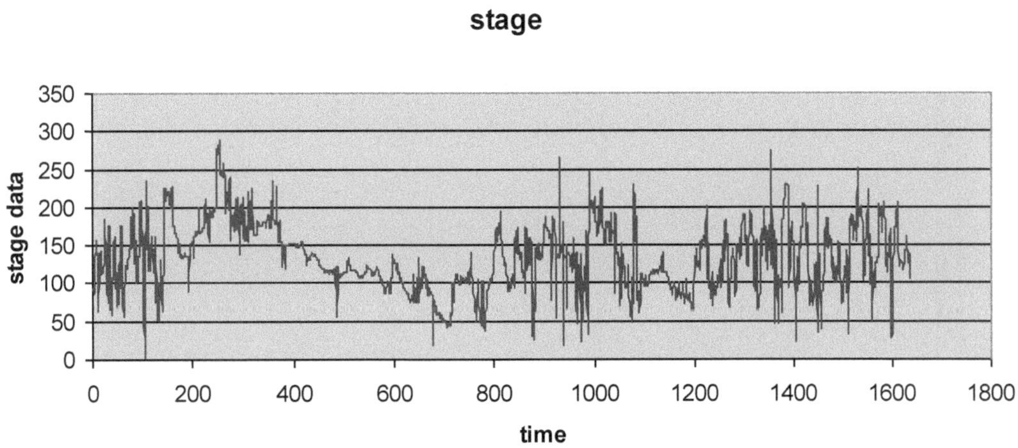

Figure 2. Time series of stage data.

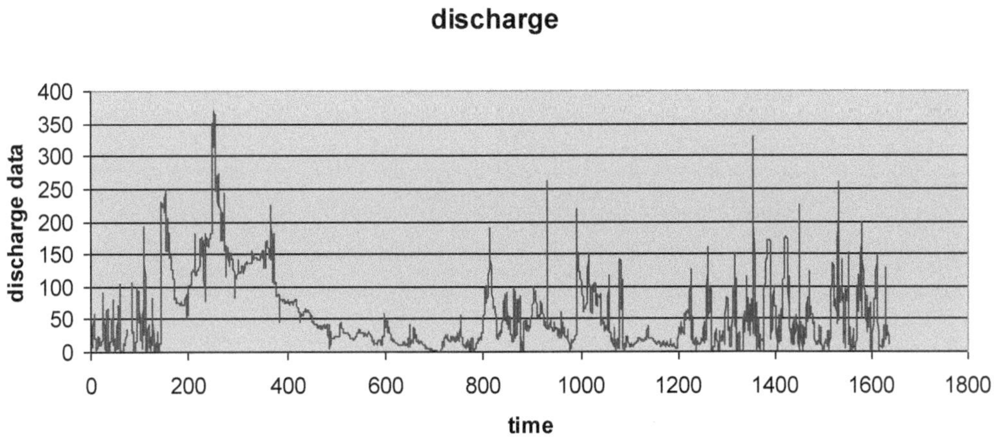

Figure 3. Time series of discharge data.

Specifications of hydrometry station

Qara Su watershed basin is one of the vast in Ardabil province, Iran, its area is about 11126.3 ha. Dost Bayglu hydrometry station established in 1351 and 840 m high from the sea level. It is in 47 – 31 longitude and 38 – 32 latitude. The upstream basin station has area about 7311.1 Km and discharge and stage data are available, until 1384. The station's equipments are: Stage, Limnogragh and etc. Normalizing is one of the re-analyzing states for modeling with artificial neural networks. In this research, we used tan-sigmoid transfer function in all layers. So the data should be normalized in respect to transfer function, by following this relation (Figures 2 and 3).

$$x_i' = \frac{x_i - x_{\min}}{x_{\max} - x_{\min}} \tag{1}$$

Model efficiency scale

There are different scales to evaluate the efficiency of the model. In this research, we use Determination coefficient (2), Root Mean Square error (3) and Correlation coefficient (4).

$$R^2 = \frac{\sum_{i=1}^{n}(Q_i - \overline{Q}_i)^2 \quad \sum_{i=1}^{n}(Q_i - \hat{Q}_i)^2}{\sum_{i=1}^{n}(Q_i - \overline{Q}_i)^2} \tag{2}$$

$$RMSE = \frac{\sqrt{\sum_{i=1}^{n}(Q_i - \hat{Q}_i)^2}}{n} \tag{3}$$

$$CorrCoef = \frac{\sum_{i=1}^{n}(x_i - \overline{x})(y_i - \overline{y})}{\sum_{i=1}^{n}(x_i - \overline{x})^2(y_i - \overline{y})^2} \tag{4}$$

RESULTS

As determination coefficient tended to one, it means

Table 1. The results of modeling by artificial neural network.

Type of neural network	Number of model	input	output	Network structure	logsig				
					Correlation coefficient		R^2		RMSE
					Calibration	Validation	Calibration	Validation	Validation
FFBP	1	Stage data	Discharge data	1-7-12-6-5-1	0.85	0.8	0.81	0.77	0.0043
FFBP	2	Stage data	Discharge data	1-7-12-5-1	0.88	0.84	0.84	0.81	0.0038
FFBP	3	Stage data	Discharge data	1-10-6-1	0.9	0.86	0.87	0.84	0.0033
FFBP	4	Stage data	Discharge data	1-8-5-1	0.94	0.89	0.89	0.85	0.0031
FFBP	5	Stage data	Discharge data	1-2-3-2-1	0.94	0.91	0.9	0.87	0.0028

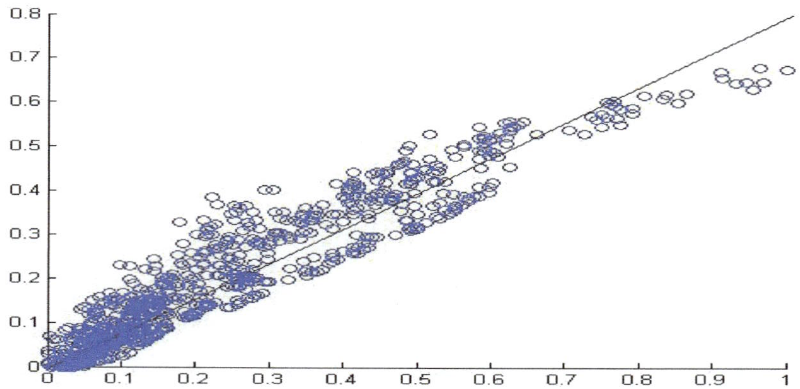

Figure 4. Scatter plot of output number 5- artificial neural network modeling, and observed data- calibration process.

there is more precise relation between stage and discharge. Data have been divided to two parts, calibrated and validating in this method. It have been used 3/4 and 1/4 of data as a calibrated and validating data, respectively, (It means that in this research data consist of 3/4 calibrated data and 1/4 validating data).

In the modeling with artificial neural network for over all models, FFBP method with Levenberg-Marquardt training algorithm (LM) as well as various states of network with different layers was evaluated that results from the best models as shown in Table 1.

Diagrams related to calibration and validation process of neural network method and, also, time series of predicted and observed data are shown in Figures 5 to 7.

In the modeling with simple regression calibrated data were used in fitting curves. So by applying regression relation on validating data, we obtained the validation correlation coefficient. The

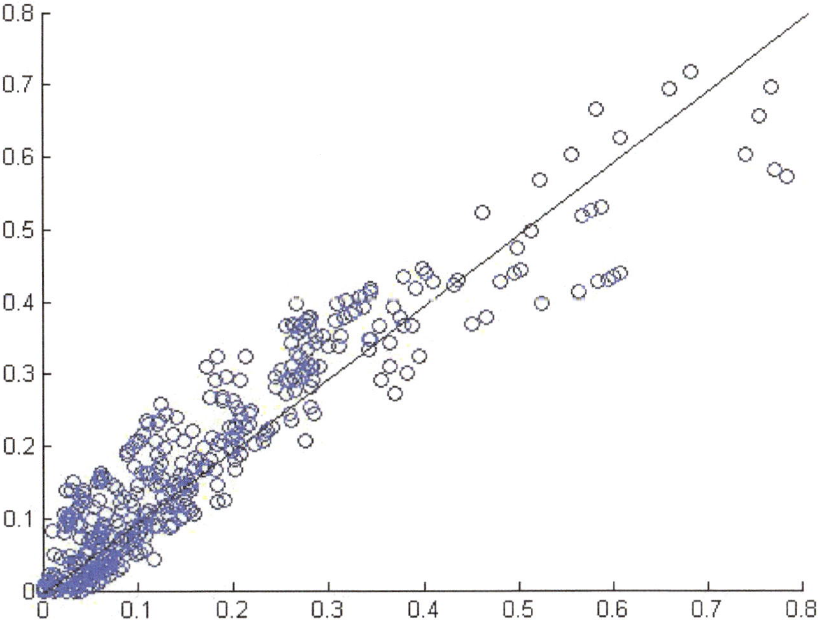

Figure 5. Scatter plot of output number 5- artificial neural network modeling, and observed data- validation process.

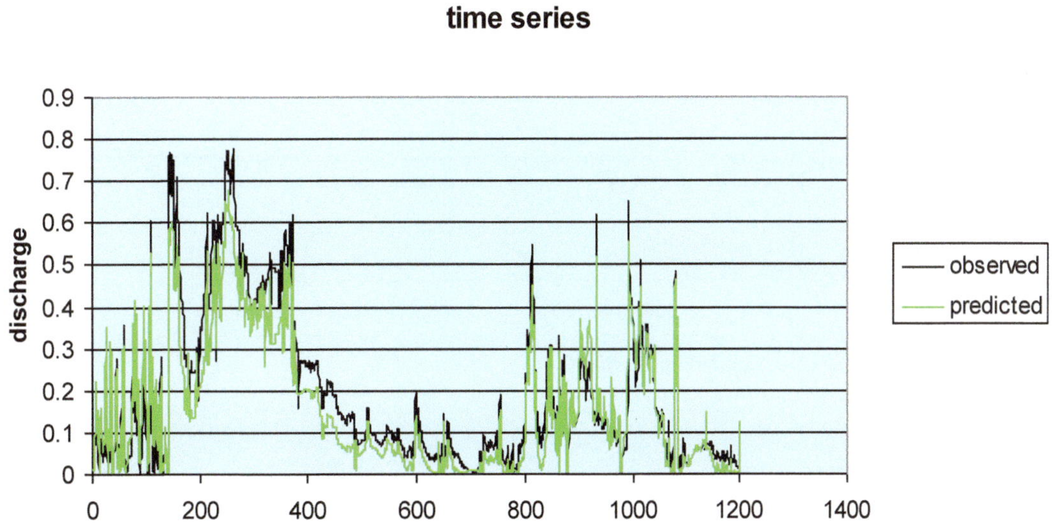

Figure 6. Time series of output number 5- artificial neural network modeling, and observed data- calibration process.

evaluation process was applied on curves and functions then, they were selected based on having the favorable determination coefficient. In this research, polynomial curve provide the best result for Dost Bayglu station. Validation and calibration correlations coefficients are shown in Table 2. Diagrams related to calibration and validation process of this method and, also, time series of predicted and observed data are shown in Figures 8 to 10.

DISCUSSION

It is finally concluded that in artificial neural network (FFBP Network with LM training algorithm), stage data as input and discharge data as output with a 1-2-3-2-1 structure with determination coefficient of 0.87 as the best model has been introduced (Table 3). It is observed that the ability of artificial neural network for stage-discharge modeling is better than other evaluated method

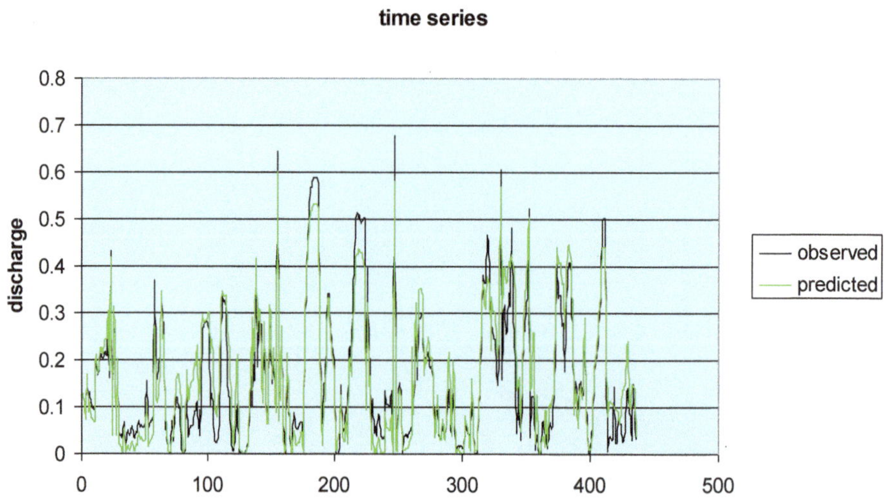

Figure 7. Time series of output number 5- artificial neural network modeling, and observed data- validation process.

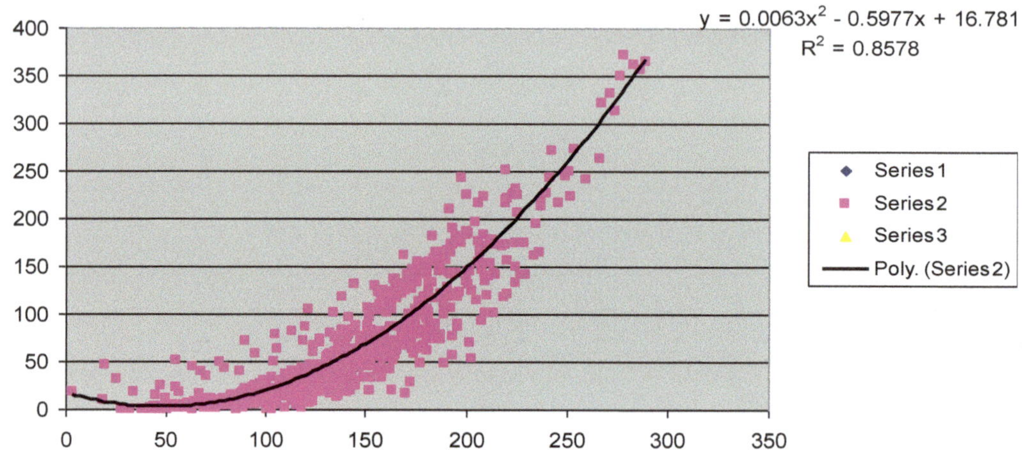

Figure 8. Rating curve - calibration process.

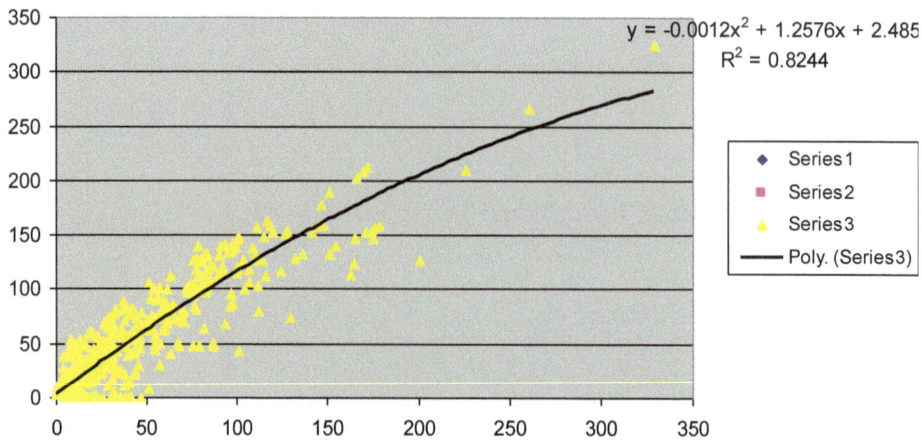

Figure 9. Rating curve - validation process.

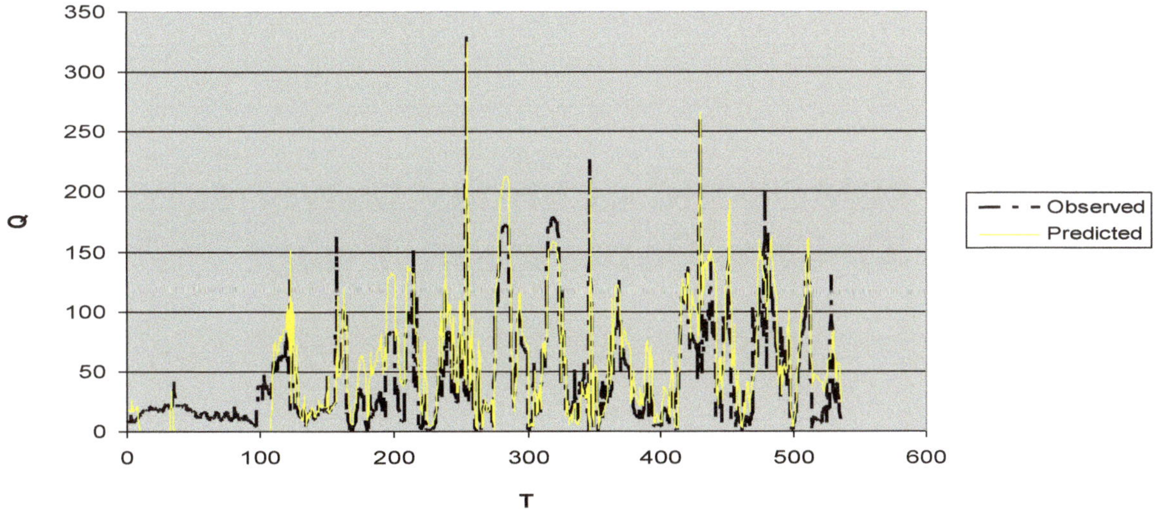

Figure 10. Time series of predicted and observed data.

Table 2. The results of modeling by simple regression.

Model	Calibration correlation coefficient	Validation correlation coefficient
Rating curve	0.85	0.82

Table 3. Comparison between results of modeling.

Modeling	Number of model	Structure	Calibration determination coefficient (R^2)	Validation determination coefficient (R^2)
Artificial neural network (ANN)	3	2-3-1	0.90	0.87
Rating curve	-		0.85	0.82

in this research (simple regression).

Conflict of Interest

The authors have not declared any conflict of interest.

REFERENCES

Alizadeh A (2002). Principles of applied hydrology. 16th edition. Emam Reza university, Mashhad.

Bhattacharya B, Solomatine DP (2000). Application of artificial neural network in stage-discharge relationship. 4th International Conference on Hydroinformatics, Iowa City, USA.

Goel A (2011). ANN-Based Approach for Predicting Rating Curve of an Indian River. International Scholarly Research Network, ISRN Civil Engineering Volume 2011, Article ID 291370, 4 pages doi:10.5402/2011/291370 http://dx.doi.org/10.5402/2011/291370

Lohani AK, Goel NK, Bhatia KKS (2007). Deriving stage-discharge-sediment concentration relationships using fuzzy logic. Hydrol. Sci. J. 52(4):793-807. http://dx.doi.org/10.1623/hysj.52.4.793

Jain SK, Chalisgaonkar (2000). Setting up stage-discharge relations using ANN. J. Hydrol. Eng. 5(4):428-433. http://dx.doi.org/10.1061/(ASCE)1084-0699(2000)5:4(428)

Thirumalaiah K, Deo MC (1998). River stage forecasting using artificial neural networks. J. Hydrol. Eng. 9(1):26-32. http://dx.doi.org/10.1061/(ASCE)1084-0699(1998)3:1(26).

19

Hydro and solar-pond-chimney power scheme for Qattara Depression, Egypt

Assem Afify

Khatib and Alami, Salman University, Al Kharj, Saudi Arabia.

This study summarizes previous proposals and studies concerning Qattara power projects and gives some details about a new concept, which aims to produce hydropower like all previous proposals, and also extra electric power and fresh water via a site specific solar-pond-chimney system. Sea water from the Mediterranean Sea and re-used water from the end of Rosetta branch would feed an artificial lake in the Qattara Depression, where salt gradient heating phenomena would be exploited. The proposed scheme entails two inlet water ways (for sea water and re-used water) and an outlet water way for the produced fresh water. Several experimental studies in a hydraulic lab will be required to clarify technical and economic feasibility aspects.

Key words: Desalination, hydraulics, hydro power, Qattara Depression, reused water, solar ponds.

INTRODUCTION

The utilization of Qattara Depression for the purpose of hydropower generation was first investigated (Ball, 1933) by Professor Penk in 1912, and later by Dr Ball in 1927. The depression is located in the north-western part of Egypt and is the world's fifth deepest natural depression (Figure 1). The depression has, at sea level, a length of about 300 km, a maximum width of 145 km, and an area of 19,500 km^2. The northern edge of the embankment is bounded by a hilly ridge with an elevation of about 200 m above sea level, with the shortest distance from the Mediterranean Sea of about 56 km. Among the development schemes studied by scientists and engineers, the following ones are singled out: Ball (1933), Bassler (De Martino, 1973), Gohar (De Martino, 1973), and lately in 2010 MIK Technology (http://miktechnology.com/powerfromseawater.html,1973; Kelada, 2010). All these proposed schemes and studies focused on hydropower production, but the scheme proposed by Kelada (2010) exploits the physical phenomenon of osmosis and allows production of both hydropower and desalinated water.

The current study reviews previous schemes and proposes a new concept which entails storing both sea water and drainage water to produce hydropower, and exploiting salt gradient heating phenomena in solar ponds (http://www.teriin.org/; http://www.teriin.org/) to produce both fresh water and extra electric power.

REVIEW OF PREVIOUS SCHEMES AND STUDIES

Qattara Depression project

As indicated by many scientists (De Martino, 1973; Kelada, 2010), the Qattara hydropower project has a primary objective to generate a sustainable source of energy corresponding to 315 MW and a secondary goal to provide an elevated storage of 45 million cubic meters to manage peak power demands.

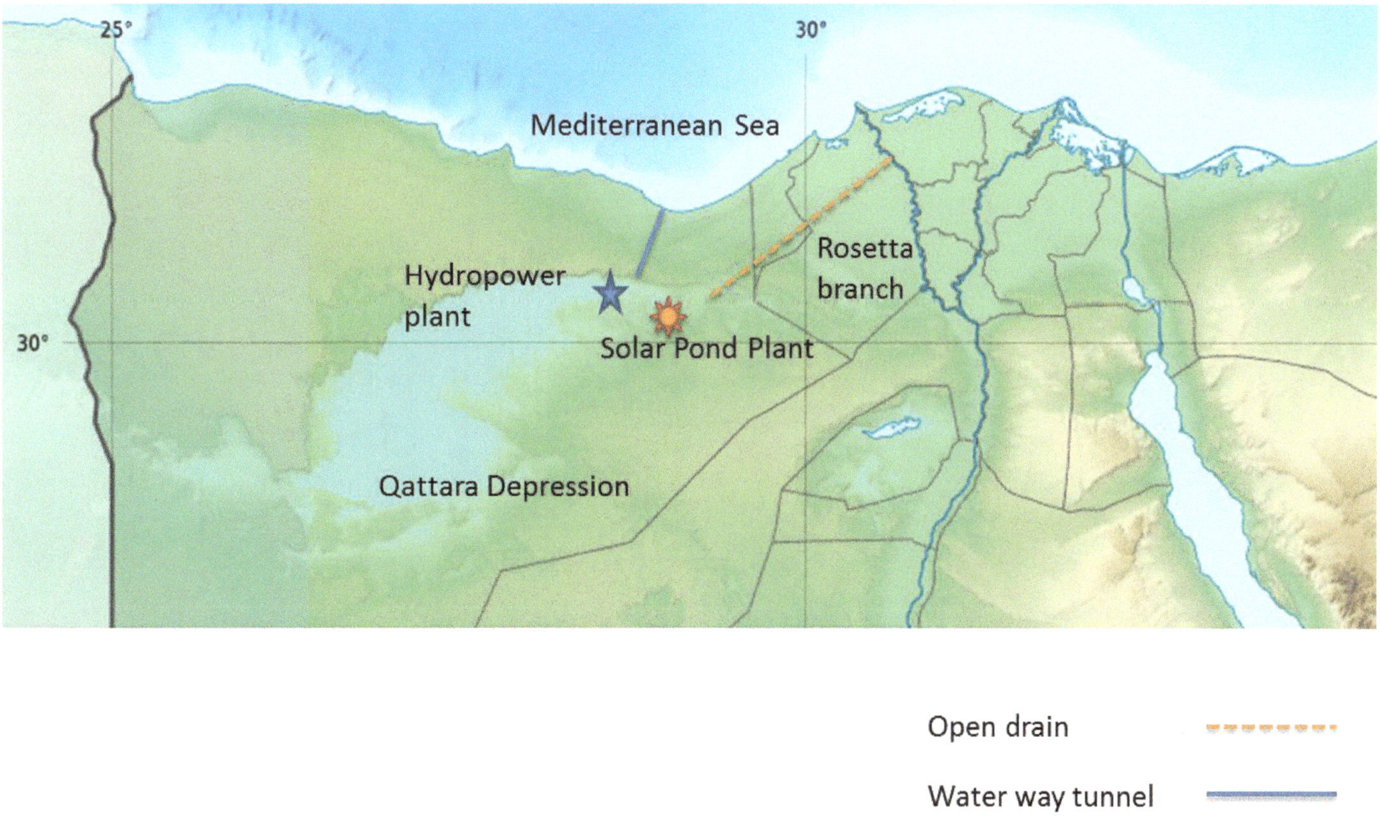

Figure 5. Aerial view for the settings of the hydro and solar-pond power plants, and the proposed water ways.

Ball's researches

Based on geological investigations, Dr. Ball (1933) studied the possibility of utilizing the Qattara Depression for hydropower generation through the formation of artificial lakes at final levels of 50, 60, or 70 m below sea level. The corresponding surface areas for the three alternatives were 13500, 12100 and 8600 km^2. Moreover, he indicated the most convenient water flow route related to each alternative. Ball considered variables such as rainfall, evaporation and seepage inflow. With respect to the rainfall, and taking into consideration the rainfall that would not precipitate directly into the lake but would find its way into the depression, Ball assumed a daily average rainfall depth of 0.15, 0.16, and 0.18 mm for the three lakes' water levels alternatives of 50, 60 or 70 m below sea level. Regarding the evaporation, and based on similar research studies at other nearby locations, he concluded that the average evaporation rates from the surfaces would be 4.6, 4.3 and 4.0 mm/day, for the same alternatives respectively. Regarding the volume of seepage inflow to the lake, Ball estimated that the daily volume would be equivalent to depths of 0.21, 0.24 and 0.33 mm, for the three alternatives, respectively. Balancing the difference between the evaporation volume and the sum of both rainfall and seepage inflow volumes, he estimated the daily volumes of feed sea water to keep the lakes' surfaces balanced at the specified levels. He could thus conclude that the daily volumes of sea water entering the depression should correspond to flow rates of 656, 546 and 348 m^3/s, respectively (Kelada, 2010).

Ball also studied the progressive increase of salinity in the artificial lakes and calculated that the maximum concentration would be reached after 160, 120, and 100 years for the three alternatives respectively (Ball, 1933). Finally, he showed that the most convenient solutions were those related to lakes' surfaces at 50 and 60 m below sea level. He suggested construction of an open channel for the first 20 km, and three tunnels of 11 m diameter to supply the required sea water. The resulting power potential would be about 175 MW for a lake surface level of -50 m. Moreover, he anticipated the possibility of using a power surplus during periods of low demands to pump part of the inflow water into a high-level reservoir and thus obtain a total head of 200 m for hydropower generation to meet peak-load requirements.

Bassler's scheme

The question of developing Qattara project was raised again at the end of the 1950s. Then Siemens proposed a scheme involving the creation of an artificial balancing

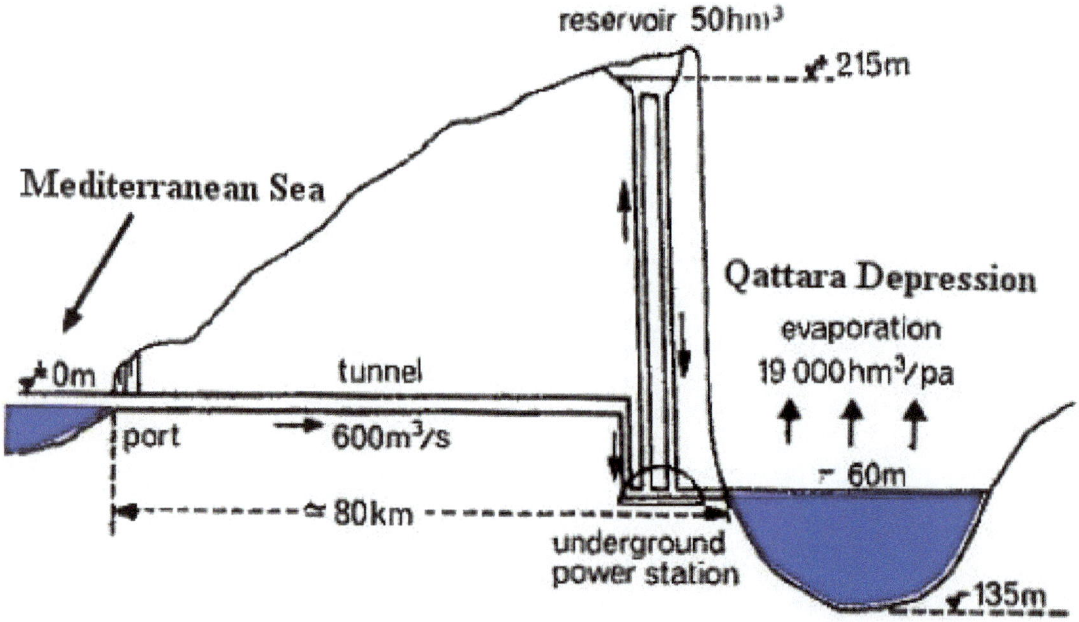

Figure 2. Cross section showing the outline of the hydropower scheme proposed by Professor Bassler.

reservoir on the edge of the depression, continuously fed by two conduits from the Mediterranean. The estimated power potential was about 100 MW, and the turbines were to function only for 6 h (De Martino, 1973). In 1964, Professor Bassler was appointed by the West German Ministry of Economics to study in greater details possible development schemes. Following geological and engineering researches, Bassler presented the scheme shown in Figure 2. The research, which was carried out in 1964 to 1965, aimed at investigating the technical and economic feasibility of a power plant in the Qattara Depression. According to Bassler (De Martino, 1973), it was also anticipated that the Qattara artificial lake would have to be fed by a sea water flow of 600 m^3/s to compensate for the evaporation volume. Similar to Ball's scheme, the construction of three tunnels having a length of 80 km would be required to feed a hydropower plant as shown in Figure 2. During the hours of low demands, the energy produced could be used to pump water into a high-level natural reservoir having a capacity of about 50 million cubic meters. Such a basin is located at the edge of the depression at an elevation of 215 m above sea level.

To summarize, this scheme combines a low head with a high head system, can produce peak power up to 450 MW, and is adaptable to a varying demand pattern.

Gohar's scheme

Gohar suggested an alternative scheme in which sea water is supplied from the Mediterranean Sea via one specific channel (route) located to the east of the routes suggested by Dr. Ball (De Martino, 1973). His point of view is that this route has more advantages than the routes suggested by Ball. He also indicated that the highest point of land between the sea and the depression was located only 11 km from the Mediterranean coast, and he suggested that sea water could be pumped up to a convenient location and then made to flow in an elevated open channel. According to Gohar (De Martino, 1973), his design could result in a simpler and more economic scheme than Ball's and Bassler's schemes which have three tunnels. At the end of the channel, the water could be carried out to supply an underground power station at a suitable location midway from the depression. However, water would need to flow in a tunnel to reach the depression. He suggested that the best scheme among various possibilities was the formation of the lake at a level of -75 m, along with a channel at a level of +80 m discharging 266 m^3/s.

The continuous net power production for his scheme would be about 100 MW, while the maximum power capacity would be 345 MW.

MIK technology scheme

The MIK Technology (http://miktechnology.com/power fromseawater.html) concept for power generation is based on the osmosis phenomenon, targeting world saline dry natural basins and salt lakes such as the Qattara Depression-Egypt, and Great Salt Lake-U.S.A. Simply, osmotic power can be generated by running salt water solutions of differing concentrations along the two sides of a suitable semi-permeable membrane (Kelada, 2010). Water tends to permeate spontaneously into the more concentrated solution and this water flow can be used to drive a power generation turbine. MIK Technology alternative scheme (Kelada, 2010) is an osmotic power generation facility that provides a source of sustainable energy at a comparable power potential, and at a considerably reduced sea water intake requirement than previously mentioned schemes. The proposed osmotic power system requires only 60 m^3/s of sea water to generate 360 MW of power. Since osmotic power requires high salinity brine, then seawater has to be concentrated by solar evaporation to reach the required salinity. Therefore, an area of low land approximately 630 km^2 will be required as proposed by Kelada (2010) to generate the brine for this process as shown in Figure 3.

HYDRO AND SOLAR-POND-CHIMNEY POWER SCHEME

Here a new concept for the Qattara Depression, conceived as a two-stage power system is introduced. The first stage concerns a conventional hydropower scheme fed by sea water and somewhat similar to previously proposed ones. The second stage concerns a hydropower scheme fed by re-used water and a site-specific solar-pond-chimney power system. The first-stage power plant would produce hydropower of about 190 MW, while the second-stage power plants would raise the total production to reach about 410 MW.

Solar-pond-chimney power system

The 'solar pond' exploits a heat storage phenomenon in a large salt water body such as a lake, which operates as a natural collector of solar radiation (http://www.greenidealive.org/; http://www.solar-chimney.biz/). For various applications, the temperature of the water should be as hot as possible but lower than 100°C to avoid boiling at the bottom of the lake. Relatively high temperatures (60 to 80°C) may be obtained, by exploiting the higher density of the more concentrated salty water of the lake, in a so called salt gradient solar pond. So below a thick insulating gradient layer, a hot bottom layer heated by the sun can store heat for several days (http://www.teriin.org/). The mean temperature of the bottom layer of the pond will be around 60 to 80°C and that of the cooler upper layer 30°C. Similar to other researches, the Solar Pond is coupled to a suitable vertical chimney to generate power by the solar chimney effect (Dhahri and Omri, 2013; Akbarzadeh et al., 2009). To this end, saturated air at the lake surface is heated by extraction of heat from the hot water bottom layer of the solar pond and fed into the Chimney at its base. After rising by the chimney effect, air gets back saturated and cooler at the lake surface.

Here, it may be diverted to chillers to catch desalinated water, and then fed back again into the chimney to repeat the cycle.

Project description

The Qattara project proposed in the present study envisages two construction stages. Stage 1 concerns the construction of a conventional hydropower plant fed by sea water, and stage 2 concerns the construction of a conventional hydropower plant fed by re-used water and a Solar-Pond-Chimney Power Plant (SPP). Figure 4 shows the cross section of the combined power system with indication of the artificial lake, the two hydropower plants and the chimney. Figure 5 shows an aerial view for the settings of the project components. The first stage hydropower plant at the left side would have a potential head of up to 70 m for the lowest lake surface level, and a sea water flow rate Q_s = 460 m^3/s. The available head would drop to 50 to 60 m at the construction end of the second stage with re-used water delivered at the top of the lake surface. The expected generated hydropower for the first stage would be around 190 MW for the lowest lake surface level of -70 m and for 60% turbine efficiency. The SPP, at the right side of the figure, should be fully operational with the completion of the second stage of the project. An additional re-used water flow rate Q_r of about 200 m^3/s would be needed to keep the lake top water level balanced at –50 m. This flow rate would be diverted from the surface drainage water of the Delta at an average salinity of 0.1% and would produce about 60 MW of hydropower with 60% turbine efficiency.

The SPP has a tall vertical chimney suitably sited above the lower lake surface. The air over this surface, and in the neighborhood of the chimney base, is first heated by hydrophobic heat wheels that convey heat from the hot water bottom layer of the lake and then enters the chimney base. The air flowing over the lake surface would be saturated by water vapor at a temperature of 30°C and, after being heated; it would have a temperature of 80°C and a density of 1 kg/m^3. With a density of the saturated air around the chimney of 1.2 kg/m^3, a pressure difference of (1.2 - 1) × 9.81 × 1000 = 1.962 kPa would be available for a chimney height of 1000 m. The chimney wind turbines may exploit this pressure difference to generate useful electrical power. For a chimney with a throat area of 1400 m^2, the electric power produced only by the SPP system would be around 200 MW (http://www.greenidealive.org/; http://www.solar-chimney.biz/). With the functioning of the SPP system and the hydropower plant fed by 200 m^3/s of re-used water, the available head would drop to about 55 m, and the additional hydropower produced by the re-used water plant would be around 66 MW at 60% turbine efficiency. The overall electric power produced by the

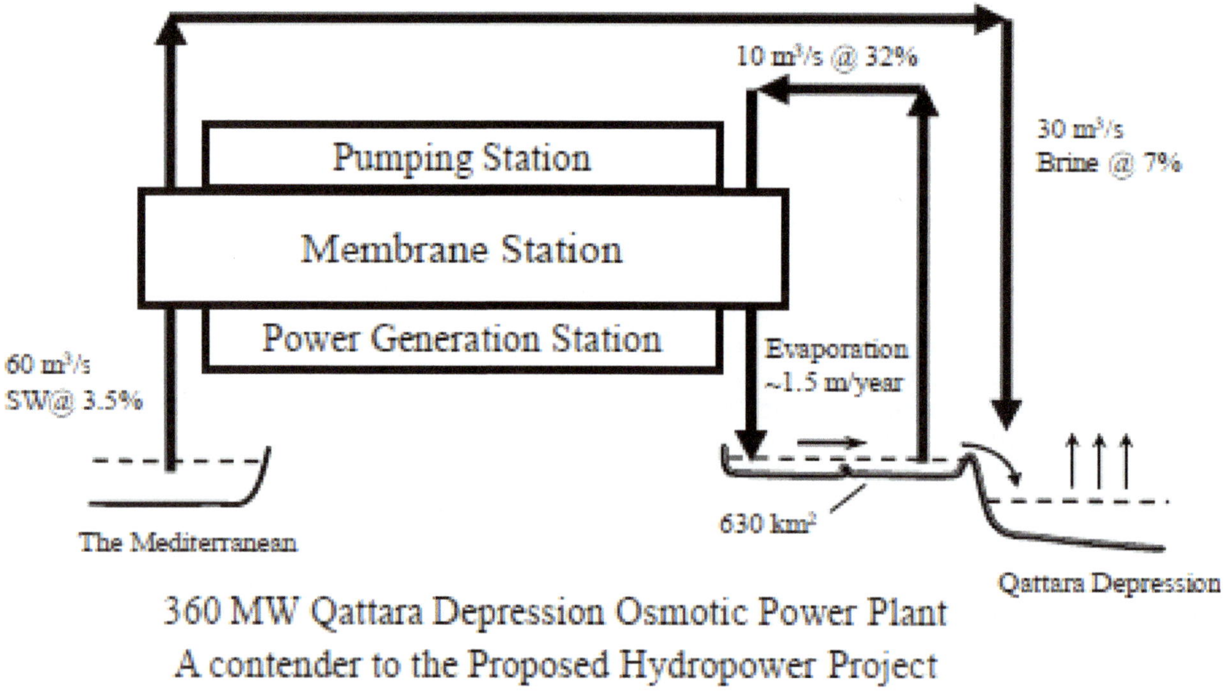

Figure 3. An alternative energy scheme for a hydro-solar system proposed by Kelada (2010).

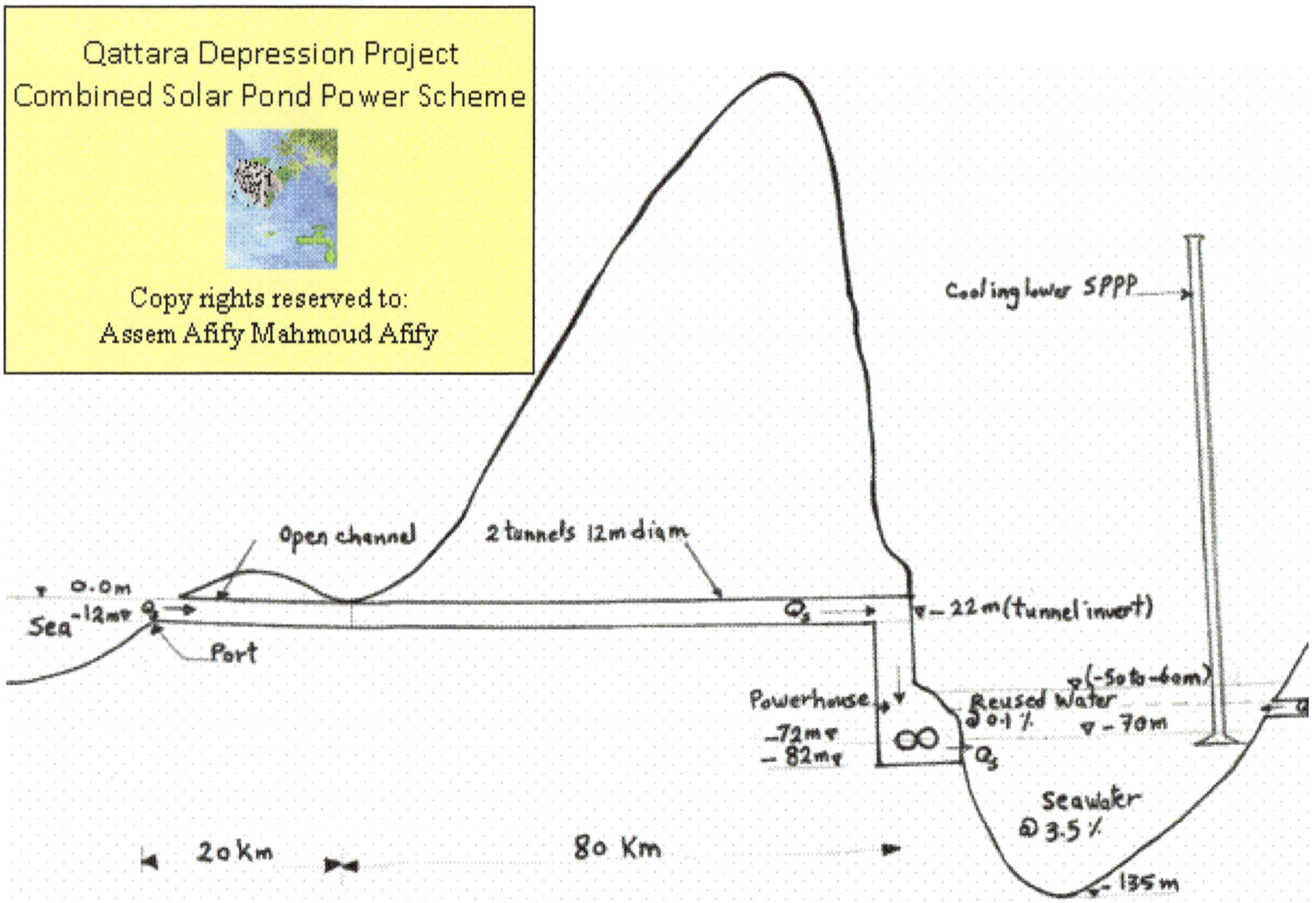

Figure 4. Cross section showing the configurations of two hydropower plants and a Solar-Pond-Chimney power plant.

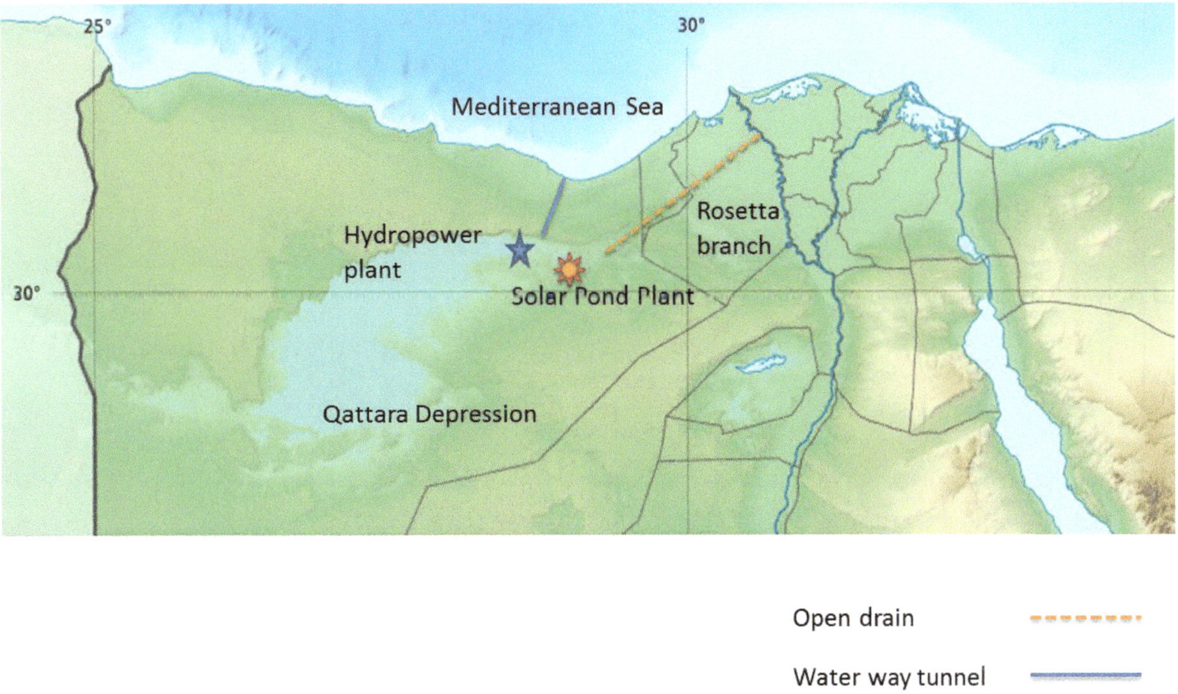

Figure 5. Aerial view for the settings of the hydro and solar-pond power plants, and the proposed water ways.

three power plants (in the first and second stages) might reach 480 MW with more efficient turbines.

Summary and comparison of schemes

Design parameters of the different schemes are summarized as listed in Table 1 for the purpose of comparison among them. The last column shows findings for the scheme proposed by the present study. The most important parameter is the artificial lake water level, which will finally govern all the remaining parameters. The lower the water level, the smaller the evaporation volume would be, and hence the needed flow rate of sea water would be less to keep the lake balanced. A clear example of this is Gohar's scheme, which proposes a water way (open channel) of reduced carrying capacity and dimensions compared to other similar schemes, such as the three tunnels proposed by Ball and Bassler. The present hydropower and Solar-Pond-Chimney scheme requires only two tunnels. However, additional water flow rates would be needed to balance the artificial lake if a partial flow rate of re-used water were diverted for the production of desalinated water. MIK Technology scheme is an exceptional case. It would require the least sea water flow rate of 60 m^3/s to produce 360 MW of hydropower, which is somewhat comparable to the peak powers expected in other schemes. However, if the sea water flow rate is relatively small with respect to the huge amount of evaporation loss, problems related with salt accumulation and unbalanced lake water surface have to be taken into account.

Project benefits and obstacles

When completed, the project would provide a yearly averaged power production of about 400 MW. In addition, it would create job opportunities in the area as well as benefit industrial and tourism developments. Excess power during periods of low demands might be used to produce desalinated water and thus help development activities. In particular, production of desalinated water may be obtained by coupling solar pond technology (Huanmin et al., 2002) with a thermal desalination plant.

Many obstacles face the Qattara project, and among them is the presence of dangerous explosives from the Second World War. As for site investigations, remote sensing can offer a solution to this problem. The problem of groundwater losses in the area has already been investigated through a hydro-geological study (Bastiaanssen and Menenti, 1990) using satellite data. Another critical issue for the project concerns the national strategic defense from the western side borders. The claims that the sea water channel would be a water barrier can be resolved, since a major part of this water way would be tunneled.

Groundwater salinity is also a big concern and has to be further studied with the help of hydraulic models so as to come up with the most suitable water level for the salty lake and thus prevent outward seepage flow.

Table 1. Design parameters in some schemes proposed for the Qattara project.

Design parameters	Schemes				
	Ball	Bassler	Gohar	MIK Technology	Present hydro SPP
Flow rate (m^3/s)	656	600	266	60	$Q_s = 460$ $Q_r = 200$
Lake water level (m)	-50	-60	-75	-40	Top = -55 Saline = -70
Mean power (MW)	175	190	100	360	190 (1st stage) 410 (from both 1st and 2nd stage)
Peak power (MW)	175	450	345	360	480
Project main features	Open channel for the first 20 km, followed by 3 tunnels with 11 m diameter.		Elevated open channel at a level of 80 m and reservoir capacity not mentioned	Brine lake to be formed with an area around 630 km^2, another lake "Qattara sea" at a level of -20 m	Lake is connected to the sea via two tunnels of 12 m diameter, and to the Delta via an open drain.
	Reservoir capacity not mentioned.	Elevated reservoir capacity of 50 million m^3 at a level of +215 m.			Solar-pond-chimney plant and two conventional hydropower plants.

CONCLUSIONS AND RECOMMENDATIONS

The proposed scheme promises some advantages with respect to previous schemes. In particular, for the hydropower plant fed by sea water (first stage), only two tunnels are needed rather than three as suggested by Bassler. Also, no elevated reservoir is needed since the produced power during low demand periods can be utilized in the desalination process of re-used water (drainage). Feeding a hydropower plant with re-used water (2nd stage) allows storing re-used water on top of sea water in the artificial lake and thus minimizes risks of increasing groundwater salinity. Funds for the construction of the SPP power plant (2nd stage), with suitable chimney dimensions, will be available by selling extra electric power generated and produced desalinated water for the development of tourism and industry. An overall feasibility study encompassing all aspects of project components is strongly recommended. Groundwater modeling should be specially investigated to find out the most suitable level of the artificial lake. A physical model may also be needed to find proper locations for the chimney and for the drain inlet of the re-used water so as to avoid hydraulic mixing with the insulating salty layer of the lake.

REFERENCES

Akbarzadeh A, Johnson P, Singh R (2009). Examining potential benefits of combining a chimney with a salinity gradient solar pond for production of power in salt affected areas. Solar Energy 83(8):1345-1359.

Ball J (1933). The Qattara Depression of the Libyan Desert and the possibility of its utilization for power-Production. Geogr. J. 82(4):289-314.

Bastiaanssen WGM, Menenti M (1990). Mapping Groundwater Losses in the Western Desert of Egypt with Satellite Measurements of Surface Reflectance and Surface Temperature. Water Management and Remote Sensing: Proceedings and information No 42, The Netherlands.

De Martino G (1973). The Qattara Depression. Technical repot, Hydraulics Institute, Naples, Italy. MIK Technology Website. http://miktechnology.com/powerfromseawater.html.

Dhahri A, Omri A (2013). A review of solar chimney power generation technology. Int. J. Eng. Adv. Technol. 2(3):1-17.

Homepage of Solar Chimneys Website. http://www.solar-chimney.biz/.

Huanmin L, John CW, Hein H (2002). Thermal desalination using MEMS and salinity-gradient solar pond technology. Desalination Research and Development Program, Report No. 80.

Kelada M (2010). Global hyper saline power generation Qattara Depression potential. Fourteenth International Middle East Power Systems Conference, Cairo, Egypt, December, 19-21.

The Energy and Resources Institute Website. http://www.teriin.org/ GreenIdeaLive - Find, discuss and publish ideas. http://www.greenidealive.org/

Environmental impact assessment of the potability of water from bore-hole, hand dug well and stream at Itagunmodi gold deposits Southwestern, Nigeria using FORTRAN algorithm for monitoring leachates and interpreting physicochemical data of contaminants in groundwater

Adetoyinbo A. A.[1], Adelegan F. T.[1] and Bello A. K.[2]

[1]Department of Physics, University of Ibadan, Ibadan, Nigeria
[2]Department of Physical Sciences, Bells University of technology, Ota, Nigeria

This study determined the quality of surface and underground water in Itagunmodi in Nigeria. Physicochemical analysis of water samples collected within gold mining area and the water consumed in the town was carried out to determine their hygienic conditions. The major objective of this study was assessment degree of groundwater pollution around the mining sites and the available water consumed in the town. Four major sources of water were analyzed. Several physical and chemical parameters were tested in groundwater, these include pH, total solids (TS), total dissolved solids (TDS), conductivity, total alkalinity, total hardness (TH), chloride (Cl^{-1}), sulphur (S), heavy metals (Pb, Zn, Cu), major cation (Mg). The results show that the parameters are within the standard acceptable levels which are required for drinking water adapted from World Health Organization. The pH of the samples was suitable for drinking according to the World Health Organization standard. For a safe drinking water, the pH should be between 6.5 and 8.5); it can be said that the water quality of the samples are fairly good based on the monitored elements and physicochemical characteristics.

Key words: Groundwater, conductivity, heavy metals, gold mining site, pollution.

INTRODUCTION

Groundwater is water located beneath the ground surface in soil pore space and in the fractures of lithologic. It makes up about 20% of the world's fresh water supply which is about 0.61% of the entire world's water including oceans and permanent ice. Groundwater is naturally replenished by surface water from precipitation, streams and rivers when this recharge reaches the water table. Rain water dissolves soluble salts from vegetations, topsoil, river bed, lake bed into water bodies, hence most ions (Ca^{2+}, Mg^{2+}, Na^{+}, K^{+} and NH_4^{+}) in rain water are also found in surface and groundwater.

Groundwater is an important water resource in both the urban and rural areas of Nigeria but in the cities, pipe borne water is also available. Rural dwellers rely basically on hand-dug wells for potable water supply as the streams usually dry up in dry season. These resources are under threat from pollution either from human life style manifested by the low level of hygiene practiced in the developing nations (Punmia and Jain, 1998; Akujieze et al., 2003). Environmental health involves all the factors, circumstances and conditions in the environment or surroundings of humans that can influence health and well being. The neglect of rural areas in most developing countries in terms of basic infrastructures such as pipe-borne water and sanitation facilities, expose the villagers to a variety of health related problems such as water - borne diseases (Sridha, 2000).

Recent research conducted by Yisa and Jimoh (2010) showed that there was an increase in the demand for freshwater due to rapid growth of population as well as the accelerated pace of industrialization in the last few decades. Groundwater has long been considered as one of the purest forms of water available in nature and meets the overall demand for rural and semi-rural people (Tyagi et al., 2002). This was considered as the major source of water for human activities (consumption inclusive) especially in the rural area (Fasunwon et al., 2008).

Inadequate amounts of water for basic hygiene can contribute to poor hygiene practices, which in turn can lead to skin and eye diseases, and act as a key factor in the transmission of many diarrhoeal diseases (Ramakrishnaiah et al., 2009) reported that once ground water is contaminated, its quality cannot be restored. It therefore becomes imperative to protect it in other to prevent water borne diseases such as typhoid, cholera, diarrhoea and dysentery which are potentially communicable (Okeke and Igboanua, 2003; Suleiman, 2006).

Water from various sources contains varying forms of dissolved gases, minerals, organic and inorganic substances, chemical compounds and suspended matter such as organisms and dirt (Erah et al., 2002). Pure water is never found in nature, but contains impurities which vary from dissolved gases, chemical compounds to suspended matter such as organism and dirt especially in this part of the world (Sandhu et al., 1976). The most important physical characteristics are turbidity, colour, odour, taste, temperature and solids, while the chemical characteristics are acidity, alkalinity, hardness and corrosiveness (Ademoroti, 1996a). Water of good quality is necessary if it is to be acceptable by the people.

Palatability of water describes the characteristics of being pleasing to sense of taste. However, palatable water in real sense is not necessarily safe to drink or potable (Packham, 1996). Trace metals of different types can be found in water. Very low amount of trace metals are required by living things, but in excess concentration, metals can be harmful (Rand and Petriocelli, 1985). Potassium and sodium are very soluble salts and most of them leached from rocks, soils and plants tend to remain in solution. High concentration of potassium and sodium cause foaming in boiling water and have toxic effect on fish (Vermani, 1989).

Some water bodies may contain pathogens that are harmful in the body. The water sources in the area were either from bore-hole, streams or other sources, in which one may not be sure if analysis were carried out or not. Access to potable water is still a major concern in all the developing nation of the world as millions of people died of water-borne diseases annually.

In this study, the levels of some physical, chemical, water quality parameters in hand-dug wells, streams, borehole, deep well located in the residential areas and in the vicinities of Itagunmodi, a rural settlement in southwest Nigeria was carried out. In this work water from gold mining area was included in the analysis. Since gold is extremely resistant to weathering, when freed from enclosing rocks, is carried downstream as metallic particles consisting of "dust," flakes, grains, or nuggets.

Geology of the study area

Itagunmodi is located on latitude (DMS) 7031'60" longitude (DMS) 4039'0" and altitude (meters) 347 the time zone is east (est). The approximate population for 7 km radius from this point is 12655. Itagunmodi is a town very close to towns such as: Ile-Ife, Ilesa in Osun State, Nigeria. The sediment-hosted disseminated gold deposits (SHDGD) in Itagunmodi, Osun State, Nigeria is located in the clayey soil types derived from variably migmatised gneiss, biotite-and-biotite hornblende-gneiss and weathered amphibolites respectively (Figure 1).

MATERIALS AND METHODS

Sample collection

The analyses were carried out between the months of February to July, 2013, and their average values were recorded in Table 1. The water sources include: (i) Deep well; (ii) Shallow well; (iii) stream, and (iv) Borehole. Samples were taken to the laboratory within two weeks. It was found out in the course of the work that most of the inhabitants get most of their domestic water from the well and drink directly without purification. The mean concentrations of the samples were calculated and used for the study.

Parameters analyzed

Physical and chemical parameters were analyzed in groundwater samples. The physical parameters include pH and electrical conductivity (EC). The chemical parameters include: Nitrite (NO_3),

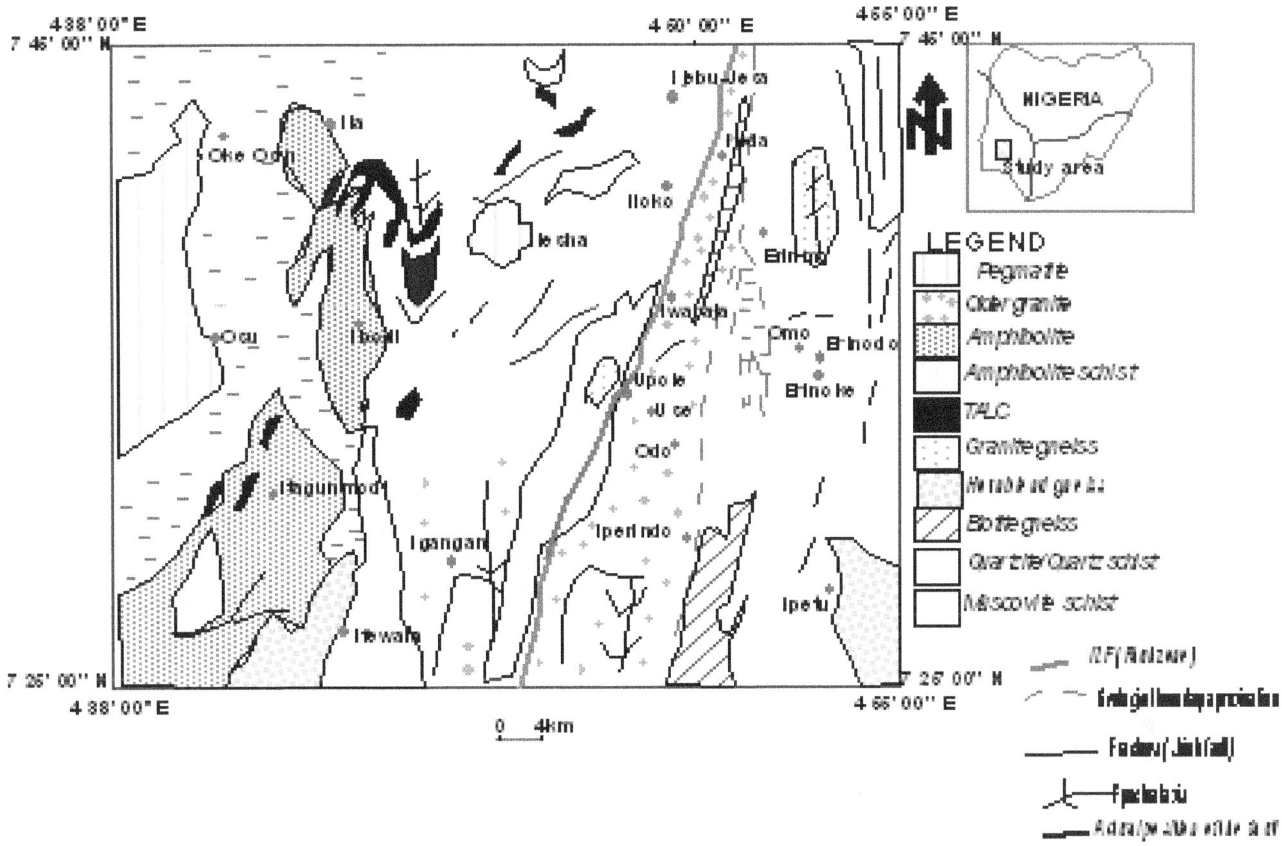

Figure 1. Generalized geological map of Ilesa schist belt SW Nigeria (after Elieze, 1982).

Table 1. Average values of physiochemical analysis on five sources of water in Itagunmodi.

Parameters	Deep water	Stream water	Bore hole	Hand dug well	Mining site	WHO Max. permitted
pH	7.12	7.83	7.36	7.40	6.50	6.5-8.5
Total solids mg/L	100.00	200.00	275.00	250.00	13.20	
Total dissolved solids mg/L	200.00	400.00	300.00	300.00	90.00	500
Conductivity $\mu S/cm$	143.00	159.00	457.00	417.00	120.00	1000
Total alkalinity mg/L	2.00	2.00	3.00	2.00	74.50	4.5
Total hardness $CaCO_3$ mg/L	230.00	355.00	360.00	285.00	20.46	150
Magnesium, Mg mg/L	8.50	11.80	26.00	27.30	13.27	0.2
Chlorine, Cl^- mg/L	4.80	2.50	7.80	4.30	3.05	5.0
Sulphide, mg/L	0.239	0.239	0.2997	0.393	115.22	0.2
Lead, Pb mg/L	0.00	1.10	0.00	0.00	0.384	0.01
Copper, Cu mg/L	0.070	0.100	0.000	0.040	0.77	1.0
Zinc, Zn mg/L	0.030	0.060	0.030	0.040	0.032	2.0
Sodium, Na mg/L	32.00	12.00	31.00	29.00	29.00	100
Manganese, Mn mg/L	0.01	0.03	0.03	0.01	0.10	0.05
Cadmium, Cd mg/L	0.058	0.085	0.018	0.069	0.097	0.003
Calcium, Ca mg/L	68.80	32.40	84.00	93.40	59.30	100-300
Iron Fe mg/L	0.10	0.28	0.06	0.07	0.15	0.3
Phosphate Phs mg/L	0.045	0.048	0.043	0.068	0.052	0.4
Nitrate, Ntr mg/L	4.93	3.29	5.26	5.15	18.53	45
Arsenic, As mg/L	0.26	0.12	0.83	0.75	8.60	0.01

chloride (Cl), sulphate (SO_4), lead (Pb), iron (Fe), zinc (Zn), copper (Cu), biological oxygen demand (BOD), and groundwater, these include pH, total solids (TS), total dissolved solids (TDS), conductivity, total alkalinity, total hardness (TH), chloride (Cl^{-1}), sulphur (S), heavy metals (Pb, Zn, Cu), major cation (Mg).

Theory

The following methods were applied in the calculations:

$$Acidity(mg/l) = \frac{V \times M \times 100{,}000}{ml\ of\ sample} \tag{1}$$

Where V = volume of NaOH used, and M = molarity of NaOH used.

$$Total\ alkalinity(mg/l) = \frac{V \times M \times 100{,}000}{ml\ of\ sample} \tag{2}$$

Where V = volume of acid used, and M = molarity of acid used.

$$Total\ solids\ (mg/l) = \frac{(W_2 - W_1)mg \times 1000}{ml\ of\ sample} \tag{3}$$

Where W_1 = initial weight of evaporating dish, and W_2 = final weight of the dish (evaporating dish + residue).

$$Total\ dissolved\ solids\ (mg/l) = \frac{(W_2 - W_1)mg \times 1000}{ml\ of\ filyrate\ used} \tag{4}$$

Where W_1 = initial weight of evaporating dish, and W_2 = final weight of the dish (evaporating dish + residue).

$$Hardness\ in\ mg/l\ CaCO_3 = \frac{V \times M \times 1000}{ml\ of\ sample\ used} \tag{5}$$

Where M = Molarity of EDTA used and V = volume of EDTA used.

Fe and Cr were measured with S series atomic absorption spectrophotometer (AAS)

RESULTS AND DISCUSSION

The pH values of the samples were shown in Table 1. The highest average value of 7.83 was measured in sample 2, and the lowest value of 6.5 was measured in sample 5, this was shown in Figure 1. The results obtained were within the standard range of 6.5 to 8.5 required for drinking water according to W.H.O and the Nigeria standard. This occurs in most natural waters. The lower the pH, the more corrosive the water will be. The pH of sample collected from the gold mining site is acidic while the other samples are alkaline. Those samples within the town of Itagunmodi are also regarded as safe drinking water; the average pH values lies between 6.5 and 7.4, these values lie between the permitted ranges in drinking water quality standard used in Nigeria.

Low values are most often caused by lack of carbonate minerals, such as calcium and magnesium found in limestone. Water leaking from a landfill may also lower pH. Other analysis such as conductivity, total dissolved solids, chlorine, copper and zinc, falls within the maximum permitted level with reference to Nigeria drinking water standard which are: 1000.00 $\mu S/cm$, 500.00 mg/L, 250.00 mg/L, 0.01 mg/L, 1.00 mg/L and 3.00 mg/L.

The total solids analyzed, the highest value was recorded in sample 3, and lowest in sample 5. It was high in sample 4 and the lowest was recorded in sample 5 while in total dissolved solids, the highest was recorded in sample 2 and the lowest was in sample 5, samples 3 and 4 were higher than that in sample 5. The conductivity was higher in sample 3 than in sample 4 while those of samples 1 and 2 were lower and sample 5 has the lowest value. The alkalinity was higher in sample 5, while those of samples1, 2, 3 and 4 were low. The concentration of $CaCO_3$ in samples 2 and 3 were higher than the value obtained sample 1 and 4 with the lowest value in sample 5. All the samples have higher concentration in Mg because they exceed the WHO standard for drinking.

The average concentration of chlorine was higher in sample 3 then sample 1, 5 and 4, with the lowest value in sample 2. Average sulphate concentration was highest in sample 5, high in sample 4 and low in samples: 1, 2 and 3. For Lead, sample1, 3, 4 has zero concentration and sample 2 and 3 concentrations exceed WHO standard. The average concentration of Cu was highest in sample 5 than in other sample. The average concentration of Zn was higher in sample 2 than in other samples, The maximum permitted level for magnesium in Nigeria is 0.20 mg/L, this value depend on consumer acceptability e.t.c., can be used to confirm the water samples as been hygienic for human consumption. This does not pose any significant water quality problem, because these results are within the standard acceptable levels of drinking water used in Nigeria. For Arsenic, all the samples are polluted with Arsenic because the arsenic concentrations exceed WHO standard. Arsenic can cause damage to the nervous system. Examples of diseases caused by arsenic are: Cancer of the lung, skin, prostate, kidney and liver, diabetes mellitus and hypertension. The WHO's guideline value for arsenic in drinking water is 0.05 mg/L (given a provisional guideline value of 0.01 mg/L in 2006).

Phosphate in all the samples exceeds the WHO standard which means that water is polluted with phosphate in all the samples. In the sample Iron in sample 2 has the highest concentration, samples 3, 4, and 5 have higher concentration and sample 1 has the lowest concentration. Iron in all the samples is within WHO standard for drinking water. Although, iron is an important dietary requirement in humans needed by haemoglobin and good for several other functions, when high concentrations of iron are absorbed, iron is stored in the pancreas, the liver, the spleen and the heart. This may damage those vital organs. The presence of excess iron in water imparts the taste and it also promotes growth of iron bacteria that hasten rusting process of all the ferrous metals that come in contact with the water (Chukwu et al., 2008).

A FORTRAN program was also used to determine the present state of water sources in the town, the data for each set of parameters were insert into program at various stage and run. The results show that the samples were within the intake of the people.

Conclusion

Water is the essence of basic survival. Without it, life on Earth would cease to exist. In order to ensure that human life continues to exist, we must work together and do our part to improve the quality of drinking water. Researchers must do their part in the laboratory to come up with treatment methods to improve the quality of drinking water. When the quality of drinking water is good, human health is also good. The non detection of pollution or small concentration of these metals implies that the groundwater in the study area were good. Pollution has not taken place and could be a result of the distant from the pollutants. It could also be due to the direction of groundwater flow, that is the pattern and movement, it could be due to the quality of recharge water as well as the complicated reaction between the water and minerals contained in the rocks. Therefore water sources from the study area were suitable for domestic use and are not likely to pose a major health risk to consumers. This research may serve as a reference for future studies on the assessment of water quality in the study area and its surrounding.

RECOMMENDATIONS

Human activities such as the use of chemicals in agriculture, effluent from homes, and sewage disposal, industrial waste discharged are all factors that contaminate water. After running the initial set of tests, well users should continue to test for bacteria once in a year. It's also a good idea to test for nitrate annually for several years. However, a nitrate test should always be conducted if an infant is drinking the water. Hence, it was recommended that physicochemical analysis should be carried out from time to time. At least once every three years. And on site drinking water systems shall be checked at least every 3 years (Nigerian Industrial Standard NIS 554: 2007).

Water is susceptible to various toxic pathogens and chemicals. Therefore, it is of extreme importance that the quality of water is tested through frequent monitoring. Water sources must be protected from potential source of contamination.

For on-site drinking water system, a minimum distance of 15 m shall be kept between the water system and potential source of contamination. Communities shall keep clean the protected area surrounding on-site drinking water systems.

The Federal Ministry of Environment in consultation with the States should declare special protection zones for Chemical Elements Sensitive Areas (such as Nitrates, heavy metals), wetlands based on such local peculiarities. For mechanized centralized drinking water, systems (high yield), broader protection zones shall be established and enforced by the Ministry of Environment.

There should be a long term sustained programme of monitoring the quality of groundwater in the community, to have a complete understanding of the ground water resources. Awareness should be created among the people on the possible danger of polluted water consumption and the diseases associated with it if not treated.

Conflict of Interest

The authors have not declared any conflict of interests.

ACKNOWLEDGEMENT

The authors appreciate the contributions made by Geophysics Group at the University of Ibadan, Nigeria.

REFERENCES

Ademoroti CMA (1996a). Environmental Chemistry and Toxicology, Foludex press Ltd. Ibadan, Nigeria pp. 215-218.

Akujieze CN, Coker S, Oteze GE (2003). Groundwater in Nigeria – A millennium experience: distribution, practice, problems and solutions. Hydrogeol. J. 1:259-274.

Chukwu O (2008). Analysis of Groundwater pollution from abattoir waste in Minna, Nigeria. Res. J. Dairy Sci. 2(4):74-77.

Erah PO, Akujieze, CN, Oteze GE (2002). Quality of groundwater in Benin City: A baseline study on inorganic chemicals and microbial contaminants of health importance in boreholes and open wells. Trop. J. Pharm. Res. 1:75-82.

Fasunwon O, Olowofola J, Akinyemi O, Fasunwon B, Akintoku O (2008). Contaminants Evaluation as Water Quality Indicator in Ago-Iwoye, South-western, Nigeria. Afr. Phys. Res. 2:12.

Nigerian Industrial Standard (NIS) (2007). 554:1-30.

Okeke CO, Igboanua AH (2003). Characteristics and quality assessment of surface water and groundwater recourses of Awka Town, Southeast, Nigeria. J. Nig. Assoc. Hydrol. Geol. 71-77.

Packham RF (1996). Drinking water quality and health: In pollution causes, Effects and control. Third Edition. (Harrison RM Ed.) Royal Society of Chemical, U.K pp. 52-65.

Punmia BC, Jain AK (1998). Wastewater Engineering. Laxmi Publications (P) Ltd, New Delhi.

Ramakrishnaiah CR, Sadashivaiah C, Ranganna G (2009). Assessment of water quality index for the groundwater in Tumkur Taluk, Karnataka State India 6(2):523-530.

Rand GN, Petrocelli SR (1985). Fundamentals of aquatic toxicity Hemisphere Publishing Corporation, pp. 1-46.

Sandhu SS, Waren WJ, Nelson SP (1976). Traces of inorganic in rural potable water and their correlation to possible sources. J. Water Res. 12:257-261.

Sridha MKC (2000). Ground water in Nigerian urban centers: problems and options. Schriftenr VerWasser Boden Luftyg. (105):393-397.
Suleiman FB (2006). Analysis of Some Sachet Water Samples in Katsina, Nigeria. Chem. Class J. Chem. Soc. Nig. Zaria Chapter 3:42-44.
Tyagi PD, Buddhi R, Chaudhary KC, Sawhney RL (2002). Degradation of ground water quality in industrial area in India. Ind. J. Environ. Prot. 20:174-181.
Vermani OPN (1989). Applied Chemistry, Theory and Practice by Inily Eastern Ltd, New Delhi, pp. 34-60.
Yisa J, Jimoh T (2010). Analytical Studies on Water Quality Index of River Landzu. Am. J. Appl. Sci. 7(4):453-458.

APPENDIX

This program is designed in the Department of Physics University of Ibadan to interprete geochemical data from leachates-polluted groundwater with main focus on 19 physio-chemical priority parameters (mainly inorganic) in accordance with WHO (2006) Standard and the United States Geological Survey and the United State Environmental Protection Agency Standard (2002) on secondary drinking water regulation.

TMP: Temprature
CLR: Colour
CDTY:Conductivity
ODR: Odour
TDS: TOTAL DISSO
LVED SOLIDS
PH: HYDROGEN ION CONCENTRATION
NA: SODIUM
CA: CALCIUM
MG: MAGNESIUM
FE: IRON
MN: MANGANESE
AL: ALUMINIUM
CU: COPPER
ZN: ZINC
AG: SILVER
CHL: CHLORIDE
SUL: SULPHATE
PHS: PHOSPHATE
FL: FLUORIDE
NTR: NITRATE
AS: ARSENIC
CD: CADMIUM
PB: LEAD
Z: SUMMATION OF THE DATA OF THE SEVEN MAJOR CATIONS: CA,MG,FE,MN,AL,AS,CD,PB FOR THE HARDNESS TEST
A: SUMMATION OF ANIONS
C: SUMMATION OF CATIONS

```
7 FORMAT(T1,A25,F9.3,A20)
10 FORMAT(T1,A160)
17 FORMAT(T1,A35,T1/T1,32('='),1X)
REAL::
TMP,CLR,CDTY,ODR,TDS,PH,NA,CA,MG,FE,MN,AL,C
U,ZN,AG,CHL,PHS,FL,NTR,AS,CD,PB,Z
OPEN(UNIT=1111,FILE='RESULTS')

PRINT*, 'THE INTERPRETATION OF RESULT'
PRINT*,''

WRITE(1,7)'DATE: 15/05/2013'
WRITE(1,10)
WRITE(1,7)'SAMPLE ID: MINING SITE'
WRITE(1,10)
WRITE(1,7)'SITE/LOCATION: ITAGUNMODI'
WRITE(1,10)
PRINT*,'INSERT THE DATA OF THE PHYSICAL
PARAMETERS:TMP,CLR,CDTY,ODR,TDS'
READ(*,*) TMP,CLR,CDTY,ODR,TDS
PRINT*,''
WRITE(1,17)'INTERPRETATION AND TEST RESULT'
WRITE(1,10)
WRITE(1,17)'PHYSICAL PARAMETERS'
WRITE(1,10)
WRITE(1,7)'TEMPERATURE =',TMP,'C'
IF(TMP.GE.12.AND.TMP.LE.25)THEN
WRITE(1,10)'TEMPERATURE WITHIN STANDARD'
ELSEIF(TMP.LT.12)THEN
WRITE(1,10)'TEMPERATURE BELOW
STANDARD,VISCOCITY AND IONIZATION MAY
INCREASE'
ELSE
WRITE(1,10)'TEMPERATURE EXCEEDS
STANDARD,POSSIBLITIES OF INCREASE IN
VAPOUR,PRESSURE AND SOLUBILITY'
ENDIF
WRITE(1,10)
WRITE(1,7)'COLOUR =',CLR,'UNIT'
IF(CLR.GT.15)THEN
WRITE(1,10)'COLOUR EXCEEDS STANDARD'
ELSEIF(CLR.LT.15)THEN
WRITE(1,10)'COLOUR WITHIN STANDARD'
ELSE
WRITE(1,10)'COLOUR EQUALS STANDARD'
ENDIF
```

The use of high resolution monitoring in the management of water quality failures caused by discolouration

Anonye, D. and Nwachukwu, A. N.

Williamson Research Centre for Molecular Environmental Sciences, School of Earth, Atmospheric and Environmental Sciences, University of Manchester, Manchester, UK M13 9PL.

Water quality failures caused by discolouration has become a serious problem for water service providers as they seek measures to comply with increasingly stringent standards from regulatory authorities. Such measures include tools that would enable them predict discolouration events before they occur, so that they can act in a more proactive manner. This paper demonstrates how high temporal monitoring of the distribution networks could be a useful tool for better prediction and management of discolouration. The turbidity results from high temporal measurements show that the bulk of sediment movement across networks are at low concentration; 75% of sediment movement caused by hydraulic disturbance was of low quantity (< 2 NTU), why 13% was of higher quantity (> 4 NTU). This suggests that during normal operations, distribution networks are in most cases, self-cleaning. Monitoring of sediment movement at low concentration proved to be useful in predicting discolouration risks as an increase of the former by 37 to 50% resulted in a quality failure in most cases. Such observation of sediment movement in a network was possible with measurements at high temporal resolution. High temporal resolution data was also used to map sediment budget of 0.032 kg in a section of a 400 m length network, which was equivalent to 1.04 gm^{-1}y^{-1}. It was also possible to monitor turbidity with residual chlorine concentration which gave a weak positive r^2 value of 0.107. Overall, the results suggest that a great deal of water quality information could come from high resolution monitoring of networks and therefore could help to proactively reducing discolouration risk.

Key words: Water quality, turbidity, sediment movement, discolouration, hydraulic disturbance.

INTRODUCTION

Failures in water quality can be defined as an event of exceedance in one or more quality standards set by regulatory authority. Management of these failures in drinking water quality have become a subject of strong concern for regulatory authorities in many parts of the world (Sadiq et al., 2008; Andreas et al., 2008). Among the various customer complaints about water quality failures, those with regards to its aesthetic qualities are in

Figure 1. Showing discoloured water before (A) and after (B) it had been allowed to stand for a while with a partitioning of the water into its clear (upper) and sediment (bottom) phases.

most cases, highest on the list of reoccurring issues. This is because many of such water quality failures, for example discolouration and chlorine-induced odour are largely unpredictable and measures to manage them are only carried out after it has been called for by several complaints from customers. The risk of the unpredictable re-occurrence of these events is likely to put more pressure on companies in the water sector, or water service providers (WSPs) as they are called in the UK, given that they are required to ensure the provision of water of high quality to customers at all time. Furthermore, the Water Service Regulatory Authority (WSRA) in the UK requires WSPs to provide these services to customers at minimum cost. WSPs that do not comply with regulatory water quality standards could however, face fines and requirements to make commitments to improve on the quality of their water. Despite the huge investment within the water sector, the risk of failures in critical water quality parameters like discolouration (Figure 1) still remains a key cause for concern within the UK water sector (Cook et al., 2005).

Consequently, there is a likely problem of increasing operational and administrative cost as regulatory authorities become stricter on their quality requirements. Hence, there is an urgent need on the part of WSPs for proper management of the issue of water quality failures. However, for the issue to be properly managed, it would be necessary to be able to predict its occurrence with greater confidence. Accurate prediction would require a better scientific understanding of the causes of these failures which are at the moment high on the complaint list. In a recent study carried out to explain the possible mechanism of discolouration, Husband and Boxall (2011) proposed a two-phase process requirement for discolouration events. These are:

1. A supply of material to the distribution network (which could come from the treatment process) and material accumulation (which could be likened to a depositional zone in hydrological terms).
2. Material mobilisation which at sufficient quantity, could ultimately result to a discolouration event.

The proposal therefore suggests that for discolouration to occur, there must be a sufficient amount of particulates (or sediments) available within the distribution system ("sediment availability"), whereas mobilisation or suspension would require hydraulic disturbances like an abrupt change in flow or pressure (Boxall et al., 2001; Prince et al., 2003). This became the underlying concept on which some established models used in predicting discolouration risks were based. An example is the Resuspension Potential Model (RPM) which uses changes in unidirectional flow as obtainable in water-distributing pipeworks in studying the risk of particulate resuspension (Vreeburg et al., 2004) and the risk of discolouration using the discolouration risk model (DRM) by Dewis and Randall -Smith (2005).

Another model known as PODDS (Prediction of Discolouration in Distribution Systems) which was developed by Boxall et al. (2001) adopts a similar "hydraulic" approach to predict discolouration. Whilst it has been possible for many decades to record hydraulic disturbance at high spatial and temporal resolution within the inaccessible pipework of the water distribution system there has been no way to get similar information about sediment availability. Thus far, management practice has relied solely on hydraulic changes in predicting discolouration events. There is little doubt about the possibility of hydraulic disturbance having some control over sediment movement which could result in discolouration

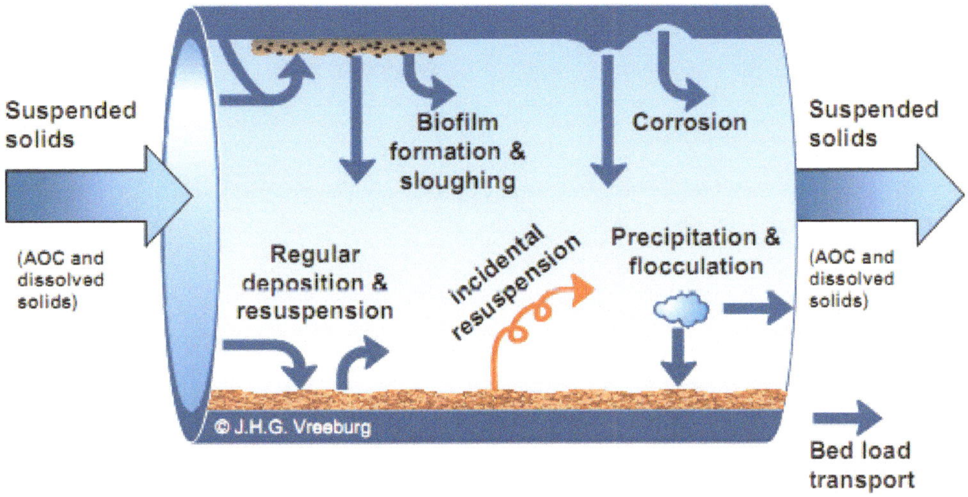

Figure 2. Showing the different types of sediment movements. Adapted from Newcombe and Dixon (2006), p281.

(Jānis et al., 2008; Vreeburg and Boxall, 2007). Some studies have even monitored the different mechanisms of sediment movement and sizes within a pipe under unidirectional flow (Figure 2) with much success (Subhasish and Athanasios, 2008; Cheng, 2006; Verberk et al., 2006; Seth et al., 2004; Gauthier et al., 2001; Paphitilis, 2001).

However, it is also likely that hydraulic disturbance alone may not give a consistently accurate prediction of discolouration. For example, when there is a hydraulic disturbance but insufficient sediments in the resulting suspension to cause discolouration. It therefore follows that the knowledge of sediment availability within a distribution network would be required in order to better understand and predict the likelihood and size of discolouration events after a hydraulic disturbance. However, the current method of managing discolouration falls short of this critical requirement. Currently, the process of managing discolouration has depended on information from customer complaints to verify the location and the subsequent sampling of water from taps within those locations. Relying on customer complaint or tap sampling, apart from the bare risk of fines from regulatory authorities, cannot provide sufficient data that could help in understand the processes that leads to discolouration. This is because discolouration events in most cases are high magnitude and short period events. Hence, low temporal resolution sampling, as currently obtainable is likely to miss those critical events that could render a more robust scientific understanding of the process.

Hence, the need for high temporal resolution monitoring of the process therefore arises. Since discolouration events still remains largely unpredictable even with a good understanding of high resolution hydraulic data across the network, it therefore suggests that knowledge of hydraulic disturbances may have a limited control on discolouration. Hence, the knowledge of sediment availability and extent to which hydraulic disturbances may affect its movement would give a better scientific understanding of the controls of discolouration events. However, such high resolution data are hitherto scarce. The reason for the paucity of high temporal resolution field data for turbidity (or sediment availability) is not farfetched since there is a practical difficulty in assessing water within the pipelines in a distribution network.

Also, the technology to carry out high temporal resolution monitoring required to better understand this process was previously not in existence. Such technology now exists and observing sediment movement at high temporal resolution can now be made possible. It is therefore important to demonstrate whether these new instruments can be used to improve the prediction of water quality failures by monitoring turbidity along with hydraulic disturbance at high resolution. An initial requirement for the success of such study would then be to understand the hydraulics of the water distribution system so as to note the likely areas where hydraulic disturbances could be more frequent. Water, on leaving the treatment plant is normally transported to the end users through a system of trunk mains, reservoirs and pumps (or booster stations). Trunk mains are generally used to transport water to several districts which are known in the UK as district meter area (DMA). From the DMAs, water is distributed via smaller mains and finally to service pipes that goes to end users. Trunk mains tend to maintain a high and relatively stable (or undisturbed) flow when compared to distribution networks in a DMA.

Some studies have suggested that the difference in flow between trunk mains and normal distribution pipes could be as high as a factor of a hundred (Blokker et al., 2008). The lower velocity makes distribution networks

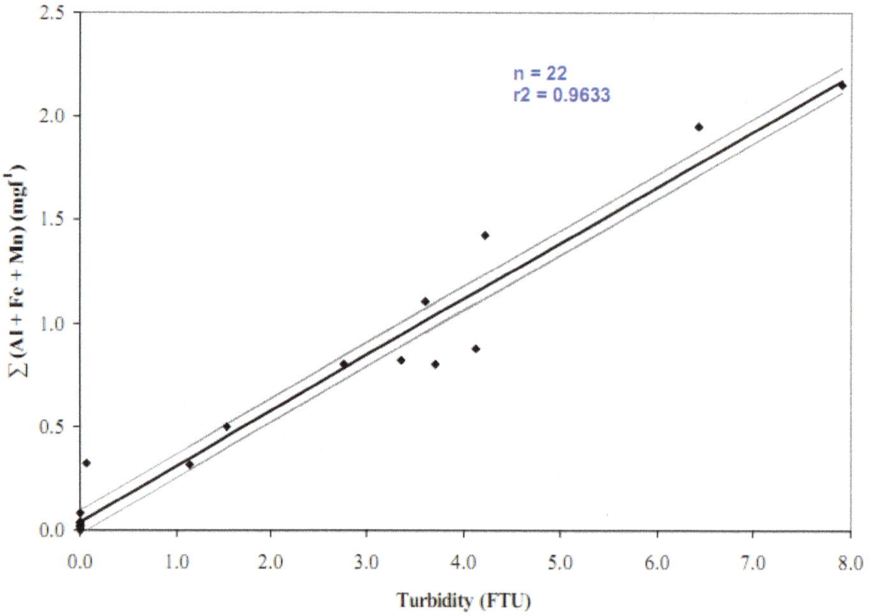

Figure 3. Calibration curve for turbidity events with sediment concentration. Adapted from Gaffney and Boult, 2011.

an ideal location for sediments to accumulate and therefore a possibility of resuspension during, say a sudden hydraulic change. Also, these areas normally witness sharp changes in flow or pressure (which are the only hydraulic disturbance within the network) during peak and low demands; a phenomenon known as the 'diurnal effect' (Gary, 2008) which could help in understanding sediment movement even better. This then suggests that distribution pipes within DMAs could be a very useful location to study the behaviour and movement of sediments in relation to hydraulic changes.

Hence, with the use of these new instruments, the likely relationship between discolouration (caused by sediment movement and measured as turbidity) and hydraulic disturbance can be observed at high resolution. High resolution turbidity monitoring will allow the relative importance to discolouration events of hydraulic disturbance and sediment availability to be determined. This knowledge is required in order to determine the amount of resource that should be allocated to the collection of each type of data in order to best be able to predict discolouration risk. However, it has been observed that there is a continuous movement of sediments at low concentration within the distribution network (Gaffney and Boult, 2012).

It therefore follows that a deeper understanding of the likely relationship between sediment movement and hydraulic disturbance could be established if observations are made based on the size of the turbidity event per hydraulic disturbance. Furthermore, turbidity monitoring at high temporal resolution would provide ample data that could be used to monitor sediment mass balance and spatial resolution were high enough to map sediment sources and sinks. This would also help improve prediction of discolouration since it affects the aesthetic quality of water and reduces confidence of customers. However, discolouration may not be solely an aesthetics concern as several studies have suggested, discolouration could also render aid to microbial activities (Bishop, 2007; Bachmann and Edyvean, 2006; Ainsworth, 2004).

Gaffney and Boult (2012) demonstrated the possibility of using high resolution data to estimate sediment budgets within a DMA. More importantly with regards to this work, was the development of a working curve which was used in the conversion of turbidity into particulate matter (or sediment) as shown in Figure 3.

Since this work is an extension of the previous study carried out by Gaffney and Boult (2012), the calibration curve above was also used in conjunction with flow (or 'discharge') data to determine sediment budget within a section of a DMA. In determining the sediment budget, it is important to ascertain whether a mass balance approach, that is, input and output fluxes, similar to the concept used in production engineering can be used to calculate sediment flux by using the working curve above as the medium for converting continuous turbidity data into sediments. To achieve this, high resolution monitoring would be carried out using the Hydraclam (Siemens WT, UK) which can provide high resolution measurements turbidity data. The aim of this study is to determine whether high resolution water quality data could be useful in improving the management of water quality particularly discolouration. Achieving this aim would be the first step towards the future success of proactive management of water quality failures for WSPs which would reduce the potential risk of fines from regulatory authorities if such failures persist.

METHODOLOGY AND METHOD

The study was carried out in two stages. The first stage involved collection of continuous turbidity and pressure datasets from the hydraclam. This took a period of eighty days. Data collected from this first stage was used to analyse correlation between sediment movement and hydraulic disturbance and the likely influence on discolouration. The second stage then involved collection of turbidity and flow data in other to calculate sediment flux for the particular section of the DMA which was used to map the sediment in that section. Collection of this data took twenty eight days. Both data from stages one and two of the research were however amalgamated and the influence of hydraulic disturbances on discolouration was analysed. Unlike previous studies which had seek to observe for discolouration events that could be tied to hydraulic changes alone, this study also focused on those sediment movements that may not result to any noticeable discolouration and check if such low concentration sediment movement may have any role to play in an eventual discolouration. The Hydraclam was the principal instrument deployed for this study. Also, the period of the study was carefully selected so that datasets were collected during normal operations when no maintenance was schedule.

The Hydraclam has four sensors that are capable of capturing four different water quality parameters at fifteen minutes interval. The parameters measured are turbidity, pressure, temperature and conductivity. It has a range between 0-14 FTU (formazine turbidity unit) for turbidity with a resolution of 0.01 FTU, and can be screwed on a hydrant line within the network. As water from the network gets into the hydraclam, it passes through an internal column where the nephelometric turbidity sensor is positioned to measure turbidity every 15 min. Data is normally offloaded from the Hydraclam via software on the Palm ™ PDA platform which could also be synchronised with a PC. The sensors in the Hydraclam that measures turbidity was calibrated with a 0 FTU and 4 FTU standard solutions, freshly prepared daily before it was sent for field measurements. The 0 FTU standard suspensions were 15 MΩ DIW. To prepare the 4 FTU standard suspension the remainder of the 0 FTU standard suspensions were divided into two portion of 1 L each. 10 g of Hydrazinium Sulphate (VWR International) was dissolved in one of the portion and 100 g of Hexamine (VWR International) was dissolved in the other portion. Both suspensions were then mixed and allowed to stand for 24 h in the dark, which resulted in a 4000 FTU standard suspension. The 4 FTU turbidity standard was then prepared by diluting a portion of the 4000 FTU standard suspension. Recalibration of the Hydraclam turbidity sensors was done each time it was removed from the distribution network using freshly prepared standard suspensions. Another instrument used was a Chloroclam. The calibration of the chloroclam sensor was done in the Siemens Water Technologies laboratory using standard chlorine solutions prepared from water in the network.

High temporal resolution data was used to analyse the correlation between discolouration and hydraulic disturbance. However, since sediments continuously move at low concentration along the distribution networks, the size of each turbidity event was compared to the hydraulic disturbance prior to that event so as to evaluate if the higher turbidity events were actually influenced by hydraulic disturbance. The size of turbidity events was divided into five size categories, which are: sizes of events >0.5NTU to <1NTU, >1NTU to <2NTU, >2NTU to <4NTU, >4NTU to <10NTU, and >10NTU (NTU is an acronym for Nephelometric Turbidity Units). The size of each event was determined by calculating the maximum hourly difference in turbidity concentration in a given location. The determination of the sediment budget was done within a pipeline that was unbranched as shown in Figure 6. The section of the pipe was carefully selected so as to ensure that no sediment 'leaks out' of the system under study thereby ensuring that the mass balance determined was as accurate as possible. The sediment budget was then calculated as the difference in sediment discharged into and out of the system.

In calculating the sediment flux, discharge data was collected at 15 min interval using a hydraulic model (SynerGEE, GL Water) and this was multiplied by the corresponding turbidity data after appropriate conversion to sediments in mg/l using the calibrated curve. The SynerGEE hydraulic model uses an input data that included household numbers connected to the network, consumption, both flow and pressure measurements. To account for slight systemic or operational changes, continuous data was collected over a large period of time (80 days for the first stage and 28 days for the second stage).

The calibrated hydraclams were deployed to the hydrant node locations ex-reservoir, each one of them was flushed to ensure the sensor is in full contact with the water in the pipes as shown in Figure 4. Data of pressure and turbidity were then collected at high temporal resolution at several locations within the distribution network. The nephelometric sensors in the Hydraclam were carefully monitored for fouling. When fouling of the sensor was suspected, cleaning of the sensor lens was done using 0.01M HNO_3 (Fischer chemicals) and re-calibrated using the procedure explained earlier. The turbidity in NTU was determined by calculating the difference between successive measurements. This approach ensures the elimination of errors associated with the measuring device.

Each turbidity event was counted and categorised by size of event in each location. Hourly hydraulic changes prior to each turbidity event were observed to check if it was causal or acausal to subsequent discolouration event. Hydraulic changes analysed and compared with turbidity was that of pressure and flow. Data collected from the second stage of the study was used to determine the sediment budget for that section of the DMA as displayed in Figure 4 using the working curve in Figure 2 to convert turbidity to sediment. The data from this stage was also used to develop a time series sediment balance graph for the section of the DMA. Also, continuous data of residual chlorine was collected alongside turbidity data at one of the locations where the Hydraclam is located with the Chloroclam. The flow of the water from the pipes into the tubing of the chloroclam was maintained between 120 to 160 ml/min. to ensure optimal chloroclam performance.

RESULTS

Overall monitoring period was 108 days with a total of 15,550 and 4922 datasets collected during the first and second stages of the research, respectively. There was no planned flushing of the networks all through the period of the study in order to ensure a favourable opportunity to study sediment movements during "normal operations". Study of the data collected across all sites reveals a temporal variance of turbidity between 0 NTU and 14.6 NTU during the duration of the research. It was also found that 75% of suspended particles (measured as turbidity) ranged between 0.5 NTU and 2 NTU this can be compared with those with magnitude greater than 4NTU that was only 13%, which gives a spread of 62%. This then suggests that the bulk of suspended material fluxes within networks are being transported at low concentration which could be difficult to detect by mere customer observation. The hourly correlation between hydraulic

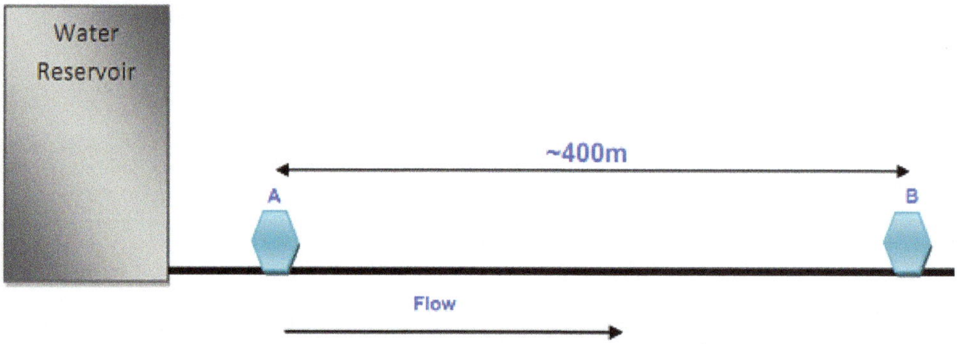

Figure 4. The set-up for the determination of sediment budget, necessary to demonstrate the possibility of sediment mapping within a DMA. A and B are Hydraclams (Siemens WT, UK).

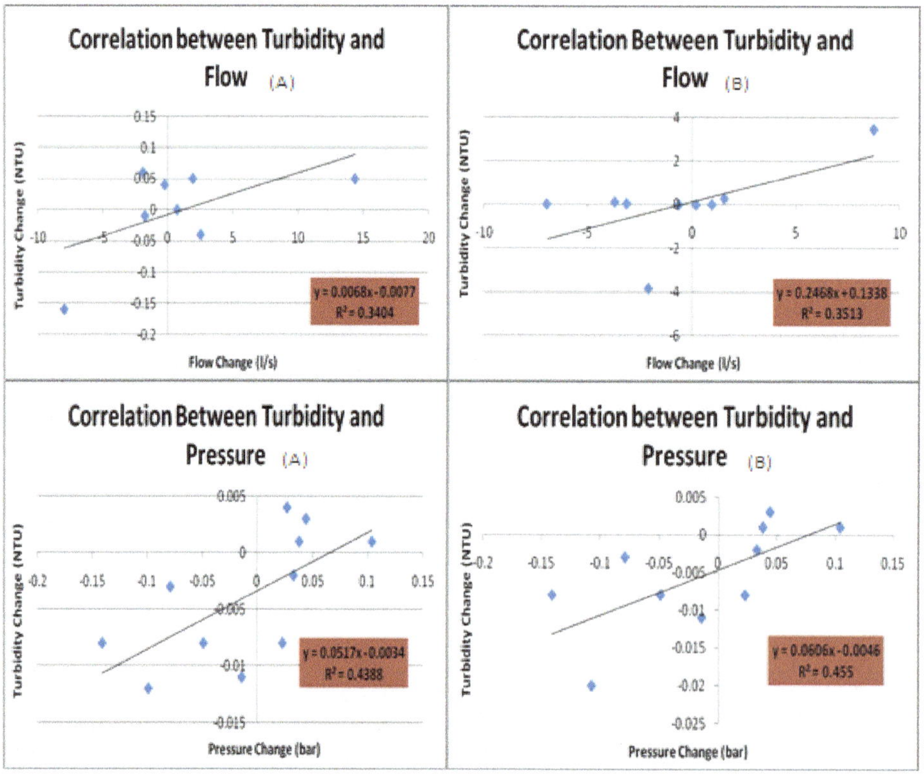

Figure 5. Showing general hourly graphs of hydraulic changes and turbidity.

changes and turbidity was found to be generally positive, but weak (Figure 5). Hydraulic disturbance caused by changes in flow was found to have an r^2 value between 0.34 (for changes occurring during periods of relatively low turbidity) and 0.35 (for changes occurring during periods of high turbidity). A similarly positive but weak correlation was also found with hydraulic disturbances caused by pressure changes with an r^2 of 0.44 (for changes occurring during periods of relatively low turbidity) and 0.46 (for changes occurring during periods of high turbidity), respectively. This weak correlation is likely to be as a result of turbidity events occurring at periods when there was no prior hydraulic change and *vice versa*. It then suggests that during "normal operating conditions" hydraulic changes may be playing a weak or passive role in causing or determining discolouration risk than initially observed.

The weak correlation between hydraulic disturbance and turbidity was also made apparent with respect to magnitude of sediment movement within the distribution network (Figure 6). Since there was a limited amount of data on sediment movement from a single site, datasets from all sites was combined so as to make a meaningful assertion on the data that is being analysed. The data

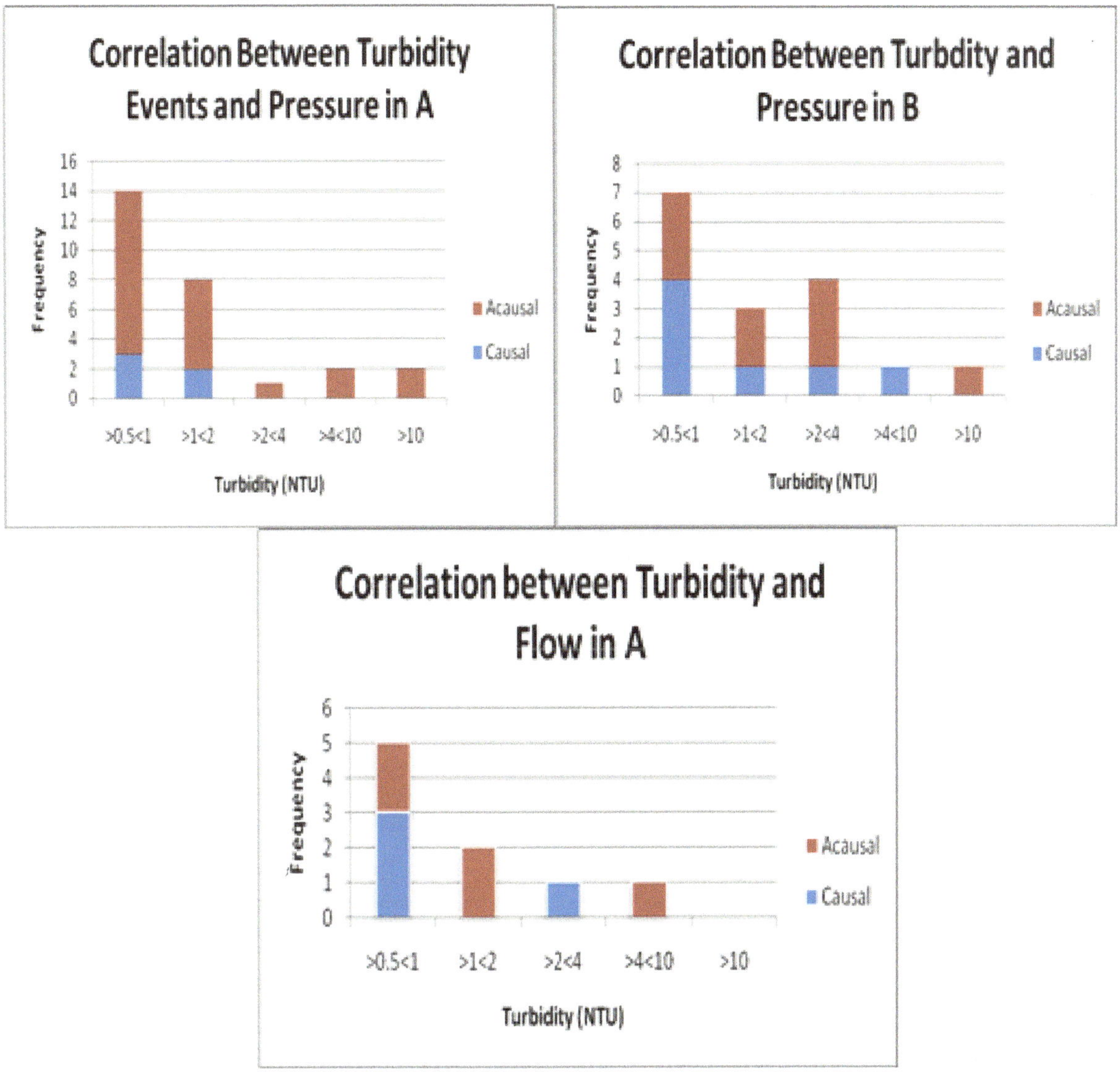

Figure 6. Causal and non-causal turbidity events occurring in the distribution networks.

collected across all sites showed that only 14% of the entire sediment movement was greater than the regulatory standard of 4 NTU (representing just 2% of the whole turbidity events of interest in this research) was actually caused by a hydraulic disturbance.

In a sharp contrast, the percentage of sediment movement below the magnitude of 4 NTU that was caused by hydraulic disturbance was 33% which represents 29% of the whole ranges of sediment movement events of interest in this study. It therefore follows that 69% of all the sediment movements analysed were not caused by any hydraulic disturbance. However, an interesting finding in the result was the manner by which turbidity varied over time.

The number of quality failures (that is, measurements above 4 NTU) was five (5) which is minor as seen from the percentage time exceedance graphs below (Figure 7). Although, continuous turbidity measurement during the period of the study revealed a high range, temporal variability (Figure 7), the temporal differences in the baseline turbidity (lowest turbidity measured) all through the period of the study was low and more "predictable". For example, for every 10% rise in the baseline turbidity, an occurrence of a turbidity event within the range of interest in this study was likely. At one of the sites, a 50% increase in the baseline turbidity was immediately followed by a quality failure while in another site similar water quality failures were observed when the baseline turbidity increased by 37 and 49% without any prior hydraulic disturbance to account for them.

The period within which the baseline turbidity accumulated to the amount capable of causing a quality failure averaged 10 days. It was also found however that turbidity events that was more than the regulatory

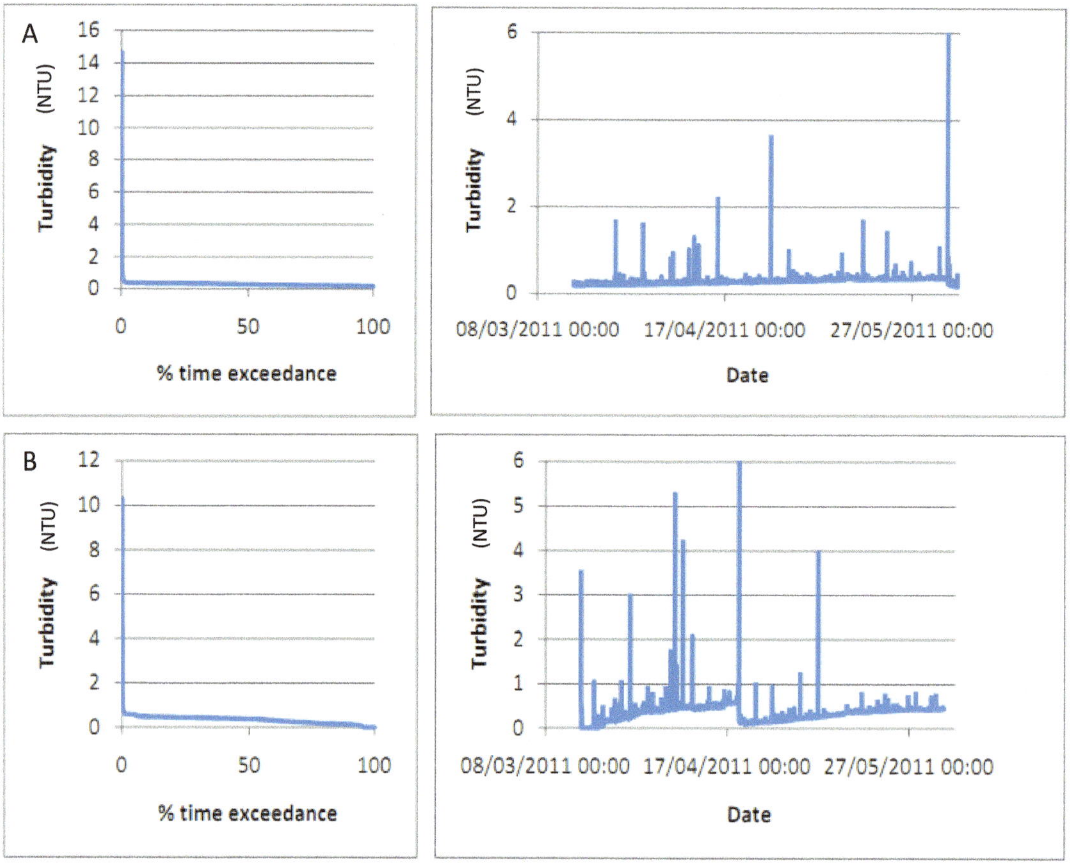

Figure 7 (A-B). Showing high temporal turbidity data collected from different sites, with the corresponding % time exceedance curves.

Table 1. Showing the sediment budget of a section of the DMA.

Sediment input		Sediment output		Balance		$gm^{-1}y^{-1}$
Mass (kg)	Error (kg)	Mass (kg)	Error (kg)	Mass (kg)	Error (kg)	
2.660	3.0×10^{-5}	2.628	3.0×10^{-5}	0.032	4.24×10^{-5}	1.04

standard was followed by a sharp drop in the baseline turbidity value. For example, at one site the baseline turbidity dropped by 44% immediately after a turbidity event of 14.6 NTU. At another site, a drop of 56% was observed after an event of 10.3 NTU. This behaviour suggests a self-cleaning process that goes on within the distribution network which can only be properly observed at such high temporal resolution as carried out in this study.

The section of the distribution network that was analysed in this study was found to be net accumulating with 0.0320 kg ± 4.24×10^{-5} kg of sediment budget accumulated over the period of the study (Table 1). The standard error in Table 1 is expressed at 95% confidence limit and includes only the regression line error from Figure 3, used in converting turbidity to concentration of metal oxides (which is a surrogate for sediments).

It is apparent from the results that sediment movement within the distribution network also proved to be temporally dependent (Figure 8). At some periods, the sediment balance (that is output-input budget) was depositional (that is net accumulating), whereby the input was higher than the output. In other periods the sediment output flux was much higher with an apparent mobilisation of sediments into other sections within the networks. However, when such mobilisation occurred, it was not as a result of any sudden increase in flow but just an increase in concentration of sediments within the system. The result also revealed the superimposition of output and input fluxes in most instances, which suggests a "steady-state" process going on within the network.

The results from the chloroclam displayed a sinusoidal pattern of residual chlorine concentration (Figure 9), with the maximum at 0.33 mg/l and minimum at 0.14 mg/l

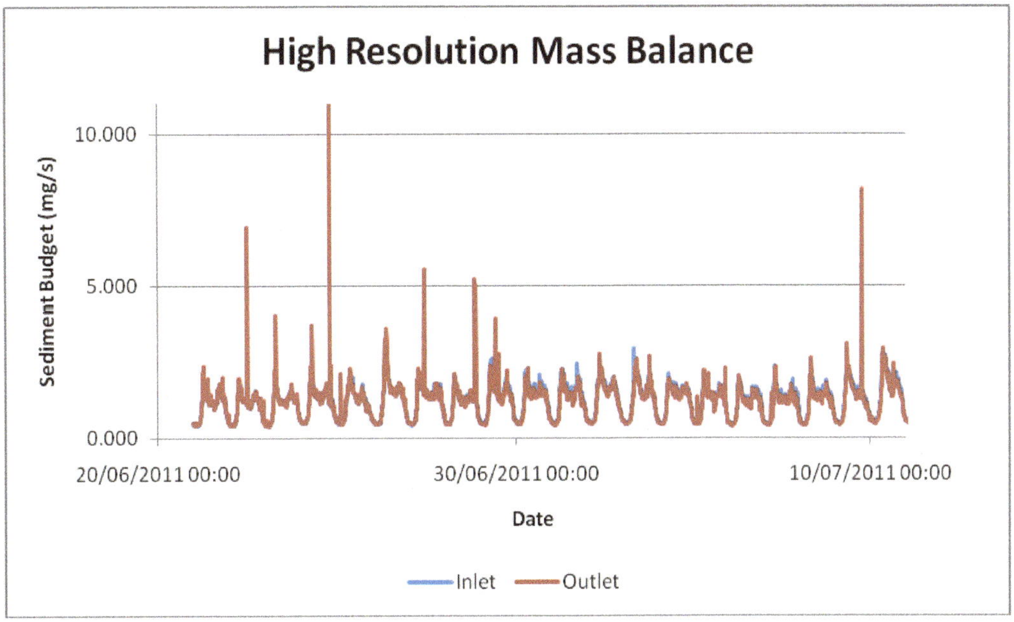

Figure 8. Showing high temporal sediment balance within a section of the DMA.

(Figure 10). However, when the changing chlorine concentration was compared with the varying turbidity there was no clear indication of decreasing residual chlorine caused by any change in turbidity. However, when the data sets for periods when residual chlorine concentration was decreasing and when it was increasing were compared with the corresponding turbidity (Figure 10), it was found that the correlation was positive. The positive correlation was also found to be order of magnitude stronger during the periods of increasing residual chlorine concentration than the period when chlorine concentration was decreasing.

DISCUSSION

As water service providers (WSPs) seek a more proactive approach in resolving issues of water quality failure, like those with regards to discolouration, the results from this study show that high resolution monitoring would play a very important role in achieving this aim. High resolution monitoring showed specific behaviours of sediments within distribution networks that would not have been otherwise observed. Such information would be useful in adopting a more proactive quality failure management and could provide an "early warning signal" for management to act upon before such failures gets to the customer's or regulatory authority's notice. How continuous monitoring could help to achieve this aim would be discussed below.

It is clear from the results of hydraulic changes and sediment movement during "normal operations" that there were a lot of instances where sediment movement resulted in a water quality failure without any apparent or significant hydraulic disturbance and *vice versa*. This could however either be as a result of the time lag between measurements in this study (which was every 15 min), which would have missed a hydraulic change that might have lasted only for some few seconds. It could also be that there was no mobilisation after a hydraulic disturbance because there were no sediments to be mobilised. Thirdly, it could be that sediments were just dislodged from an accumulation point without any requirement for a hydraulic disturbance. However important the first two possibilities are, since there were significant amount of sediment mobilisation that were unrelated to hydraulic disturbance, the latter case seems more likely. It suggests that the system is likely to be self-cleaning, a possibility that had also been proposed in some earlier studies (Vreeburg and Boxall, 2007; Boxall and Prince, 2006). To confirm this viewpoint would however require a longer period of monitoring and more data collection of high temporal and spatial resolution. However, the implication of such undisturbed sediment mobilisation across distribution networks is that better prediction of discolouration risks would require more resource to be channelled towards collecting data for sediment availability.

Nevertheless, the likelihood of a self-cleaning process as made evident with the varying turbidity results is suggestive of the possibility of sediments within distribution networks tending to behave like sand piles under "monotonic" or unidirectional loadings (Liu et al., 2011; Roul and Schinner, 2010). One explanation to this with regards to discolouration would be that; as the bulk sediment material increases due to accumulation, the increased "internal pressure" of the bulk (or "sand pile") results in a "crumbling effect". This crumbling of bulk

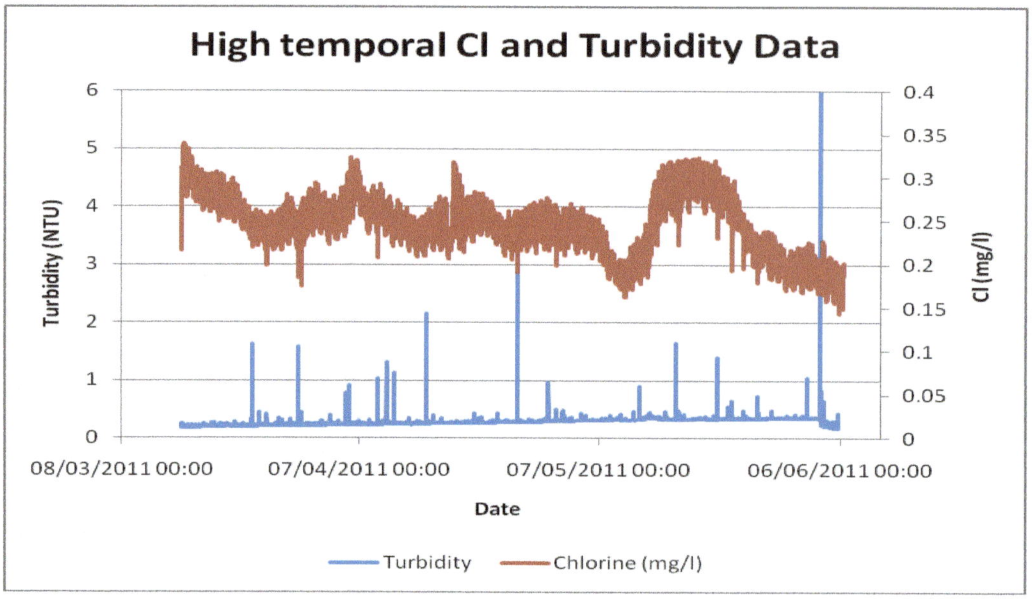

Figure 9. Showing high temporal changes in turbidity and residual chlorine concentration.

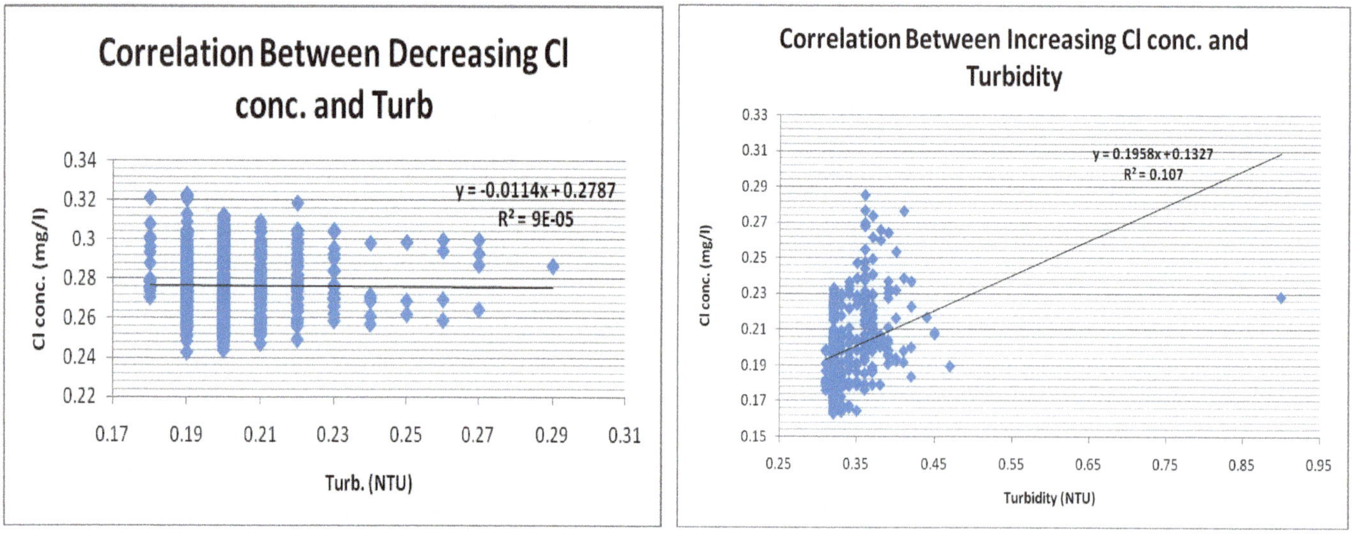

Figure 10. Showing the correlation between turbidity and changing pattern (that is, instances of decreasing and increasing) of residual chlorine concentration.

sediment would probably be observed as sediment mobilisation without any need for an additional external shear (or hydraulic disturbance). If this is true, as suggested by the weak correlation between hydraulic disturbance and turbidity, it would then explain why discolouration risk tends to be less predictable, as desired, since the risk of quality failures still lingers even after "maintenance" by flushing or pigging as observed by previous studies (Gaffney and Boult, 2011; Barbeau et al., 2005; Carriere et al., 2005).

This then suggests that bulk sediments within the pipes may withstand external hydraulic disturbance as well as internal pressure during normal operations until they build-up to a size that they can no longer continue growing and are therefore mobilised. For example, at one of the sites studied, a continuous period of low sediment movement was observed, which was immediately followed by a sudden water quality failure (high turbidity) without any prior hydraulic disturbance. Although, such possibilities are likely to occur in areas where sediment layers exists (Husband and Boxall, 2011) or conditions where the hydraulic status are relatively quiescent (Boxall and Saul, 2005; Prince et al., 2003) even during flushing, like dead-ends within the networks. It therefore follows that during normal operations significant mobilisation of sediments occurs only when sufficient amount has been accumulated. But this likely condition could change quickly, say, during a turbulent flow as obtainable in flushing, when mobilisation then takes preference over accumulation, which is also supported by previous studies (Blokker et al., 2010).

From the analysis of the results, it can be suggests that under "normal operations", as obtained during the period of the research, the role of hydraulic disturbances may be passive and only becomes strong when sufficient amount of shear is induced on the system, such as obtainable during "flushing". Whether sediment mobilisation would result in a water quality failure would only depend on the amount of material accumulated and mobilised at any point in time. This result also suggests that a high hydraulic disturbance would be important but not necessary for sediment mobilisation to result in a quality failure. It therefore follows that reducing or eliminating the risk of discolouration would require knowledge of areas where sediments are or could possibly be and an estimation of the quantity present at that period. This would then require a system of sediment mapping to ascertain the locations of highest risks within a DMA. The calculation of a section of the DMA would be discussed later.

An interesting finding from the results analysed was the strong correlation between increasing and decreasing baseline turbidity and corresponding risk of water quality failures. The baseline turbidity datasets were low enough not to be detectable by customers and drops sharply immediately after a significant turbidity event, that they are not likely to be observable by low temporal sampling. However, the possibility of incorporating the pattern of increase and decrease of these low turbidity data (or baseline turbidity) in risk models could provide a promising alternative in predicting discolouration risk. For example, in all the turbidity events that resulted in a water quality failure, an increase of the baseline turbidity by 37 to 50% was observed. This failure occurred without any observable hydraulic disturbance. Also in one of the sites analysed, the baseline turbidity drop by 30% immediately after a turbidity event that resulted in water quality failure, still suggesting the possibility of a self-cleaning process as earlier explained.

Whether this observed behaviour is a "rule" or just an exception cannot be established by the data from this study. It would therefore require a more analysis that would contain both high temporal and high spatial data-sets to fully explain this behaviour. However, the query by Gaffney and Boult (2011) of whether a self-cleaning process (which this observation proposes), could result in a compliance failure could now be clearly seen as a likely possibility as 17% of all the events considered as self-cleaning actually resulted in a quality failure.

In the 28 days of continuous monitoring of a section of the DMA, it was demonstrated that data from high resolution monitoring can be used to calculate the sediment budget of a section in a DMA. The sediment budget calculated by adopting the concept of material balance into and out of the selected section of DMA was 0.032 kg, which is equivalent to 1.040 $gm^{-1}y^{-1}$, estimated as metal oxides from the calibration curve of Figure 5. This result was within the range of previous studies as it can be compared with 0.112 $gm^{-1}y^{-1}$ and 5.212 $gm^{-1}y^{-1}$ from the work done by Gaffney and Boult (2011) using a similar method, and that carried out by Vreeburg et al. (2008) which estimated the weight of sediments from a flushed section of a DMA to be 1.071 $gm^{-1}y^{-1}$.

It is however important to note the variability that could occur in such mapping exercise. For example, during the first 10 days of the mapping study, data collected showed that the system was not net accumulating but rather "erosional" (a term used by hydrologists to describe a river system that dislodges more material than it takes). During those first 10 days the sediment deficit was roughly about the same figure it turns out to have accumulated by the end of the 28 days of monitoring (that is 1.047 $gm^{-1}y^{-1}$). The reason for the "reversal" in the character of this section of the DMA could be attributed to a likely event of sediment mobilisation, as supported by the results which showed that the highest sediment output from the system also occurred within this period. It however suggests that for any sediment mapping result to provide an accurate picture of the character of any section of a DMA, that is either net accumulating or sediment source, an extensive study would be required for that section of the DMA.

From the results, the chlorine concentration was also variable just like the turbidity data with a maximum measurement of 0.33 mg/l and a minimum of 0.14 mg/l. The correlation with turbidity was generally weak. However, a careful observation of the pattern of scatter on both graphs reveals that the density of the data was within the area of low turbidity and at high turbidity there was little or no residual chlorine. This therefore suggests that high turbidity is likely to affect residual chlorine across distribution networks, although this may not be instantaneous possibly because of time required for any chlorine-demanding reaction to occur. For example, on day 15 of the first stage of the study, an increase in turbidity from 0.22 NTU to 1.06 NTU (380% increase), resulted in a drop of residual chlorine by 17%, thirty minutes after the turbidity event. Also on day 30, an increase of turbidity from 0.24 NTU to 2.16 NTU (800% increase) resulted in a drop of residual chlorine by 12%, also thirty minutes after the turbidity event. Such drops were not however observed at other times, even with similar increase in turbidity, which suggests the complexity of the controls and the need for more studies.

It was also interesting to note that residual chlorine concentration could display a sinusoidal pattern within the network, since it is only possible to add chlorine in treatment plant. However, a recent laboratory simulation study done by Ryan and Jayaratne (2003) suggested that such changes in concentration as observed in this study could be as a result of the age or transient time of the water and/or the pipe material used in its distribution.

The huge difference in the r^2 values of the results of increasing and decreasing residual chlorine with respect to turbidity is suggestive of the possibility that high chlorine concentration could also possibly play a role in particulates generation. Earlier studies have also suggested this, with respect to wall reactions and reactions with other organic compounds (Zhang et al., 2011; Castro and Neves, 2003). However, the data from the study is insufficient to make certain any such claim, but further investigation could help to shed more light on this finding.

Another potential use of high temporal data of residual chlorine concentration would be the prospects of tracking odour complaint, which would be indicated by an unusually high chlorine concentration. This has never been previously observed and could only be made possible by high temporal measurements.

Conclusion

The monitoring of water quality at high temporal resolution has given tangible proof of high magnitude but short period changes that occurs in water quality as it travels through distribution networks. The results from high temporal measurements also confirm, with greater confidence, the following:

i. The bulk of sediment movement occurs at low concentration,
ii. Under normal operating conditions, the hydraulic control on the risk of discolouration is low and sediment mobilisation would depend more on availability. The implication of this is that more resource would be required for the collection of data about sediment availability so as to ensure a more accurate prediction of discolouration,
iii. Distribution networks are likely to be self cleaning with regards to discolouration. However, it is recommended that the density per DMA of such continuous monitoring devices like those used in this study be increased so as to determine the sinks or fate of the sediments from the section that seem to be self cleaning,
iv. Mapping of sediments within a DMA is possible and can become useful in predicting areas of high risk, by conducting a more extensive mapping exercise with the method used in this study.

Further recommendations

Another usefulness of high temporal monitoring is the ability for it to track the baseline turbidity within a DMA. As shown in this study, this could be an alternative method in predicting discolouration risks under normal operating conditions, and can be monitored only with high temporal measurements. Hence, further study on this is recommended. It is also recommended that sediment mapping, as demonstrated in this study be extended to other sections of the DMA or distribution networks in general. This would help in the development of sediment sinks that would aid proactive management and reduce the uncertainties of general periodic flushing.

It is important for any proactive approach to managing water quality failures to adopt continuous monitoring across the networks as more extensive data from such monitoring would be useful for further research towards understanding the behaviour of sediments in distribution networks.

Conflict of interests

The author(s) have not declared any conflict of interests.

REFERENCES

Ainsworth R (2004). Safe Piped Water: Managing Microbial Water Quality in Piped Distribution Systems. Published on behalf of the World Health Organisation by IWA Publishing London, UK.

Andreas K, Jure L, Carlos G, Jeanne V, Christos F (2008). Efficient Sensor Placement Optimization for Securing Large Water Distribution Networks. J. Water Res. Plan. Mgt. 134(6):516 -526. http://dx.doi.org/10.1061/(ASCE)0733-9496(2008)134:6(516)

Bachmann R, Edyvean R (2006). Biofouling: an historic and contemporary review of its causes, consequences and control in drinking water distribution systems. Biofilms 2(3):197–227. http://dx.doi.org/10.1017/S1479050506001979

Barbeau B, Gauthier V, Julienne K, Carriere A (2005). Dead-end flushing of a distribution system: Short and long-term effects on water quality. J. Water Suppl. Res. Technol.-Aqua. 54(6):371-383.

Bishop P (2007). The role of biofilms in water reclamation and use. Water Sci. Technol. 55(1–2):19–26. http://dx.doi.org/10.2166/wst.2007.005 PMid:17305119

Blokker E, Vreeburg J, Buchberger S, van Dijk C (2008). Importance of demand modelling in network water quality models: a review. Drink. Water Eng. Sci. 1:27–38. http://dx.doi.org/10.5194/dwes-1-27-2008

Boxall, JB, Prince RA (2006) Modelling discolouration in a Melbourne (Australia) potable water distribution system. J. Water Suppl. Res. Technol.-Aqua, 55(3):207-219.

Boxall JB, Saul AJ (2005). Modelling Discolouration in Potable Water Distribution Systems. J. Environ. Eng. 131(5):716-725. http://dx.doi.org/10.1061/(ASCE)0733-9372(2005)131:5(716)

Boxall JB, Skipworth PJ, Saul AJ (2001). A novel approach to modelling sediment movement in distribution mains based on particle characteristics. Water Software Systems: v. 1: Theory and Applications (Water Engineering & Management). B. Ulanicki, B. Coulbeck and J. P. Rance, Research Studies Press, Hertfordshire, UK. 1:263-273.

Carriere A, Gauthier V, Desjardins R, Barbeau B (2005). Evaluation of loose deposits in distribution systems through unidirectional flushing. Am. Water Works Assoc. 97(9):82-92.

Castro P, Neves M (2003). Chlorine decay in water distribution systems case study – lousada network. Electron. J. Environ. Agric. Food Chem. 2(2):261-266.

Cheng NS (2006). Influence of shear stress fluctuation on bed particle instability. Physics of Fluids. 18(9):1-28: Art. 096602.

Cook DM, Boxall JB, Hall SA, Styan E (2005). Structural Intergrity and water quality in water distribution networks. Proceedings of CCWI Water Management for the 21st Century, Centre for Water Systems, University of Exeter, Exeter, UK.

Dewis N, Randall-Smith M (2005). Discolouration risk modelling. Water Asset Mgt. Int'l. 1.3:16-18.
Gaffney J, Boult S (2012). The need for and use of high resolution turbidity monitoring in managing discolouration in distribution. J. Environ. Eng. 138(6):637–644. http://dx.doi.org/10.1061/(ASCE)EE.1943-7870.0000521
Gary NF (2008). Drinking Water Quality, Problems and Solutions, John Wiley and Sons, Ltd. Second Edition, Chapters 20 and 21.
Gauthier V, Barbeau B, Millette R, Block JC, Prevost M (2001). Suspended particles in the drinking water of two distribution systems. Water Sci. Technol. Water Suppl. 1(4):237-245.
Husband PS, Boxall JB (2011). Asset deterioration and discolouration in water distribution systems. Water Res. 45(1):113-124. http://dx.doi.org/10.1016/j.watres.2010.08.021PMid:20813391
Jānis R, Verberk J, Vreeburg J, Kamila G, Tālis J (2008). Chemical and microbial composition of loose deposits in drinking water distribution systems. Environ. Eng.: The 7th International Conference, pp. 695-702.
Liu W, Li Y, Cao L (2011). The Qualitative Research of Macro-Micro Association of Sand under the Monotonic Loading. Advan. Mat. Res. 261-263:1544-1548. http://dx.doi.org/10.4028/www.scientific.net/AMR.261-263.1544
Newcombe G, Dixon D (2006). Interface Science in Drinking Water Treatment: Theory and Applications. Academic Press, 10:14-16.
Paphitilis D (2001). Sediment movement under unidirectional flows: an assessment of empirical threshold curves. Coastal Eng. 43(3-4):227-245.http://dx.doi.org/10.1016/S0378-3839(01)00015-1
Prince R, Ryan G, Goulter I (2003). Role of operational changes in forming discoloured water in a water distribution system. Advances in water supply management: Proceedings of International Conference on Computing and Control in the Water Industry, Swets & Zeitlinger, London, U.K.
Roul P, Schinner A (2010). Simulation study on micro and macro mechanical behaviour of sand piles. Powder Technol. 204(1):113-123.http://dx.doi.org/10.1016/j.powtec.2010.07.026
Ryan G, Jayaratne A (2003). Particles in distribution systems and assessment of discoloured water. In: Maintenance and Assessment of Distribution Systems to Improve Water Quality. C.G. Workshop, Sydney.
Sadiq R, Saint-Martin E, Kleiner Y (2008). Predicting risk of water quality failures in distribution networks under uncertainties using fault-tree analysis. Urban Wat. 5(4):287-304. http://dx.doi.org/10.1080/15730620802213504
Seth A, Bachmann R, Boxall J, Saul A, Edyvean R (2004). Characterisation of materials causing discolouration in potable water systems. Water Sci. Technol. 49(2):27-32. PMid:14982160
Subhasish D, Athanasios P (2008). Sediment Threshold under Stream Flow: A State-of-the-Art Rev. Water Eng. 12(1):45-60.
Verberk J, Hamilton L, O'Halloran K, Van Der Horst W, Vreeburg J (2006). Analysis of particle numbers, size and composition in drinking water transportation pipelines: results of online measurements. Water Sci. Technol. 6 (4):35-43.http://dx.doi.org/10.2166/ws.2006.902
Vreeburg JHG (2007). Discolouration in Drinking Water Systems: A Particular Approach. Delft 978-90-74741-91-0.
Vreeburg JHG, Schaap PG, van Dijk JC (2004a). Particles in the drinking water system: from source to discolouration. Water Sci. Technol. 4(5–6):431–438.
Vreeburg JHG, Schippers D, Verberk J, van Dijk JC (2008). Impact of particles on sediment accumulation in a drinking water distribution system. Water Res. 42(16):4233-4242. http://dx.doi.org/10.1016/j.watres.2008.05.024 PMid:18789809
Zhang Y, Yi-Pin L (2011). Adsorption of Free Chlorine on Tetravalent Lead Corrosion Product (PbO2). Environ. Eng. Sci. 29(1):1-8.

Simulation of rainfall runoff process for Khartoum State (Sudan) using remote sensing and geographic information systems (GIS)

A. Aldoma and M. Y. Mohamed

Faculty of agriculture AlzaiemAlazhary University, Sudan

Using geographic information systems (GIS) and remote sensing integration to determine runoff caused due to rainfall from watershed has performed increasingly attention in recent years. This study was conducted in the Khartoum State. Curve Number (CN) method was applied for estimating the runoff depth in the watershed. Hydrologic soil group and land use maps were generated in GIS Environment. Hydrologic soil groups and land use maps were used to generate the CN map. Soil Conservation Services-Curve number (SCS-CN) method was followed to estimate runoff for the watershed. It was found that the model can predict runoff reasonably well. It could be concluded that the SCS-CN method can be applied to predict runoff volumes for planning of various conservation measures and for other water resources applications.

Key words: Soil Conservation Services-Curve number (SCS-CN), geographic information systems (GIS), remote sensing, runoff estimation

INTRODUCTION

Surface runoff is the water flow that occurs when soil is saturated to full capacity and excess water from rain over the land. This is a major component of the water cycle. Rainfall generates runoff, and its occurrence and quantity are dependent on the characteristics of the rainfall event, that is, the intensity, duration and distribution. Apart from these rainfall characteristics, there are numbers of catchment specific factors, which have a direct effect on the occurrence and volume of runoff.

Measured rainfall is one of the most significant input data in applying the hydrological models for runoff estimations. Unfortunately, the distribution of rainfall usually varies significantly in both space and time. Therefore, the limited number of rainfall stations in the catchment can have a major impact on the accuracy of runoff estimations. The accurate estimation of the spatial distribution of rainfall therefore requires a very dense rainfall network, which involves high installation and operational costs.

Remote sensing can provide measurements of many of the hydrologic variables used in hydrologic applications, either as direct measurements comparable to traditional forms, or as entirely new data set. The pixel format of digital remote sensing data makes it suitable to merge it with geographic information system (GIS).

Most of the previous work on adapting remote sensing to hydrologic modeling has involved the Natural Resources Conservation Service runoff Curve Number

(CN) model. This involvement used remote sensing data as a substitute for land cover maps which had been obtained by conventional means.

The GIS technology provides suitable alternatives for efficient management of large and complex databases; it is used in hydrologic modeling to facilitate processing, management and interpretation of hydrologic data.

Several studies have been done to incorporate GIS in to hydrologic modeling of watersheds. The traditional method for establishing CN on small watersheds includes field surveys and interpretations of aerial photographs. For large drainage basins, field surveys are prohibitively expensive and an excessive number of aerial photographs may be required for complete coverage. A further disadvantage of conventional techniques may be the infrequency of the surveys and the consequent failure to account for changes in vegetative cover and land use.

Land use is an important characteristic of the runoff process that affects infiltration, erosion, and evapotranspiration. Hydrologic models, distributed models in particular, need specific data on land use and its location within the basin. Remote sensing can provide measurements of many of the hydrologic variables used in hydrologic and environmental model.

Rainfall-runoff modeling is a very important topic of research that can be utilized for management and control of water resources. Where there is lack of important elements such as digital land use map, digital curve number map, hydrological soil groups map and digital runoff map, these types of maps are very useful in hydrological investigations. In addition, future predictions with different scenarios can be undertaken by rainfall-runoff modeling.

The integration of remote sensing and GIS has been widely applied and has been recognized as a powerful and effective tool in estimating and evaluating rainfall; therefore the main aim of the present study is to apply a rainfall-runoff model using GIS and remote sensing techniques to simulate the rainfall-runoff process and create digital runoff map for the study area using SCS-CN method.

Nowadays, modeling has become a common practice in every field of endeavor, and runoff modeling is no exception (Donigian, 1995). The main reason behind the using of modeling in general is the limitations of the techniques used in measuring and observing the various components of hydrological systems (Beven, 2001). Also using hydrologic models will increase our understanding and explanation of the natural phenomena and its dynamic interactions with the surrounding systems (that is, climatic terrestrial, pedologic, lithologic and hydrologic systems) (Chiew, 1994). However, under some conditions, predictions can be made in deterministic or probabilistic sense.

Another use of modeling is to predict how the system will respond to the future alternative conditions and actions (Donigian et al., 1995; Linsley, 1982) summarized the principal purposes for which hydrological model have or can be employed. In general they can be used for hydrologic research purposes, for forecasting and prediction of stream flow, and for engineering and statistical applications (record extension, operational simulation, data fill-in, and data revision).

Detailed analysis of flood hydrograph has special importance in flood mitigation, flood forecasting and/or establishing flows for many structures which must convey floodwaters (Linsley, 1982). Rainfall-Runoff modeling covers a wide range of applications and practices. This can be divided into two main groups, viz: flood studies (planning and designing new hydraulic structure, operating and/or evaluatingexisting hydraulic structures, preparing for and responding to flood, flood damage reduction, and regulating flood plain activities), and storage studies (catchment and reservoir yield analysis, and water resource potential) (Blandford and Meadows, 1990; Chiew, 1994; Connolly, 1995).

The Soil Conservation Service - curve number (SCS-CN) method has its origins in the unit hydrograph approach to rainfall-runoff modeling. The unit hydrograph approach always requires a method for predicting how much of the rainfall contributes to the storm runoff. The SCS-CN method arose out of the empirical analysis of runoff from small catchments and hill slope plots monitored by the USDA. The Curve Number method (SCS, 1972),also known as the Hydrologic Soil Cover Complex Method, is a versatile and widely used procedure for runoff estimation. In this method, runoff producing capability is expressed by a numerical value varying between 0 to 100.

METHODOLOGY

This study is conducted in Khartoum State, the study area located in the middle of Sudan, lies between latitudes 15.8° N to 16.45° N and longitudes 31.5° E to 34.45° E, surrounded by five states, divided to seven localities. It has an area of 22,122 km^2 (Figure 1).

The topography of the area is undulating. Khartoum featured as hot arid climate, with only months of July and August seeing significant precipitation. Khartoum averages a little over 155 mm (6.1 in) of precipitation per year. Temperatures may exceed 45°C in summer time.

Rainfall data is collected from National Meteorological Corporation. Data from five gauges were used here, namely: Al Sahafa, Jabalawlia, Shambat, Soba and Omdurman; the rainfall data were obtained for 20 years.

Soil data are obtained from the Ministry of Agriculture of Khartoum State. Land use map was obtained from remote sensing authority. Arc GIS version 9.3 produced by ESRI, 2008 was used for creating, managing and generating of different layers and maps.

Soils were classified into A, B, and D groups according to their minimum infiltration rate, which obtained for a bare soil after prolonged wetting as shown in Table 1.

To prepare the land use layers in the study area, eight major land use classes were identified. Classified land use map was integrated with the hydrological soil groups map to produce the CN. To estimate curve numbers for the entire basin, vector coverage of soil showing the HSG, and use map were overlaid. The soil layer and classified land use layer were used to calculate curve numbers

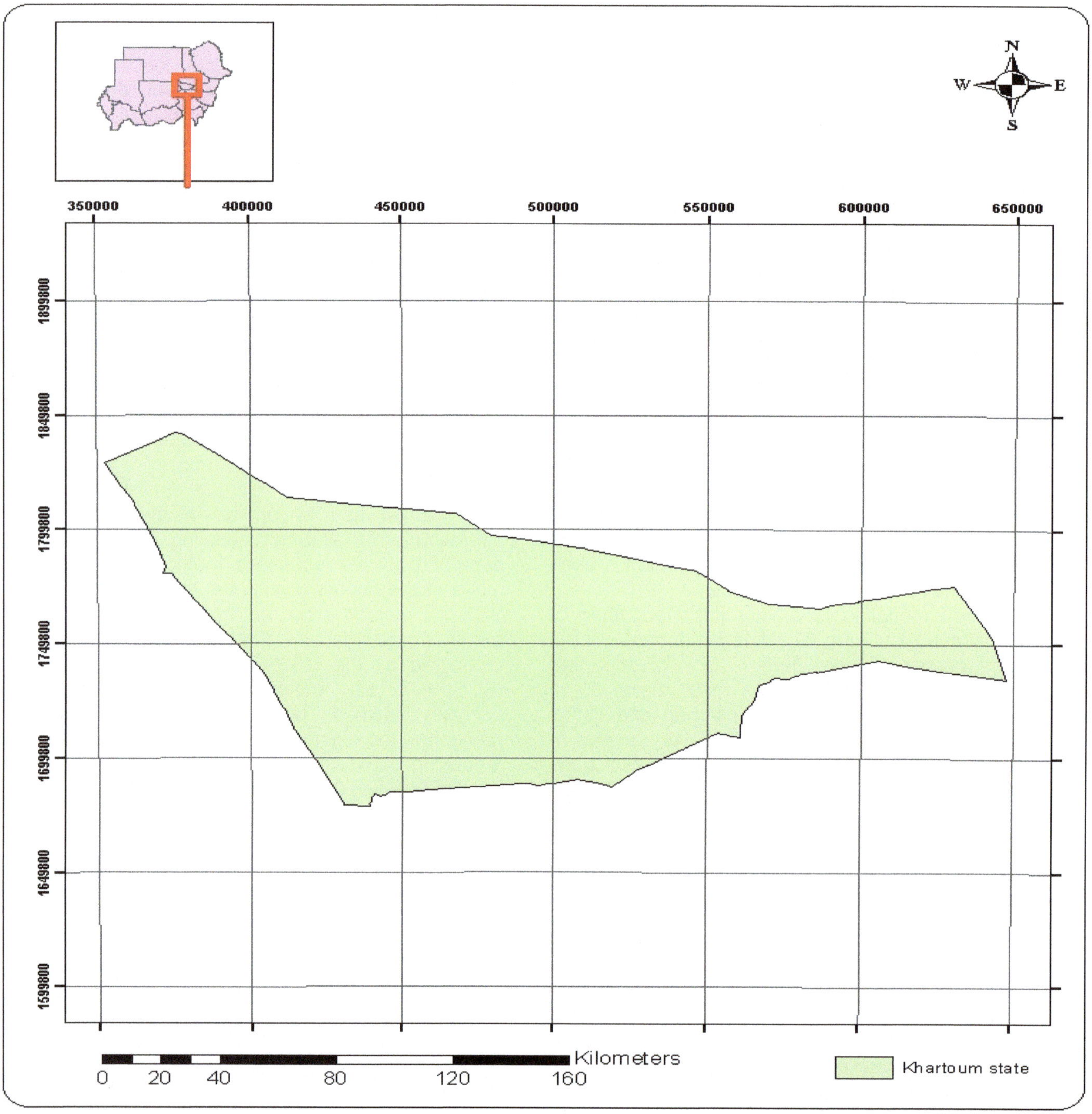

Figure 1. Study area.

value.

The rainfall-runoff equation used for estimating depth of direct runoff from storm rainfall (USDA, 1972):

$$Q=(P-0.2S)^2/(P+0.8S)$$

The CN is a dimensionless runoff index determined based on hydrologic soilgroup (HSG) and land use. The CN method is able to reflect the effect of land use on runoff. The NRCS runoff equation is widely used in estimating directRunoff because of its simplicity, flexibility and versatility.

RESULTS AND DISCUSSION

Remote sensing and GIS technologies are suitable tools for rainfall-runoff estimation which were used to determine CN distribution in the study area. It could be used to calculate runoff depth of the basin area. The results of this study may be very useful for flood forecasting and the development of hydrologic and hydraulic structure. These results are useful to understand

Table 1. Hydrological soil groups.

Soil Group	Characteristics
A	Low overland flow potential, high minimum infiltration capacity even when thoroughly wetted (>0.76 cm/h), deep, well to excessively drained sands and gravel.
B	Moderate minimum infiltration capacity when thoroughly wetted (0.13 to 0.76 cm/h), moderately deep to deep, moderately to well drained, moderately fine to moderately coarse grained (e.g. sandy loam).
D	I ligh overland flow potential: Very low minimum infiltration capacity when thoroughly wetted (<0.13 cm/h) clay soils with high swelling potential, soils with permanent high water table. Soils with a clay layer near the surface, shallow soils over impervious bedrock.

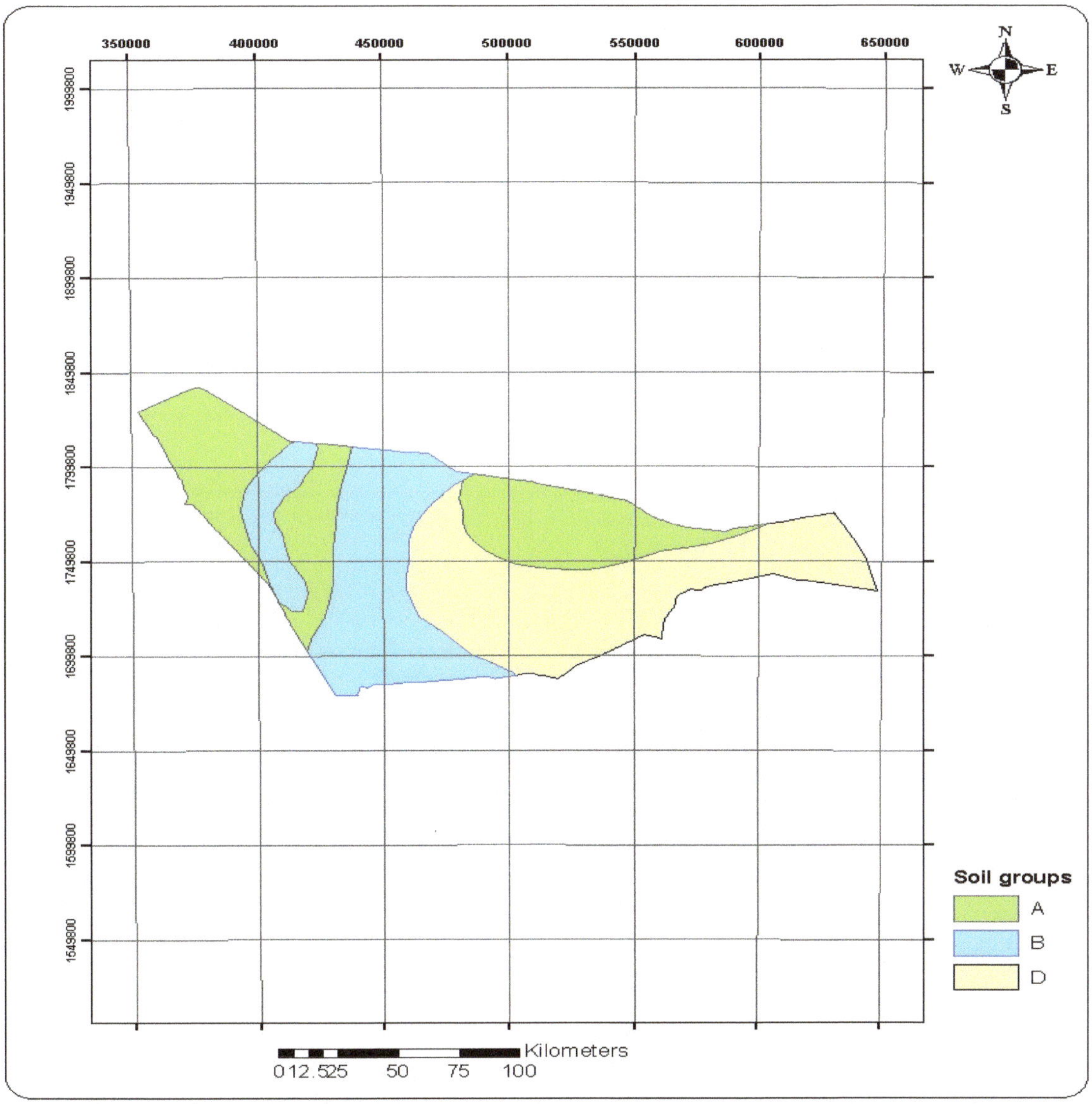

Figure 2. Hydrological soil groups (HSG).

hydrological characteristics of basin area.

From classified soil map three hydrological soil groups (HSG), it has been found that three HSG covered the study area, namely A, B, and D. Figure 2 represents the distribution of these groups.

From Figure 2 it can be observed that 30% of the study

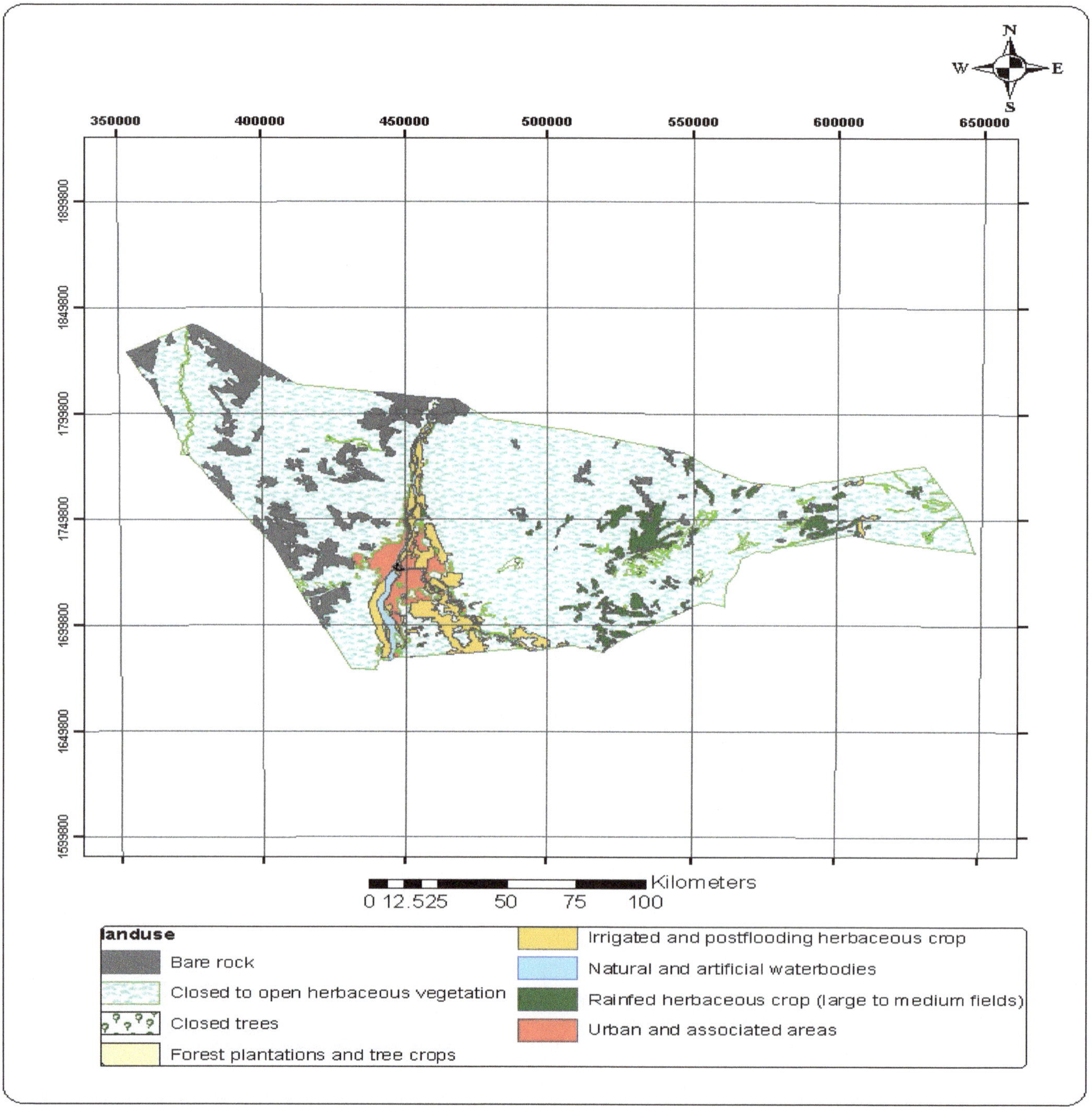

Figure 3. Land use classed distribution in the study area.

area is covered by group B and 34% is covered by group A while the rest of the area covered by group D (36%). The classified land use map is shown in Figure 3. The study area has been classified into eight major land use classes. Table 2 represents the percentage of each class.

From Table 2 it can be observed that most of the study area was covered by closed to open herbaceous vegetation while the forest plantation and tree crops have least coverage.

The CN value for each soil hydrologic group and corresponding land use class were calculated. Hydrologic soil groups A and B leads to low CN value while the hydrologic group D leads to high CN value in the study area. The lowest CN value was found to be 30 in closed trees while the highest CN value was found to be 100 in water bodies; watershed generates more runoff for a given rainfall in areas having greater CN values. The CN

Table 2. Percentage of land use classes.

Land use	% Area
Bare rock	9.09
Urban and associated area	2.67
Rainfed herbaceous crops	3.53
Irrigated and post flooding herbaceous crops	4.29
Natural and artificial water body	0.99
Forest plantation and tree crops	0.014
Closed trees	1.73
Closed to open herbaceous vegetation	77.66

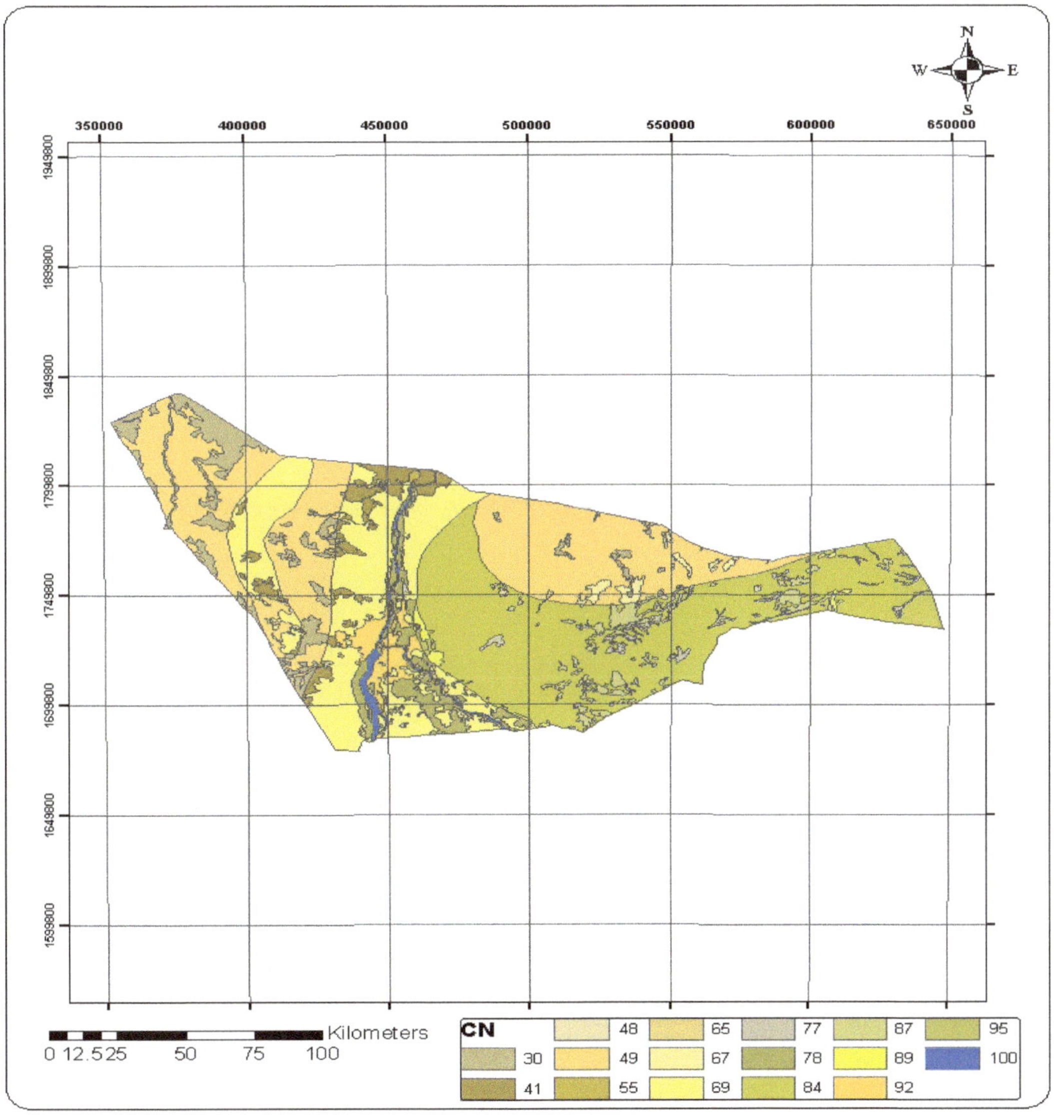

Figure 4. Distribution of CN values in the study area.

values were mapped and displayed in Figure 4.

Runoff depths of the basin were estimated using CN values in the SCS-CN equation as shown in Figure 5.

Conclusion

Remote sensing and GIS technologies can be used to

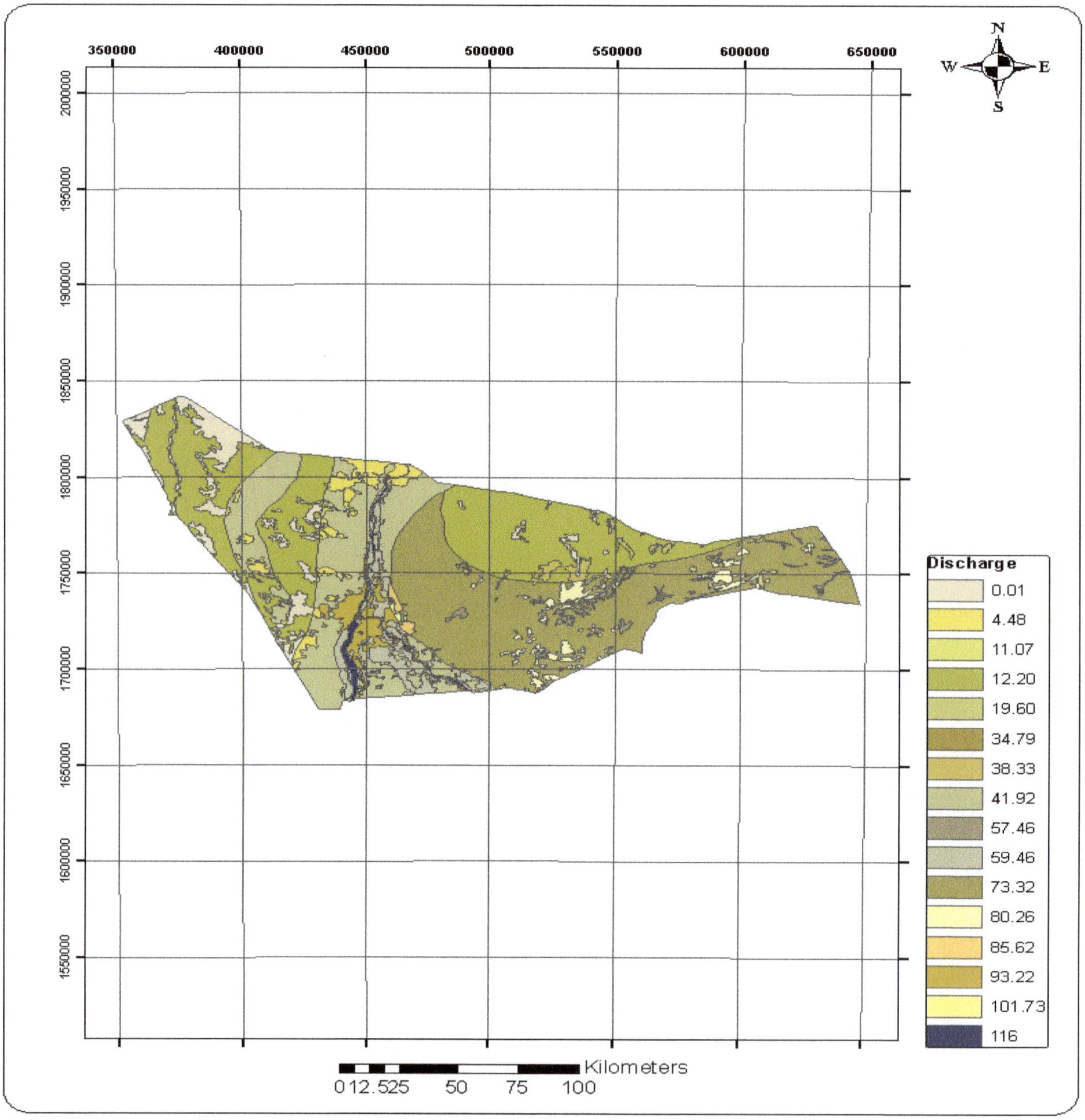

Figure 5. Runoff estimation based on CN values.

determine curve number values which could be used to calculate runoff depth. In this study, SCS-CN method was applied to simulate the rainfall-runoff process and create digital runoff map using GIS and remote sensing techniques. Three HSG covered the study area. The study area has been classified into eight major land use classes. Curve numbers values were estimated and thereafter the runoff amounts were determined for different areas. The results of the study would be very useful for flood forecasts and future land use changes scenarios.

Conflict of Interests

The author(s) have not declared any conflict of interests.

REFERENCES

Beven KJ (2001). Rainfall-Runoff Modelling. UK, the Primer, Wiley, Chichester.

Blandford GE, Meadows M (1990). Finite element simulation of non linear kinematic surface runoff. J. Hydrol. 119:335-356.

Chiew FH, SM, TA, (1994). Application of the daily rainfall-runoff modelMODHYDROLOG to twenty-eight Australian catchments. J.

Hydrol. 153:383-416.
Connolly RD, SDM (1995). Distributed parameter hydrology model (ANSWERS) applied to a range of catchment scales using rainfall simulator data I: Infiltration modelling and parameter measurement.
Donigian AS, Bicknell BR, Imhof JC (1995). Hydrological Simulation Program – Fortran (HSPF). Colorado, Water Resources Pubs
Linsley RK, Kohler MA, JL, Paulhus H (1982). Hydrology for Engineers New York, McGraw-Hill Book Co.
SCS (1972). National Engineering Handbook. Washington, D.C., Department of Agriculture, U.S.

Flood hazard assessment of River Dep floodplains in North-Central Nigeria

Rose E. Daffi[1] , Johnson A. Otun[2] and Abubakar Ismail[2]

[1]Department of Civil Engineering, Plateau State Polytechnic, Barkin Ladi, Nigeria.
[2]Department of Water Resources and Environmental Engineering, Ahmadu Bello University, Zaria, Nigeria.

Flood is a recurring event that leads to hazards. The probability of a flood occurring is normally investigated followed by flood hazard mapping which defines the areas that are at risk of flood inundation. This study carries out flood hazard assessment for the flood prone areas within the low-lying flat river valley of the River Dep watershed using Remote Sensing (RS) and geographic information system (GIS) for 2-year to 1000-year return periods, with regards to inhabited areas and other land uses that will be affected. Result shows that the most affected land use within the floodplain is agriculture with inundated area ranging from 68.82 to 146.10 km^2. Low to medium flood hazards predominant dominate the floodplain with area extent increasing from 112.2 to 140.75 km^2 for low hazard and 35.65 to 163.65 km^2 for medium hazard. High hazard is mainly within the deep part of the floodplain with minimal area extent of 4.11 km^2. The study recommends low hazard areas to be used for irrigation farming and early rainy season farming, medium and high hazard areas for irrigation farming only while low, medium and high hazard areas for the 100-year flood should be avoided with respect to construction of residential or commercial structures. Generally areas close to rivers should be avoided for rainy season farming and residential or commercial development.

Key words: Flood hazard, remote sensing, geographic information system (GIS), land use, land cover.

INTRODUCTION

Hazard is defined as a 'potentially damaging physical phenomenon or human activity that may cause the loss of life or injury, property damage, social and economic disruption or environmental degradation' (Damayanti, 2011). Alkema and Westen (2005) defined flood hazard as 'the chance that a flood event of a certain magnitude will occur in a given area within a given period of time'. Each hazard is characterised by its location, intensity, frequency and probability. Flood hazard can be described by different parameters including flood extent, water depth, flow velocity, duration and the rate at which the water rises, where flood depth, velocity and duration are important factors in flood damage (de Moel et al, 2009). While flood hazard is the impact of flooding on development and people, the velocity and depth of floodwaters greatly affect personal safety and damage to infrastructure and agricultural lands (Floodplain Development Manual). The manual suggests that at velocities of over excess of 2 m/s grass and earth surfaces begin to scour affecting stability of foundations, and depths of over 2 m with less velocity can cause damage to light-framed buildings.

Catchment flood hazard assessment focuses on the catchment as a whole and looks at how the different characteristics of the catchment integrate to contribute to flooding within the floodplains. There is a yearly occurrence of floods of different severity within the River Dep floodplains resulting in loss of human lives, destruction of flood infrastructure, livestock and crops. Flood depth and velocity affect agricultural activities within the Dep floodplain including harvest and transportation of crops within the area. During the rainy season, the farms become submerged in floodwater which can be over 1 m in depth, sometimes taking several days to recede and resulting in destruction of submerged crops.

The inhabitation and use of the floodplains results in encroachment into a land that should be left to the natural elements, posing a great danger to the inhabitants and users, and high risk of destruction and loss of investments and lives. People living within the River Dep floodplains can still enjoy the agricultural benefits that it provides and yet avoid the hazards that come with floods. This can be achieved by ensuring that man's activities do not conflict with the natural occurrence of flooding by adequately planning the use and management of the floodplain. Though conventional traditional methods can be used for flood hazard assessment, the use of remote sensing and geographic information system techniques have been suggested to provide quick, efficient and effective results as investigated and documented by Balanova and Vassilev (2010), Damayanti (2011), Kafle et al. (2006), Salimi et al. (2008), Ahmed et al. (2010) and others. This study is aimed at carrying out flood hazard assessment for the flood prone areas within the low-lying or flat river valley of the watershed with regards to land uses that will be affected. Flood hazard maps for some extreme flood events will be analysed and the affected land uses estimated.

Study area

The Dep River Basin lies between latitudes 8°00'00"N to 9°20'00"N and longitudes 8°20'00"E to 9°35'00"E as shown in Figure 1. It falls within the Lower Benue River Basin Development Authority in North Central Nigeria. The flood plain is within the relatively flat river valley with elevations between 78 to 200 masl. The land within the area is predominantly used for agriculture.

MATERIALS AND METHODS

Softwares

1. ERDAS Imagine 9.2 for remote sensing analysis.
2. ArcGIS 9.3 for GIS analysis. It has the capability of accepting compatible add-on extensions such as the HEC-GeoRAS.

Satellite data

30 m × 30 m resolution Landsat ETM image of Dep River watershed was downloaded from Global Land Cover Facility. It is the only one available for the complete study area. Date of acquisition is 02/11/2001.

Flood hazard assessment

Flood hazard assessment was carried out by assessing the impact of flood depth and flood velocity where deep inundating water and high velocity were classified as the destructive force, deep inundating flood with low velocity as less destructive and shallow inundating water with high velocity as more destructive (Damayanti, 2011). The classification of flood hazard according to depth by Cooper and Opadeyi (2006) and the relationship between flood depth and flood velocity established in the Floodplain Development Manual were taken into consideration.

The following criteria were used by multiplying the depths and the velocities to assess flood hazard:

Level 1: Shallow water and low velocity
Level 2: Shallow water and medium velocity or deeper water and low velocity
Level 3: Shallow to medium or deep water with high velocity or deep water with medium velocity.

The different categories of velocities and depths are tabulated below for the 100-year flood:

Table 1 implies that the threshold for high hazard was set at 1.4m of flood inundation depth and 2m/s of flood velocity.

Flood hazard analyses were carried out for 2, 5, 10, 20, 50, 100, 200, 500 and 1000-year floods, with more emphasis on the 100-year flood, using flood inundation maps previously obtained by Daffi (2013). The map was overlain on the settlement map of the study area to view those that will be affected by flood of these return periods. It was also overlaid on the Landsat land use classified map to view and analyse the land uses that would be inundated by the floods.

RESULTS AND DISCUSSION

Assessment of flood inundation on settlements

Figure 2 shows the flood inundation extent on a 3-D map of the study area. It shows the floodplain is within the low lying river valleys of the Dep River.

Overlaying the 100-year flood inundation map on the settlement map showed that about four settlements are likely to be affected by the 100-year flood as seen in Figure 3.

Ground validation however, showed that of the four settlements seen to be within the flood area, only one settlement called 'Wuse' is under risk of inundation from the flood because of its proximity to the adjacent river.

Assessment of flood inundation on land uses

Overlaying the flood inundation extent map on the land use land cover classification map gave an indication of the area and percentage area inundated by the 100-year

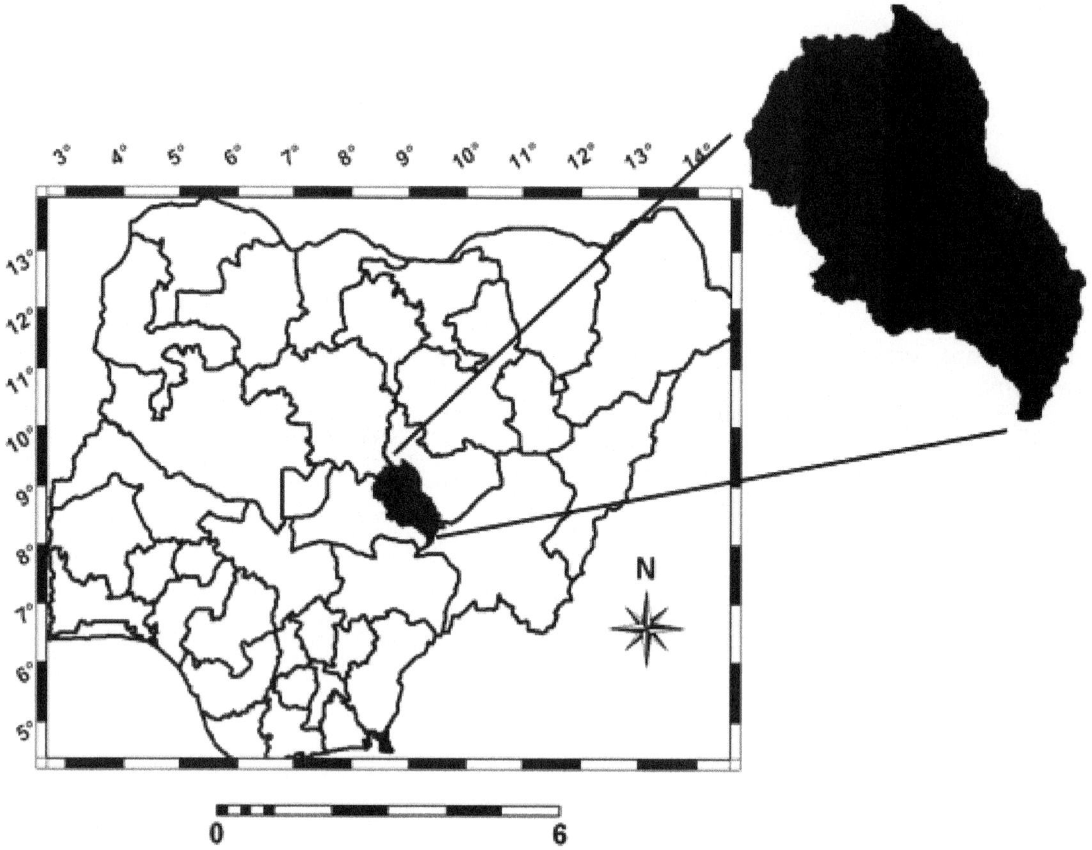

Figure 1. Map of Nigeria showing river dep basin.

Table 1. Flood hazard classification based on flood depths and flood velocities.

Flood depth (m)	Flood velocity (m/s)	Hazard classification	Indicators
0 – 1	0 – 1	Low Hazard	Unlikely loss of life, minor increase to existing flood levels, able-bodied adults can wade to safety with little difficulty
1 – 1.4	1 – 2	Medium Hazard	Possible loss of life, significant increase in flood levels with respect to crops, roads and buildings.
1.4 – 11.86	2 – 3.52	High Hazard	Probable loss of life, major increase to existing flood levels with respect to crops, roads and buildings.

flood shown in Table 2.

Of the total land area of the floodplain, generally shrubs/scattered trees make up the highest land cover class inundated by floodwaters with percentage area of between 46.4 to 46.5% for the 2-year and 1000-year floods respectively. This is because the areas closest to the river, most especially around the confluence of River Dep and River Benue, are made up of dense shrubs and trees and some of the areas are even impenetrable. There was a general increase of 77.28 km^2 or 52.9% for the agricultural land affected by flood from the 2-year to the 1000-year floods.

Also, for all floods agricultural cultivated land is the land use that is significantly affected with an area of 119.86 km^2 for the 100-year flood. A lot of dry season (irrigation) farming takes place within the floodplain which is commonly referred to as 'fadama'. This is the most important land use within the floodplain.

Bare surface/degraded land is the land cover that is least affected by the flood with area of 1.57 km^2 or 0.6% for the 100-year flood. This is because there are hardly any bare surfaces available within the floodplain; the land is mostly cultivated or covered with shrubs or trees because of the swampy nature of the area.

The remaining 6.4% or 16.62 km^2 is made up of water bodies and rock outcrops where the rock outcrops are mainly within the upper parts of the Dep River catchment which mainly make up the source of the river.

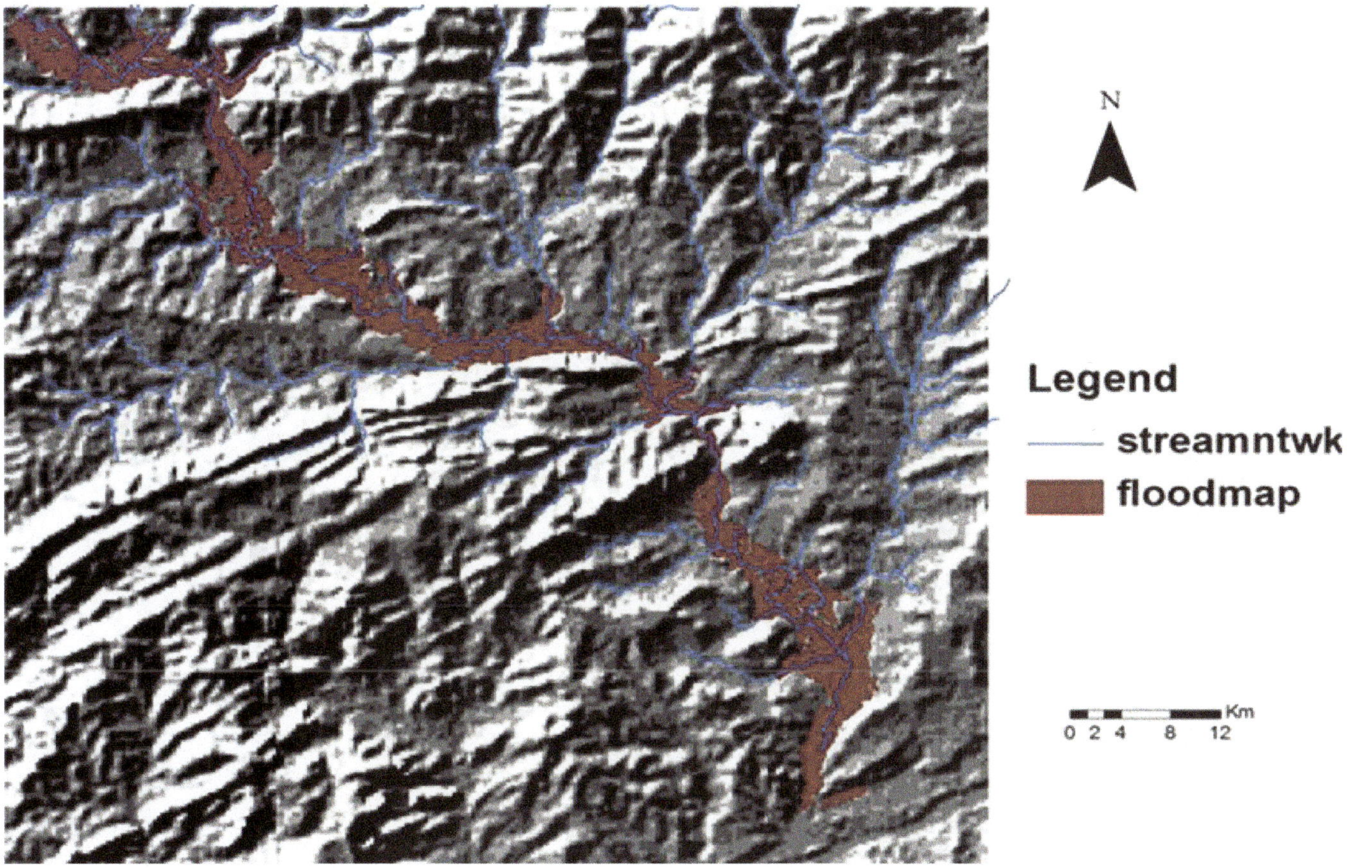

Figure 2. Flood inundation extent on 3-D map.

Generally, the land use that is more prone to flood hazard is agriculture and the worst affected land cover is shrubs and scattered trees.

Flood hazard with respect to flood velocities and depths

Figure 4 shows the flood hazard areas within the Dep River catchment with sections of flood areas for different return periods. These are within the low lying and flat parts of the catchment. The flood hazard classification for the different return periods is shown in Table 3.

The floodplain is majorly inundated by low to medium hazard flood with an area of 105.4 and 150.1 km^2 respectively for the 100-year return period while high hazard flood covers 5.82 km^2. This agrees with the result of a similar study by Damayanti (2011) which showed the area studied to be mostly of high to medium hazard. A marked increase of 145.7 km^2 was observed for medium hazard (71.28%) and high hazard increased by 12.87 km^2 (93.94%). This means that as the return periods increased, the hazard level of the floods was also increasing significantly.

Considering the problems of lack of institutional and infrastructural measures to solve the problem of flooding within the Dep River floodplains, the flood hazard analysis carried out from this study will greatly enhance the suggestion of solutions to the incessant flooding and its effects within the basin. This is because the extent of the area that flooding will affect has been determined by this process thereby locating the critical areas which should be avoided and left for flood waters or used with proper management with regards to how and when to use them.

CONCLUSION AND RECOMMENDATIONS

The most affected activity within the floodplain is agriculture with area inundated ranging from 68.82 to 146.10 km^2 for 2-year to 1000-year return periods. Low to medium flood hazards predominate the floodplain with maximum area of 105.7 km^2 for low hazard and 204.4 km^2 for medium flood. High hazard is mainly within the deepest parts of the floodplain.

It is recommended that:

1. The low hazard areas can be used for irrigation farming and early rainy season farming starting between

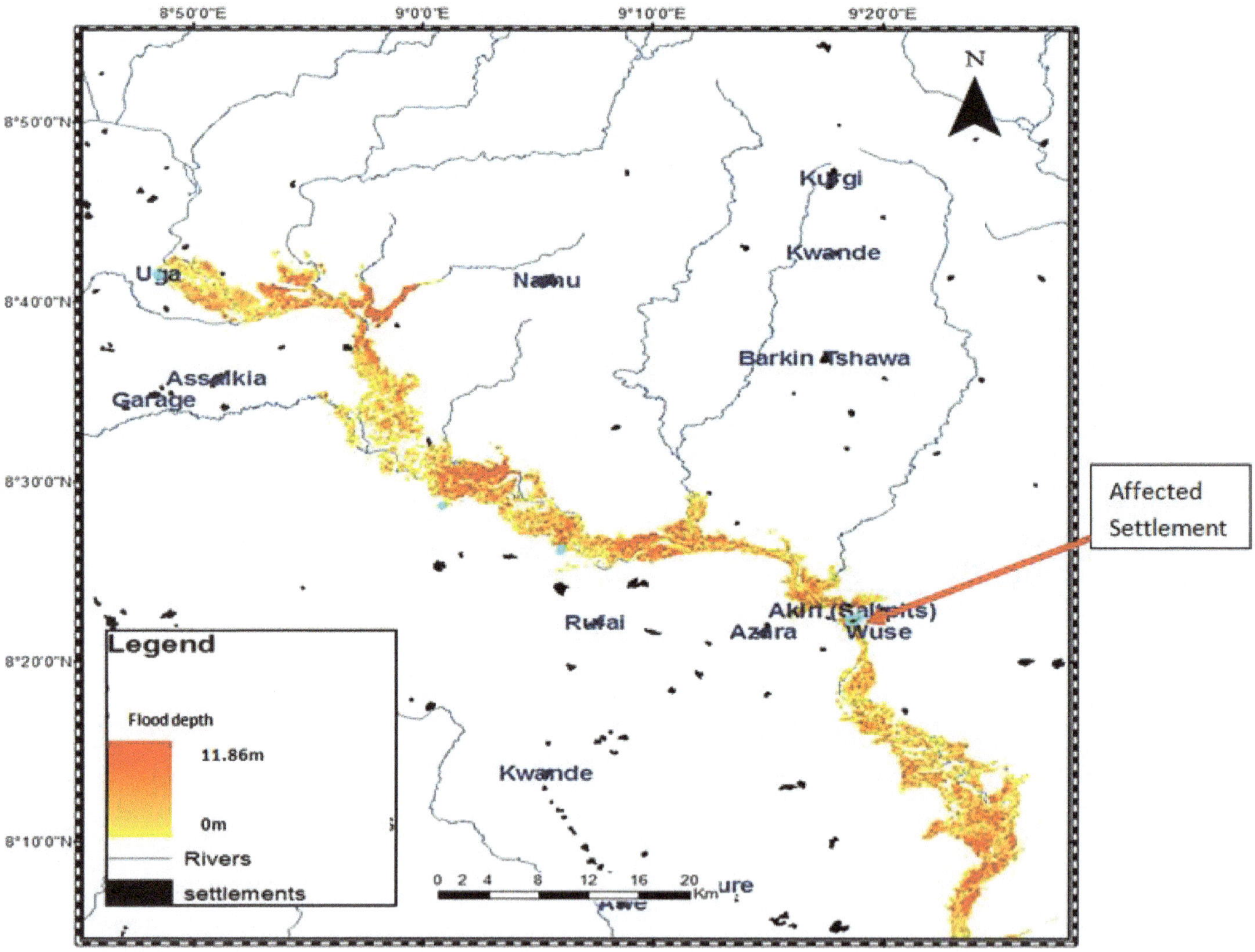

Figure 3. Flood inundation map showing affected settlement.

Table 2. Percentage and area coverage of land use land cover classes inundated by different floods.

Land use land cover	Area of Inundation (km²)								
	TR-2	**TR-5**	**TR-10**	**TR-20**	**TR-50**	**TR-100**	**TR-200**	**TR-500**	**TR-1000**
Bare/degraded lands	1.01	1.12	1.26	1.37	1.49	1.57	1.67	1.71	1.76
Water bodies	9.28	10.71	11.68	12.54	13.62	14.50	15.37	16.44	17.09
Rock outcrop	1.26	1.44	1.56	1.78	2.00	2.12	2.27	2.42	2.56
Scattered cultivation	68.82	83.19	92.53	101.35	111.80	119.86	128.13	138.23	146.10
Shrubs/scattered trees	69.55	84.25	94.01	102.83	113.90	122.00	129.46	139.35	145.88
Total	149.9	180.7	201.03	219.87	242.8	260.06	276.9	298.15	313.39

March and first week of April or as indicated by the time of rainfall for any particular year.

2. The medium and high hazard areas can be used for irrigation farming only.

3. All the hazard areas for the 100-year flood should be avoided with respect to construction of residential or commercial structures.

4. Generally, areas close to rivers should be avoided for rainy season farming or any other activities during the rainy season.

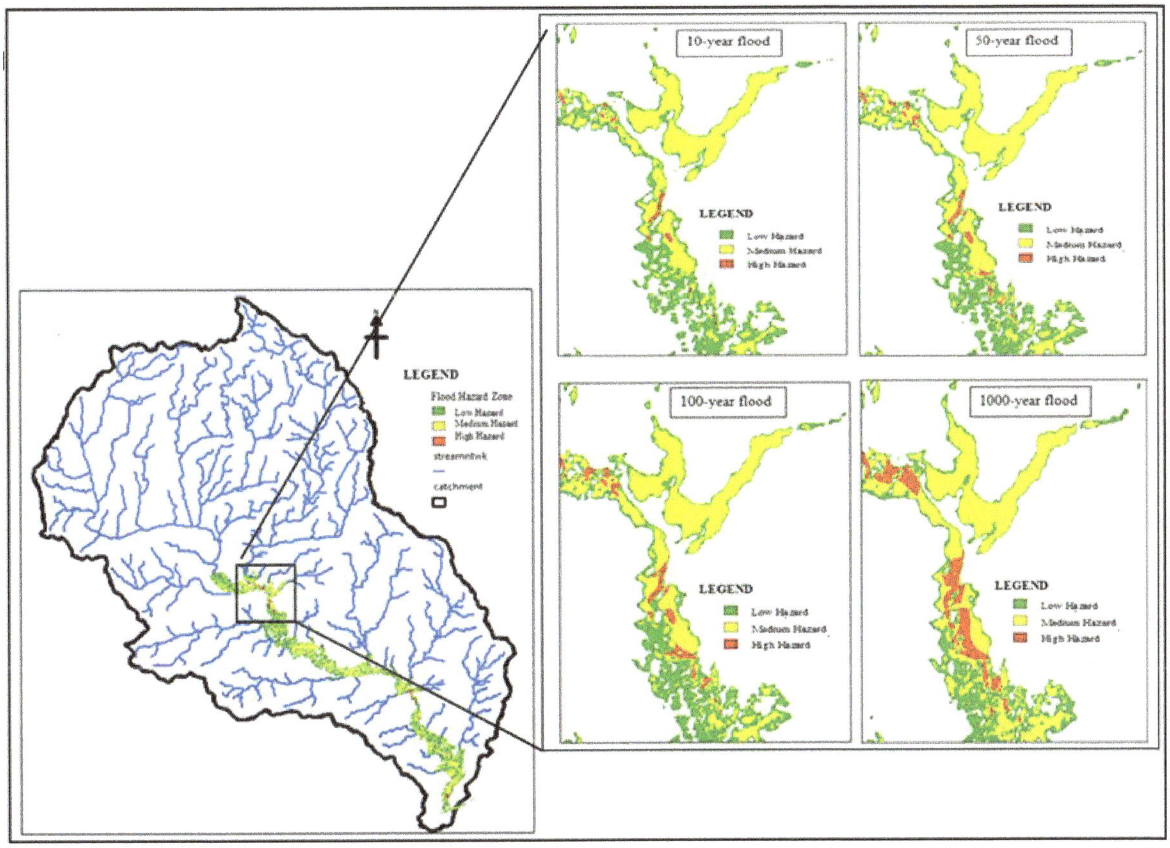

Figure 4. Flood hazard classification for different floods.

Table 3. Flood hazard classification for different return periods.

Hazard	TR-2	TR-5	TR-10	TR-20	TR-50	TR-100	TR-200	TR-500	TR-1000
Low hazard (km^2)	88.7	99.2	103.5	105.5	105.7	105.4	104.1	100.7	97.2
Medium hazard (km^2)	58.7	79.8	95.9	112.6	133.9	150.1	166.4	188.2	204.4
High hazard (km^2)	0.83	1.22	1.49	2.09	3.98	5.82	7.79	10.88	13.70

5. More detailed flood hazard mapping can be carried out with high resolution satellite data like QuickBird or IKONOS.

Conflict of Interests

The author(s) have not declared any conflict of interests.

REFERENCES

Ahmed B, Kaleem MS, Butt MJ and Dahri ZH (2010). Hydrological modelling and flood hazard mapping of Nullah Lai. Proc. Pak. Acad. Sci. 47(4):215-226.

Alkema D, Westen C (2005). Flood Hazard and Risk Assessment. Refresher Course on Geo-Information for natural Disaster Reduction in Easter Africa. Makerere University, 12- 23 September.

Balabanova S, Vassilev V (2010). Creation of Flood Hazard Maps. BALWOIS 2010 - Ohrid, Republic of Macedonia.

Cooper V, Opadeyi J (2006). Flood Hazard Mapping of St. Lucia. Final Report for the Caribbean Development Bank.

Daffi RE (2013). Flood Hazard Assessment for River Dep Catchment Using Remote Sensing and Geographic Information System Techniques. Ahmadu Bello University Zaria. PhD Dissertation. Unpublished.

Damayanti F (2011). Hydrodynamic Modeling for Flood Hazard Assessment inTelemoyo Catchment Central Java, Indonesia. Unpublished. M.Sc Thesis. Gadjah Mada University and University of Twente. The Netherlands.

De Moel H, van Alphen J, Aerts JCJH (2009). Flood maps in Europe – methods, availability and use. Nat. Hazards Earth Syst. Sci. 9:289-301.

Floodplain Development Manual: The management of flood liable land. Appendix L - Hydraulic and hazard categorisation. www.environment.nsw.gov.au/.../**flood**plains/18_**flood**_manual.pdf

Kafle TP, Hazarika MK, Shrestha KG, Prathumchai K, Samarakoon L (2006). Integration of remote sensing and GIS with flood simulation model for flood hazard mapping in the Bagmati River, Nepal. New Technol. Urban Saf. Mega Cities Asia, pp. 257-267.

Salimi S, Ghanbarpour MR, Solaimani K, Ahmadi MZ (2008). Floodplain mapping using hydraulic simulation model in GIS. J. Appl. Sci. 8(4):660-665.

Stratigraphic mapping of subsurface structures and groundwater potentials from electrical resistivity soundings in Onicha Olona, Atuma Iga and Akwukwu-Igbo, Delta State Nigeria

OKOLIE, Emmanuel Chukwuemeka

Department of Physics, Delta State University Abraka, Nigeria.

Onicha Olona, Atuma Iga and Akwukwu-Igbo have experienced severe hardship in getting reliable boreholes over many years due to their high topography. They are transition towns as it is from here that marked variations in subsurface structures and topography commence. Although the entry point at onicha Olona which is tied to Issele Uku is a flat land by contouring, the western end of the town has evidence of intermittently occurring deep valleys. It is important to study and ascertain the formation strata and groundwater occurrence in these towns. Hence, this study was made in Onicha Olona, Atuma Iga and Akwukwu-Igbo Delta State Nigeria. Seven vertical electrical sounding (VES) stations were sounded using Schlumberger configuration. The apparent resistivity values obtained in the field were plotted against half current electrode spacings. Interpretations of data were done qualitatively and quantitatively from which seven to eight layers with complex curve types of HAK, HAH, HK, QHK, QKHK, HKHKQ and QHKHA were mapped. Parches of springs from faults in the weathered rocks within the valleys at depths of about 30 m and low yield aquifers at depths of 125 m and beyond are delineated in Onicha Olona. Akwukwu-Igbo and Atuma Iga have sustainable aquifers at depths of 85 to 105 m due to topographic variation.

Key words: Formation strata, Schlumberger, apparent resistivity, aquifer, topographic variation.

INTRODUCTION

Subsurface structures in Delta State vary remarkably as the state stretches from near the ocean at Forcados and Warri to the hinterland at Agbor and Issele Uku from where it again slopes into River Niger. Many towns in the high altitude zone of Delta State have inadequate water resources whereas availability of groundwater/portable water is a major determinant of settlement and population growth in rural and urban areas (Okolie, 2010). Onicha Olona, Akwukwu-Igbo and Atuma Iga are towns worst hit in this zone where availability of portable water is obviously a serious problem. Onicha Olona seems to be the transition zone by contouring as the topography gradually drops from here with respect to Issele Uku (Figure 1). Many attempts of citing viable boreholes and deep water wells have woefully failed. These difficulties may be attributed to the structural distributions of the subsurface and neglect of elaborate geophysical investigations capable of delineating the subsurface lithology, prolific aquifers and static water levels (Kogbe, 1976) and (Olorunfemi et al., 1999). The increasing population and aspiration for rapid technological advancement here call for urgent empirical study aimed at providing lasting solutions to this problem. Hence, this empirical investigation was made to stratify the subsurface,

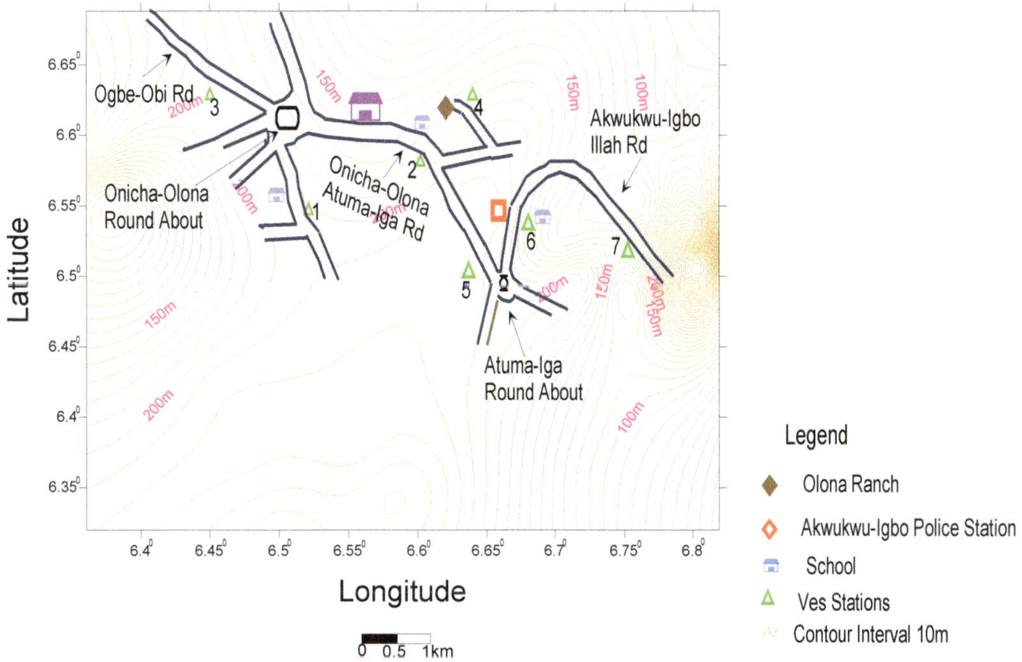

Figure 1. Base Map of Study Area – Onicha Olona, Atuma Iga and Akwukwu-Igbo.

provide geologic and geophysical features and determine the aquifer distributions in this area in order to eliminate unsuccessful well drills. The electrical resistivity method was applied using Schlumberger configuration. The obtained field data was interpreted qualitatively by curve matching to obtain the trend of the strata and quantitatively from Vander resist version (1988 and 2004) iteration technique to deduce the structural lithology.

Location of study area

Onicha-Olona is located in Aniocha North LGA while Atuma Iga and Akwukwu-Igbo towns are located in Oshimili North L.G.A of Delta State. Although the study area is in two different LGAs, the road stretch is only about four kilometres. It is within E006°.34' and Latitude N06° 22' in Onicha Olona and E006° 35' and Latitude N06° 21' in Akwukwu Igbo on the Western bank of River Niger (Figure 1). It has heavy rainfall for a period of about seven months (April-October) in the year. It has generally flat topography while with patches of undulating terrain and is accessible by network of roads and footpaths.

MATERIALS AND METHODS

Seven vertical electrical sounding (VES) stations were sounded in the area and the topography of each station was recorded with a sensitive GPS and plotted (Figure 1). Electrical resistivity technique is used in delineating subsurface structures. It is based on the fact that the sub-surface structures possess varying resistivities. These differences in electrical properties are probed or investigated and properly analyzed to delineate underground structures and groundwater occurrence. The vertical electrical sounding method measures ground potential differences between suitable implanted electrodes resulting from variations in some important physical properties of the underlain rock formations such as porosity, permittivity and resistivity with respect to a fixed array center, the station. The current and potential electrodes are pegged collinearly with the inter-electrode spacing expanding progressively about the station with increasing sounding number (Okolie, 2011). The VES sounding measures internal variations in *apparent* resistivity since it is an indirect approach. The apparent resistivity is dependent on pressure and porosity which is a function of the pore space resulting from aqueous electrolytes, the fluid viscosity which decreases as the mobility of the ions increases and water saturation in accordance to ARCHIE'S LAW for rock resistivity. Archie's law states that the resistivity of a rock is directly proportional to resistivity of groundwater present in a formation but inversely proportional to the formation porosity and water saturation. Archies law for formation resistivity is therefore given by

$$\rho_r = a\phi^{-m} S_w^{-n} \rho_w \tag{1}$$

Where ρ_r = resistivity of rock also, ρ_w = resistivity of groundwater S_w = water saturation, $0 \leq S \leq 1$ and $\varnothing$ = porosity, $0 \leq \phi \leq 1$ a, m, and n are predetermined empirical constants given by $0.5 \leq a \geq 25$, $1.3 \leq m \geq 2.5$ and n = 2.

In Schlumberger arrays, four electrodes are collinearly arranged with the current electrode spacing which is much greater than the potential electrode $^{AB}/_2 \geq 5^{MN}/_2$ where AB is current electrodes spacing and MN is potential electrodes spacing as in Figure 2. The dimensional implication of the array is given by Equations (2 to 6).

$$r_1 = r_4 = \frac{1}{2}[L - b]$$

$$r_2 - r_3 = \frac{1}{2}[L + b] \tag{2}$$

Hence, from theory (Okolie, 2010) and (Lowrie, 1997)

$$\ell = 2\pi\left[\frac{1}{\left(\frac{1}{r_1}-\frac{1}{r_3}\right)-\left(\frac{1}{r_2}-\frac{1}{r_4}\right)}\right]\frac{V}{I} \quad (3)$$

So that

$$\ell_a = 2\pi\left[\frac{1}{\left(\frac{2}{L-b}-\frac{2}{L+b}\right)-\left(\frac{2}{L+b}-\frac{2}{l-b}\right)}\right]\frac{V}{I} \quad (4)$$

$$\ell_a = \frac{\pi\left(\frac{L}{2}\right)^2-\left(\frac{b}{2}\right)^2 R}{b} \quad (5)$$

In which case, the instrument operational Geometric factor "G" is given by

$$G = \frac{\pi\left(\frac{L}{2}\right)^2-\left(\frac{b}{2}\right)^2}{b} \quad (6)$$

The Schlumberger configuration of VES is a choice because of ease of laying field equipment and minimal field logistic of less manpower need since potential electrodes are not frequently moved (Okolie, 2011). Moreover, it has greater probing depths and resolving power than other arrays because of its large current electrodes spacing. Also, Schlumberger curves are generally smoother and step free (Reinhard, 1974). Schlumberger array uses exert logarithmic spacing of the current electrodes from six readings per decade in sequence of 1.00, 1.47, 2.15, 3.14, 4.64, 6.81, 10, 14.7, etc. (Milson, 1992; Reynolds, 1997; Ako and Olorunfemi, 1989).

A total of seven VES stations were sounded. While four stations were sounded in Onicha Olona three stations were sounded at Atuma Iga and Akwukwu-Igbo. In each station, the four electrodes were hammered down collinearly to make good contact with the earth for effective current to flow in the subsurface. Thereafter, the two current electrodes were tightly connected to the current terminals of the terrameter while the other two potential electrodes were similarly connected to the corresponding potential terminals of the equipment. The Signal Averaging System (SAS) 1000 terrameter was powered by a 12 Volts accumulator (Ayolabi, 2005). With this set up the subsurface under investigation was sounded, measured and the mean apparent resistivity value computed by the terrameter was digitally displayed on the instrument screen (Okolie, 2011) and recorded for use on application of relevant theories (Table 1). The terrameter was switched off after each reading to avoid shock on handling the electrodes.

DATA ANALYSIS

The data obtain from the electrical resistivity survey in Table 1 above was plotted on a log-log graph paper with the electrode separation ($^{AB}/_2$) on the abscissa and apparent resistivity (ρ_a) value on the ordinate. The qualitative analysis was made to determine the curve types exhibited by each site. The quantitative analysis was made by employing partial curve matching technique using the relevant Schlumberger standard curves A, Q, K, or H and their corresponding auxiliary curves (Zohdy, 1974). The field curves on the log-log graph were matched segment by segment with the two layer master curves and corresponding auxiliary curves to obtain the actual resistivity and thickness of the first subsurface layer in each station as well as the resistivity replacements (ρ_r), reflection factors and Depth Indices (DI) of the other subsurface layers.

With the above parameter obtained from curve matching, the apparent resistivity (ρ_a) and the corresponding thickness of each of the other layers were calculated for from

$$\rho_{a=k_{(n+1)}} \times \rho_{(n-1)r} \quad (7)$$

where n is an integer greater than one (n >1), and the real layer depth is obtained from

$$h_n = h_{(n-1)} \times DI \quad (8)$$

and used as starting models for the computer iteration technique in which the Win Resist Software was applied based on the work of Vander Velper (1988, 2004) to delineate the actual apparent resistivities and thicknesses of the other layers. This computer iteration involved entering the field spread type (Schlumberger array), plotting half electrode separation 'AB/2' against computed apparent resistivity in the columns provided by the software. Thereafter the total number of layers, the resistivity and corresponding thickness of each layer modeled from the qualitative interpretation in each station were fed into the computer and automatically iterated to its lowest Root-Mean Square (RMS) percentage error. In this application, the thickness and apparent resistivity of the first medium are real. Moreover, although the resistivity of the last layer can be computed for, it has seemly undefined thickness/depth as h_n cannot be evaluated. The results of the computer iterated curves for the seven sites are explicitly shown in the Figures 3 to 7 and Table 2.

DISCUSSION

The qualitative analysis shows that the area consists of heterogeneous lithology with seven to eight layers made of HAK, HAH, HK, QHK, QKHK, HKHKQ, QHKHA curve types (Figure 3 to 7).

The quantitative analysis and computer iteration reveals that the top soil has resistivity of 160 to 1970 Ωm and thickness of 0.5 to 1.0 m and consisting of loose sand except at the ranch area where the top soil consists of weathered rock of anomalous resistivity values with the same thickness range. The anomalous nature (Very high resistivity) around the ranch at Onicha Olona is as a result of the hugh presence of near surface weathered

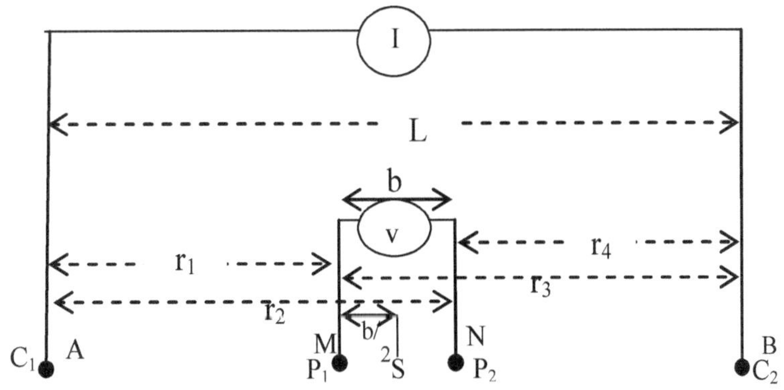

Figure 2. Schlumberger field electrodes configuration. Where "L" is the distance between the current electrodes and "b" is the distance between potential electrodes in station S.

Table 1. Apparent Resistivity field Data as recorded from ABEM SAS 1000 Terrameter.

Current electrode separation AB/2 (m)	Potential electrode separation MN/2 (m)	VES 1 ρ (Ωm) Onicha Olona	VES 2 ρ (Ωm) Onicha Olona	VES 3 ρ (Ωm) Onicha Olona	VES 4 ρ (Ωm) Onicha Olona (Ranch)	VES 5 ρ (Ωm) Atuma Iga (Palace)	VES 6 ρ (Ωm) Akwukwu/Atuma Iga market	VES 7 ρ (Ωm) Akwukwu-Igbo
1.0	0.2	135.20	140.0	1794.8	85760.0	250.90	1500.2	3500.61
1.47	0.2	92.017	116.85	828.80	72110.0	330.29	1121.6	2623.90
2.15	0.2	69.046	130.87	477.94	50033.0	443.20	956.83	1789.22
3.16	0.2	55.24	165.05	340.23	21700.0	469.83	1080.2	1096.38
4.64	0.2	65.272	225.14	344.85	10900.0	462.30	1250.67	875.00
4.64	1.0	66.35	240.14	287.86	11000.0	483.08	1260.64	830.57
6.81	1.0	106.88	290.76	317.16	11500.0	499.90	1406.57	911.28
10.0	1.0	140.88	351.25	420.4	13050.0	623.16	1430.55	700.09
10.0	3.0	150.27	380.6	469.62	13100.0	650.78	1450.17	750.88
14.70	3.0	185.56	560.6	628.06	17500.0	840.74	1062.9	685.17
21.50	3.0	241.37	575.0	1250.4	18279.0	1330.5	907.02	635.9
31.6	3.0	350.72	700.7	1087.5	18900.0	1500.16	950.70	434.6
31.6	8.0	400.12	768.71	1273.49	19049.0	1600.60	974.86	439.8
46.40	8.0	402.49	970.0	1314.2	17240.0	1305.70	1201.7	404.313
68.10	16.0	454.19	1045.5	1783.6	15070.0	950.80	1414.31	424.765
100.00	16.0	435.44	900.7	1968.1	8570.0	1206.10	1289.3	743.14
100.00	30.0	440.36	950.0	1907.6	8600.0	1300.4	1334.6	755.40
147.00	30.0	320.797	802.9	1660.9	4290.0	2000.5	1007.51	1060.13
147.00	50.0	333.92	850.2	1855.7	4306.0	2010.6	1050.76	1065.2
215.00	50.0	320.95	1000.4	1196.98	2174.47	1905.20	830.49	1537.9
316.00	50.0	435.2	1351.6	750.30	1577.7	1445.33	684.39	2057.5

rock and stones as well as human activities on the top soil within and around Olona ranch. This is followed by generally low resistive formation ranging from 30.3 to 50 Ωm at Onicha Olona suggestive of clay stratum with thickness of 1.7 m. The second layer in Atuma Iga and Akwukwu-Igbo has resistivity of 630 to 820 Ωm and consists of clayey sand of thickness of about 1 m but at VES 7 it has a thickness of about 10 m (Figure 3 to 7). The third layer consists of sandy clay of resistivity values ranging from 200 to 480 Ωm. Thus, only VES 5 has sandy clay formation before clay formation. The fourth layer in all sites have moderately high resistivity of 1,300 to 2588 Ωm suggestive of compact weathered rock and granulated sand grain formation with thickness of 8 to 13 m to a depth of about 31 m. At the ranch this layer is to a depth of 65 m. The fifth layer follows with low resistive layer with resistivity range of 390 to 700 Ωm except at VES 7 in Akwukwu-Igbo which has resistivity of 2700 Ωm

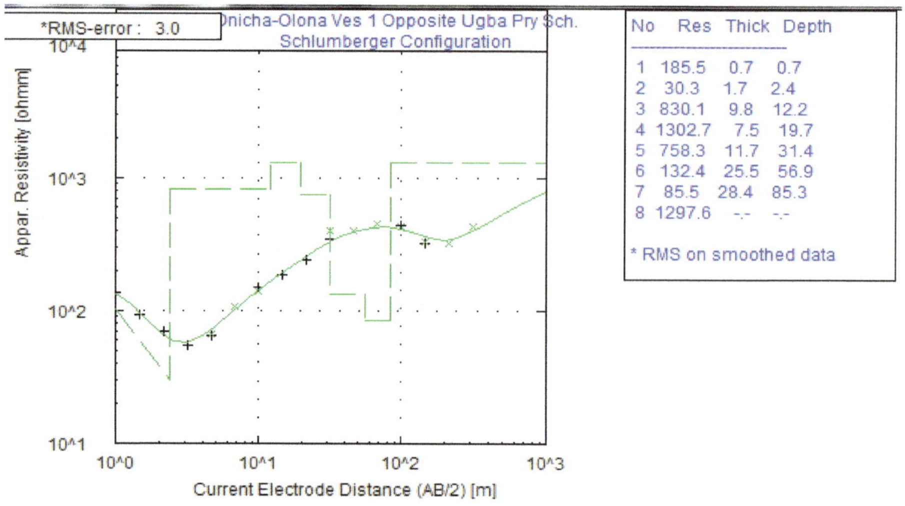

Figure 3. Typical Sounding Curve for Onicha Olona (VES 1) HAK - curve type.

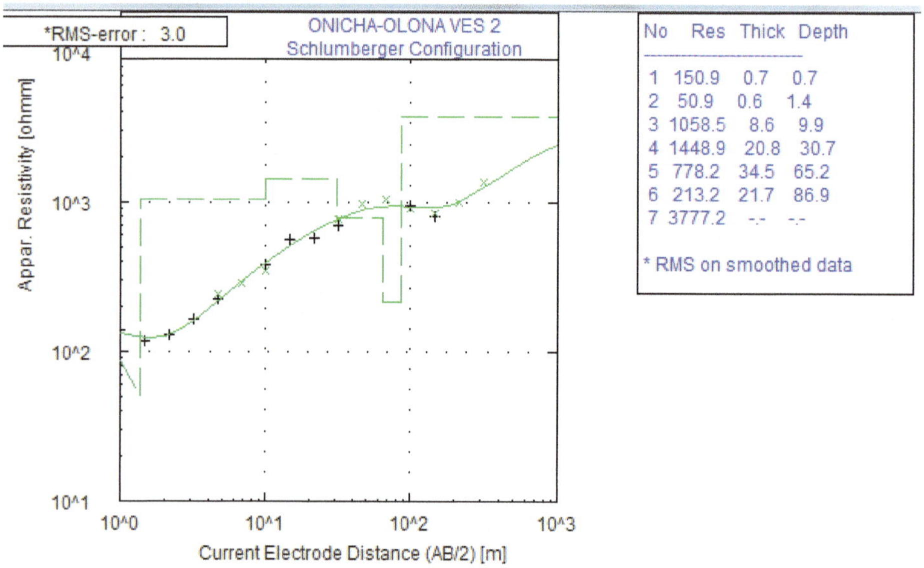

Figure 4. Typical Sounding Curve for Onicha Olona (VES 2) HAH - curve type.

due to drastic topographic change. The sixth layers at Atuma Iga and Akwukwu-Igbo have remarkable high resistivity range of 2244 to 7955 m evident of weathered and granulated rock formations. The seventh and eight layers have low resistivity values which suggest the occurrence of medium to coarse grain sand and gravelly stones respectively indicative of near water bearing medium. Thus, by this study, appreciable groundwater yield at Atuma Iga and Akwukwu-Igbo is at depths of 85 to 100 m with static water level within and about 75 m. Also, the water bearing medium (aquifer) is found to be shielded by the weathered rock formation at Akwukwu-Igbo so it is advisable that wells are drilled to about 95 to 100 m and even deeper around the Ministry of works to obtain appreciable aquifer yield and avoid contamination. The subsurface strata at Atuma Iga consist generally of low resistive formations suggestive of sandy clay, clay and clayey sand formations to far depths. This is indicative of confined aquifer.

At the ranch, the upper layer consists of high resistive weathered rocks. However, at depths below, the resistivities drop drastically suggestive of rock faults and some groundwater. Thus, there is evidence of perched aquifer or near surface groundwater accumulation beneath resulting from continuous flow from rock faults at about 50 m. Here also, there are springs emanating from the faults which eventually form the slow running streams like the 'Iyi Ishiekpe' stream within the valleys. These

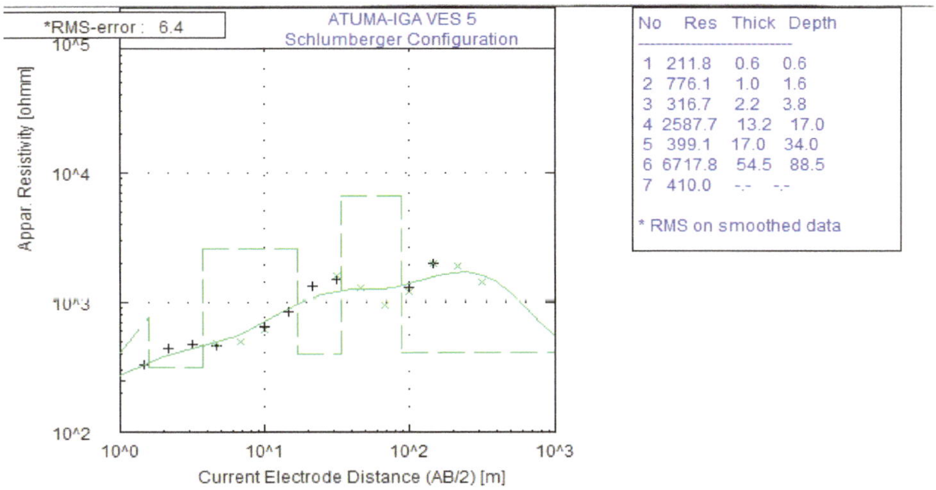

Figure 5. Typical Sounding Curve for Atuma Iga (VES 5) QKHK - curve type.

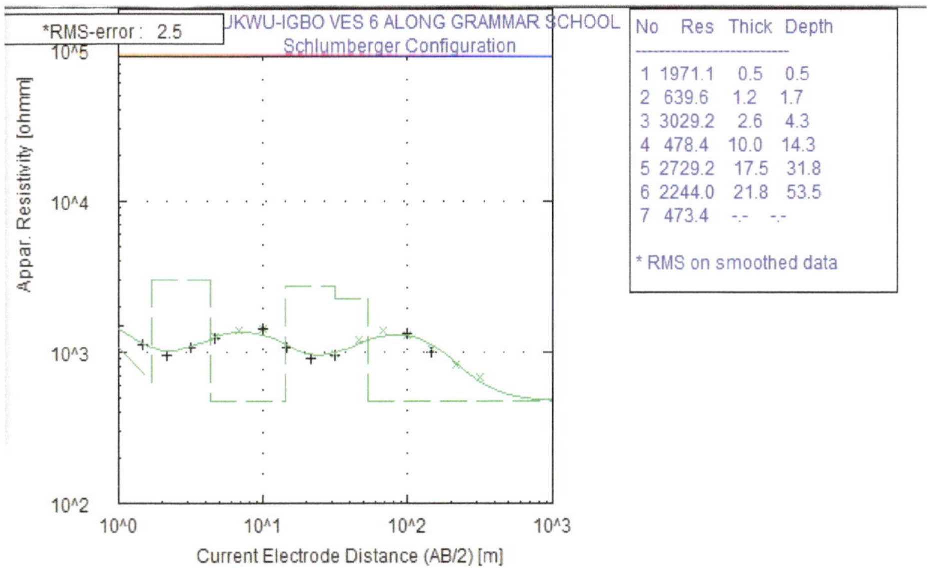

Figure 6. Typical Sounding Curve for Akwukwu-Igbo (VES 6) HKHKQ - curve type.

valley streams stand out as the major sources of the domestic water for the residence of Onicha Olona, Atuma Iga and Akwukwu-Igbo for many years.

The geoelectric parameters and delineated lithology are presented in Table 2.

Conclusion

Onicha Olona, Atuma Iga and Akwukwu-Igbo are true Niger Delta towns with a typical feature of massive subsurface structural deposits. The study shows that the area contains sediments of varying lithology with complex curve types of HAK, HAH, HK, QHK, QKHK, HKHKQ and QHKHA. It consists of a mix of clayey sand with springs from faults in the weathered rocks within the valleys at Onicha Olona. It also shows that groundwater flow originates from faults within the rocky valleys and culminates into running surface waters around the ranch. A section of these running waters form the slow running Iyi Ishiekpe stream at Onicha Olona which has been one of the main sources of water for domestics use for many a people. Aquifers are delineated in Onicha Olona at depths of 120 m and more, while at Atuma Iga and Akwukwu-Igbo sustainable aquifers exist at depths of about 85 to 105 m due to reduction in topography from Onicha Olona to Akwukwu-Igbo. The static water level in Atuma Iga and Akwukwu-Igbo is about 80 m. Perched

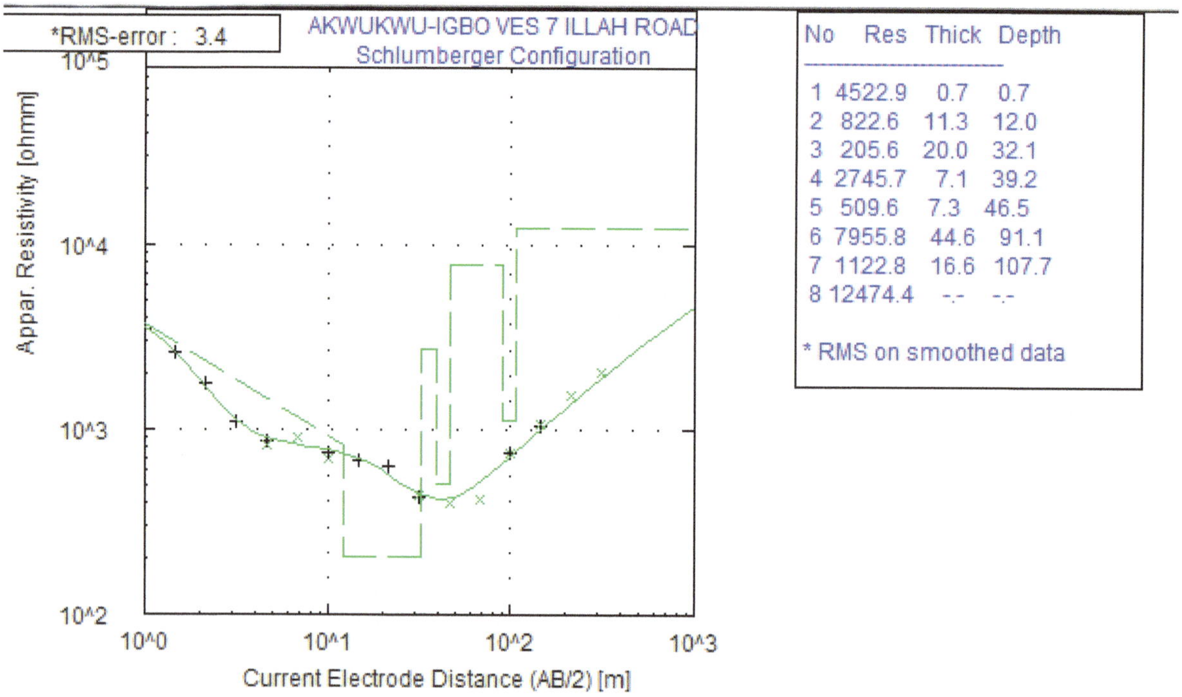

Figure 7. Typical Sounding Curve for Akwukwu-Igbo (VES 7) QHKHA - curve type.

Table 2. Geoelectric parameters and delineated lithology in the study area.

VES	Layer No.	Resistivity (Ωm)	Thickness (m)	Depth (m)	Delineated lithology	VES station and type curve
	1	185.5	0.7	0.7	Topsoil	
	2	30.3	1.7	2.4	Clayey Sand	**Onicha Olona**
	3	830.1	9.8	12.2	Clay	(Along Mkpitime Rd)
VES 1	4	1302.7	7.5	19.7	Sandy Sand	
	5	758.3	11.7	31.4	Fine Sand	$\ell_1>\ell_2<\ell_3<\ell_4>\ell_5>\ell_6>\ell_7<\ell_8$
	6	132.4	25.5	56.9	Brownish Sandy Clay	HAK
	7	85.5	28.4	85.3	Fine to Medium Grain sand	
	8	1297.6	---	-----	Medium to Coarse Grain Sand	
	1	150.9	0.7	0.7	Topsoil	
	2	50.9	0.6	1.3	Clayey Sand	**Onicha Olona**
	3	1058.5	8.6	9.9	Clay	(Atuma Iga Rd)
VES 2	4	1448.9	20.8	30.7	Sandy Sand	
	5	778.2	34.5	65.2	Fine Sand	$\ell_1>\ell_2<\ell_3<\ell_4>\ell_5>\ell_6<\ell_7$
	6	213.2	21.7	86.9	Brownish Sandy Clay	HAH
	7	3777.2	---	-----	Fine to Medium Grain sand	
	1	2956.5	0.5	0.5	Topsoil	
	2	250.3	4.5	5.0	Clayey Sand	**Onicha olona**
	3	3659.6	14.9	19.9	Clay	(Palm Guest House)
VES 3	4	1724.0	9.7	29.6	Sandy Sand	
	5	4523.9	15.9	45.5	Fine Sand	$\ell_1>\ell_2<\ell_3>\ell_4<\ell_5>\ell_6>\ell_7$
	6	3748.2	18.4	63.9	Brownish Sandy Clay	HK
	7	231.9	---	------	Fine to Medium Grain sand	

Table 2. Contd.

	1	1100000.0	1.0	1.0	Topsoil	
	2	5379.9	1.9	2.9	Granulated Sand & stones	**Onicha Olona**
	3	22817.9	39.3	41.2	Weathered rocks	(Olona Ranch)
VES 4	4	1498.0	22.5	63.7	Sand stones	
	5	751.0	13.8	77.5	Fine Sand	$\ell_1>\ell_2<\ell_3>\ell_4>\ell_5>\ell_6<\ell_7$
	6	346.5	9.5	87.0	Brownish Sandy Clay	QHK
	7	2402.0	---	------	Fine to Medium Grain sand	
	1	211.8	0.6	0.6	Topsoil	
	2	776.1	1.0	1.6	Clayey Sand	**Atuma Iga**
	3	316.7	2.2	3.8	Clay	(Obi's Palace)
VES 5	4	2587.7	13.2	17.0	Medium grain sand	
	5	399.1	17.0	24.0	Sandy Clay	$\ell_1<\ell_2>\ell_3<\ell_4>\ell_5<\ell_6>\ell_7$
	6	6717.8	54.5	78.5	Brownish Sandy Clay	QKHK
	7	410.0	---	----	Fine to Medium Grain sand	
	1	1971.1	0.5	0.5	Topsoil	
	2	639.6	1.2	1.7	Clayey Sand	**Atuma Iga/Akwuwu-Igbo**
	3	3029.2	2.6	4.3	Weathered rock/ Sand stones	(Market)
VES 6	4	478.4	10.0	14.3	Sandy Sand	
	5	2729.2	17.5	31.8	Fine Sand	$\ell_1>\ell_2<\ell_3>\ell_4<\ell_5>\ell_6>\ell_7$
	6	2244.0	21.8	53.6	Brownish Sandy Clay	HKHKQ
	7	473.4	---	-----	Fine to Medium Grain sand	
	1	4522.9	0.7	0.7	Topsoil	
	2	822.6	11.3	12.0	Clayey Sand	
	3	205.6	20.0	32.0	Clay	**Akwukwu-Igbo**
	4	2745.7	7.1	39.1	Sandy Sand	(Ministry of Works)
VES 7	5	509.6	7.3	46.4	Fine Sand	
	6	7955.8	44.6	91.0	Brownish Weathered Sand stones	$\ell_1>\ell_2>\ell_3<\ell_4>\ell_5<\ell_6>\ell_7<\ell_8$
	7	1122.8	16.6	107.6	Fine to Medium Grain sand	QHKHA
	8	12474.4	---	-----	Medium to Coarse Grain Sand	

aquifers are also delineated at 50 m in Atuma Iga and at 40 m in Akwukwu-Igbo. The study has thus, been successfully used to delineate the subsurface structure and increase the probability of drilling successful and sustainable boreholes in the area.

RECOMMENDATION

From the study, the subsurface formations contain rocks and the prolific aquifers exist only at very far depths. It is obvious that obtaining uncontaminated groundwater from viable aquifers in these towns, Onicha Olona, Atuma Iga and Akwukwu-Igbo requires government intervention as deep wells in the area can only be obtain by professional drillers and special drilling equipment. Indeed, only very few individuals may afford this feat. It is therefore recommended that Government and private sectors should as a matter of urgency assist these agro-based and fast developing towns in the provision of reliable and sustainable boreholes that will stand the test of time to improve the quality of life of the people.

ACKNOWLEDGEMENTS

Author gratefully acknowledge Prof Andy Egwunyenga and geophysics students for their concern, field commitments and immense contributions; The Vice Chancellor Prof. Eric Arubayi and Dean of Faculty of Science Prof S.H.O. Egboh both of Delta State University, Abraka for granting official permission to go for this study; His Royal Majesty, Obi Chukwuka Ogbudo of Atuma Iga, His Royal Highness, Obi David Azuka of Akwukwu-Igbo and the Obi of Onicha Olona for the hospitality and unflinching support received during the study.

REFERENCES

Ako AO, Olorunfemi MO (1998). Geoeletric survey for groundwater in the newer Basalts of Vom Plateau State. J. Min. Geol. 25(1&2):247-450.

Ayolabi EA (2005). Geoelectric Evaluation of Groundwater Potential: A case study of Alagba Primary School, Akure, S.W. Nig. J. Geol. Soc. India 66:491-495.

Kogbe CA (1976). The Stratigraphy and Sedimentation of Niger Delta. Geol. Nig. pp. 311-323.

Lowrie W (1997). Fundamentals of Geophysics. Cambridge University Press, pp. 212-218.

Milson J (1992). Field Geophysics. Geophysical Society of London. John Wiley & Sons NY, pp. 93-97.

Okolie EC (2010). Geophysical Investigation of Effects of Topographic Complexities on Groundwater Potential in Ibusa, Delta State Nigeria. Scholars Research Library: Arch. Phys. Res. 1(3):62-71.

Okolie EC (2011). Geophysical Investigation of treasured formation strata and groundwater potential in Ogume Delta State, Nig. Intl. J. Phys. Sci. 6(2):1152-1162.

Olorunfemi MO, Ojo JS, Akintunde OM (1999). Hydrogeophysical evaluation of groundwater potentials of Akure metropolis, Southwestern Nigeria. J. Min. Geol. 35(2):207-228.

Reinhard KF (1974). Combined Geoelectrical and Drill-Hole investigations for detecting Fresh-water aquifers in Northern Missouri. Geophysics 39(3):340-352.

Reynolds JM (1997). An Introduction to Applied and Environmental Geophysics: John Wiley and Sons Ltd, England.

Vander VBPA (1988, 2004). Win Resist Version 1.0. M. Sc Research Project ITC, Deft, Netherlands.

Vander VBPA (2004). Win Resist Version 1.0. M.Sc Research Project ITC, Deft, Netherlands.

Zohdy AA (1974). Automatic Interpretation of Schlumberger Sounding Curves. Geological Survey Bulletin, 1313-E US Government Printing Office, Washington, p. 71.

25

Critical analysis of existing economic tools available for assessing river water quality

Sudevi Basu[1] and Lokesh K. S[2]

[1]Department of Biotechnology, Sir M Visvesvaraya Institute of Technology, Bangalore – 562157, Karnataka, India.
[2]Department of Environmental Engineering, Sri Jayachamarajendra College of Engineering (Autonomous), Mysore – 570006, Karnataka, India.

This review paper critically analyzes the economic literature on the approaches of measuring the environmental benefits. It focuses on the economic methodologies that are available for the evaluation of the effects (social costs and benefits) of environmental changes (degradation/preservation) on river water quality. Further, it shows how the monetary valuations of these effects can have an impact in making of economic policy for creating more efficient water quality management for environmentally sustainable aspects. Over 85 papers were reviewed and it was found that the economic assessment tools were studied independently without comparing the impact of one method over the other. The literature does not provide information on economics of the interventions to protect the river water quality and relate it to the increase in local flora and fauna and decrease in averting costs incurred by local people. Furthermore, the reviewed papers have not economically quantified various pollution control measures to improve water quality in rivers.

Key words: River water quality, river pollution control, monetary valuation, economic policy, environmental benefits.

INTRODUCTION

River systems are one of the most important natural resources which form the basis of human livelihoods. Rivers exhibit extraordinary phenomena, with physical, cultural and psychological expression in human societies; they bring life and death, civilization and devastation, opportunity and risk. These river systems have been an important source for irrigation, potable water, cheap transportation, electricity and other facilities and play an important role in human development. However, huge economic development and population growth result in continuing environmental degradation. Intensification of agriculture, industrialization and increasing urbanization are the most severe driving forces of water quality deterioration in rivers and these river systems are one of the most important vulnerable natural resource.

To value natural environmental resources like river water, it necessitated the need for non-market valuation methods because these resources are neither bought nor sold in the market but nonetheless have significant value (Boyer et al., 2004). Valuation of environmental impacts using nonmarket valuation methods has evolved from being completely a U.S activity in 1960s and 1970s to become a very important field also in Europe in 1980s and 1990s. At the same time, nonmarket valuation methods have been applied at an increasing rate in developing countries in Asia, Latin America and Africa (Navrud et al., 1997). Thus monetization of the benefits of water quality improvements is a central component of evaluating the cost-benefit performance of water quality regulations.

The basic concept of economic value incorporates a wide range of measurement techniques than use of market prices. Value theory begins by examining people in situations where they must make choices involving a trade-off, and there are basically two concepts that choice situations present: one in which people give up something to obtain an object of choice (that is, they pay for it), called Willingness to Pay (WTP) and another where they receive compensation in return for giving up an object of choice (that is, they sell it), called Willingness to Accept (WTA) (Hanemann, 1991).

As concern over the deterioration of water quality grows, regular analysis of water quality, diversion of sewerage lines, setting up of industrial wastewater treatment plants to prevent discharge of untreated wastes are some of the positive measures to prevent water pollution. Further effective management of river water quality requires the evaluation of benefits derived from water quality improvements. But as river restoration grows into huge investments a year effort, certainly there will be individuals who will ask whether the benefits of such efforts are worth the costs and hence it is necessary to establish their full value and incorporate this into the decision making process. Widely accepted and often used framework for decision making is Cost Benefit Analysis (CBA). It is an analytical tool conducted by aggregating the total costs and benefits of a project or policy, which represent welfare improvement only if net benefits of costs are positive and an option with highest net benefits is the optimal one, over space and time (Birol et al., 2006). Since 2000, World Health Organization (WHO) has been putting its efforts behind developing and applying approaches to cost–benefit analysis on issues of water, sanitation, hygiene and health. Work is now focusing on the development of methods appropriate for application at the country level to assist in analysis of the cost effectiveness and benefit–cost ratios of water, sanitation and hygiene interventions (Prüss-Üstün et al., 2008). However, values of water resources are not straightforward to estimate for CBA purposes because water bodies are public goods in nature and their values are more complex compared to private goods since water bodies are composed of both use and non-use values (Birol et al., 2006). Capturing the total economic value of water resources is crucial to policy and management decisions, thus enabling society to allocate its scarce economic and environmental resources efficiently.

Various economic methods have been developed to capture the total economic value of environmental resources. Most commonly used are revealed or indirect preference methods (Hedonic Pricing Method (HPM), Travel Cost Method (TCM), Averting Expenditure Method (AEM), and Cost of Illness (COI)) and stated or direct preference methods (Contingent Valuation Method (CVM) and Choice Experiment Method (CEM)).

Aim and objectives

i) Conduct an in depth review of various existing economic tools and techniques for assessing river water quality.
ii) Explain the limitations of these techniques.
iii) Describe the need for economically quantifying various pollution control measures to prevent pollution and improve water quality in rivers.

EXISTING ECONOMIC METHODS

Various researchers have made an attempt to measure the economic value of water quality (Steinnes, 1992; Rogers et al., 2002; Ward et al., 2002; Elsin et al., 2010; Olmstead et al., 2010). This paper is an attempt to review the existing tools the economists use for valuing river water quality for pollution control.

i) Physical Linkage Method [Damage Function Method] and
ii) Behavioural Linkage Method [Direct Methods - Contingent Valuation Method (CVM) and Choice Experiment Method (CEM)] and [Indirect Methods - Averting Expenditure Method (AEM), Travel Cost Method (TCM), Hedonic Pricing (HP) and Cost of Illness (COI)].

Physical linkage method

This measure the benefits based on technical relationship between an environmental resource and user of that resource. Common estimation method is damage function method (Callan et al., 2010).

Damage function method

This uses a functional relationship to capture the link between contaminant and any associated damages. Using this method, incremental benefits are measured as reduction in damages arising from policy-induced decrease in contaminant. This damage reduction is then monetized to obtain the value of benefits brought about by the policy. (Callan et al., 2010).

Behavioural linkage method

Direct methods

These are also called Stated Preference Methods. These valuation methods have been developed to determine the economic benefits of valuing those environmental resources that are not traded in any market, including surrogate ones like water bodies. The commonly used direct methods are Contingent Valuation Method (CVM) and Choice Experiment Method (CEM).

a) Contingent valuation method: Contingent valuation method (CVM) has become a popular way of placing a monetary value on various aspects of the environment with the aim of determining whether the benefits of a proposed project outweigh the costs. The CVM has risen to prominence among these tools (Spash, 2000). With CVM, valuation is dependent or 'contingent' upon a hypothetical situation or scenario whereby a sample of the population is interviewed and individuals are asked to state their maximum WTP (or minimum willingness to accept (WTA) compensation) for an increase, or decrease, in the level of environmental quantity or quality (Birol et al., 2006). The contingent valuation surveys have been conducted by several researchers to see how people were willing to pay for improvements in river water quality (Carson et al., 1993; Kwak et al., 1994; Douglas et al., 1999; Loomis et al., 2000; Turpie et al., 2001; Alam, 2006; Imandoust et al., 2007; Monarchova et al., 2009; Nallathiga et al., 2010; Tu, 2013). CV studies have also been conducted on coastal water quality improvements by Soderqvist (1998), Hokby et al. (2003), Hanley et al. (2003) and Zhai et al. (2009).

Kristrom (1993) explored two commonly used methods to elicit an individual's WTP for a public good in CV studies. According to him, the most preferred method is discrete valuation question where the respondent accepts or rejects a suggested cost for the good. The other method, the traditional one is the CV question which simply asks an individual to state his WTP for the suggested change in the provision of a public good.

An interesting survey carried out by Lindhjem et al. (2009) on the aggregate welfare measures for change in the provision of public goods found it would be higher if the same elicited mean willingness to pay was added up over individuals rather than households. It was revealed that when people were prompted to answer for response unit, an average of 43% decided to state higher willingness to pay than individual willingness to pay, while 52% stated the same willingness to pay. Also more people stated higher household willingness to pay if individual willingness to pay were asked first.

Despite the strengths of CVM regarding its ability to estimate non-use values and evaluate irreversible changes, this method has been criticised for its lack of validity and reliability (Diamond et al., 1994; Carson et al., 2001; Whittington, 2002; Cooper et al., 2004). This is on account of potential problems including information bias (Park et al., 1991; Whitehead et al., 1991; Poe et al., 1997), design bias -starting point bias and vehicle bias (Johansson, 1996; Morrison et al., 2000; Ivehammar, 2009), Yea-saying bias (Remoundou et al., 2009), hypothetical bias (Balistreri et al., 2001; Vossler et al., 2006; Murphy et al., 2010), selection bias (Svento, 1993; Yoo et al., 2001), protest bias (Jorgensen et al., 1999; Strazzer et al., 2003), sequencing bias (Halvorsen 1996), elicitation bias (Loomis, 1997; Loomis et al., 1997; Bohara et al., 2001; Crooker et al., 2004; Farmer et al., 2008; Watanbe et al., 2009), anchoring bias (Frykblom et al., 2000; Arana et al., 2007) and embedding effects (Diamond et al., 1994; Hanemann, 1994). Hypothetical bias contends that respondents may be prepared to reveal their true values but are not capable of knowing these values without participating in a market in the first place. Strategic bias occurs when respondents deliberately under- or overstate their WTP. Respondents may understate their WTP if they believe that the actual fees they will pay for provision of the environmental resources will be influenced by their response to the CV question. Conversely, realizing that payments expressed in a CV exercise are purely hypothetical, respondents may overstate their true WTP in the hope that this may increase the likelihood of a policy being accepted. Yea-saying bias indicates that respondents may express a positive WTP because they feel good about the act of giving for a social good although they believe that the good itself is unimportant while embedding bias implies that WTP is not affected by the scale of the good being offered (Remoundou et al., 2009).

b) Choice experiment method: Choice Experiments (CE) involve eliciting responses from individuals in constructed, hypothetical markets, rather than the study of actual behaviour. The Choice Experiment technique is based on random utility theory and the characteristics theory of value, where environmental goods are valued in terms of their attributes and by making one of these attributes a price or cost term, marginal utility estimates can be converted into willingness-to-pay estimates for changes in attribute levels, and welfare estimates obtained for combinations of attribute changes (Hanley et al., 2006). Researchers (Carlsson et al., 2003; Hanley et al., 2006; Viscusi et al., 2008; Borg et al., 2009; Birol et al., 2010) have applied CE in valuing the improvements in water quality.

Carlsson et al. (2003) conducted CE for valuing wetland attributes in Staffanstorp, southern Sweden and found that attributes like biodiversity and walking facilities were the two greatest contributors to welfare while a fenced waterline and introduction of crayfish decreased welfare.

Birol et al. (2010) used the Choice Experiment method to estimate around 150 randomly selected local public's willingness to pay (WTP) for improvements in the capacity and technology of a pilot scale sewage treatment plant (STP) in Chandernagore municipality, located on the banks of the River Ganga in India. The benefit estimates reported in this study reveal that an average household in the sample would be willing to pay Rs 8.36 per month (Rs 4.82 for high quality of treated water plus 3.54 for high quantity of treated water) in municipal taxes, in order to improve the capacity and technology of the STP. This would amount to Rs 100.32 per annum in additional municipal taxes per household. When aggregated over the entire population (32,939 households), Chandernagore municipality residents' WTP for increasing the capacity of the STP amounts to Rs 3,304,441 per annum. Thus the results reported in this paper are indicative of local public's demand for higher quality and quantity of treated wastewater to minimize the high levels of environmental and health risks in the Ganga.

Alpizar et al. (2001) highlighted the advantages of choice experiments stating that values for each attribute as well as marginal rates between non-monetary attributes can be obtained. According to Johnston (2007), CE results are well suited for benefits transfer because CE are designed to account for variations in environmental resources and site characteristics, as well as potential implications of these variations for willingness to pay. The main objective of the study by Brouwer et al. (2010) was to examine how repeated choice affects preference learning in stated preference experiments. Choice consistency tests suggested that preferences in the choice experiment were stable and coherent. Thus respondents felt significantly more confident and certain about their choice at the end of the choice experiment than they were at the beginning. The research by Taylor et al. (2010) explored the incentive properties of repeated, attribute-based choice questions when subjects are provided with an explicit connection between choices and outcomes. Their study results indicated that the choice-modeling studies that have no explicit provision rule can have a relative error that is more than double the contingent-valuation average error when applied to public goods (96%/39% = 2.5). They also found that when an explicit provision rule discussion is included, the average error in the choice experiments decreased to 57%, which is still larger than that found in contingent valuation studies (57% vs. 39%). These results clearly indicated that the inclusion of a provision rule is necessary for a credible choice-modeling study.

Limitations of CE were discussed by Hanley et al. (1998), Meyerhoff et al. (2008) and Morkbak et al. (2010). Hanley et al. (1998) stated that the principle problems in using the CE method are often the complex nature of statistical/experimental design and the selection of appropriate attributes and levels. Meyerhoff et al. (2008) studied the protest responses in a CE and CV. They used an attitude scale based on respondents' protest beliefs and found a significant negative effect of this attitude on willingness to pay in both methods. However, in one of the two study regions, the effect was found to be weaker in CE than in CV. Morkbak et al. (2010) addressed the issue of defining the levels of the cost variable in Choice Experiments. The main focus was on changes in the maximum price level—the expected choke price affecting consumers' preferences and WTP. The results showed that increasing the maximum price level gave rise to statistically significant increases in the WTP estimates for all attributes. So the high maximum price would indicate to the respondents that the good in question is more valuable, and that they should pay more money to obtain the good. Their results showed that increasing the maximum price level 50%, gave rise to increased WTP estimates of up to 68%, which very well could alter the outcomes of a cost-benefit analysis. So they concluded that setting the choke price can be crucial and hence it deserved attention in the experimental design stage.

Indirect methods

These are also called Revealed Preference Methods. These methods look for related or surrogate markets in which the environmental good is implicitly traded. These methods are suitable for valuing those water resources that are marketed indirectly and are thus only able to estimate their direct and indirect use values. The commonly used indirect methods are Averting Expenditure Method (AEM), Travel Cost Method (TCM), Hedonic Pricing (HP) and Cost of Illness (COI).

a) Averting expenditure method: This method is based on function theory of consumer behaviour and is used to indirectly estimate the willingness to pay for non-marketed commodities like clean water. In the context of water resources, households may respond to increased degradation of water quality in various ways that are generally referred to as averting or defensive behaviours so as to avoid the adverse impacts of water contaminants (Birol et al., 2006). According to Courant et al., (1981), between two differently located but otherwise identical individuals, the difference in their averting expenditures may or may not be a close estimate of their willingness to pay for the preferred location.

There are however important limitations to this method. Individuals may undertake more than one form of averting behaviour in response to an environmental change and the averting behaviour may have other beneficial effects that are not considered explicitly. Furthermore, averting behaviour is often not a continuous decision but a discrete one, depending on the situation. Generally, the averting expenditures does not measure all the costs related to pollution that affect household utility and are therefore only able to provide a lower bound estimate of the true cost of increased pollution (Birol et al., 2006).

Thus if general environment is improved by certain policy initiatives, individual can spend less on substitute goods and this gives an indirect estimate of individual's willingness to pay for associated incremental benefits (Callan et al., 2010).

b) Travel cost method: Travel cost approach is mainly applied to study the recreational value of sites like water bodies, for boating, fishing, watching birds. It uses information about number of trips to particular sites and cost of those trips to infer people's willingness to pay for access to the sites (Boyer et al., 2004). Several researchers have employed travel cost method to measure the welfare effects to changes in water quality and include the work done by Caulkins et al. (1986), Kealy et al. (1986), Hellerstein (1993), Englin et al. (1996), Ortacesme et al. (2002), Carr et al. (2003), McKean et al. (2005), Shrestha et al. (2007) and Hosking (2011). The travel cost method was applied to evaluate the recreational value of the RAMSAR site of the estuary of Massa River (El-Bekkay et al., 2013).

Even though the TCM have been regularly used to determine the value of recreation, a key site attribute often omitted is that of congestion, which describe the number of other individuals encountered during the recreation experience. Researchers (Michael et al., 1997; Boxall et al., 2003; Timmins et al., 2007) studied the recreational congestion. The results of Michael et al. (1997) indicated that failing to account for heterogeneous preferences for congestion by time of visit led to overestimates of the benefits of relieving peak-time congestion, while accounting for expectations raised questions about the validity of the standard optimal use model. Further, the results of this study indicated that the effect of congestion on recreation benefits was best modeled by the difference between actual and expected congestion rather than by simple objective measures of actual congestion.

c) Hedonic pricing: This is a measure of variations in housing prices that reflect the value of local environmental resources like water quality, aesthetics, local flora and fauna. The relationship between land prices and surface water access (both in quantity and quality terms) has been studied in the hedonic framework by Epp et al. (1979), D'arge et al. (1989), Garrod et al. (1994), Lansford et al. (1995), Doss et al. (1996), Mahan et al. (2000), Poor et al. (2007) and Higgins et al. (2009). Mahan et al. (2000) used data on more than 14000 home sales in Portland, Oregon metropolitan area to estimate the effect of proximity to wetlands on property values and found that a decrease in the distance to the nearest wetland by 304.8 m from an initial distance of 1 mile resulted in an increase in property value of €371.6.

A limitation of the HPM is that it only measures direct use values of water resources as perceived by the consumers' of the good in which it is implicitly traded. Services such as flood control, water quality improvement, habitat provision for species, and groundwater recharge may provide values that benefit individuals far away, beyond the consumers of the good, which the HPM is unable to capture (Boyer et al., 2004). Leggett et al. (2000) found that hedonic studies of environmental quality are particularly vulnerable to omitted variables bias: the emitters of pollution often have direct effects on the value of nearby properties-for reasons completely unrelated to water quality.

Another limitation discussed by Koundouri et al. (2003) is the effect of selectivity bias on hedonic price analysis. They stated that these valuations can be misleading when the quality characteristics intended for the valuation have sample selection implications and considered the case of land close to seaside that could be used either as an input in agricultural production or for tourist development. The proximity to the sea could reduce the quality of land as an input in agricultural production due to salinization of groundwater supplies, but increased the probability of switching the land usage from agriculture to the lucrative tourism market. Thereafter, the deterioration of groundwater supplies could appear to have a positive effect on the price of agricultural land. They have cautioned that this technique can give rise to misleading conclusions about the effect of an environmental attribute on producers' or consumers' welfare if potential biases from inappropriate sample selection criteria are ignored.

d) Cost of illness method: This method measures the direct (medical costs, nursing care, drugs) and indirect (opportunity) economic costs associated with a disease and estimate the potential savings from the eradication of the disease. This approach also values loss of life based on the foregone earnings associated with premature mortality. The notion is that people should be willing to pay at least as much as the value of the income they would lose by dying prematurely (Remoundou et al., 2009).

Two important limitations of this approach is that it does not consider the actual disutility of those who are ill, nor does it account for the defensive or averting expenditures that individuals may have taken to protect themselves

RESULTS AND DISCUSSION

For this review, the journal papers were collected, collated, reviewed and analyzed intensely to find out the gaps and limitations of various methods were discussed and presented.

In the wake of rampant pollution of water sources, the need of the hour is proper management of water bodies to maintain their purity and sanctity. This paper is an attempt to critically assess and bridge the loopholes of the economic tools for evaluating river water quality and use it to quantify the environmental degradation by monitoring the water quality of the rivers.

The researchers have used the economic assessment tools independently without finding the impact of one method on another. For instance, the water bodies used for recreation are valued using TCM only (McKean et al., 2005; Shrestha et al., 2007; El-Bekkay et al., 2013) without considering the Damage Function and Cost of Illness of local people, which might still result due to microbial contamination of water. Similarly, Hedonic Pricing measures the land prices near water bodies without considering the local flora and fauna in the vicinity of water bodies (Higgins et al., 2009; Mahan et al., 2000). Also the literature does not provide information on economics of the interventions to protect the river water quality and relate it to the increase in local flora and fauna and decrease in averting costs incurred by local people. Further, the literature does not quantify economically various pollution control measures to improve water quality in rivers.

The following discussions are drawn based on the critical review of papers related to Economic tools.

i) Select social discount rates to assess future benefits from policy interventions and carry out a cost - benefit analysis of the interventions.
ii) Develop a mathematical relationship considering various parameters of concern to quantify (in terms of monetary valuation) the pollution abatement investments of the rivers.
iii) Justify investments made for pollution control in the riverine systems through economic quantification.

Conclusions

This review is certainly a step forward in understanding the principles of Environmental Economics by quantifying the impacts of pollution in the rivers and studying the implications of various pollution prevention projects undertaken by the Government, industries and private organizations to improve the quality of water in these rivers. Economic quantification is the basis for assessment of the investments in the form of interventions to protect the river water quality and social benefits like improvements in health of people, local flora and fauna and water quality of rivers. This research is unique because the flaws of existing economic tools are studied for assessing river water quality and this assumes significance in the light of the investments to improve the water quality of these rivers. A major limitation of existing economic tools is the lack of mathematical models for quantifying investments on river water protection. Future work is to combine different economic tools and study the effects of investments made for water protection and pollution control.

Conflict of Interest

The authors have not declared any conflict of interests.

REFERENCES

Alam K (2006). Valuing the Environment in Developing Countries: Problems and Potentials. Asia Pacific J. Environ. Dev. 13(1&2):27-44.

Alpizar F, Carlsson F, Martinsson P (2001). Using Choice Experiments for Non-Market Valuation. Econ. Issues 8(1):83.110.

Arana JE, Leon CJ (2007). Repeated Dichotomous Choice Formats for Elicitation of Willingness to Pay: Simultaneous Estimation and Anchoring Effect. Environ. Res. Econ. 36(4):475-497.

Balistreri E, McClelland G, Poe G, Schulze W (2001). Can Hypothetical Questions Reveal True Values? A Laboratory Comparison of Dichotomous Choice and Open-Ended Contingent Values with Auction Values. Environ. Res. Econ. 18(3):275-292.

Birol E, Karousakis K, Koundouri P (2006). Using Economic Valuation Techniques to Inform Water Resources Management: A Survey and Critical Appraisal of Available Techniques and an Application. Sci. Total Environ. 365(1-3):105-122.

Birol E, Das S (2010). Estimating the Value of Improved Wastewater Treatment: The Case of River Ganga, India. J. Environ. Manage. 91:2163-2171.

Bohara AK, Kerkvliet J, Berrens RP (2001). Addressing Negative Willingness to Pay in Dichotomous Choice Contingent Valuation: A Monte Carlo Simulation. Environ. Res. Econ. 20(3):173-195.

Borg NB, Hensher DA, Scarpa R (2009). An Analytical Framework for Joint vs Separate Decisions by Couples in Choice Experiments: The Case of Coastal Water Quality in Tobago. Environ. Res. Econ. 43(1):95-117.

Boxall P, Rollins K, Englin J (2003). Heterogenous Preferences for Congestion During a Wilderness Experience. Res. Energy Econ. 25:177-195.

Boyer T, Polasky S (2004). Valuing Urban Wetlands: A Review of Non – Market Valuation Studies. Wetlands 24(4):744-755.

Brouwer R, Dekker T, Rolfe J, Windle J (2010). Choice Certainty and Consistency in Repeated Choice Experiments. Environ. Res. Econ. 46(1):93-109.

Callan SJ, Thomas JM (2010). Environmental Economics and Management: Theory, Policy and applications, 5th Edition, Nelson Education Ltd., pp 155-168.

Carr L, Mendelsohn R (2003). Valuing Coral Reefs: A Travel Cost Analysis of the Great Barrier Reef. Ambio 32(5):353-357.

Carlsson F, Frykblom P, Liljenstolpe C (2003). Valuing Wetland Attributes: An Application of Choice Experiments. Ecol. Econ. 47(1):95-103.

Carson RT, Mitchell RC (1993). The Value of Clean Water: The Public's Willingness to Pay for Boatable, Fishable and Swimmable Quality Water. Water Resour. Res. 29(7):2445-2454.

Carson RT, Flores NE, Meade NF (2001). Contingent Valuation: Controversies and Evidence. Environ. Resour. Econ. 19(2):173:210.

Caulkins PP, Bishop RC, Bouwes NW (1986). The Travel Cost Model for Lake Recreation: A Comparison of Two Methods for Incorporating Site Quality and Substitution Effects. Am. J. Agric. Econ. 68(2):291-297.

Cooper P, Poe GL, Bateman IJ (2004). The Structure of Motivation for Contingent Values: A Case Study of Lake Water Quality Improvement. Ecol. Econ. 50:69-82.

Courant PN. Porter RC (1981). Averting Expenditure and Cost of Pollution. J. Environ. Econ. Manage. 8:321-329.

Crooker JR, Herriges JA (2004). Parametric and Semi–Nonparametric Estimation of Wilingness–to–Pay in the Dichotomous Choice Contingent Valuation Framework. Environ. Resour. Econ. 27(4):451-480.

D'arge RC, Shogren JF (1989). Okoboji Experiment: Comparing Non–Market Valuation Techniques in an Unusually Well–Defined Market for Water Quality. Ecol. Econ. 1:251-259.

Diamond PA, Hausman JA (1994). Contingent Valuation: Is Some Number Better than No Number? J. Econ. Perspect. 8(4):45-64.

Doss CR, Taff SJ (1996). The Influence of Wetland Type and Wetland Proximity on Residential Property Values. J. Agric. Resour. Econ. 21(1):120-129.

Douglas AJ, Taylor JG (1999). The Economic Value of Trinity River Water. Water Resour. Dev. 15(3):309-322.

El-Bekkay M, Moukrim A, Benchakroun F (2013). An Economic Assessment of the Ramsar Site of Massa (Morocco) with Travel Cost and Contingent Valuation Methods. Afr. J. Environ. Sci. Technol. 7(6):441-447.
Elsin YK, Kramer RA, Jenkins WA (2010). Valuing Drinking Water Provision as Ecosystem Service in the Neuse River Basin. J. Water Resour. Plan. Manage. 136(4):474-482.
Englin J, Cameron TA (1996). Augmenting Travel Cost Models with Contingent Behaviour Data – Poisson Regression Analyses with Individual Panel Data. Environ. Resour. Econ. 7(2):133-147.
Epp DJ, Al-Ani KS (1979). The Effect of Water Quality on Rural Nonfarm Residential Property Values. Am. J. Agric. Econ. 61:529-534.
Farmer MC, Lipscomb CA (2008). Conservative Dichotomous Choice Responses in the Active Policy Setting: DC Rejections below WTP. Environ. Resour. Econ. 39(3):223-246.
Frykblom P, Shogren JF (2000). An Experimental Testing of Anchoring Effects in Discrete Choice Questions. Environ. Resour. Econ. 16(3):329-341.
Garrod G, Willis K (1994). An Economic Estimate of the Effect of a Waterside Location on Property Values, Environ. Resour. Econ. 4(2):209-217.
Halvorsen B (1996). Ordering Effects in Contingent Valuation Surveys – Willingness to Pay for Reduced Health Damage from Air Pollution. Environ. Resour. Econ. 8(4):485-499.
Hanley N, Wright RE, Adamowicz V (1998). Using Choice Experiments to Value the Environment, Environ. Resour. Econ. 11(3-4):413-428.
Hanley N., Bell D, Farizo BA (2003). Valuing the Benefits of Coastal Water Quality Improvements Using Contingent and Real Behaviour. Environ. Resour. Econ. 24(3):273-285.
Hanley N, Wright RE, Alvarez-Farizo B (2006). Estimating the Economic Value of Improvements in River Ecology Using Choice Experiments: An Application to the Water Framework Directive. J. Environ. Manage. 78:183-193.
Hanemann WM (1994). Valuing the Environment through Contingent Valuation. J. Econ. Perspect. 8(4):19-43.
Hanemann WM (1991). Willingness to Pay and Willingness to Accept – How Much Can They Differ? Am. Econ. Rev. 81(3):635-647.
Hellerstein DM (1993). Intertemporal Data and Travel Cost Analysis. Environ. Resour. Econ. 3(2):193-207.
Higgins A, Pearson L, Laredo L (2009). Modeling the Financial Value of the Maroochy River to Property Values: An Application of Neural Networks. J. Water Resour. Prot. 4:237-248.
Hokby S, Soderqvist T (2003). Elasticities of Demand and Willingness to Pay for Environmental Services in Sweden. Environ. Resour. Econ. 26(3):361-383.
Hosking S (2011). The Recreational Value of River Inflows into South African Estuaries. Water SA 37(5):711-717.
Imandoust SB, Gadam SN (2007). Are People Willing to Pay for River Water Quality, Contingent Valuation. Int. J. Environ. Sci. Technol. 4(3):401-408.
Ivehammar P (2009). The Payment Vehicle Used in CV Studies of Environmental Goods Does Matter. J. Agric. Resour. Econ. 34(3):450-463.
Johansson P (1996). Commodity Prices as Payment Vehicles in Valuation Experiments, Environ. Resour. Econ. 8(2):247-254.
Johnston RJ (2007). Choice Experiments, Site Similarity and Benefits Transfer. Environ. Resour. Econ. 38(3):331-351.
Jorgensen BS, Syme GJ, Bishop BJ, Nancarrow BE (1999). Protest Responses in Contingent Valuation. Environ. Resour. Econ. 14(1):131-150.
Kealy MJ, Bishop RC (1986). Theoretical and Empirical Specifications Issues in Travel Cost Demand Studies. Am. J. Agric. Econ. pp. 660-667.
Koundouri P, Pashardes P (2003). Hedonic Price Analysis and Selectivity Bias: Water Salinity and Demand for Land. Environ. Resour. Econ. 26(1):45-56.
Kristrom B (1993). Comparing Continuous and Discrete Contingent Valuation Questions. Environ. Resour. Econ. 3(1):63-71.
Kwak SJ, Russel CS (1994). Contingent Valuation in Korean Environmental Planning: A Pilot Application to the Protection of Drinking Water Quality in Seoul. Environ. Resour. Econ. 4(5): 511-526.
Lansford NH, Jones LL (1995). Recreational and Aesthetic Value of Water using Hedonic Price Analysis. J. Agric. Resour. Econ. 20(2):341-355.
Legget CG, Bockstael NE (2000). Evidence of the Effects of Water Quality on Residential Land Prices. J. Environ. Econ. Manage. 39:121-144.
Lindhjem H, Navrud S (2009). Asking for Individual or Household Willingness to Pay for Environmental Goods: Implication for Aggregate Welfare Measures. Environ. Resour. Econ. 43(1):11-29.
Loomis JB (1997). Panel Estimators to Combine Revealed and Stated Preference Dichotomous Choice Data. J. Agric. Resour. Econ. 22(2):233-245.
Loomis J, Brown T, Lucero B, Peterson G (1997). Evaluating the Validity of the Dichotomous Choice Question Format in Contingent Valuation. Environ. Resour. Econ. 10(2):109-123.
Loomis J, Kent P, Strange L, Fausch K, Covich A (2000). Measuring the Total Economic Value of Restoring Ecosystem Services in an Impaired River Basin: Results from a Contingent Valuation Survey. Ecol. Econ. 33(1):103-117.
Mahan BL, Polasky S, Adams RM (2000). Valuing Urban Wetlands: A Property Price Approach. Land Econ. 76(1):100-113.
McKean JR, Johnson D, Taylor RG, Johnson RL (2005). Willingness to Pay for Non Angler Recreation at the Lower Snake River Reservoirs. J. Leis. Res. 37(2):178-194.
Meyerhoff J, Liebe U (2008). Do Protest Responses to a Contingent Valuation Question and a Choice Experiment Differ? Environ. Resour. Econ. 39(4):433-446.
Michael JA, Reiling SD (1997). The Role of Expectations and Heterogeneous Preferences for Congestion in the Valuation of Recreation Benefits. Agric. Resour. Econ. Rev. 166-173.
Monarchova J, Gudas M (2009). Contingent Valuation Approach for Estimating the Benefits of Water Quality Improvement in the Baltic Seas. Environ. Res. Eng. Manage. 47(1):5-12.
Morkbak MR, Christensen T, Gyrd-Hansen D (2010). Choke Price Bias in Choice Experiments. Environ. Resour. Econ. 45(4):537-551.
Morrison MD, Blamey RK, Bennett JW (2000). Minimizing Payment Vehicle Bias in Contingent Valuation Studies. Environ. Resour. Econ. 16(4):407-422.
Murphy JJ, Stevens TH, Yadav L (2010). A Comparison of Induced Value and Home–Grown Experiments to Test for Hypothetical Bias in Contingent Valuation. Environ. Resour. Econ. 47(1):111-123.
Nallathiga R, Paravasthu R (2010). Economic value of conserving river water quality: Results from a contingent valuation survey in Yamuna River Basin, India. Water Pol. 12(2):260-271.
Navrud S, Pruckner GJ (1997). Environmental valuation – To use or not to use? A comparative study of the United States and Europe. Environ. Resour. Econ. 10:1-26.
Olmstead SM (2010). The Economics of water quality. Rev. Environ. Econ. Pol. 4(1):44-62.
Ortacesme V, Ozkan B, Karaguzel O (2002). An estimation of the recreational use value of Kursunlu waterfall nature park by the individual travel cost method. Turk. J. Agric. For. 26(1):57-62.
Park T, Loomis JB, Creel M (1991). Confidence intervals for evaluating benefits estimates from dichotomous choice contingent valuation studies. Land Econ. 67(1):64-73.
Poe GL, Welsh MP, Champ PA (1997). Measuring the difference in mean willingness to pay when dichotomous choice contingent valuation responses are not independent. Land Econ. 73(2):255-267.
Poor PJ, Pessagno KL, Paul RW (2007). Exploring the hedonic value of ambient water quality: A local watershed–based study. Ecol. Econ. 60(1):797-806.
Pruss-Ustun A, Bos R, Gore F, Bartram J (2008). Safer water, better health: costs, benefits and sustainability of interventions to protect and promote health. World Health Organization, Geneva, pp. 1-53.
Remoundou K, Koundouri P (2009). Environmental effects on public health: An economic perspective. Int. J. Environ. Res. Public Hlth. 6:2160-2178.
Rogers P, de Silva R, Bhatia R (2002). Water is an economic good: How to use prices to promote equity, efficiency, and sustainability. Water Pol. 4:1-17.
Shrestha RK, Stein TV, Clark J (2007). Valuing nature–based recreation

in public natural areas of the Apalachicola River Region, Florida. J. Environ. Manage. 85(4):977-985.

Soderqvist T (1998). Why give up money for the baltic seas? Motives for people's willingness (or reluctance) to pay. Environ. Resour. Econ. 12(2):249-254.

Spash CL (2000). Ecosystems, contingent valuation and ethics: The case of wetland recreation. Ecol. Econ. 34(2):195-215.

Spash CL (2000). Multiple value expression in contingent valuation: Economics and ethics. Environ. Sci. Technol. 34:1433-1438.

Steinnes DN (1992). Measuring the economic value of water quality: The case of Lakeshore land. Ann. Reg. Sci. 26:171-176.

Strazzer E, Genius M, Scarpa R, Hutchinson G (2003). The effect of protest votes on the estimates of WTP for use values of recreational sites. Environ. Resour. Econ. 25(4):461-476.

Svento R (1993). Some notes on trichotomous choice discrete valuation. Environ. Resour. Econ. 3(6):533-543.

Taylor LO, Morrison MD, Boyle KJ (2010). Exchange rules and the incentive compatibility of choice experiments. Environ. Resour. Econ. 47(2):197-220.

Timmins C, Murdock J (2007). A revealed preference approach to the measurement of congestion in travel cost models. J. Environ. Econ. Manage. 53:230-249.

Turpie J, Joubert A (2001). Estimating potential impacts of a change in river quality on the tourism value of Kruger National Park: An application of travel cost, contingent and conjoint valuation methods. Water SA 27(3):387-398.

Viscusi WK, Huber J, Bell J (2008). The economic value of water quality. Environ. Resour. Econ. 41(2):169-187.

Vossler CA, McKee M (2006). Induced value tests of contingent valuation elicitation mechanisms. Environ. Resour. Econ. 35(2):137-168.

Ward FA, Michelsen A (2002). The economic value of water in agriculture: Concepts and policy applications. Water Pol. 4:423-446.

Watanabe M, Asano K (2009). Distribution free consistent estimation of mean WTP in dichotomous choice contingent valuation. Environ. Resour. Econ. 44(1):1-10.

Whitehead JC, Blomquist GC (1991). Measuring contingent values for wetlands: Effects of information about related environmental goods. Water Resour. Res. 27(10):2523-2531.

Whittington D (2002). Improving the performance of contingent valuation studies in developing countries. Environ. Resour. Econ. 22(1-2):323-367.

Tu Z (2013). Contingent valuation methodology: Evaluation of benefits of improving water quality in the Lake Tai Region. Undergrad. Econ. Rev. 10:1, 8:1-11.

Yoo SH, Yang HG (2001). Application of sample selection model to double–bounded dichotomous choice contingent valuation studies. Environ. Resour. Econ. 20(2):147-163.

Zhai G, Suzuki T (2009). Evaluating economic value of coastal waterfront in Tokyo Bay, Japan with willingness–to–accept measure. Water Resour. Manage. 23(4):633-645.

Prospective techniques for *in-situ* treatment and protection of aquifers: A sustainable hydrology review

Michael A. Nwachukwu

Department of Environmental Technology, Federal University of Technology, Owerri, Nigeria.

Most water table aquifers are polluted, yet they are the primary sources of urban and rural water supply. Sustainable hydrology in this millennium will capture *in-situ* treatment and protection of shallow aquifers particularly in developing countries. Permeable reactive barrier (PRB), biological and electrochemical methods with air and steam injection techniques will advance. *Ex-situ* pump and treat method is slow, costly and unsustainable. Many private treatment plants for pump-treat-use have been abandoned due to unsatisfactory output and operational problems. *In-situ* treatment of polluted aquifer starts with mapping pollution source(s), identifying stressors and migration pathway, estimating quantum of stress, and terminating release of stressors from the source. Sustainable hydrology shall also include developing hydrologic models code that can predict pollution, treatment method; amount and period of treatment. This shall be based on the characteristics of the aquifers; the pollution stress and the subsurface.

Key words: Aquifer cleanup, permeable reactive barrier (PRB), biological, chemical, hydrodynamic model, sustainability

INTRODUCTION

An aquifer may be polluted by a number of pollution stressors from one or more sources in a continuous or discontinuous manner. The most active stressor(s) produce the *pollution stress* that characterizes the aquifer. Pollution occurs only when the accumulated stress exceed the aquifer containment level. Sustainability of water resources is not only for the purpose of fulfilling needs for water usage but also for bringing people into harmony with their ambient natural environment. To achieve this, Melloul and Collin (2002) made two important suggestions: (1) develop a global understanding of an aquifer's hydrological and environmental properties in order to delineate appropriate eco-hydrological scenarios and recommend corresponding operational management activities; and (2) emphasize the importance of educating and increasing the awareness of citizens involved as to the need for and viability of socially acceptable measures for sustainable management of groundwater and other resources. Nandakumar (2012) reported that groundwater resources in Kerala–India have come under increasing stress from rising levels of exploitation and pollution. Nwachukwu et al. (2013) reported the water table aquifer of the lower Imo River basin as being increasingly polluted due surface pollutants arising from poor waste management.

International and national action at all levels will be needed to improve water access in those regions lacking water such as North Africa and West Asia and to improve the efficiency of water use in the regions that have water. This is the aspect of sustainable hydrology that challenges hydrologist in this 21st century. The goal is to achieve sustainable development of groundwater so that water supply could be sustained for future generations. Though the aspect of sustainable hydrology is of global concern, it is yet to be properly recognized with respect to groundwater management in Africa. Key to sustainable management of water resources is having the knowledge needed to make the right decisions towards preservation of the accessible aquifer systems worldwide. Melloul and Collin (2000) described migration and accumulation of contaminants into groundwater, rate of water abstraction and land use as the three major aquifer stress vehicles. This paper focuses on how a polluted aquifer can be cleaned up, recovered, and preserved for future use, while considering safety of public health.

Accumulation of pollutants following improvement of urban and rural infrastructure such as road construction in particular in many developing countries has increased pollution stress on water table aquifers, affecting water quality. For example, borrow pits from where materials were collected for road projects are usually abandoned instead of reclaimed. Soon after, these abandoned borrow pits become urban waste disposal pits. Emissions from these pits accumulate to pollution stress until the underlying water aquifer becomes polluted (Nwachukwu and Osoro, 2013). Again, lack of storm water management during road construction and other open engineering sites such as mechanic villages causes continuous enrichment of pollutants to surface water. This in turn, causes groundwater pollution. As a result, near surface aquifers in many regions of the world are highly polluted and most shallow wells in these regions produce water of poor quality. The consequence is endemic water related diseases, poverty, and poor quality of life (Nwachukwu et al., 2010).

Cases of water table aquifer under pollution stress have been discussed by many scholars. For example, Benaabidate and Cholli (2011) reported groundwater stress and vulnerability to pollution of Saiss basin shallow aquifer, Morocco. According to them, the vulnerability of this shallow aquifer to pollution diagnosed by DRASTIC method occurs in different degrees. Not all aquifers are equally vulnerable to pollution. Those where fractures or cavities permit rapid flow tend to be more vulnerable than those where water flows slowly through pore spaces and more opportunities exist for attenuation of pollutants. However, vulnerability to pollution has an inverse relationship to the difficulty of remediation. Once polluted, slow movement of groundwater through a porous aquifer generally makes cleanup difficult, expensive, and in some cases impossible. Three main sources of groundwater pollution are: agricultural, urban and industrial developments.

In-situ remediation is often the method of choice as it destroys contaminants in place, rather than merely transferring them to another environmental medium, as occurs with Pump and Treat method. Thus, industrialized nations like USA, different parts of Europe and Australia are investing on *in-situ* clean up of their polluted aquifers. Unlike most developing countries, they have continuously invoked various techniques in their efforts towards sustainable hydrology. The idea of sustainable hydrology this millennium shall be the primary concern of all hydrologists, NGOs and governments worldwide. Case examples of these applications are hereby reviewed. For example, Key et al. (2013) demonstrated how two pilot tests of an aerobic *in situ* bioreactor (ISBR) have been conducted at field sites contaminated with petroleum hydrocarbons. The two sites differed with respect to hydrocarbon concentrations. At one site, concentrations were low but persistent, and at the other site concentrations were high enough to be inhibitory to biodegradation. Brown et al. (2010) applied surfactant-enhanced aquifer remediation (SEAR) as alternatives to conventional pump-and-treat remediation for aquifers contaminated by dense non aqueous phase organic liquids.

Suthersan et al. (2010) reported a full-scale non–aqueous phase liquid (NAPL) remediation of Area A of the Northeast Site at the Young-Rainey STAR Center, Largo, Florida. Area A covered an area of 930 m^2 (10,000 ft^2) and extended to a depth of 10.7 m (35 ft), representing a total cleanup volume of 9930 m^3 (12,960 cubic yards). The site was contaminated with ~2500 kg (5500 lb) of NAPL constituents such as trichloroethylene, *cis*-1,2-dichloroethylene, methylene chloride, toluene, and petroleum hydrocarbons. The site consists of a fine-grained sand aquifer underlain by Hawthorn clay at 9 m (30 ft) depth. The upper 1.5 m (5 ft) of this clay formed part of the remediation volume, as dense non–aqueous phase liquid was present in this layer. The site was remediated using a combination of steam-enhanced extraction and electrical resistance heating. Operations lasted 4.5 months. The site was heated to the target temperatures within 6 weeks, at which time the mass removal rate increased more than 1000-fold. The post operational sampling showed that ~0.5 kg (1 pound) of VOCs remained in the remedial volume, and showed remedial efficiencies between 99.85% and 99.99% for the four chemicals of concern.

Piotrowski et al. (2006) presented the case of soil and two aquifers under an active lumber mill in Libby, Montana, which had been contaminated from 1946 to 1969 by uncontrolled releases of creosote and pentachlorophenol (PCP). In 1983, because the contaminated surface soil and the shallower aquifer posed immediate risks to human health and the natural environment, the U.S. Environmental Protection Agency placed the site on its National Priorities List. Feasibility studies in 1987 and 1988 determined that *in situ* bioremediation would help clean up this aquifer and that biological treatment would help clean up the contaminated soils.

Rao et al. (2010) showed a comprehensive, field-scale evaluation of *in situ* co-solvent flushing for enhanced remediation of non aqueous phase liquid (NAPL) in a field site at Hill Air Force Base, Utah. This sand-gravel-cobble subsurface aquifer, underlain by a deep clay confining unit at about 6 m below ground surface, was contaminated with a multi component NAPL as a result of jet fuel and chlorinated solvent disposal during the 1940s and 1950s. The co- solvent flushing test consisted of pumping about 40,000 L (approximately nine pore volumes) of a ternary co-solvent mixture (70% ethanol, 12% *n*-pentanol, and 18% water) through the test cell over a period of 10 days, followed by flushing with water for another 20 days. Several methods for assessing site remediation yielded consistent results, indicating that on the average >85% mass of the several target contaminants was removed as a result of the co solvent flushing; NAPL constituent removal effectiveness was greater (90 to 99^{+}%) in the upper 1 m zone, in comparison to about 70 to 80% in the bottom 0.5 m zone near the clay confining unit.

A popular type of *in situ* remediation method currently used to clean up contaminated groundwater is the permeable reactive barrier (PRB). PRBs are treatment zones composed of materials that degrade or immobilize contaminants as the groundwater passes through the barrier. They can be installed as permanent, semi permanent or replaceable barriers within the flow path of a contaminant plume. The material chosen for the barrier is based on the contaminant(s) of concern (U.S. EPA 2001). Liu (2013) investigated the biodegradation of three selected benzo triazoles (BTs), namely benzotriazole (BT), 5-methyl-benzotriazole (5-TTri) and 5-chloro-benzotriazole (CBT), in aquifer materials. Under anaerobic conditions, CBT was found to be degraded better with its half-life of 21 days under nitrate reducing conditions than under aerobic conditions with its half-life of 47 days.

Nano-technology

Nano-remediation methods entail the application of reactive nano-materials for transformation and detoxification of pollutants. These nano particles have properties that enable both chemical reduction and catalysis to mitigate the pollutants of concern. For nano-remediation *in situ*, no groundwater is pumped out for above-ground treatment, and no soil is transported to other places for treatment and disposal (Otto et al., 2008). Nano-materials have highly desired properties for *in situ* applications. Because of their minute size and innovative surface coatings, nano-particles may be able to pervade very small spaces in the subsurface and remain suspended in groundwater, allowing the particles to travel farther than larger, macro-sized particles and achieve wider distribution. However, in practice, current nano-materials used for remediation do not move very far from their injection point (Tratnyek and Johnson, 2006). For a comprehensive overview of the chemistry and engineering of various nanotechnology applications addressed in Supplemental Material, Table 1, and used for remediation, see Theron et al. (2008) and Zhang (2003). For a comprehensive overview of supplemental materials of nano-particles and their remediation applications adapted from Theron et al. (2008) and Zhang (2003), see Table 1.

Nano particles range from 10 to 100 nm in diameter, although some vendors sell micrometer-scale iron powders as "nano-particles." The second metal creates a catalytic synergy between itself and Fe and also aids in the nano-particles' distribution and mobility once injected into the ground (Saleh et al., 2007; Tratnyek and Johnson, 2006; U.S. EPA, 2008b). Some noble metals, particularly palladium, catalyze dechlorination and hydrogenation and can make the remediation more efficient (U.S. EPA, 2008b; Zhang and Elliott, 2006). The underlying chemistry of the reaction of Fe with environmental pollutants (particularly chlorinated solvents) has been extensively studied and applied in micrometer-scale zero-valent iron (ZVI) PRBs (Matheson and Tratnyek, 1994). There are two main degradation pathways for chlorinated solvents: beta elimination and reductive chlorination. Beta elimination occurs most frequently when the contaminant comes into direct contact with the Fe particle (Elliott and Zhang, 2001; Glazier et al., 2003). Zhang (2003) also showed that modifying Fe nano-particles could enhance the speed and efficiency of the remediation process. Some of the particles flow with the groundwater and remain in suspension for various amounts of time, whereas others are filtered out and bind to soil particles, providing an *in situ* treatment zone that could hold back emanating plumes (Henn and Waddill, 2006).

Laboratory tests have shown 99% removal of As using 12-nm-diameter Fe oxide nano-particles (Rickerby and Morrison, 2007). Kanel et al. (2006) concluded that nZVI can reduce As(V) to As(III) in a short period of time at neutral pH. Macé et al. (2006) found that nanoscale zero-valent iron (nZVI) moved with groundwater away from the injection site. Based on this, they hypothesized that nZVI could treat larger areas of the affected aquifers. They found dramatic but short-lived reductions of volatile organic compounds (VOCs) in fractured bedrock and a slower, steadier decrease of VOCs in primary porosity aquifers. The same study suggested that the degradation of VOCs and travel velocity are indirectly proportional to the hydraulic conductivity. In an extensive study, the Navy conducted field tests using nZVI to remediate two of its contaminated sites (Naval Air Engineering Station, Lakehurst, NJ, and Naval Air Station, Jacksonville, FL). In the Jacksonville study, TCE concentrations in a well approximately 20 ft from the source zone were reduced up to 99%, suggesting that some of the nZVI migrated

Table 1. List of manufactured nanoparticles and the pollutants potentially remediated

Number	Nano particle	Application
1	Nanocrystalline zeolites	Toluene, nitrogen dioxide
2	Carbon nanotubes (CNTs)	Benzene, toluene, xylene, ethylbenzene, heavy metal
3	Activated carbon fibers,	Benzene, toluene, xylene, ethylbenzene, heavy metal ions
4	CNTs functionalized with polymers and Fe	p-nitrophenol Benzene,toluene, dimethylbenzene, heavy metal ions
5	Multi-walled CNTs	Heavy metal ions, THMs, Chlorophenols, Herbicides, Microcystin toxins
6	Self-assembled monolayer on mesoporous supports	Inorganic ions, Heavy metal ions, Actinides, Lanthanides, Radionuclides TiO2
7	Zero-valent iron nanoparticles	Chlorinated methanes, Trihalomethanes, Chlorinated benzenes, Chlorinated ethenes, Pesticides, Polychlorinated hydrocarbons, Organic dyes, Heavy metal ions, Inorganic ions, Chlorinated organic compounds.
8	Photo catalysts	Heavy metal ions, Azo dyes, Phenol, Aromatic pollutants, toluene
9	Bimetallic: Pd/Fe nanoparticles	PCBs, Chlorinated ethane and methanes
10	Ni/Fe nanoparticles Pd/Au nanoparticles	TCE, PCBs, Dichlorophenol, Triclorobenzene, Chlorinated ethene, Brominated organic compounds

Adapted from Theron et al. (2008) and Zhang (2003)

outside of the treatment zone through preferential pathways (Gavaskar et al., 2005).

In addition to groundwater remediation, nanotechnology holds promise in reducing the presence of NAPLs.

Recently, a material using nano-sized oxides (mostly calcium) was used *in situ* to clean up heating oil spills from underground oil tanks. Preliminary results from this redox-based technology suggest faster, cheaper methods and, ultimately, lower overall contaminant levels compared with previous remediation methods. Most of these sites have been in New Jersey, with cleanup conducted in consultation with the New Jersey Department of Environmental Protection (Continental Remediation LLC, 2009). These fate processes depend on both characteristics of the particle and that of the environmental system (Boxall et al., 2007). The use of nano-particles in environmental remediation will inevitably lead to the release of nano-particles into the environment and subsequent ecosystems. To understand and quantify the potential risks, the mobility, bioavailability, toxicity, and persistence of manufactured nano-particles need to be studied (Nowack, 2008). The Royal Commission on Environmental Pollution (2008) summed up the current approach to potential implications from nano-material.

METHODS AND SPECIFIC ADVANCEMENTS

To accomplish this study, several journal articles, text books, technical reports and regulatory documents from different parts of the world concerning groundwater pollution and management have been reviewed. Recent opinions going by technological advancement are most considered. Pollution stress can be quantified and measured as pollution index and stress enrichment factor. These are the key elements that can be measured in order to determine the pollution level of aquifer and the quantum of treatment required. The challenge is to develop a hydrodynamic model code for polluted shallow aquifers. Such a model code should indicate shallow aquifer pollution stress (SAPS). The code shall predict pollution, treatment method; pathway, amount and period of treatment. Quantifying SAPS is important in the present day hydrology and hydro-geological studies. SAPS hydrodynamic model software can be adaptable in all cases to monitor shallow aquifer pollution even as the pollution stressors are being removed.

In designing aquifer treatment, a primary assignment is to determine the direction of groundwater flow. This involves the use of advanced numerical modeling software such as DRASTIC, RADMOD, and ZONEBUDGET, PHAST and SHARP etc. MODFLOW code (McDonald and Harbough, 1988) and MODDATH module Pollock (1994) of the GMS package are widely applicable. These model software codes will by graphical illustration explain the direction of groundwater flow along with contaminants migration pathway, fate and transport.

The understanding of this will enable effective design of treatment method, quantum of treatment and procedure. A case example is the study conducted by Nwachukwu et al. (2010b) on south-west end of the lower Imo River Basin. They used MODFOW and MODPATH codes to show Groundwater Flow Model and Particle Track Analysis across Owerri metropolis (Figure 1). Based on

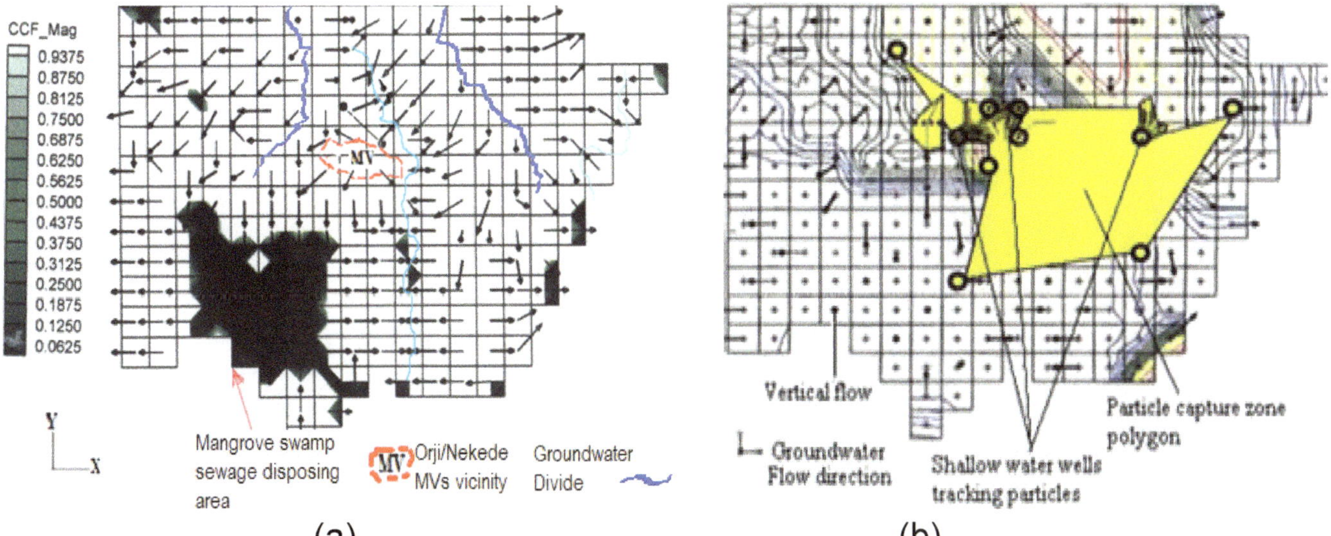

Figure 1. (a) Groundwater flow pattern based on model layer 5. In the SW portion is the highly contaminated sewage dump mangrove swamp area of Umuapu and Egbeda (b) Particle capture zone across Owerri metropolis (Source: Nwachukwu et al., 2010b).

the result of this study, the polluted water table aquifer which remains the very source of drinking water in the area could receive *in-situ* treatment as per result of monitoring wells. This aquifer is easily accessible at shallow depth, even with cheap manual drilling tools, making more than 95% of wells in the area to be shallow. Most of these wells terminate at depths from 34 to 49 m, and are private/commercial water wells. As a result, about 60% of wells in the area are producing water of poor quality (Nwachukwu et al., 2012).

Isotopes can be used to investigate groundwater sources of water to determine their source, how they are recharged, whether they are at risk of saltwater intrusion or pollution, and whether they can be used in a sustainable manner. The isotopes of pollutants, such as trace metals, or chemical compounds dissolved in water, also offer clues about its origin and applicable treatment. Jaisi (2013) conducted a study aimed at determining whether abiotic reactions of phosphate during its transport involve fractionation of oxygen isotopes in phosphate (δ18Op). The results collectively suggest that abiotic reactive transport processes exert minimal influence on the δ18Op composition of groundwater systems.

New advances in geophysics have enabled characterization of sites based on time and cost. The hydrology and basic geochemistry at the sites are established, and used as input data to a hydrologic model code. Parameters required include average depth and thickness of aquifer, drainage area, and distance to pollution source. Others include aquifer pollution stress maximum containment limit (MCL), concentration of primary pollutant, hydraulic conductivity and porosity. These are obtainable from geophysical field study and other hydro geological methods to improve the prospects of sustainable hydrology in this 21st century.

A baseline characterization of the microbiology is essential to establish that the right microorganisms are present for biological treatment of a polluted aquifer. It must be established that the micro-organisms can be stimulated, and that no undesirable reactions with the stimulants or daughter products from the stimulation will occur. This usually requires some treatability and soil compatibility studies and monitoring of microbial community structure and function to establish the base conditions prior to stimulation (Plaza et al., 2001). With new advances in geophysics, aquifer heterogeneity: hydraulic conductivity, aquifer depth, thickness and drainage area are easily obtained. Groundwater flow model and changes in biogeochemistry as well as injection pathway are determined (Faybishenko et al., 2008; Hubbard et al., 2008).

Characterization studies should determine which general remedial approach is appropriate for each sub-area of the site. In many cases, the appropriate remedy for source areas will be containment, and the appropriate remedial level for pollution plume areas will be active restoration. Natural attenuation may be acceptable for some zones where it can show that natural biological, chemical and physical processes will lower pollutant concentrations to cleanup levels (Cherry, 1992). Once the level of remediation is established, site characterization studies should be designed to collect the data required in order to design a technically effective and cost effective remedial action to obtain the desired end result. *In-situ* treatment of polluted aquifer starts with mapping the pollution source(s), identifying stressors and migration pathway, estimating quantum of stress, and terminating release of

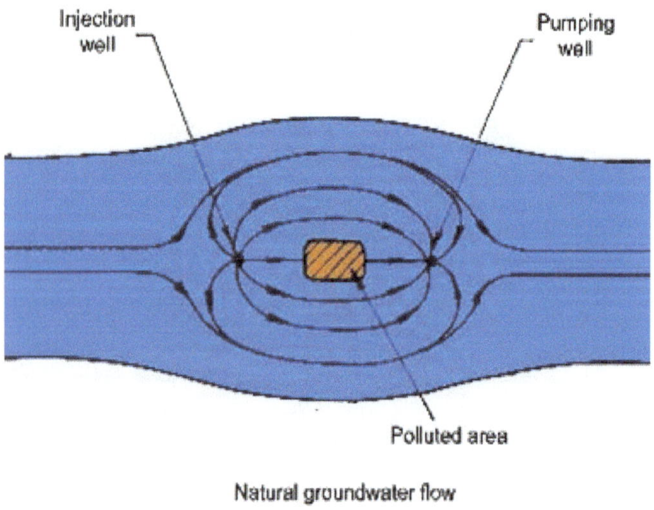

(a)

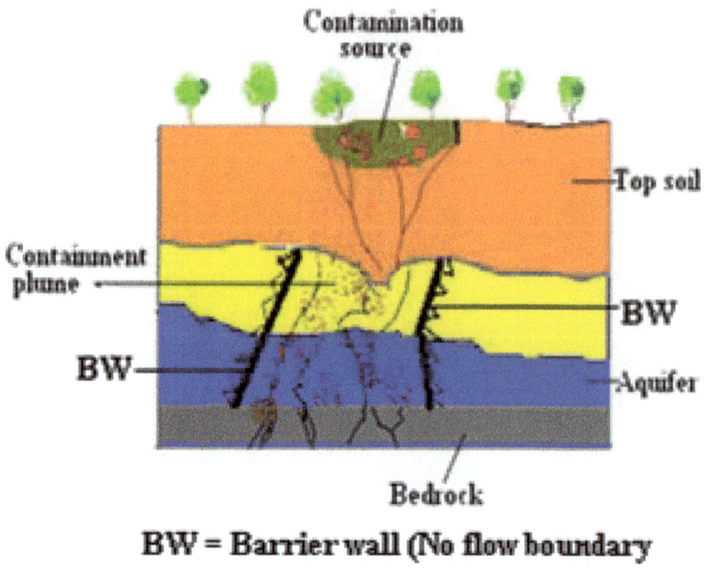

(b)

Figure 2. Scheme of containment systems (a) Hydraulic containment (Source: Fetter, 1993); (b) Containment barrier walls.

stressors from the source.

Aquifer treatment field techniques

Containment techniques

One of the most used alternatives to manage with groundwater pollution areas is to contain the future spread of the pollutant. This task may be achieved by subsurface barriers (Fetter, 1993) which are designed to prevent or control the groundwater polluted flow into the desired location. Two main types of subsurface barriers are used now: physical and hydraulic barriers. Physical barriers include slurry trench walls, grout curtains and cutoff walls. These barriers are no flow boundaries, made with materials that have a lower permeability than the aquifer. For example, compacted clay, geomembranes, soil and bentonite mixtures can be used.

Hydraulic barriers are used to isolate a plume of pollutant from the natural water flow pattern in order to prevent the increase of the polluted area. Another solution of the hydraulic barrier may be realized by controlling the hydraulic gradient of groundwater around and inside the polluted plume by using a pair of injection and pumping wells (Figure 2). The main disadvantages of a well system includes higher operation and maintenance costs, system failures, due to the breakdown of equipment or power outages, flexibility is reduced in fine soils and incorrect pumping rates can draw a significant part of the plume into the wells making treatment necessary before recharge into the aquifer (Barcelona et al., 1990).

Pump and treat technique

Barely a few years ago, pump and treat was one of the most often used processes for polluted groundwater remediation. The basic concept of this technology is very simple; the polluted groundwater is extracted from the subsurface and treated using one of the methods that is currently applied for removing pollutants from industrial or domestic waste water. The clean water may be discharged into a surface water body, such as a stream, or it may be injected into the subsurface. Remediation by pump and treat is a slow process. The calculations for a variety of typical situations show that predicted cleanup times range from a few years to hundreds or even thousands of years. Because of the slow rates of pollutants' desorption and dissolution, pump and treat systems must displace many volumes of aquifer water to flush out pollutants. Pump and treat systems are generally an inefficient method for removing pollutants from the aquifer, but they may be used to clean up plumes of dissolved pollutants. A common private application of this method is the pump-treat-use. A number of treatment plants are available for different treatment purposes. Often treatment plants are abandoned for inefficiency in removing a number of pollutants, compelling operational stress and maintenance.

A very efficient and recommended application of pump and treat systems is to remove light non-aqueous-phase-liquids (LANPLs) from the subsurface. LNAPLs delineate the pollutant that is not miscible with water and is easier than it, so that it will accumulate as a mobile layer on the water table or the top of capillary fringes. These

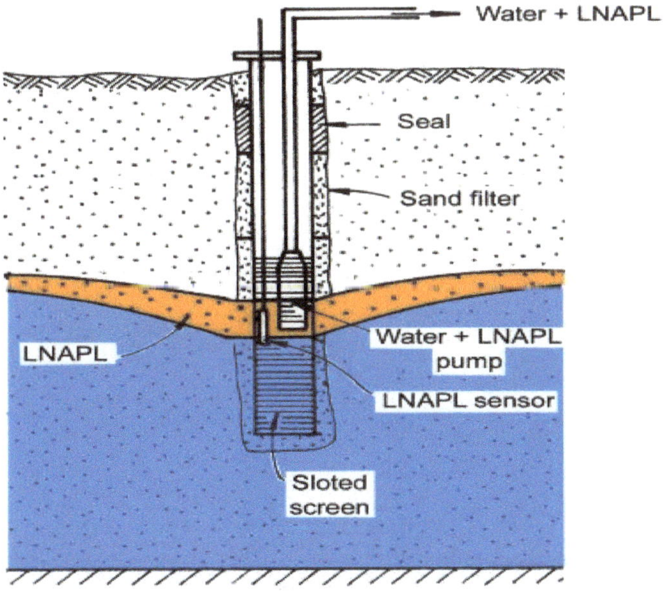

Figure 3. Examples of NAPL recovery by pump and treat system. Source: Beddient et al. (1993).

pollutants can be recovered by depressing the groundwater table with extraction wells in order to facilitate their accumulation in the well (Figure 3). For this application of pump and treat system, two alternatives can be employed: pumping the combined groundwater and LNAPL mixture with a single pump or using two separate pumps working under a control system to remove water and NAPL separately.

Soil vapor extraction technique

This technology is primarily applied to extract vapor of organic compound from an unsaturated zone by applying a vacuum at a sealed wellhead (Figure 4). Soil vapor extraction can be more efficient and cost effective in comparison to others remediation strategies like pump and treat. Soil vapor extraction involves passing large volumes of air through a polluted spill. The organic compounds or various fractions of a mixture of organic compounds volatilize or evaporate into air and are transported to the surface. The system may consist of one or more extraction wells, vacuum pump and a treatment system for the extracted vapors.

Bioremediation technique

The main feature of this method is to stimulate subsurface microorganisms to biodegrade pollutants (Figure 5a). In some proper conditions microorganisms can transform the chemical compounds pollutants of the groundwater to carbon dioxide and water (Figure 5b). Typical *in situ* bioremediation systems inject electron acceptors and nutrients through the polluted area so as to enhance the bioactivity of the microorganisms. *In situ* bioremediation has been successfully applied for treating the soil and the groundwater polluted with hydrocarbons, phenols, acetone and cellulosic waste. *In situ* bioremediation has some advantages; it treats the pollutant on the spot and transforms it into an innocuous compound; so this method reduces the requirement for surface water treatment and disposal of recovered water.

In-situ bioremediation is looked upon as the best approach to deal with the underground water contamination. In order to increase microbial activity and to ensure the continuous growth of the bacteria in the Biologically Active Zone (BAZ), the limiting substances such as macro nutrient (nitrogen or phosphorus), electron acceptor (such as oxygen, nitrate) or carbon source has to be continuously supplied and injected to the system. Success of the whole program is largely dependent on the soil properties, selection of the right bacteria to use matching the type of contaminants present, hydrogeology and other factors. In the line of bioremediation is bioimmobilization. *In situ* bioimmobilization has recently gained attention as a potentially effective remediation strategy for metal- or radionuclide-contaminated groundwater (Tokunaga et al., 2003; Istok et al., 2004). During *in situ* bioimmobilization, electron donor additions are used to stimulate iron- and sulfate-reducing conditions, which promote the reductive precipitation of redox-sensitive metals and radionuclides from groundwater. Diverse or extreme geochemical conditions common to radionuclide-contaminated sites present unique challenges to successful implementation of bioimmobilization.

In situ chemical treatment technique

By this method, chemical compounds are used to transform pollutants in the subsurface. *In situ* chemical treatment processes generally consist of installing a series of injection wells for delivering chemical compounds at the head or within the plume of polluted groundwater. Chemicals that are added can oxidize or reduce pollutants, converting them to non toxic forms or immobilizing them to minimize their migration ability. Possible oxidizing compounds include hydrogen peroxide, ozone and potassium permanganate. The most often used reducing agent includes sulfur dioxide, sulfide salts, ferrous sulfate, metallic iron and zinc.

This technique can be considered in cases where specific pollutants, their concentrations and the extent of pollution plume in the aquifer are well-defined. The treatment chemical compounds specific to the class of pollutants are delivered in the subsurface by injection wells, as shown in Figure 6a (Barcelona et al., 1990). Figure 6b also illustrates a simple field technique of *in-situ* oxidation treatment method for contaminants like

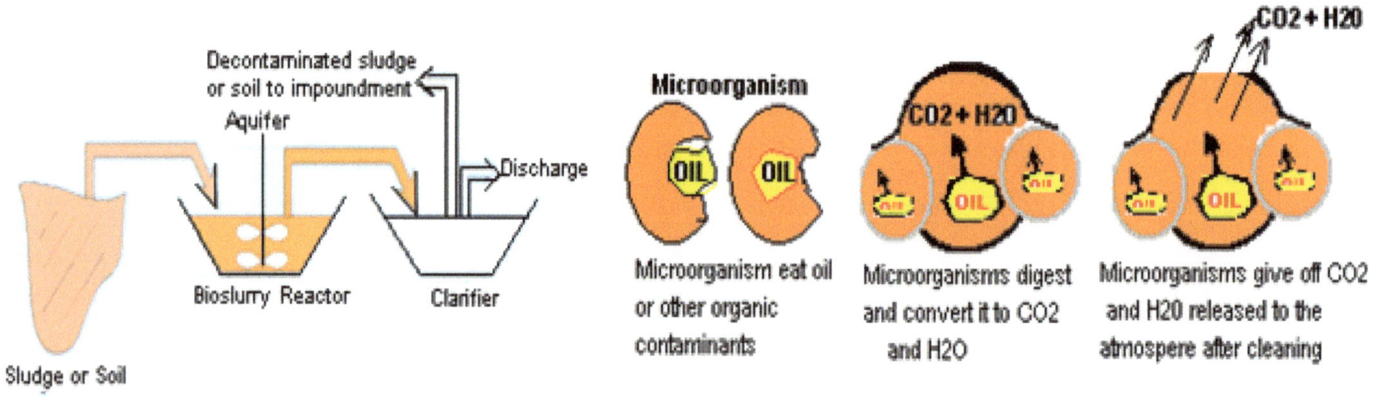

Figure 5. (a) Bioslurry bioremediation system (USEPA, 1991) (b) Bioremediation processes.

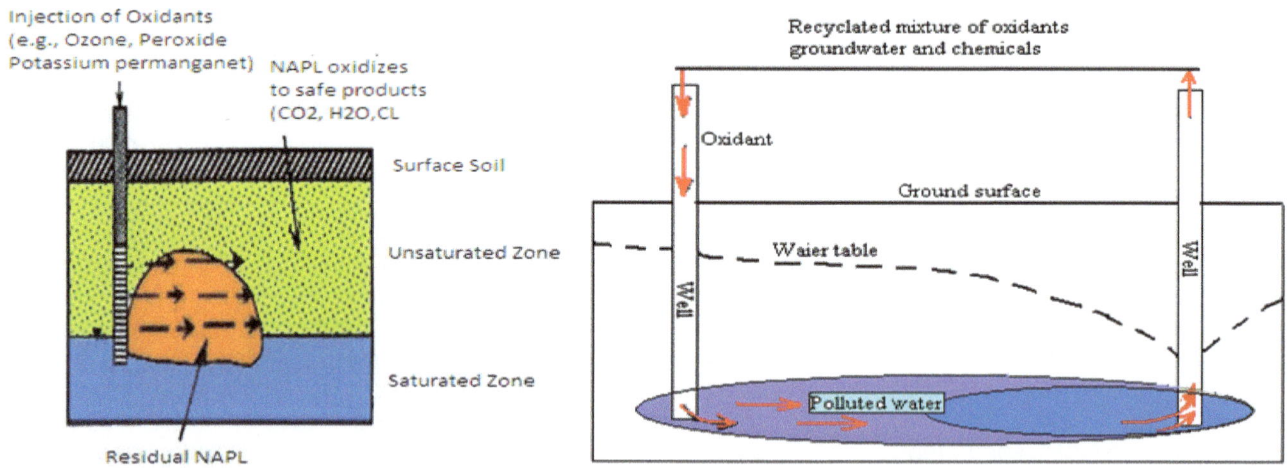

(a) (b)

Figure 6. Techniques of chemical treatment system: (a) Barcelona et al. (1990); (b) Simple oxidation field technique.

amines, phenols, cyanides, halogenated aliphatic compounds, and certain pesticides in liquid waste streams.

In situ thermal technologies

In situ thermal technologies are designed to increase the temperature, of the subsurface to improve and accelerate the removal mechanisms of the pollutants. At the increasing of temperature the organic pollutants will mobilize more rapidly due to enhanced volatilization, enhanced desorption from soil, increased water solubility and increased fluid flow rates. Higher temperatures can also accelerate biodegradation and other certain abiotic transformation or decomposition reactions. Temperature can be increased by injection of heated fluids such as air, water or steam. Other means of heat addition include installation of vertical electrodes that provide resistance heating by passing electrical current through the soil between the electrodes, with the amount of heat controlled by the amount of electrical current. *In situ* steam injection – extraction removes volatile and semi-volatile pollutants from soil and groundwater without requiring excavation. Application of steam injection and extraction (Figure 7) relies on the ability to deliver, control flow and recover the heating fluid.

Permeable reactive barriers technique

Permeable reactive barrier (PRB) is an *in situ* treatment zone that passively captures a plume of contaminants and removes or breaks down the contaminants, releasing uncontaminated water (Gillham et al., 2010). The primary removal methods include: (1) sorption and precipitation, 2) chemical reaction, and (3) reactions involving biological

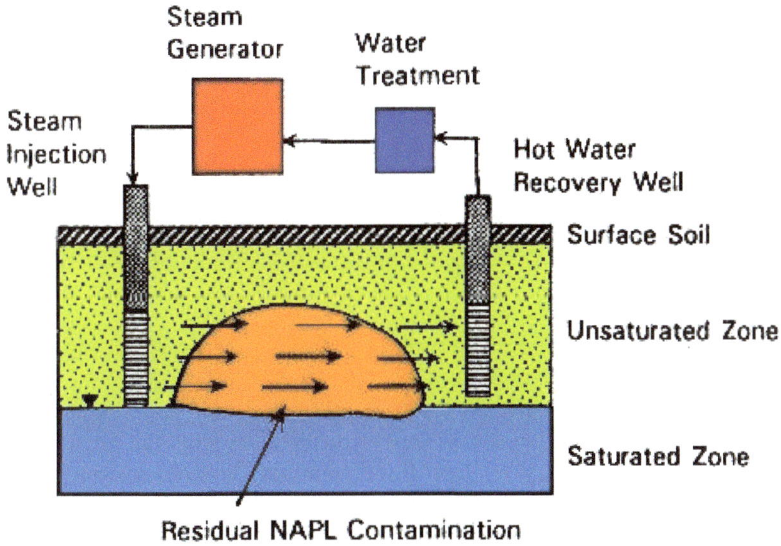

Figure 7. Scheme of a steam injection system (Barcelona et al., 1990).

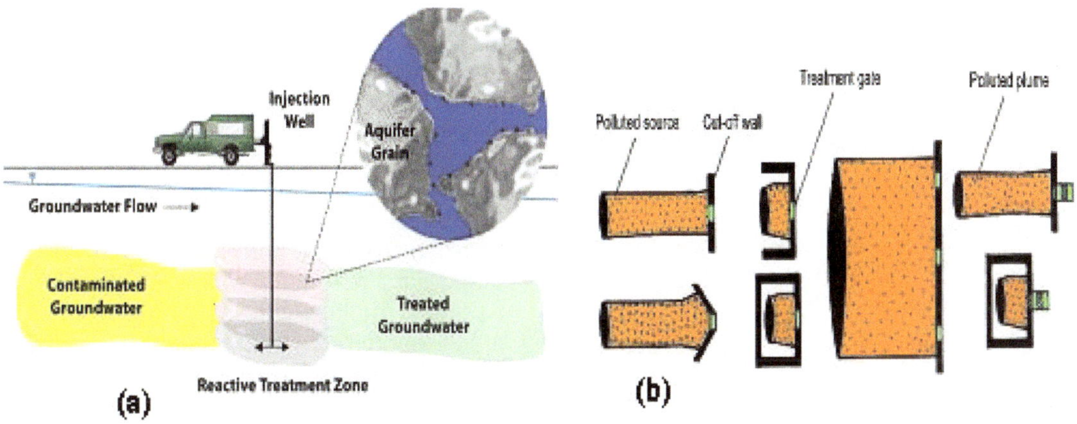

Figure 8. (a) Funnel and gate configuration (b) Field technique of PRB.

mechanisms involve construction of permanent or semi-permanent wall across the flow path of pollutant plume (Figure 8a). As the polluted groundwater moves passively through the wall, the pollutants are removed. Funnel and gate systems are used to channel the contaminant plume into a gate which contains the reactive material Figure 8b). The funnels are non-permeable, and the simplest design consists of a single gate with walls extending from both sides. An advantage of the funnel and gate system is that a smaller reactive region can be used for treating the plume, resulting in lower cost.

Denitrification technique

Most methods of nitrate removal that have been applied for *in situ* groundwater treatment are based on chemical and/or biological denitrification. The methods apply redox reactions, often with biological catalysis, to reduce the nitrogen (N) oxides, nitrate (NO_3) and nitrite (NO_2), to the gases nitric oxide (NO), nitrous oxide (N_2O), and dinitrogen (N_2) to nitrogen gas. The appeal of using denitrification reactions for *in situ* application lies mainlyin the fact that the main products of the reactions are gaseous and do not accumulate as hazardous by-products in the subsurface. Some of the techniques also do not require highly sophisticated technology. A series of factors influence the denitrification reactions and the most important ones for biological denitrification are listed below:

i) Temperature, preferably near 35°C; and pH, preferably

neutral to alkaline;
ii) Organic carbon present as substrate for bacteria; and Carbon: nitrogen ratio of 1:25;
iii) Presence of nutrients such as phosphate; and sufficient denitrifying bacteria;
iv) Low oxygen, that is, anoxic or anaerobic conditions; and high soil water content;
v) Anthropogenic disturbance of the soil and Porosity and permeability of aquifer.

Biological denitrification plants in the USA, France and parts of Europe generally use ethanol or methanol as carbon substrate. In one pilot plant in Israel, sucrose was used. Iron and manganese that are mobilized from the aquifer during the denitrification process are re-precipitated in the oxidation step.

Other field techniques

Natural attenuation

Natural attenuation defines the natural occurring processes in the subsurface environment such as dilution, volatilization, biodegradation, adsorption and chemical reactions with subsurface compounds that contain the spread of pollution and reduce the concentration and amount of pollutants at polluted sites. Consideration of this option requires modeling and evaluation of pollutant degradation rates and pathways. The processes contributing to natural attenuation are typically acting at many sites, but at varying rates and degrees of effectiveness, depending on the types of pollutants present and physical, chemical and biological characteristics of the soil and groundwater. To estimate how well natural attenuation will work and how long it will take requires a detailed study of the contaminated site.

Hydraulic and pneumatic fracturing

The pneumatic fracturing process involves injection of highly pressurized air into consolidated sediments that are polluted to extend existing fractures and create a secondary network of fissures and channels. In this way, the hydraulic permeability of the soil increases and the removal of the pollutants will be accelerated particularly by vapor extraction, biodegradation and thermal treatment. Hydraulic fracturing creates distinct fractures in low permeability and over-consolidated clays or sediments. High-pressure water is first injected into the bottom of a borehole to cut a notch that will serve as the starting point for the fracture. Slurry of water, sand and thick gel are pumped at high pressure into the borehole to propagate the fracture. These fractures will serve for steam or hot air injection or pollutant recovery and they can also improve pumping efficiency and the delivery for other *in situ* processes.

Air injection

This involves the injection of air with high pressure in a trench or in a borehole to form bubbles in a groundwater column. The bubbles will strip volatile organic compounds from the dissolved phase from NAPls present along the path of the bubbles; they will add oxygen to the water to enhance the *in situ* bioremediation process and they will establish large circulation in the subsurface which tends to move polluted water to the surrounding wells for extraction. After the bubbles make their way to the unsaturated zone, a soil vapor extraction system is used to remove the vapors for treatment prior to release to the atmosphere. One major drawback of this technique consists in the risk of the potential off-site migration of vapors pollutant plumes so that an adequate monitoring system will be necessary.

Solidification and stabilization

Solidification and stabilization refers to treatment processes that mix or inject treatment agents into the polluted material to accomplish one of the following objectives: to improve the physical characteristics of the water by producing a solid form liquid or semi-liquid wastes; to reduce pollutant solubility; to decrease the exposed surface area across which mass transfer of pollutants may occur. Solidification refers to techniques that encapsulate hazardous waste into a solid material of high structural integrity. Stabilization refers to techniques that treat hazardous waste by converting it into a less soluble, mobile or toxic form. Solidification and stabilization technologies can immobilize many heavy metals, certain radionuclide and selected organic compounds while decreasing waste surface areas and permeability for many types of sludge, contaminated soils and solid wastes. Common solidification and stabilization agents include cement, lime, limestone, and various mixtures of these materials and various organic binders.

DISCUSSION OF PROSPECTS

Advantages

In situ treatment of contaminated groundwater uses the aquifer as a subsurface "treatment plant" to improve the quality of groundwater supplies. This has several advantages over conventional *ex situ* treatment technologies (pump-and-treat systems). The advantages are;

i) Cost and time savings
ii) Exposure to chemical reagents.
iii) There is a long period of operation (anything from 5 to 30 years) with less stringent control of operating conditions.

iv) Robust and require less plumbing.
v) Redox-based systems can often remove other contaminants e.g. chromate and organic chemicals, if these are also present.

Side effects

Some *in situ* treatment methods for removing groundwater contaminants may cause undesirable side effects, of which the most common problem is clogging in the subsurface. In the case of removal of organic compounds, the products of the treatment reaction may cause clogging of the aquifer due to biofilm build-up. If metals are removed by precipitation, the solid precipitates in the aquifer matrix may reduce the permeability of the aquifer over the long term. The specific hydrogeological conditions and the contaminant load will determine the extent of these phenomena. Side effects need to be managed to maintain the efficiency of the scheme and increase the treatment lifetime. Biofilm build-up in the aquifer, e.g. in biological systems still constitute a major disadvantage.

Design criteria

For the design of any treatment system, both the hydrodynamics of the flow system and the source of the pollution need to be exceptionally well characterized to optimize the system. The nature of the pollution source will affect the choice of system design. Trench and fill barrier methods are only suitable for shallow flow systems where the barrier can be constructed down to the impermeable bedrock. In deeper flow systems, treatment barriers can be created by borehole injection. *In situ* treatment methods are more likely to be successful in primary aquifers, where hydrodynamics are more easily understood and greater control over the treatment zone can be exercised than in fractured complex flow environments. For nitrate removal systems, management of nitrogen inputs to the subsurface is still required, as treatment of the nitrate in the aquifer does not remove the contamination source. Although the treatment systems generally have a long lifetime, it is still important that the source of contamination be eliminated wherever possible.

Prospect case example

In Nigeria are great potentials for *in situ* treatment of shallow aquifers that constitute the accessible and economic sources of water supply to the ever growing population. From the basement complex rocks of the North and Western Nigeria to south-east and to the south-south sedimentary basins are shallow aquifers. About 85% of the population depends on shallow wells producing water from these water table aquifers significantly polluted. The configuration of these wells range from 40 ft hand-dug tube wells to 100 to 120 ft hand driven mono pump wells and to between 120 to 200 ft hand-driven minor motorized pump wells. These are predominantly the private/commercial wells. Deep wells are usually the few public or government motorized rig drilled wells that are often non functional or abandoned. Aquifer contamination in most parts of Nigeria is mainly the product of poor domestic and industrial waste management and chemical fertilizer agricultural input across the country.

In the Niger Delta significantly, groundwater may have variable levels of pollution by petroleum products and wastes, and oil and gas spills. Another major pollution issue of the Niger Delta region of Nigeria is the saline intrusion to aquifer. The Niger Delta covers 20,000 km² within wetlands of 70,000 km² formed primarily by sediment deposition. It is home to 20 million people and 40 different ethnic groups. It is the largest wetland and maintains the third-largest drainage basin in Africa. The Niger Delta environment can be broken down into four ecological zones: coastal barrier islands, mangrove swamp forests, freshwater swamps, and lowland rainforests. A UNDP report states that there have been a total of 6,817 oil spills between 1976 and 2001, which account for a loss of three million barrels of oil, of which more than 70% was not recovered. 69% of these spills occurred off-shore, a quarter was in swamps and 6% spilled on land (UNDP, 2006). Some spills are caused by sabotage and stealing, while most are due to poor maintenance by oil companies such as Shell.

Presently, there is no significant research effort towards *in situ* treatment of groundwater in this region. A prospect model suggested for treating groundwater polluted by oil or gas in the Niger delta based on air sparging technique is illustrated in Figure 9.

Wells A, B, C are the compressed air injection wells, with injection nozzles extended from the water table. 1 and 2 are the air lifting wells connected to a blower that passes the aerated oil and gas to the atmosphere after some cleaning. X is the monitoring well with pump. This method is prospective in the Niger Delta where water table depth fluctuates to an average of 18 m or as shallow as 0 to 5 m in the swamps. A pilot study is necessary in order to determine the effectiveness of this technique in the area.

Prospect comparison

Capital investment for the *ex situ* plant may be between three and seven times higher than for the *in situ* systems. *In situ* methods have virtually no operating costs, and the permeable reactive barrier system have the lowest capital costs. This explains why *in situ* methods have already gained wide acceptance. *In situ* treatment systems for iron and manganese removal have been used successfully and economically for many decades, while *in*

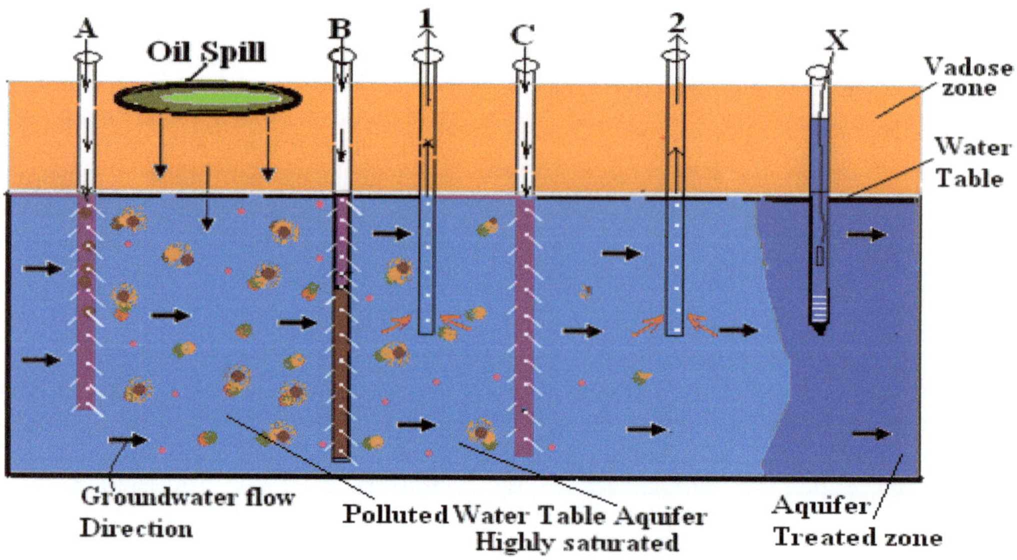

Figure 9. Air sparging field technique prospective for treating oil polluted areas of Niger Delta where the vadose and the shallow aquifer water consists of coastal plain sands.

situ nitrate removal has been in place for about two decades, with successful application in South Africa.

Summary

Although most water table aquifers are polluted, they are the easily accessible primary sources of urban and rural water supply. Sustainable hydrology this millennium will capture *in-situ* treatment of polluted aquifers particularly in developing countries. The most important *in situ* groundwater treatment methods can be divided into the following three main groups: Permeable Reactive Barrier (PRB), Biological and Electrochemical methods with Air and Steam injections techniques. *Ex-situ* pump and treat method is slow, costly and unsustainable. *In-situ* treatment of polluted aquifer starts with mapping pollution source(s), identifying stressors and migration pathway, estimating quantum of stress, and terminating release of stressors from the source. Sustainable hydrology shall include developing a hydrologic model code that can predict pollution, treatment method; amount and period of treatment. The model code shall be based on the characteristics of the aquifers; the pollution stress and the subsurface. *In situ* treatment however has shown potential for total groundwater cleanup and considered as a 21st century aspect of sustainable hydrology. Inorganic contaminants susceptible to this cleanup include Arsenic (As), Cadmium (Cd), Chromium (Cr), Copper (Cu), Mercury (Hg), Iron (Fe), Manganese (Mn), Molybdenum (Mo), Nickel (Ni), Lead (Pb), Selenium (Se), Technetium (Te), Uranium (U), Vanadium (V), Nitrate (NO_3^-), Phosphate (PO_4^-), and Sulphate (SO_4^{2-}). Organic compounds can also be removed by *in situ* treatment.

Conclusion

Permeable reactive barriers can be constructed from cheap, readily available materials and would be relatively simple for sub urban communities in Nigeria to install with limited training. *In situ redox manipulation* (ISRM) should be tested for deeper primary aquifers in South and West African countries. The dithionite chemical reagent should be readily available, since it is used in the pulp and paper industry. The electrokinetic method is, however, the only one found that has been claimed to be suitable for fractured aquifers with high technical skill. *Vyredox* method for Iron and manganese removal is imperative in many high volume abstraction situations to avoid or minimize borehole clogging effects. Advancement in groundwater geophysics, to advanced hydrological or hydrodynamic model codes is inevitable for sustainable hydrology this 21st century. Though global, the aspect of sustainable hydrology needs to be well infused into groundwater management in Africa, and now is the time.

REFERENCES

Barcelona M, Keely J, Pettijohn W, Wehrman A (1990). Contamination of ground water: Prevention, assessment, and restoration. Pollution Technology Review No. 184. Noyes, Park

Benaabidate L, Cholli M (2011). Groundwater stress and vulnerability to pollution of Saiss basin shallow aquifer, Morocco. 15th International Water Technology Conference, IWTC-15 2011, Alexandria, Egypt.

Brown CL, Pope GA, Abriola LM, Sepehrnoori K (2010). Simulation of surfactant-enhanced aquifer remediation. Water Resour. Res. 30(11):2959-2977.

Key KC, Sublette KL, Johannes TW, Raes E, Sullivan E, Duba J, Ogles D, Baldwin BR, Biernacki A (2013). An *In Situ* Bioreactor for the Treatment of Petroleum Hydrocarbons in Groundwater. Remed. J. 23(3):55-84.

http://journals.ohiolink.edu/ejc/article.cgi?issn=10515658&issue=v23i0003&article=55_aisbfttophig
Cherry AJ (1992). Developing rational goals for *in situ* remedial technologies. Subsurface Restor Conf, Dallas.
Elliott DW, Zhang WX (2001). Field assessment of nanoscale bimetallic particles for groundwater treatment. Environ. Sci. Technol. 35:4922-4926
Fetter CW (1993). Contaminant Hydrogeology. Macmillan Publishing Company, New York.
Faybishenko B, Hazen TC, Long PE, Brodie EL, Conrad ME, Hubbard SS, Christensen JN, Gavaskar A, Tatar L, Condit W (2008). Cost and Performance Report Nanoscale Zero-Valent Iron Technologies for Source Remediation. Port Hueneme, CA. Naval Facilities Engineering Command; CR-05-007-ENV: at www.clu-in.org/download/techdrct/td_CR-05-007-ENV.pdf.
Glazier R, Venkatakrishnan R, Gheorghiu F, Walata L, Nash R, Zhang W (2003). Nanotechnology takes root. Civ. Eng. 73(5):64-69.
Gillham R, Vogan J, Gui L, Duchene M, Son J (2010). Iron barrier walls for chlorinated solvent remediation. In: Stroo, H. F.; Ward, C. H. (eds.), *In Situ* Remediation of Chlorinated Solvent Plumes. Springer Science+Business Media, NY, p. 537. doi:10.1007/978-1-4419-1401-9
Henn KW, Waddill DW (2006). Utilization of nanoscale zerovalent iron for source remediation - A case study. Remediation 16(2):57-77.
Hubbard SS, Williams KH, Conrad ME, Faybishenko B, Peterson J, Chen J, Long PE, Hazen TC (2008). Geophysics for monitoring of hydrological and bio-geochemical transformations associated with contaminant remediation. Environ. Sci. Technol. 42:3757-3765.
Istok, JD, Senko JM, Krumholz LR, Watson D, Bogle MA, Peacock A, Chang YJ, White DC (2004). In situ bioreduction of Technetium and Uranium in a nitrate-contaminated aquifer. Environ. Sci. Technol. 38:468-475.
Jaisi DP (2013). Stable isotope fractionations during reactive transport of phosphate in packed-bed sediment columns. J. Contam. Hydrol. 154:10-16.
Macé C, Desrocher S, Gheorghiu F, Kane A, Pupeza M, Cernik M, Kvapil P, Venkatakrishnan R, Zhang W-x (2006). Nanotechnology and groundwater remediation: A step forward in technology understanding. Rem. J. 16(2):23-33.
Matheson LJ, Tratnyek PG (1994). Reductive dehalogenation of chlorinated methanes by iron metal. Environ. Sci Technol. 28(12):2045-2053.
Melloul AJ., Collin ML (2002). Prioritization of Sustainable Groundwater Management Needs: Case of Israel's Stressed Coastal Aquifer. Environ. Dev. Sustain. 4(4): 347-360.
Melloul AJ, Collin ML (2000). Sustainable groundwater management of the stressed Coastal aquifer in the Gaza Region. Hydrol. Sci-J-des Sci. Hydrologiques 45(1):1-47.
McDonald MG, Harbough AW (1988). A modular three-dimensional finite-difference groundwater flow model, MODFLOW. U.S. Geo Survey Open-file Report 83-875.
Nwachukwu MA, Osoro E (2013). Assessment of environ hazards: Linking borrow pits, gully erosion, and road failure. J. Geotech. Eng. Construct. Tech. 4(6):199-210,
Nwachukwu MA, Huan F (2012). *In-situ* Remediation of Shallow Aquifer. Pollut. J. Hydrol. Eng. Editorial, 1:1.
Nwachukwu MA, Ohuawunwa C, Nwachukwu I, Nnorom U (2012). Sustainable water supply from boreholes in neighborhood communities besieged by off-campus students. Int. J. Water Res. Environ. Eng. 4(11):352-362.
Nwachukwu MA, Huan F, Ophori D (2010). Groundwater model and particle track analysis, for water quality monitoring and pollution assess around contaminated sites. J. Spatial Hydrol. 10(1).
Nwachukwu MA, Feng H, Amadi MI, Umunna FU (2010). The Causes and the Control of Selective Pollution of Shallow Wells by Coliform Bacteria: Imo River Basin Nigeria. Water Qual. Expo. Health 2:75-84.
Nowack B (2008). Pollution prevention and treatment using nanotechnology. In: Krug H, Nanotechn, Environ Aspects. Weinheim: Wiley-VCH Verlag. 2:1-15.
Nandakumar T (2012). Aquifer atlas reveals groundwater stress. The Hindu, Today's Paper (October, 12) Otto M, Floyd M, Bajpai S. Nanotechnology for site remediation. Rem. J. 19(1):99-108.
Plaza G, Ulfig K, Hazen TC, Brigmon RL (2001). Use of Molecular techniques in bioremediation Acta. Microbiol. Pol. 50:205-218.
Pollock DW (1994). User's guide for MODPATH/MODPATH PLOT, version 3: A particle tracking post processing package for MODFLOW: The U.S. Geological Survey finite-difference groundwater flow model
Piotrowski MR, Doyle JR, Carraway JW (2006). Integrated bioremediation of soil and groundwater at a superfund site. Rem. J. 2(3):293-309.
Rickerby DG, Morrison M (2007) Nano-technology and the environment: A European perspective. Sci. Technol. Adv. Mater. 8:19-24.
Rao PS, Annable MD, Sillan RK, Dongping D, Hatfield K, Graham WD, Wood LA, Enfield CG (2010). Field-scale evaluation of *in situ* cosolvent flushing for enhanced aquifer remediation. Water Res. 33:2673-2686,
Royal Commission on Environmental Pollution (2008). Novel Materials in the Environment: The Case of Nanotechnology. London: Royal Committee on Environ Pollution. Accessed at: http://www.official-documents.gov.uk/document/cm74/7468/7468.pdf
Saleh N, Sirk K, Liu Y, Phenrat T, Dufour B, Maty-jaszewski K., Tilton RD, Lowry GV (2007). Surface modifications enhance nanoiron transport and NAPL target-ing in saturated porous media. Environ. Eng. Sci. 24:45.
Suthersan S, John H, Denice N, Scott P (2010). *In situ* air sparging remediation engineering design. Ground water Monit. Rem. 30(3):39-44,
Theron J, Walker JA, Cloete TE (2008). Nanotechnology and water treatment: Applications and emerging opportunities. Crit. Rev. Microbiol. 34:43-69.
Tratnyek PG, Johnson RL (2006). Nanotechnology for environmental cleanup. Nanotoday 1(2):44-48.
Tokunaga TK, Wan JM, Firestone MK, Hazen TC, Olson KR, Herman DJ, Sutton SR, Lanzirotti A (2003). *In situ* reduction of chromium (VI) in heavily contaminated soils through organic carbon amendment. J. Environ. Qual. 32:1641-1649.
U.S. EPA. (2001). Cost Analysis for Selected Groundwater Cleanup Projects: Pump and Treat Systems and Permeable Reactive Barriers. Washington, DC: Environ Protect Agency 542-R-00-013.
U.S. EPA (2008b). Nanotechnology for Site Remediation: Fact Sheet. Washington, DC: EPA 542-F-08-009 at http://www.epa.gov/tio/download/remed/542-f-08-09.pdf
U.S. EPA (1991). "Sampling and analysis plan for a bench-scale bioslurry test of soil from the 9th Ave Dump Superfund site, U.S. Army Engineering Waterways Experiment Station, Vicksburg, Washington, DC.
UNDP (2006). "Niger Delta Human Development Report." UNDP p. 76. Retrieved 19 June, 2013
Zhang W-X (2003). Nanoscale iron particles for environmental remediation: An overview. J. Nanopart. Res. 5:323-332.
Zhang W-X (2005). Frontiers in Nanotechnology: U.S. EPA Millenium Lecture Series. Nanotechnology for Water Purification and Waste Treatment July 18, 2005 Washington D.C.: http://www.epa.gov/ncer/nano/lectures/zhang0705.pdf
Zhang W-X, Elliott DW (2006). Application of iron nanoparticles for groundwater remediation. Rem. J. 16(2):7-21.

Spatial-temporal estimation of evapotranspiration over Black Volta of West Africa

Amos T. Kabo-bah[1], G. K. Anornu[2], Eric Ofosu[1], Robert Andoh[3] and Kamila Justyna Lis[4]

[1]Department of Energy and Environmental Engineering, School of Engineering, University of Energy and Natural Resources, P. O. Box 214, Sunyani, Ghana.
[2]Department of Civil Engineering, Kwame Nkrumah University of Science and Technology, P. M. B., Kumasi, Ghana.
[3]Hydro International, 94 Hutchins Dr., Portland, USA.
[4]Department of Geography, School of Geosciences, University of Energy and Natural Resources, P. O. Box 214, Sunyani, Ghana.

Evapotranspiration is an important component necessary for ensuring efficient management of water resources for every river basin. The Bui Hydroelectric Plant completed early 2014 to provide electricity for the people of Ghana is located within the Black Volta River Basin of West Africa. With the construction of this plant and related developments, there is a need for attention to be paid to the conservation of water resources within this basin. This paper therefore assesses the spatial temporal behaviour of evapotranspiration within the basin from 2001 to 2013 and compares this with the average monthly rainfall estimates from 1983 to 2012. The assessment shows that the average monthly evapotranspiration is within 0 to 150 mm/month for the period assessed. However, significant variation exists between the years with a highly erratic pattern observed. The pattern of the rainfall indicates that the upstream portion of the basin receives lower rainfall compared to the downstream portion of the basin. This pattern is consistent with the overall variation of the evapotranspiration within the basin. This spatial-temporal behaviour is an important factor that needs to be considered by all stakeholders given the anticipated land-use changes expected within the basin coupled with likely future increases in temperatures resulting from climate change.

Key words: Evapotranspiration, moderate resolution imaging spectroradiometer 16 (MODIS16), satellite application facility (SAF), Tropical Applications of Meteorology using SATellite (TAMSAT) rainfall, Bui Hydroelectric plant.

INTRODUCTION

Evapotranspiration (ET) is one of the difficult hydrological cycle parameters to estimate. It is impossible to estimate directly both *in-situ* and through satellite data. As a result, there have been several studies over the past decade using various derived data/products to estimate evapotranspiration (Bastiaanssen, 2000; Minacapilli et al., 2009; Or and Hanks, 1992; Rana and Katerji, 2000; Rwasoka et al., 2011; Su et al., 2003, 2005; Sucksdorff and Ottle, 1990). The large number of studies highlights the significance of this parameter and it relevance to water resource managers and agriculturist as it informs decisions on water budgets. The ET dynamics cannot be directly quantified from satellite observations; as a result it is derived based on the assumption, that surface variables have a direct influence on evapotranspiration (Choudhury, 1991).

In the recent past, satellite data has provided a leap forward in the ability to indirectly estimate evapotranspiration fluxes in space and time, further boosting our ability to monitor the ET over large areas (Allen et al., 2007; Bastiaanssen, 2000; Su, 2002). The surface energy balance algorithm for land (SEBAL) is a remote sensing based model that can be used to estimate large-scale ET over an area (Bastiaanssen, 2000; Bastiaanssen et al., 1998a, b). The SEBAL model uses satellite derived data for the land surface characteristics such as the surface albedo, land surface temperature and vegetation index. This is then coupled with meteorological data (that is, wind speed, humidity, solar radiation and air temperature) to resolve and estimate the instantaneous energy balance and then extrapolated to obtain the daily evapotranspiration. Another remote sensing based model is the surface energy balance system (SEBS) which also uses a similar approach to SEBAL by combining satellite derived products and meteorological data to estimate evapotranspiration (Su, 2002, 2009). Several improved versions by different researchers have been proposed however these are all fundamentally derived from the basic concepts indicated by SEBAL and SEBS. Other methods which have also been used include Vegetation Index/Temperature Trapezoid (VITT) (Moran et al., 1994) and Mapping Evapotranspiration at high resolution with internalized calibration (METRIC) (Allen et al., 2007).

Building on these fundamentals of remotely sensed ET products, the National Aeronautics and Space Administration (NASA) developed the moderate resolution imaging spectroradiometer (MODIS). The remotely sensed MODIS has both vegetative and land surface products. Mu et al. (2007) developed global evapotranspiration algorithm based on the remotely sensed MODIS data and global meteorological data. This algorithm was further improved by Mu et al. (2011) to improve the calculation of vegetation cover fraction, soil heat flux estimation and resistances for stomatal, aerodynamic and boundary layer. On the other hand, European Organisation for the Exploitation of Meteorological Satellites (EUMETSAT) has also developed products under the land surface analysis satellite applications facility (LSA SAF) initiatives. LSA-SAF ET algorithm is one of the first kind to derive quasi-real time ET over large areas such as Europe, Africa and South America using remote sensing. It uses the meteosat second generation (MSG) to derive ET at high temporal scale over large areas.

For the Black Volta Basin of West Africa that is data scarce, the availability of these remotely sensed data provide a critical gap filling for understanding water budgets in the basin. Generally, *in-situ* stations within the basin are limited, and hence for real time decision making; it is imperative to support this with remotely sensed data.

In the Volta Basin, a number of researches have been conducted on the estimation of evapotranspiration using remote sensing methods (Hafeez et al., 2007; Mu et al., 2011b; Opoku-Duah et al., 2008) and others with water balance approach (Friesen et al., 2005; Kunstmann and Jung, 2007; Oguntunde et al., 2006). The Volta Basin consists of the Black, White and Oti River Basins. Most work has been conducted in the White Basin and Oti River Basin, there is no clear studies on the application of remote sensing techniques for estimation of ET in the Black Volta Basin. With the increasing relevance of the Black Volta to the economies of Ghana and Cote D'Ivoire, the water budget analysis in terms of the losses – ET become crucial.

The availability of MODIS ET and LSA SAF ET provides adequate spatially and temporally information that practitioners can use to inform decision making as it relates to water budgets in river basins. The objective of the paper is to estimate the spatial temporal ET over the Black River Basin using remote sensing approaches with data from MODIS ET and LSA SAF ET from 2001 to 2014. The products adopted are to assess and gain an understanding of the spatial-temporal ET variability between the 2001 and 2014 for the Black Volta Basin of West Africa; providing useful information for this important basin of West Africa given its socio-economic contribution to Ghana and neighbouring countries.

MATERIALS AND METHODS

Study area

The Black Volta River which is a tributary of the White Volta River flows through Ghana for about 1352 km before it joins the White Volta Basin. The Black Volta River Basin covers an area of 147000 km^2 and forms part of the international border between the La Cote D'Ivoire and Ghana, and Burkina Faso and Ghana. The basin has a tropical climate with temperatures averaging 24°C and a double maxima rainfall pattern. Average annual rainfalls for the northern and southern parts of the basin are 1000 and 1400 mm, respectively.

With the completion of the 400 MW Hydroelectric Project at Bui Gorge at the south end of the Bui National Park in Ghana in 2014, the basin has become very important on the national and international agenda as there is now a heightened need to monitor closely the water budgets in this basin to ensure sustainable management of the water resources. This is because the Bui Hydro plant is the second largest hydroelectric plant in the country after the Akosombo Dam. Figure 1 shows the basin including the Bui reservoir.

Datasets

MODIS global ET

The MODIS is aboard the Terra and Aqua satellites. The Terra's

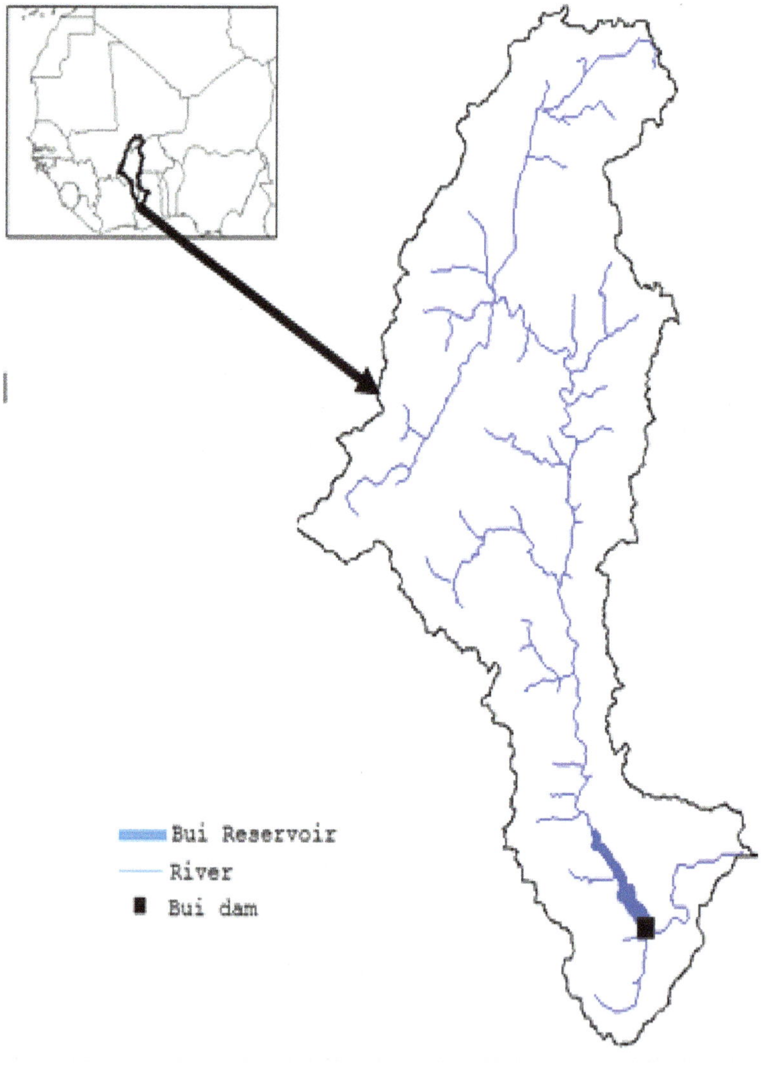

Figure 1. Black Volta Basin showing Bui Reservoir.

orbit around the Earth is programmed to pass from north to south and across the equator in the morning while the Aqua from south to north over the equator in the afternoon. The Terra and Aqua views the complete Earth in every 1 to 2 days and hence provide adequate information to monitor environmental fluxes. The MODIS16 ET datasets are estimated using the improved ET algorithm from Mu et al. (2011) over the previous (Mu et al., 2007). The ET algorithm is based on the Penman-Monteith equation with the concept that surface resistance is an effective resistance to evaporation from land surface and transpiration plant canopy. The MOD16 monthly global ET dataset are 5600 m spatial resolution at the equator. Due to its coarse resolution of the product, it is appropriate for the study of large areas or continental scale but not suitable for smaller watersheds. It would also be interesting if the regular production of such products was implemented so that on a timely basis, MODIS ET products could be obtained based on the algorithm in real time. The dataset covers a time period of 2000 to 2012, and, it is expected that in the future the dataset will be produced and posted periodically but this is not in real-time. Data for the period 2000 to 2012 was downloaded from ftp://ftp.ntsg.umt.edu/pub/MODIS/NTSG_ Products/MOD16/.

LSASAF Global ET

The EUMETSAT LSA SAF is part of the SAF Network, a set of specialized development and processing centres, serving as EUMETSAT distributed Applications Ground Segment (LSA SAF, 2011). The SAF network supplements the product-oriented activities at the EUMETSAT Central Facility in Darmstadt. The prime purpose of the LSA SAF is to take full advantage of remotely sensed data to measure land surface variables that are of relevance; for supporting applications in meteorology (http://landsaf.meteo.pt/). ET is therefore one of the products developed by the LSA SAF.

The ET algorithm developed by the LSA SAF uses the quantification of flux of water vapour from the ground surface (that is, soil and canopy) into the atmosphere using derived products from the Meteosat Second Generation (MSG) satellites. The method mimics the physical approach and can be simplified as a simplified soil-vegetation-atmosphere transfer (SVAT) module adapted to integrate forcing satellite remote sensing (SRS) derived data with data from numerical weather products (NWP). The physics of the model on the Tiled European Centre for Medium-Range

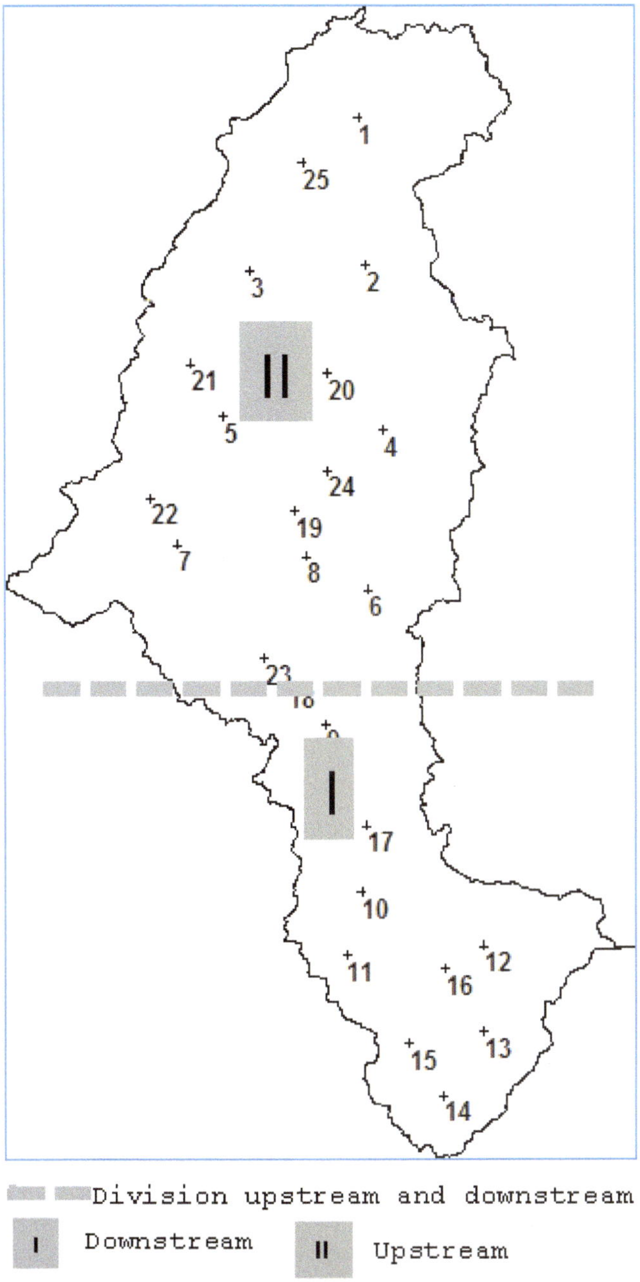

Figure 2. Randomly selected locations with the basin.

Weather Forecasts (ECMRWF) surface scheme for exchange processes over land, TESSEL-SVAT scheme are documented in (Van den Hurk et al., 2000; Viterbo and Beljaars, 1995). Daily ET products were downloaded for 2013 from ftp://landsaf.ipma.pt/. The greatest advantage of this product is the real-time production of the ET measurements on continental scale. However, the coarse resolution of the dataset is still a challenge for smaller watersheds assessment.

TAMSAT rainfall data

The Tropical Applications of Meteorology using SATellite (TAMSAT) research group at the Reading University in UK has produced climatological rainfall estimates for the period 1983-2012. This data is currently available at: http://www.met.reading.ac.uk/~tamsat/public_data/TAMSAT_clim_monthly.zip. The TAMSAT rainfall estimates are produced for Africa using the Meteosat thermal infrared observations. The TAMSAT rainfall estimation algorithm is calibrated using historic rain gauge records providing monthly and regional calibration parameters in both real and historical time scales.

Procedure

The ET products for both LSA SAF and MODIS16 were correlated to check their comparability. The spatial correlation between the two datasets was relatively good (that is, R^2 is 0.77). This suggested that the datasets could be used to evaluate and understand the spatial-temporal behaviour for the study area. The monthly MODIS ET products have been validated (Mu et al., 2011a); however these products are only available from 2000 to 2012. The LSA SAF daily ET products on the other hand were only available from July 2010. Therefore, the LSA SAF ET daily products were first resampled to the same spatial resolution as the MODIS16 ET and then, used to derive the monthly ET for the period 2013 to 2014. In Integrated Land and Water Information System (ILWIS) open, the resampling LSA SAF ET was done by choosing the georeference of the MODIS16 ET map. The integration of the two datasets provided a long-series dynamics of the Black Volta River Basin from 2000 to 2014.

To understand the spatial behaviour of the monthly ET within the basin, 25 points (Figure 2) were distributed throughout the basin with the various monthly ET quantities from each location compared for each year for the whole period. This provided evidence of the dynamic spatial and temporal behaviour within the basin.

RESULTS AND DISCUSSION

The results of the estimated monthly ET are provided in Figure 5 for the 12 year period (2001 to 2013). The estimation from 2001 to 2012 was obtained from the MODIS16 Global ET. For the year 2013, ET estimated from the LSA SAF. The 12 year period ET estimation shows an average monthly distribution of 2 to 150 mm/month from 2001 to 2013 with the exception of the year 2010, where the ET quantities were slightly higher. The upstream portion of the basin (as indicated in Figure 2) tends to show very low average monthly ET estimates (2 to 50 mm/month) whereas the downstream portion of the basin shows high of 50 to 200 mm/month. There is an apparent exception for the year 2013 where some locations in the upstream portion of the basin receive ET estimates as high as 150 mm/month. It also shows that the portion of the basin located in the Brong-Ahafo Region of Ghana has higher ET estimates for the whole period compared to the rest of the basin. The average monthly ET for this part within Ghana is within the range of 100 to 200 mm/month. The part located within the La Cote D'lvoire also receives the second highest ET estimates for the period (50 to 100 mm/month).

The landcover of the basin describes a variation of terrestrial woodlands, herbs and cultivated areas (Figure 3). This is reflective of the major economic activity in the basin as agriculturally based. Therefore, variations between

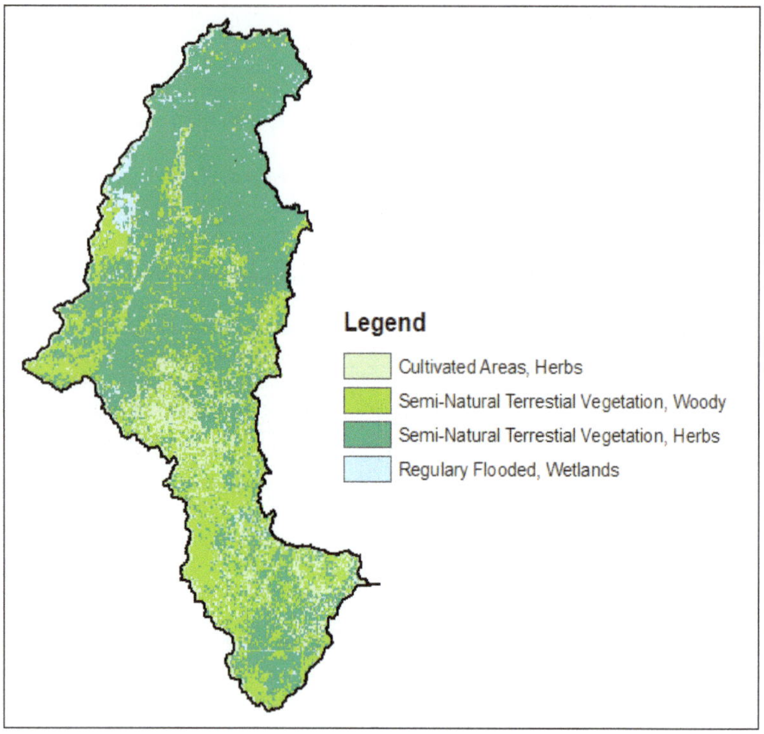

Figure 3. Landcover Map of the Black Volta. Source: (ZEF, 2014).

different landcover types in the area are expected to be changing over time. Now, there is however a clear distinction between the upstream and the downstream that the upstream is predominately covered with semi-natural terrestrial herbs while the downstream is covered with semi-natural terrestrial woodlands. Though, the density of cover in each of these classes was not determined, it can be deduced that the upstream had higher ET estimates compared to the downstream as a result of more vegetative activity in the downstream. More vegetative cover in the downstream results in higher transpiration, hence increasing the amount of ET over the area. The temperature distribution in Figure 4 shows an erratic pattern between 2001 and 2013 measuring as low as 29.5°C (2001) to 30.7°C (2013). Though, the temperature measurements over the period are not spatially distributed, it helps to explain the general variation of the ET among the various years as a result of the effect of the temperature.

Figure 4 shows a relatively stable rainfall pattern observed across the different parts of the basin. The upstream portion of the basin labelled (II) is seen to have a relatively low monthly rainfall (30-60 mm) during the last 30 years compared to the downstream portion of the basin (60 to 86 mm). This possibly accounts for the observed spatial variation in ET with ET for the basin showing a similar overall spatial pattern. As would be expected due to issues of the erratic nature of the weather and climatic conditions, there are some yearly variability.

To understand the climatological conditions and to explain the possible reason for the high ET in the southern part of the basin compared to the rest of the basin, Figure 4 was developed. Figure 6(a) shows the average rainfall obtained from TAMSAT from 1983 to 2012 at the selected locations within the basin. Figure 6(b) shows the average monthly ET for the period 2001 to 2013.

The Bui Hydroelectric dam is located at the downstream portion of the basin (Figure 1). With the average monthly ET being relatively high around this location, this means that there is the need for careful attention to be paid to the management of the water resources in the reservoir to ensure that there is a consistent and sufficient water head for the generating power at the plant. With climate change, temperatures are expected to rise (Hulme et al., 2001; Kurukulasuriya et al., 2006; Stern, 2007). It is therefore imperative to consider this rise with the existing spatial-temporal behaviour of the evapotranspiration within the basin.

Generally, the temporal trend of the ET between years is rather erratic and does not follow any consistent pattern. However, the temporal trend with the downstream and upstream parts of the basin indicates that the downstream regularly receives higher rainfall and evapotranspiration values within a given year but between years, can be a rise or a fall. The spatial trend is rather vivid, showing lower evapotranspiration in the extreme

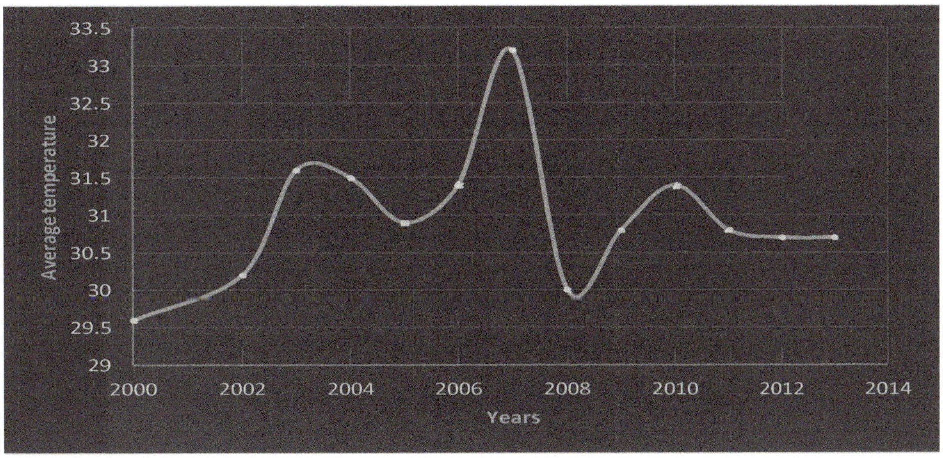

Figure 4. Average daily temperature in Black Volta River Basin. Source: GEONETCast Centre (2014).

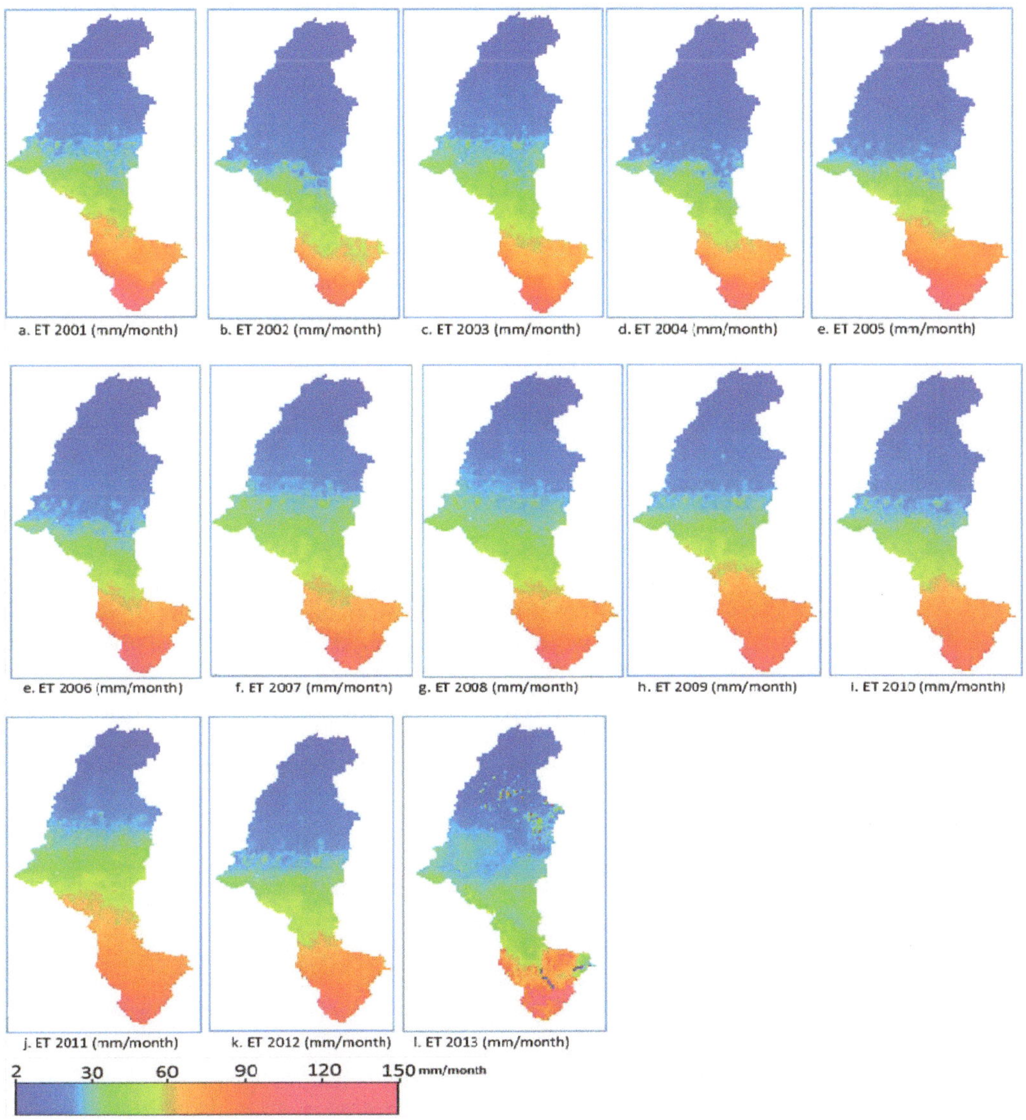

Figure 5. Spatial Average Monthly ET variation within the Black Volta Basin (2001 to 2013).

upstream and showing highest evapotranspiration in the extreme downstream, with the middle part, showing average values.

With the completion of the Bui Hydroelectric plant in 2014, it is expected that socio-economic activity within the basin is bound to rise which no doubt will result in land-use modifications that could impact the hydrology and geomorphology of the basin. For example the downstream part of the basin has seen recent developments due to the establishment of new government institutions such as Bui Community, the Catholic University College and the University of Energy and Natural Resources. These are expected to translate to the building of new buildings and related infrastructure to house the activities of these newly established institutions.

Coupling these socio-economic changes with other changes such as climate change means that the basin characteristics could be significantly impacted. Large numbers of studies have indicated the impact of land-use change on rising temperatures (Dullinger et al., 2004; Piao et al., 2007). With these changes taking place, the average monthly ET would be expected to rise higher than what is shown between the period 2001 to 2013. If that were to happen, then special measures would be needed to ensure the Bui reservoir is able to conserve the water behind the dam, to enable constant and regular supply of water to the turbines. Such measures would be necessary to ensure that purposes for which the Bui Hydroelectric plant was established are consistently met. It would also be necessary that relevant stakeholders in the basin begin to discuss and coordinate across the countries involved (Ghana, La Cote D'Ivoire, Burkina Faso and Mali) to ensure that the water resources are efficiently managed, taking close consideration of the potential large losses e.g. evapotranspiration.

Conclusion

The Black Volta is a trans-boundary basin covering four countries – Ghana, Cote D'Ivoire, Burkina Faso and Mali. With the recent construction of the Bui Hydroelectric plant and associated dam, this has increased the possible socio-economic activities within the basin, owing to the relocation of small scale farmers and the establishment of new institutions to maintain the hydroelectric plant. With such new developments, the conservation of water resources has become critical. It is important that the water resources are managed appropriately to ensure that adequate water is available for the running of the hydroelectric plant.

The paper assessed the evapotranspiration in the basin from 2001 to 2013 to determine the spatial-temporal behaviour during this time. The study revealed that the average monthly evapotranspiration is estimated to be 0-150 mm/month for the whole period with the exemption of the year 2010 where estimates were about 200 mm/month. The average monthly evapotranspiration were compared with the average monthly rainfall estimates over a 30-year period. This comparison showed that the higher evapotranspiration estimates coincide with the high rainfall pattern within the basin. The study provides baseline information for decision makers and stakeholders within the basin to consider an integrated approach to managing the losses (that is, evapotranspiration) to ensure equitable sharing of the water resources within the basin for hydropower and also small scale farming activities.

Conflict of Interest

The authors of this publication have no financial issues or related interest that have influenced the study and therefore, expresses no conflict of interest whatsoever.

ACKNOWLEDGEMENTS

The authors wish to thank Prof. Ben Maathuis of Faculty of Geo-information and Earth Observation, University of Twente, The Netherlands for his development of the ISOD toolbox and the GEONETCast toolbox. Thanks also go to staff of MODIS16 Global ET programme and LSA SAF of EUMETSAT for their continuous support to evapotranspiration studies globally. Furthermore, staff of TAMSAT research group at the Reading University are also appreciated for their contribution to spatial rainfall estimates for Africa.

REFERENCES

Alazem F, Kabo-bah AT (2012). Estimation Of Evapotranspiration In The Om Alrabeaa Region In Morroco Using The Sebs Model. In African Association of Remote Sensing of the Environment (AARSE). Morocco.

Allen RG, Tasumi M, Trezza R (2007). Satellite-based energy balance for mapping evapotranspiration with internalized calibration (METRIC)—Model. J. Irrig. Drain. Eng. 133(4):380–394. http://dx.doi.org/10.1061/(ASCE)0733-9437(2007)133:4(380)

Bastiaanssen WGM (2000). SEBAL-based sensible and latent heat fluxes in the irrigated Gediz Basin. Turk. J. Hydrol. 229(1):87–100. http://dx.doi.org/10.1016/S0022-1694(99)00202-4

Bastiaanssen WGM, Menenti M, Feddes RA, Holtslag AAM (1998). A remote sensing surface energy balance algorithm for land (SEBAL). 1. Formulation. J. Hydrol. 212-213:198–212. Retrieved from http://www.sciencedirect.com/science/article/B6V6C-4CYFRND-G/2/d7c61c460ec8ffa789dc8b2843e5b598 http://dx.doi.org/10.1016/S0022-1694(98)00253-4

Bastiaanssena WGM, Pelgruma H, Wangb J, Mab Y, Morenoc JF, Roerinka GJ, van der Walai T (1998). A remote sensing surface energy balance algorithm for land (SEBAL): Part 2: Validation. J. Hydrol. 21:213–229. http://dx.doi.org/10.1016/S0022-1694(98)00254-6

Choudhury BJ (1991). Multispectral satellite data in the context of land surface heat balance. Rev. Geophys. 29(2):217–236. http://dx.doi.org/10.1029/91RG02635

Dullinger S, Dirnböck T, Grabherr G (2004). Modelling climate change-driven treeline shifts: Relative effects of temperature increase, dispersal and invasibility. J. Ecol. 92(2):241–252. http://dx.doi.org/10.1111/j.0022-0477.2004.00872.x

Friesen J, Andreini M, Andah W (2005). Storage capacity and long-term water balance of the Volta Basin, West Africa. Regional Hydrological Impacts of Climatic Change: Hydroclim. Variabil. 2:138.

GEONETCast Centre (2014). Data Acquisition Portal. University of Energy and Natural Resources, Sunyani-Ghana.

Hafeez M, Andreini M, Liebe J, Friesen J, Marx A, Van De Giesen N (2007). Hydrological parameterization through remote sensing in Volta Basin, West Africa. Int. J. River Basin Manage. 5(1):49–56. http://dx.doi.org/10.1080/15715124.2007.9635305.

Hulme M, Doherty R, Ngara T, New M, Lister D (2001). African climate change: 1900-2100. Clim. Res. 17(2):145–168. http://dx.doi.org/10.3354/cr017145

Kunstmann H, Jung G (2007). Influence of soil-moisture and land use change on precipitation in the Volta Basin of West Africa. Int. J. River Basin Manage. 5(1):9–16. http://dx.doi.org/10.1080/15715124.2007.9635301

Kurukulasuriya P, Mendelsohn R, Hassan R, Benhin J, Deressa T, Diop M (2006). Will African agriculture survive climate change? The World Bank Econ. Rev. 20(3):367–388.http://dx.doi.org/10.1093/wber/lhl004

LANDSAF (2014). Daily ET products. Retrieved from ftp://landsaf.ipma.pt/.

LSA SAF. (2011). The EUMETSAT Satellite Application Facility on Land Surface Analysis (LSA SAF). Damstadt.

Minacapilli M, Agnese C, Blanda F, Cammalleri C, Ciraolo G, D'Urso G, Rallo G (2009). Estimation of actual evapotranspiration of Mediterranean perennial crops by means of remote-sensing based surface energy balance models. Hydrol. Earth Syst. Sci, 13:1061–1074. http://dx.doi.org/10.5194/hess-13-1061-2009

Moran MS, Clarke TR, Inoue Y, Vidal A (1994). Estimating crop water deficit using the relation between surface-air temperature and spectral vegetation index. Rem. Sens. Environ. 49(3):246–263. http://dx.doi.org/10.1016/0034-4257(94)90020-5

MODIS (2014). MODIS Global ET data. Retrieved from ftp://ftp.ntsg.umt.edu/pub/MODIS/NTSG_Products/MOD16/.

Mu Q, Heinsch FA, Zhao M, Running SW (2007). Development of a global evapotranspiration algorithm based on MODIS and global meteorology data. Rem. Sens. Environ. 111(4):519–536. http://dx.doi.org/10.1016/j.rse.2007.04.015

Mu Q, Zhao M, Running SW (2011a). Improvements to a MODIS global terrestrial evapotranspiration algorithm. Rem. Sens. Environ. 115(8):1781–1800. http://dx.doi.org/10.1016/j.rse.2011.02.019

Mu Q, Zhao M, Running SW (2011b). Improvements to a MODIS global terrestrial evapotranspiration algorithm. Rem. Sens. Environ. 115(8):1781–1800. http://dx.doi.org/10.1016/j.rse.2011.02.019

Oguntunde PG, Friesen J, van de Giesen N, Savenije HHG (2006). Hydroclimatology of the Volta River Basin in West Africa: Trends and variability from 1901 to 2002. Physics and Chemistry of the Earth, Parts A/B/C 31(18):1180–1188. http://dx.doi.org/10.1016/j.pce.2006.02.062

Opoku-Duah S, Donoghue DNM, Burt TP (2008). Intercomparison of evapotranspiration over the Savannah Volta Basin in West Africa using remote sensing data. Sensors 8(4):2736–2761. http://dx.doi.org/10.3390/s8042736PMCid:PMC3673443

Or D, Hanks RJ (1992). Spatial and temporal soil water estimation considering soil variability and evapotranspiration uncertainty. Water Resour. Res. 28(3):803–814.http://dx.doi.org/10.1029/91WR02585

Piao S, Friedlingstein P, Ciais P, de Noblet-Ducoudré N, Labat D, Zaehle S (2007). Changes in climate and land use have a larger direct impact than rising CO_2 on global river runoff trends. Proc. Natl. Acad. Sci. 104(39):15242–15247. http://dx.doi.org/10.1073/pnas.0707213104. PMid:17878298 PMCid:PMC1978487

Rana G, Katerji N (2000). Measurement and estimation of actual evapotranspiration in the field under Mediterranean climate: A review. Eur. J. Agron. 13(2-3):125–153. Retrieved from http://www.sciencedirect.com/science/article/B6T67-411PH0R-5/2/e563b4a011d62698c9154bcf28bfa87b http://dx.doi.org/10.1016/S1161-0301(00)00070-8

Rwasoka DT, Gumindoga W, Gwenzi J (2011). Estimation of actual evapotranspiration using the Surface Energy Balance System (SEBS) algorithm in the Upper Manyame catchment in Zimbabwe. Physics and Chemistry of the Earth, Parts A/B/C. http://dx.doi.org/10.1016/j.pce.2011.07.035

Stern N (2007). The economics of climate change: The Stern review. Cambridge University Press.

Su BZ (2009). Introduction to the Surface Energy Balance System (SEBS). Enschede: ITC.

Su H, McCabe MF, Wood EF, Su Z, Prueger JH (2005). Modeling evapotranspiration during SMACEX: Comparing two approaches for local-and regional-scale prediction. J. Hydrometeorol. 6(6):910–922. http://dx.doi.org/10.1175/JHM466.1

Su Z (2002). The Surface Energy Balance System (SEBS) for estimation of turbulent heat fluxes. Hydrol. Earth Syst. Sci. 6(1):85–99. http://dx.doi.org/10.5194/hess-6-85-2002

Su Z, Wen J, Wan L (2003). A methodology for the retrieval of land physical parameter and actual evaporation using NOAA/AVHRR data. J. Jilin Univ. Earth Sci. 33(1):106–118.

Sucksdorff Y, Ottle C (1990). Application of satellite remote sensing to estimate areal evapotranspiration over a watershed. J. Hydrol. 121(1-4):321–333. http://dx.doi.org/10.1016/0022-1694(90)90238-S

TAMSAT (2014). TAMSAT Rainfall Products. Retrieved http://www.met.reading.ac.uk/~tamsat/public_data/TAMSAT_clim_monthly.zip.

Van den Hurk BJJM, Viterbo P, Beljaars ACM, Betts AK (2000). Offline validation of the ERA40 surface scheme. European Centre for Medium-Range Weather Forecasts.

Viterbo P, Beljaars ACM (1995). An improved land surface parameterization scheme in the ECMWF model and its validation. J. Clim. 8(11):2716–2748. http://dx.doi.org/10.1175/1520-0442(1995)008<2716:AILSPS>2.0.CO;2

ZEF (2014). Maps. Retrieved from http://www.zef.de/publ_maps.html

28

Assessment of climate change effects on future drought levels by combining a hydrological model and Standardized Precipitation Index (SPI) index in the Nakdong river basin, Korea

Ngo Van Quan[1] and Gwangseob Kim[2]

[1]Department of Water Resources Engineering, Water Resource University, 175 Tay Son street, Dong Da district, Hanoi Vietnam.
[2]Department of Civil Engineering, Kyungpook National University, Sankyuk-Dong, Buk-Gu, Daegu, 702-701, South Korea.

The main purpose of this study is to analyze the effects of climate change on drought levels in the future by using both hydrological model (SWAT) and Standardized Precipitation Index (SPI) tools. Three benchmark periods of climate change were used for simulation such as 2010-2039, 2040-2069 and 2070-2099. These results were then compared with the baseline period (1980-2009) at nine zones in the basin. Results indicated that both the SWAT model and SPI index showed a similar correlation in duration and density of the drought occurrence levels based on shortage of soil water content and values of drought index through climate change effects. These impacts include not only temporal changes, but also spatial changes, in the future. Results also reflect that the soil water shortage and values of SPI index also showed significant reductions in April, May and November. Most of the severe droughts seem to increase in intensity in two of these months, specifically those that appeared in April in five sub-basins and in November in six sub-basins during the period of 2010-2039; drought in April appeared in four sub-basins and drought in November in six sub-basins in the period of 2040-2069; severe drought occurred in nine sub-basins in April and seven sub-basins in November during the period of 2070-2099. The results suggest that the drought occurrence levels have a trend of increased severity from the North to the South and gradually reduced from the East to the West in May, but then the drought severity increased in the middle of basin with the trend from the South to the North in November. The methods applied in this study are expected to be appropriately applicable to the evaluation of the effects of extreme hydrologic events, and this study can provide useful values for sustainable water-resource management strategies and policy in agricultural basins.

Key words: Hydrology model, standardized precipitation index, climate change, soil water content drought levels.

INTRODUCTION

Climate change has not only significantly influenced hydrologic processes during the last decades but will continue to influence processes more extremely in the future in river basins. Its impacts have affected the supply and management of water resources, especially its strong influences on disaster as floods and drought in watersheds. Several recent articles highlight concerns over climate change and changing precipitation patterns that can be particularly damaging (IPCC, 2007). In fact, drought is estimated to be the most costly natural disaster in the world (Witt, 1997). The wide range of detrimental effects associated with precipitation deficits include: Decreased crop yields, increased wildfires, death of cattle and wildlife, water shortages, and rising food prices (Witt, 1997). The estimation of direct and indirect damages makes it difficult to enumerate the exact costs, although the Federal Emergency Management Agency estimates losses on average of about US$6–8 billion in damages annually in the United States alone (Witt, 1997). Moreover, consequences of drought have long plagued ecosystems and society (Le Roy Ladurie, 1971). The consequences of drought vary greatly depending on its location, timing, extent and the type of society or societal sector impacted by the drought (Gleick, 1993). Such temporary anomaly conditions can be well represented in a number of ways by evaluating anomalous water supply (precipitation) conditions, or through other variables such as soil moisture conditions. The combination of what defines a drought that has an impact on a particular sector of society and the choice of variables to use to define drought led to over 150 definitions of drought (Wilhite and Glantz, 1985). Recognizing that set precipitation amounts could not define drought conditions, since drought period attempted to distinguish between locations that observe "permanent drought," The consequences of drought vary greatly by the length and timing of the precipitation deficit, which is often not taken into account in the interpretation of model results. More recently, Heim (2002) divides drought into four categories based on the myriad of effects meteorological, agricultural, hydrological, and socioeconomic. Meteorological drought simply refers to the atmospheric conditions that result in the absence or reduction of precipitation. Because its definition only relies on rainfall, meteorological drought can end literally overnight, as soon as sufficient precipitation falls to bring levels close to average. Agricultural drought is a short-term dryness in soil layers that can reduce crop yields. Due to its reliance on plant and soil conditions, agricultural drought usually has a lag time in response to precipitation changes (Park et al., 2005), and the impact depends greatly on the timing of the drought in relation to crop growth. Hydrological droughts have an even longer lag time, as they are defined by deficiencies in surface and subsurface water supplies, which respond more slowly to meteorological conditions.

Similarly to other countries in the world, the river basins of Korea are influenced by climate change. According to the Intergovernmental Panel on Climate Change's (IPCC, 2007) findings, developing countries like Korea will be more vulnerable from climate change. On the other hand, Korea is located in the southeastern region of the Asian Monsoon region that shows a typical seasonal pattern in its climate. Four seasons are developed due to the temperature and rainfall amount. Over 60% of the total annual precipitation is concentrated in the wet summer season, so flooding has become an annual event in the Korean Peninsula. The monsoon lasts about a month from mid-June to July, and several typhoons hit the Korean peninsula from late August to September. Convective storms also frequently develop locally to cause flash floods (Han and Byun, 1994), while the dry season is a long dry spell that continues until the monsoon season begins in spring, autumn, and winter. Therefore, drought events are inevitable in Korea in general and in particular in the Nakdong river basin. However, in contrast to floods, it is difficult to know when a drought begins and it is also difficult to determine when a drought is over and according to what criteria this determination should be made. There were many researches considering quantifying drought, most of which have been based on precipitation data analysis (Byun, 1996; Kyung et al., 2007), Drought analysis under climate current and future precipitation data (Karavitis, 2011. Although it is true that these attempts helped quantifying drought to some extent, it is expected that future climate changes as a result of increasing greenhouse gases will affect temperature or evaporation and precipitation patterns in Nakdong river basin and thus hydrologic moisture conditions will be changed. In particular, since agricultural productivity is greatly dependent on the amount of water supplied in the form of soil water, the understanding of soil water and grasping the form of its future changes become very meaningful when measures to adapt to climate changes are prepared. Therefore, it will become very difficult to compare droughts that will occur in the future with present droughts with only the analysis of precipitation data. Droughts can be said to be cases in which the amount of water going out is larger than the amount of water coming into the basin compared to ordinary times. In short, droughts can be represented by precipitation (the amount of water coming in) and the amount of evaporation (the amount of water being lost). If temperature patterns are assumed to be the same in the future too, droughts in the future can be sufficiently compared with present droughts using only the analysis of data on precipitation expected to occur in the future. However, future climates in basins in Korea predicted from several global climate models affect temperature

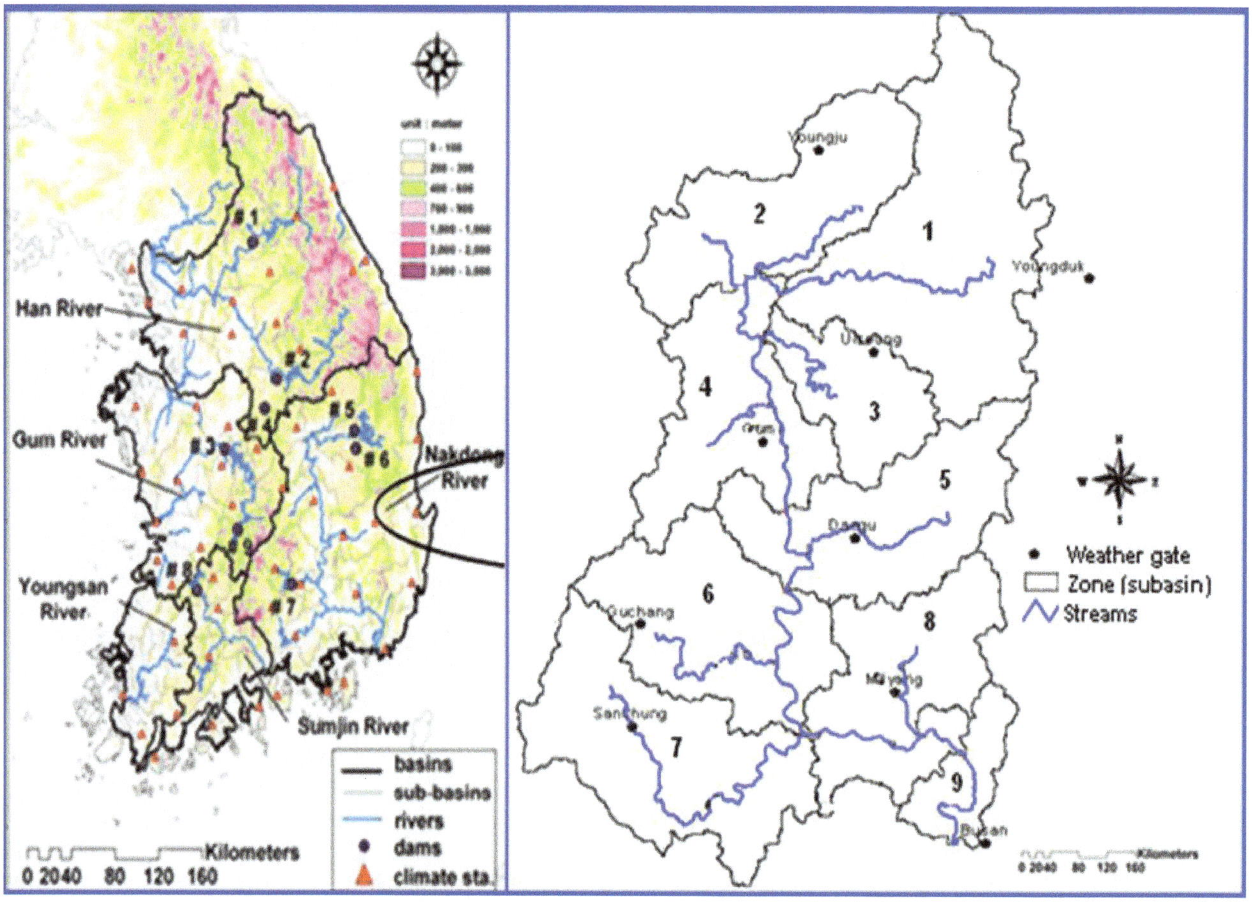

Figure 1. Location of study area, river systems.

increases (+3.2°C) or evaporation increases (+25%) more strongly than changes in precipitation (+9.3%) by the 2080s (Bae et al., 2011). Therefore, it can be said that, to predict droughts in the future, droughts should be analyzed considering not only changes in precipitation, but also temperature or evaporation increases. One of the hydrological components that best reflects changes in temperatures and precipitation should be soil water. Soil water is a hydrological component that serves the role of a link in the hydrologic cycle of precipitation-infiltration-runoff-evapotranspiration and it is perceived as a very important hydrologic component in hydrologic evaluation of climate changes because it is located in the intermediate layer that connects the earth surface with the atmosphere. On the other hand, the importance for considering the present study area is because of the following reasons: (a) The basin is situated in an underdeveloped part of the Nakdong basin, so no study was conducted earlier for drought analysis: (b) The people in the region mostly depend on agriculture, so it is very important to analyze drought in the basin; and (c) The basin was affected by severe droughts in recent years. Therefore, it was necessary for the researchers to investigate drought in the basin. In order to achieve the purpose of this study, both a hydrology model and drought index analysis were combined to assess effects of climate change on drought occurrence in the Nakdong river basin.

Study area and data description

The Nakdong river basin is one of the biggest basins in South Korea, located in the monsoon region (35-37° N, 127-129° E) (Figure 1). This region is characterized by heavy rainfall in the monsoon season in early summer from Middle June to August.

The river drains an area of 23,817 km^2 and the length of the main stream is over 525 km. The annual mean precipitation across the river basin is about 1200 mm, but more than 60% of the annual rainfall is concentrated during the summer season (June-August). The mean air temperature is 2.2°C during the coldest month (January) and 25.9°C in the warmest month (August). The Nakdong River basin is an important water resource for the southeastern area with about 7 million people residing within the basin and more than 13 million people taking drinking-water from the river. In particular, the large

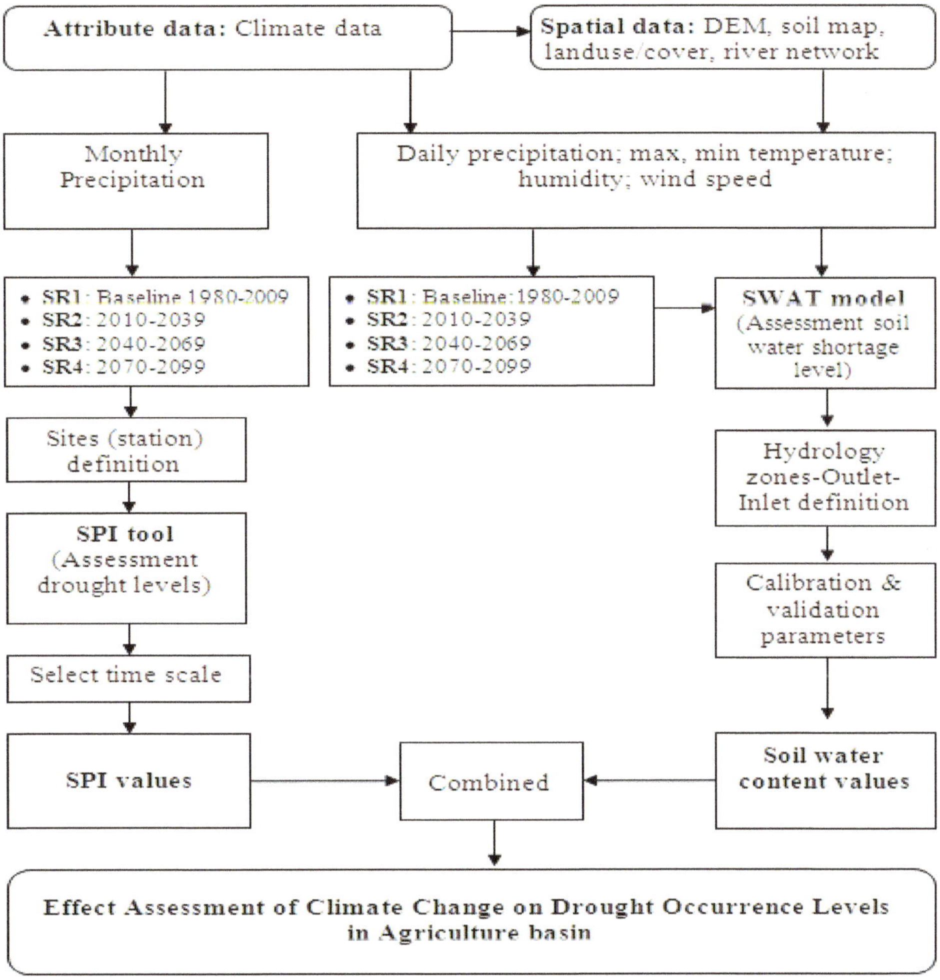

Figure 2. The flow chart of study.

amount of water demand for agricultural productivity is greatly dependent on the amount of water supplied and the cause of the water shortage in the region.

In this study, the data that were collected for both hydrological and land-use models were used in clued spatial data and time series. Spatial data include a Digital Elevation Model (DEM), a digital layer of land use/land cover, a soil map and a river system layer. Time series data include the current climate data from 1980 to 2009 and future climate change for the period 2010-2099, which included daily data of precipitation, maximum and minimum temperature, solar radiation, humidity, and wind speed and direction for fourteen weather stations. Hydrology data included monthly flow for the period of 1995-2009 around Nakdong basin stations. All of the data sources were collected by Korea Meteorological Administration (KMA) and the Water Management Information System (WAMIS), which was built for providing services including scientifically collecting, creating, and processing water-resource information.

This study of climate change is based on available climate-change data. Specifically, the data were used from the GCM output under three different emission scenarios (A2, B1, A1B) for the period of 2010-2099. In these scenarios, A2 was generated with high emission scenario and it represents a realistic worst case for climate change in Korea. Therefore, A2 scenario was selected to use in this study. On the other hand, in order to assist in comparison and assessment, as well as identification of average change in the future periods relative to the past 30-year observed baseline period of 1980-2010, the future long period was divided into three periods of 2010-2039, 2040-2069 and 2070-2099.

METHODS AND MATERIALS

Figure 2 shows the flow chart of the study methods for evaluating the effect of climate change on drought levels using both the hydrology model and Standardized Precipitation Index methods. It involves these steps: Firstly, in order to simulate the deficiency of soil water content, the Soil and Water Assessment Tool (SWAT) hydrologic model was set up. The data input into SWAT model include: Time-series (climate data), and spatial data (DEM, Landuse/land cover, soil map). In this study, the climate is change in the future, but the landuse/land cover is unchanged (as constant), and the landuse/land cover in 2000 year is used in this study, then calibrated and validated with periods of 1995-2004, respectively. Then, the SWAT model was run for climate change for three different periods (2010-2039, 2040-2069 and 2070-2099) and the baseline period (1980-2010) to simulate the soil water content in different hydrology zones (Figure 1) in the study area. Secondly, the Standardized Precipitation Index (SPI), which is based on the cumulative probability of a given rainfall event occurring at a station, was used to determine the rarity of a drought in a sub-basin in the study area. Finally, the impacts of climate change on soil water content were quantified by comparing the SWAT output for the different periods of 2010-2039, 2040-2069 and 2070-2099 with the results of simulations of the present state as of 1980-2009. In addition, a combination of the Standardized Precipitation Index (SPI) was used for describing and comparing with correlation to assess drought occurrence with different climatic conditions. In order to assist in assessment, the four scenarios were assigned to different periods of Scenario 1 (SR1): Current climate (1980-2009); Scenario 2 (SR2): Future climate period (2010-2039); Scenario 3 (SR3): Future climate period (2040-2069); Scenario 4 (SR4): Future climate period (2070-2099), and the effect of climate change on the drought levels in the future were compared between future periods with the observed period.

Hydrology model

The SWAT model was used in this study to evaluate the impacts of climate changes on soil water in the Nakdong River basin. SWAT is in the domain of semi-distributed hydrological models. It is a physically based, basin scale, continuous time model. SWAT integrates more than 30 years of model development within the US Department of Agriculture's (USDA) Agricultural Research Service (ARS) into a single model, developed to support water managers in forecasting and assessing the impacts of climate and land use management practices on water supplies, sediment, non-point source loadings, and pesticide contamination in ungauged watersheds and large, complex river basins with varying soils, land use and management conditions over long periods of time (Arnold et al., 1998; Neitsch et al., 2005a). Upland model components in the latest version include weather, hydrology, erosion, sedimentation, soil temperature, plant growth, nutrients, pesticides, agricultural management, sediment and nutrient loadings from urban areas, and bacteria growth. Stream processes allow for routing of water, sediment, nutrients and organic chemicals in the main channel and transport of bacteria from land areas to the stream network.
The hydrologic cycle is based on the water balance equation:

$$SWt = SW0 + \sum_{i=1}^{t} (Rday - Qsurf - Ea - Wdeep - Qgw) \quad (1)$$

where SWt is the final soil water content (mm H_2O), SW0 is the initial soil water content on day i (mm H_2O), t is time (days), Rday is the amount of precipitation on day i (mm H_2O), Qsurf is the amount of surface runoff on day i (mm H2O), Ea is the amount of evapotranspiration on day i (mm H_2O), Wdeep is the amount of water into the deep aquifer on day i (mm H_2O), and Qgw is the amount of return flow on day i (mm H_2O). One of the major features of SWAT is its partitioning of the study basin into sub-basins that are connected by surface flows (Neitsch et al., 2005b). Each sub-basin (Figure 1) is further divided into one or more hydrological response units (HRU) according to topography, types of land-use, and soil. In each HRU, hydrological components of the water budget for surface water, soil water content, and groundwater are calculated. In these calculations, precipitation is assumed to be intercepted by the canopy of vegetation. The amount of water held by the canopy is a function of the density of plant cover and the morphology of plant species defined by the leaf area index. Precipitation reaching the ground after interception infiltrates into soils. The infiltration rate varies according to soil water content (Neitsch et al., 2005b). In the root zone, percolation occurs when the root zone is saturated. Percolation continues to deliver soil water to the aquifer, which is very important to know the water in soil. Evapotranspiration (ET) is the primary mechanism of surface and soil water loss for HRUs. The method developed by Ritchie (1972) was used to calculate actual ET. Potential ET is available in model and it can be selected or the Penman-Monteith or Priestley-Taylor method. Runoff was predicted separately for each hydrologic response unit (HRU) and routed to obtain the total runoff for the watershed.

The standardized precipitation index

SPI is a tool that was developed primarily for defining and monitoring drought. It allows an analyst to determine the rarity of a drought at a given time scale (temporal resolution) of interest for any rainfall station data. It can also be used to determine periods of anomalously wet events. Mathematically, the SPI is based on the cumulative probability of a given rainfall event occurring at a station. The historic rainfall data of the station is fitted to a gamma distribution, as the gamma distribution has been found to fit the precipitation distribution quite well. This is done through a process of maximum likelihood estimation of the gamma distribution parameters, α and β. In simple terms, the process described above allows the rainfall distribution at the station to be effectively represented by a mathematical cumulative probability function. Therefore, based on the rainfall data, an analyst can then tell what is the probability of the rainfall which is less than or equal to a certain amount. Thus, the probability of rainfall being less than or equal to the average rainfall for that area will be about 0.5, while the probability of rainfall being less than or equal to an amount much smaller than the average will be lower (0.2, 0.1, 0.0,1 etc., depending on the amount). Therefore, if a particular rainfall event has a low probability on the cumulative probability function, then this is indicative of a likely drought event. Alternatively, a rainfall event which has a high probability on the cumulative probability function is an anomalously wet event. Moreover, the SPI can effectively represent the amount of rainfall over a given time scale, with the advantage that it provides not only information on the amount of rainfall but also gives an indication of what this amount is in relation to the normal level, thus leading to the definition of whether a station is experiencing drought or not. Drought classification levels based on SPI are shown in Table 1. The SPI method is computed by fitting a probability density function to the frequency distribution of precipitation summed over the time scale of interest. The SPI was developed by McKee et al. (1993, 1995) with the purpose of identifying and monitoring local droughts. This is performed separately for each month and for each location in space. Each probability density functions are then transformed into a standardized normal distribution. The gamma distribution is defined by its probability density function as given in Equation (1):

$$g(x) = \frac{1}{\beta^{\alpha}\Gamma(\alpha)} x^{\alpha-1} e^{-x/\beta} \quad \text{for } x > 0 \quad (2)$$

Where: $\alpha>0$ is a shape factor, $\beta>0$ is a scale factor, and $x>0$ is the amount of precipitation. $\Gamma(\alpha)$ is the gamma function. The resulting parameters are then used to find the cumulative probability of an observed precipitation event for the given month or any other time

Table 1. Drought classification levels based on SPI index.

SPI Values	Drought levels
2.0+	Extremely wet
1.5 to 1.99	Very wet
1.0 to 1.49	Moderately wet
-0.99 to 0.99	Near normal
-1.0 to -1.49	Moderately dry
-1.5 to -1.99	Severely dry
-2 and less	Extremely dry

scale using Equation (2):

$$G(x) = \int_0^x g(x)dx \quad (3)$$

Since the gamma function is undefined for x=0 and a precipitation distribution may contain zeros, the cumulative probability becomes as given in Equation (3):

$$H(x) = u + (1-u)G(x) \quad (4)$$

Where: u is the probability of zero precipitation. The cumulative probability, H(x) is then transformed to the standard normal random variable Z with a mean of zero and a variance of one, which is the value of SPI as shown in Equations (4) and (5):

$$Z = SPI = -\left(k - \frac{c_0 + c_1k + c_2k^2}{1 + d_1 + d_2k^2 + d_3k^3}\right) \quad for\ 0 < H(x) \le 0.5 \quad (5)$$

$$Z = SPI = +\left(k - \frac{c_0 + c_1k + c_2k^2}{1 + d_1 + d_2k^2 + d_3k^3}\right) \quad for\ 0 < H(x) \le 0.5 \quad (6)$$

Where:

$$k = \sqrt{\ln(\frac{1}{H(x)^2})} \quad for\ 0 < H(x) \le 0.5$$

$$k = \sqrt{\ln(\frac{1}{1-H(x)^2})} \quad for\ 0.5 < H(x) < 1$$

and c_0=2.515517, c_1=0.802853, c_2=0.010328, d_1=1.432788, d_2=0.189269, d_3=0.001308

Where the positive SPI values indicate the rainfall is greater than median rainfall and negative values indicate less than median rainfall. Dry conditions are defined by 0.00<SPI<-0.99, moderately dry by -1.00<SPI<-1.49, severely dry by -1.5<SPI<-1.99, and extremely dry by SPI <-2.00. A drought event starts when the SPI value reaches -1.00 and ends when SPI becomes positive again. Based on the SWAT model, nine sub-basins were determined for the study area (Figure 3), and the SPI time series were computed for different periods for all sub-basins.

RESULTS

In this study, the data for calibrating and validating are streamflow at the two stations of Goeangwan and Jindong in the Nakdong river basin. The SWAT model was used to calibrate against measured streamflow for a period of ten years (1995-2004) and validate by a period of five years (2005-2009). The sensitive parameters for the streamflow calibration were CN2 (the curve number); ESCO (soil evaporation compensation factor); GW_REVAP (groundwater revap coefficient); ALPHA_BF (base flow alpha factor); CH_K2 (channel hydraulic conductivity); TLAPS (temperature lapse rate) and SOL_AWC (soil available water content). These parameters were adjusted from SWAT initial parameters until acceptable matching of the simulated and observed flow. Calibration parameters for various model outputs were constrained within the range and final calibration values are shown in Table 2. Then the regression statistical parameters of Nash-Sutcliffe coefficient of efficiency (E_{NS}) (Santhi et al., 2001) and the coefficient of determination (R^2) were used to measure the model performance. In the validation process, the model is operated with input parameters set during the calibration process without any change and the results are compared to the remaining observation to evaluate the model prediction. The simulated monthly streamflow was compared with the observed streamflow values for calibration at Goengwan and Jindong stations. The results show good consistency between the simulated and measured monthly streamflow for two stations according to regression statistical parameters. The regression statistical parameters show that the validation period statistics were stronger than those computed for the calibration period for both stations, with R^2 =0.92 and E_{NS}=0.81 in the validation period versus corresponding values of 0.94 and 0.86 for the calibration period Goengwan station and R^2 =0.90 and E_{NS}=0.78 versus corresponding values of 0.92 and 0.83 for the calibration period at Jindong station. The detailed results are given in Table 3.

Climate change impact on soil water content

In this study, the result was simulated for nine of sub-basins (zones) (Figure 1). However, to order to advantage in the analysis and assessment the result were shown representative figures for four sub-basins as

Table 2. The calibrated parameters for SWAT at Goeangwan and Jindong stations.

Calibration parameter	Name	SIM1 Default	Range values	Calibrated values of Goeangwan station Final	Calibrated values of Jindong station Final
Curve number for moisture condition II	CN2	Default	±25%	+20%	+20%
Soil evaporation compensation factor	ESCO	0.95	0-1.00	0.52	0.65
Ground water Revap coefficient	GW_REVAP	0.02	0.02-0.20	0.15	0.02
Channel hydraulic conductivity	CH_K2	20.0	-0.01-150	10.00	0.00
Temperature lapse rate	TLAPS	0.00	0-1.00	0.70	1.00
Soil available water capacity	SOL_AWC	0.00	-0.04-0.04	0.04	0.02
Infiltration method used	Cure Number				
ET method used	Priestley-Taylor				

Table 3. Evaluation statistics for the observed vs. simulated monthly streamflow at Goeangwan and Jindong stations.

Stream flow compared between observed with simuated data	Calibration period (1995-2004)		Validation period (2005-2009)	
	E_{NS}	R^2	E_{NS}	R^2
Goeangwan station	0.81	0.92	0.86	0.94
Jindong station	0.78	0.90	0.83	0.92

sub2, sub4, sub6, and sub8. Therefore, the result shows the annual mean soil water simulated by the SWAT model for different periods of climate change under SRES A2 in the future. In general, results indicated annual soil water increased for all future periods, with soil water content increasing more for the period of 2040-2069 than other periods, and little decreasing tendency for the period of 2070-2099 at nine sub basins in the study area (Figure 3). In these results showed that, the annual mean soil water under the influence of climate change increased most by +1.30% for sub5 and least for+1.02% at sub6. For SR2 and SR3 compared with SR1, the highest increase in soil water content was +3.52% at sub3 and +1.95% at sub8, while the lowest change in soil water content was +1.16% and +0.85% at sub7, respectively. The calculated results revealed that the effect of climate change on soil water is not only its impact on temporal, but also on spatial, characteristics at each site in the basin.

Figure 4 shows the changes in monthly average values of soil water content. There are quite different characteristics for the changes for monthly soil water in the future. In fact, the soil water content has clearly increasing trends in the months of June, July and August, but it shows decreasing tendency in the months of April, May and November at all or most of the sub basins. In particular, at sub1, the mean soil water from June to August increased by +3.28, +4.52, and +4.07% for the period of 2010-2039; +3.38, +4.81, and +4.07% for the 2040-2069 period; and +3.20, +4.05, and +3.76% for the 2070-2099 period. In contrast, mean monthly soil water in April, May and November decreased by -2.04, -2.93 and -2.11%; -2.23, -3.12 and -2.58%; and -1.61, -2.82 and -2.77%. The changes in soil water for the other sub basins are shown in Figure 4. The soil water storage results indicated density of the maximum values that appeared in July and the highest value reached was +10.67% at sub4. However, the reduction mainly appears in May, but the lowest value is -5.28% in November at sub5 in the 2040-2069 period. In the other months, these changes are not so much. On the other hand, the effect of climate change on soil water is not only temporal but also spatial. In particular, the soil water content is high in the East-North as at sub1, sub2, and sub3 (100.42, 106.70 and 95.93 mm annually) respectively for the periods 1980-2009 and 2070-2099. But in the periods of 2010-2039 and 2040-2069, the soil water content is high both for East-North and East-South at sub1, sub2, sub3, sub8, and sub9. The determination of spatial change helps in understanding water supply capacity and future crop productivity in water-resource management strategies to adapt for agriculture development in each sub-basin. In conclusion, the above results indicate a soil water shortage resulting from reductions in April, May and November in the future. These results help to determine that the water resource supply may be more vulnerable to crop water demand in the region. Moreover, these impacts can include more extreme drought events in two months at the end of spring and the first month of the winter season.

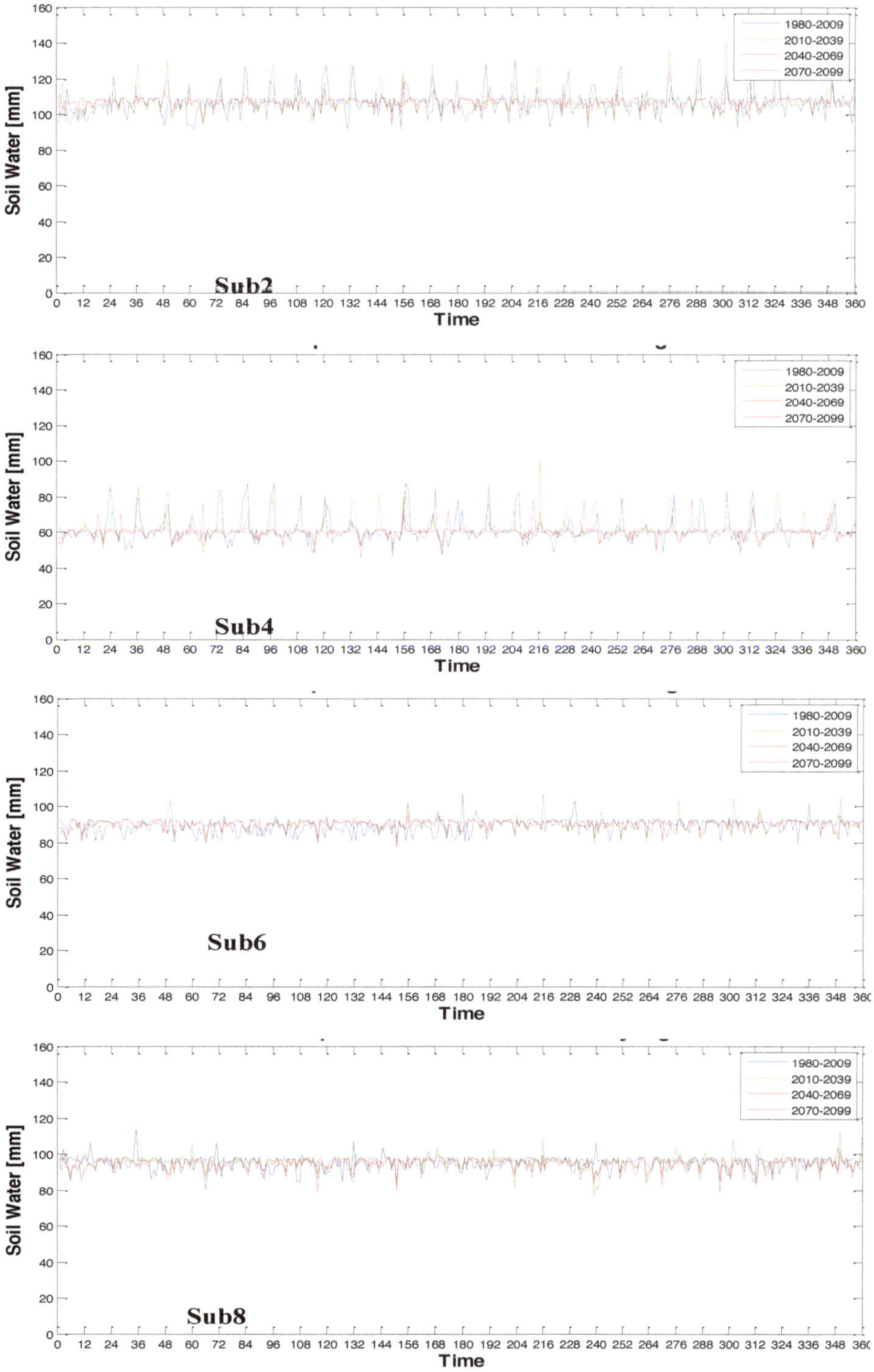

Figure 3. Impact of climate change on annual mean soil water at nine sub-basins.

Effect of climate change on drought occurrence

Figure 5 Illustrates SPI in current and future periods and Figure 6 shows the monthly mean SPI values for drought for nine sub-basins in the study area. The results clearly indicated SPI values most sharply reducing in May, then a gradually increasing tendency in June, July and August for most of the periods in the sub-basins, except in the period of 2070-2099. Result shows that the lowest SPI values occurred in April, May and November at all sub-basins (Figure 6). Moreover, the extreme drought level showed that that is mainly in May with four sub basins for

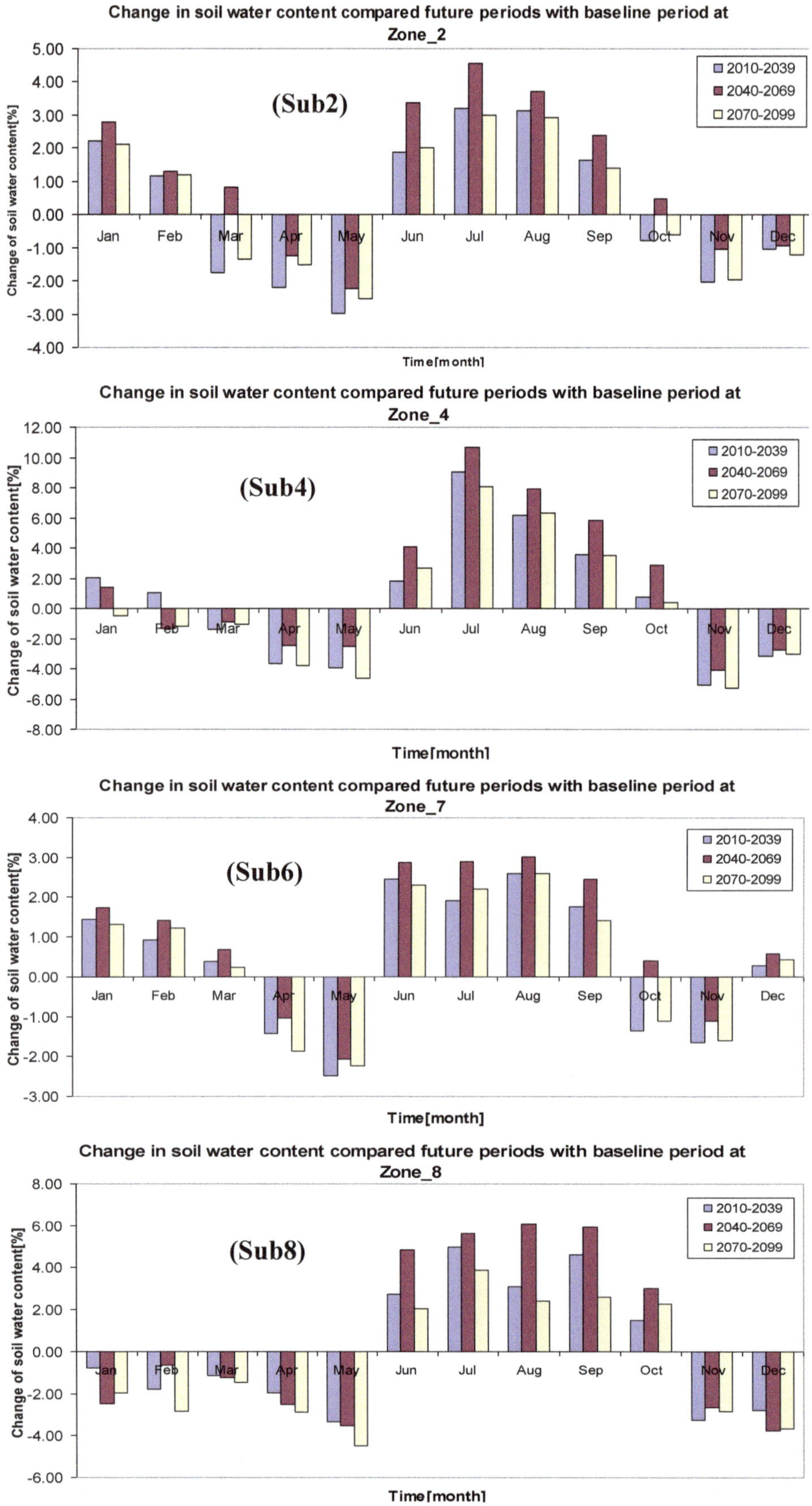

Figure 4. Effect of climate change on average month soil water at all sub basins.

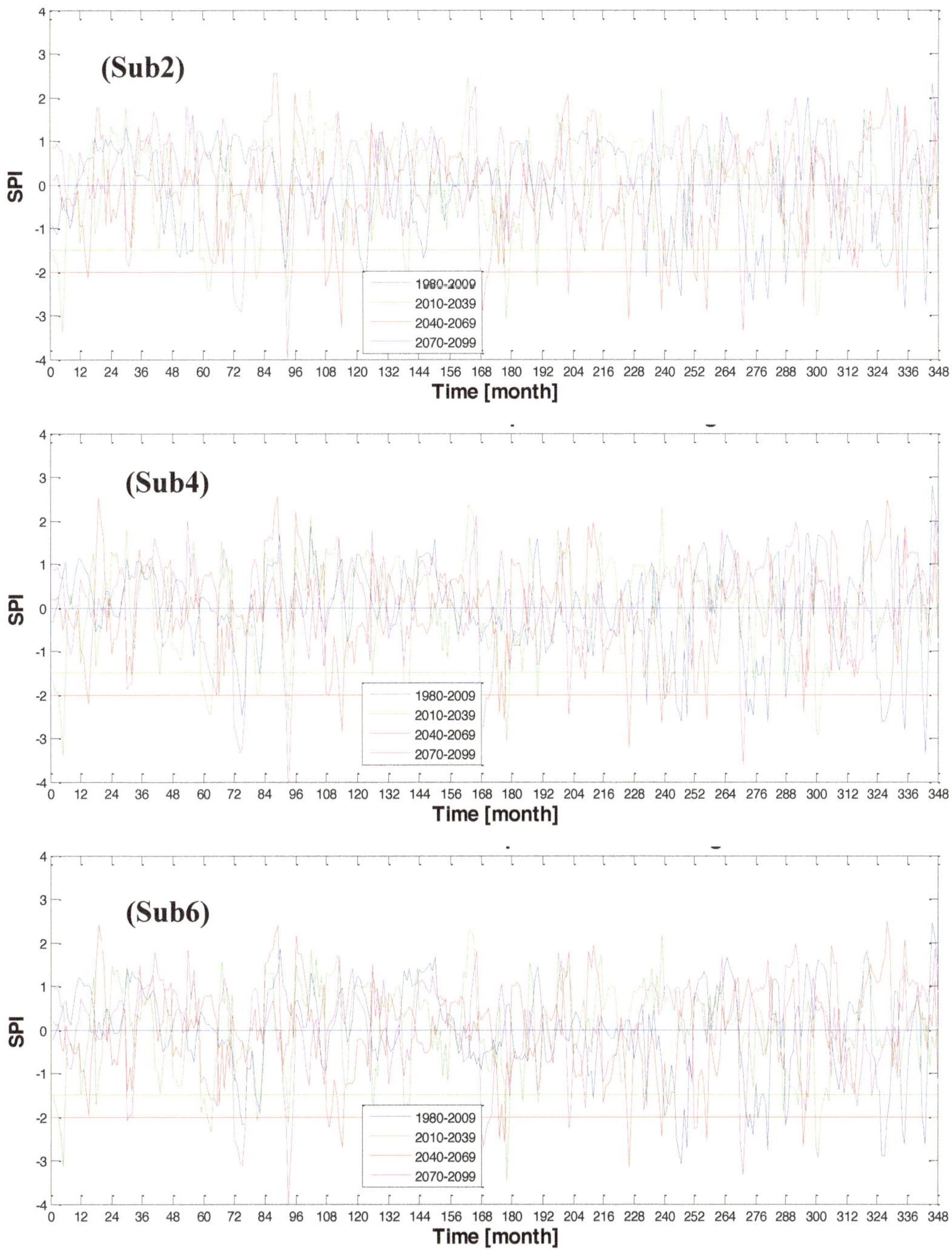

Figure 5. Illustrate SPI in current and future periods at nine sub-basins.

the period of 1980-2009; at five sub-basins for the period of 2010-2039; at three sub-basins for the period of 2040-2069; and at five sub-basins for the period of 2070-2099, and the lowest value of SPI is -2.32 (extreme drought level) in May at sub5 in the period of 2010-2039. Other months are less than May as shown in Figure 6. The extreme drought level occurred not so much in April and November, but the severe drought level happened with increasing intensity in these two months. Specifically, the severe drought level appeared at six sub-basins in April and at seven sub-basins in November in 1980-2009; in the period of 2010-2039 it appeared at five sub-basins in April and at six sub-basins in November; at four sub-basins in April and at six sub-basins in November in the 2040-2069 period; and in particular, the highest intensity of severe drought level occurred at nine sub-basins in April and seven sub-basins in November in the 2070-2099 period. Severe drought level in April is unchanged

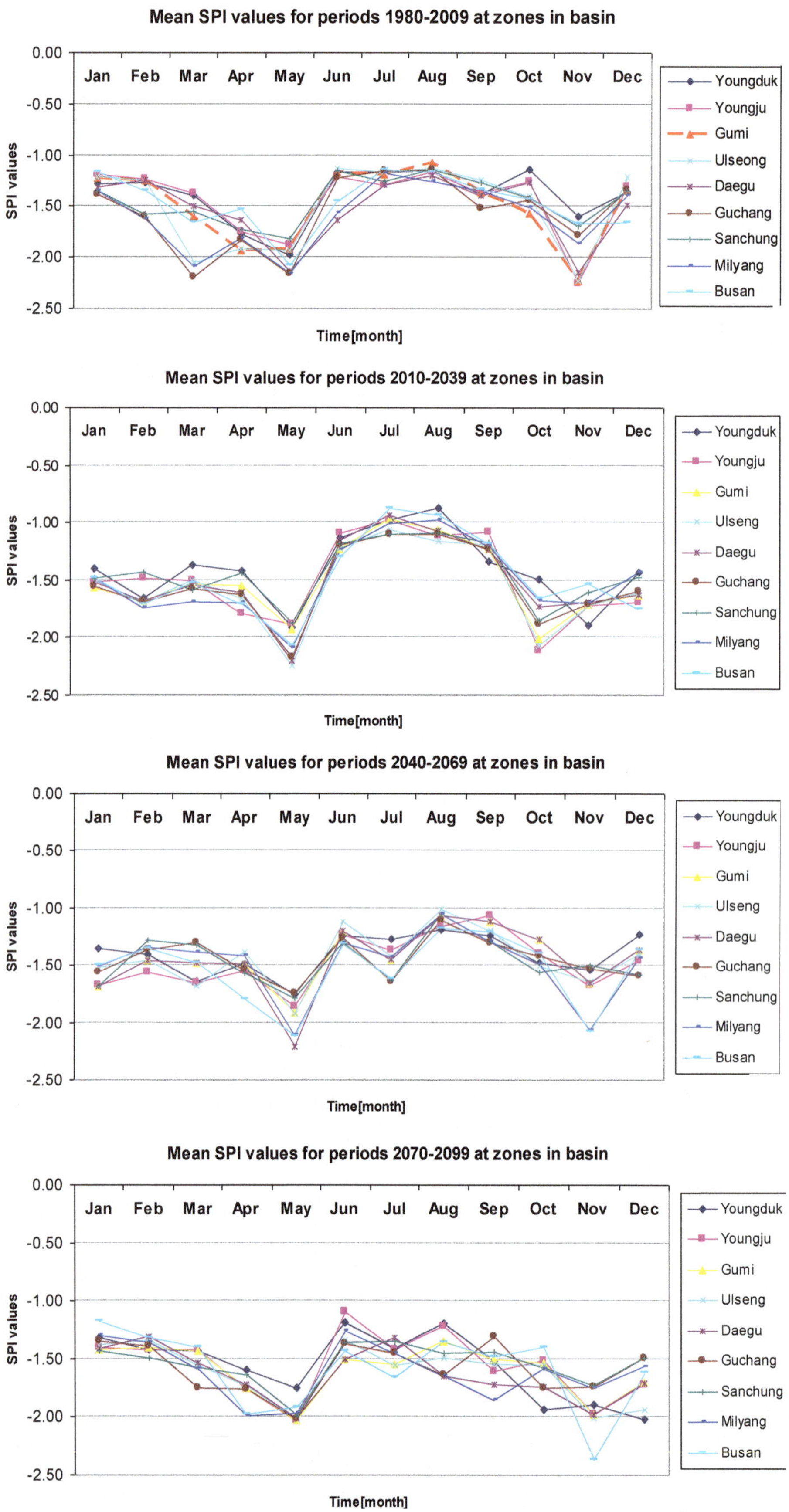

Figure 6. Mean SPI values for drought appearance for periods at sub basins.

in 2010-2039 and 2040-2069 but then sharply increasing in the 2070-2099 period, and appears in April and May which can result in drought in the spring season in the future. In addition, the results indicate that the annual drought occurrence levels have an increasing trend from West through South and gradually reducing from North to East.

DISCUSSION AND CONCLUSIONS

Since drought occurrences are not defined using absolute quantities of soil water as above-mentioned, they are defined using a combination of a hydrology model and Standardized Precipitation Index in order to achieve the purpose of this study. This study analyzed the occurrence of drought in the future under the effect of climate changes and greenhouse gas emissions under scenario A2. Based on the results of the study, the soil water showed large changes in months/seasons. The soil water content in summer is clearly distinguished from the form of soil water content in other seasons (Figure 4). Results determined the soil water content from hydrology model output for climate data, which is similar in duration and density to SPI values from using the Standardized Precipitation Index at each specific sub-basin in the study area. However, there are differences in drought duration and intensity in some sub-basins, but the differences in the changes in both soil water content and SPI in different periods are not very big. The hydrology model method applied in this study did not have any big problem expressing the behavior of soil water content combined with the Standardized Precipitation Index tool, to assess drought occurrence in the study area. Reference soil water content in sub-basins and by month were calculated and the effect of climate change on future occurrence of droughts was evaluated by using the reference soil water content. Based on the results, it was identified that, the most serious increases of soil water content and SPI value in the months of June, July and August were predicted at several sites in the basin. In contrast, the decreasing trend is more complicated than the increase in soil water content and SPI value, as the results showed reductions at sites were not only remarkable in spring (April and May) but also in the end of autumn and beginning of winter (November and December). The results determined that the lowest SPI values occurred in May and the lowest value is -2.32 at sub5 in the period of 2010-3039, and the 2070-2099 period had the highest intensity of extreme drought in five sub-basins with other periods less intense in extreme and severe drought. Extreme drought occurred not so much in April and November but severe drought happened with increasing intensity in these two months.

In summary, this study attempts to quantify the climate change impact on soil water availability by a water balance simulation modelling approach of the SWAT model and SPI values for the study of Nakdong basin in Korea. This study used a number of models for impact assessment to confer valuable outputs and at the same time introduced a number of uncertainties. For this study, two different tools were used, which give different model outputs. Results indicated that there will be high monthly variation of soil water compared to annual, as the average of simulated annual mean soil water compared between 2070-2099 and 1980-2009 has the highest increase of +1.30% in sub5 and the lowest change of +1.02% at sub6 while 2010-2039 and 2040-2069 compared with 1980-2009 reached the highest increases in soil water content of +3.52% at sub3 and +1.95% at sub8, while the soil water content had its lowest change of +1.16% at sub7 and +0.85% at sub7, respectively, while relative to the base periods the soil water in June, July, and August increased soil water storage results indicate density of maximum values in July and higher increases in the period of 2040-2069, with the highest value reached of +10.67% in sub4. In contrast, the soil water in April, May and November are strongly reduced, mainly in May, but the lowest value of -5.28% is appeared in November in sub5 of the 2040-2069 periods.

In conclusion, the results showed climate change effects on future drought levels and it also revealed that its impact is not only temporal variation, but also spatial variation on the drought levels in study region. Results indicated that the severe droughts seem to appear in April, May and November. The results suggest that the drought occurrence levels have a trend of increased severity from the North to the South and gradually reduced from the East to the West in May, but then the drought severity increased in the middle of basin with the trend from the South to the North in November. The study suggests that the drought occurrence is not only evaluated in years but also evaluate in seasons/months in the future. These impacts can be more extreme events of drought in two months in spring and the end of the autumn season and it can help to determine that the water resource supply is more vulnerable on water demand of crop in the study region. Finally, this study is expected to be appropriately applicable to the evaluation of the effects of extreme hydrologic events, and this paper can provide useful values for sustainable water-resource management strategies and policy in agricultural basins in the future.

RECOMMENDATIONS

The soil water content can be influenced not only by climate change but also land use change. Therefore, effective assessment of soil water content should be considered in model simulations of both climate change and land use change scenarios in the future. In addition, the climate change projections can use a GCM model with SRES scenarios such as A2, B1A and B1. SPI values using Standard Precipitation Index should be calculated for different time scales for effective assessment of climate change in combination with soil water content by a hydrology model.

Conflict of Interest

The authors have not declared any conflict of interest.

ACKNOWLEDGEMENTS

The author of this paper would like to thank the Korea Meteorological Administration (KMA) and Water Management Information System (WAMIS), Korea, for providing all the data for this study.

REFERENCES

Arnold JG, Srinivasan R, Muttiah RS, Williams JR (1998). Large area hydrologic modelling and assessment Part 1: Model development. American J. Water Resour. Assoc. 34(1):73-89.

Bae DH, Jung IW, Lettenmaier DP (2011). Hydrologic uncertainties in climate change from IPCC AR4 GCM simulations of the Chungju Basin. Korean J. Hydrol. 401:90-105.

Byun HR (1996). On the atmospheric circulation caused drought in Korea. Korean J. Meteorol. Soc. 32:455-469.

Gleick PH (1993). Water in Crisis: A Guide to the World's Fresh Water Resources. Oxford University Press, New York, NY.

Han YH, Byun HR (1994). On the existence of seasonal drought in the Korean Peninsula. Korean J. Meteorol. Soc. 30:457-467.

Heim RR (2002). A review of twentieth-century drought indices used in the United States. American J. Meteorol. Soc. 83(8):1149-1166.

IPCC (2007). Climate Change 2007: The Physical Science Basis. Contribution of WG I to the fourth assessment report of the Intergovernmental Panel on Climate Change. Cambridge University Press: Cambridge. Available at http://ipcc-wg1.ucar.edu/wg1/wg1-report.html.

Karavitis CA, Alexandris S, Tsesmelis DE, Athanasopoulos G (2011). Application of the Standardized Precipitation Index (SPI) in Greece. American J. Water Resour. Assoc. 3:787-805.

Kyung M, Kim S, Kim B, Kim H (2007). Construction of hydrological drought severity-area-duration curves using cluster analysis. Korean J. Soc. Civil Eng. 27:267-276.

Le Roy Ladurie E (1971). Times of feast, times of famine: A history of climate since the Year 1000 (Barbara Bray, Trans.). Doubleday, New York, NY.

McKee TB, Doesken NJ, Kleist J (1993). The relationship of drought frequency and duration to time scales, in Proceedings of the 8th Conference on Applied Climatology, Anaheim, California, USA, pp.179-184.

McKee TB, Doesken NJ, Kleist J (1995). Drought monitoring with multiple time scales, in Preprints, 9th Conference on Applied Climatology, Dallas, TX, USA. pp. 233-236.

Neitsch SL, Arnold JG, Kiniry JR, Srinivasan R, Williams JR (2005a). Soil and Water Assessment Tool Theoretical Documentation, version 2005. Temple, TX: Grassland, Soil and Water Research Laboratory, Agricultural Research Service.

Neitsch SL, Arnold JG, Kiniry JR, Srinivasan R, Williams JR (2005b). Soil and water assessment tool input/output file documentation, version 2005. Temple, Texas: Grassland, Soil and Water Research Laboratory, Agricultural Research Service. Available at: www.brc.tamus.edu/swat/doc.html. Accessed 1 November 2006.www.brc.tamus.edu/swat/doc.html. Accessed 1 November 2006.

Park S, Feddema JJ, Egbert SL (2005). MODIS land surface temperature composite data and their relationships with climatic water budget factors in the central Great Plains. Int. J. Remote Sens. 26(6):1127-1144.

Santhi C, Arnold JG, Williams, JR, Dugas WA, Srinivasan R, Hauck LM (2001). Validation of the SWAT model on a large river basin with point and nonpoint sources. Am. J. Water Resour. Assoc. 37(5):1169-1188.

Wilhite DA, Glantz MH (1985) Understanding the drought phenomenon: The role of definitions. Water International. 10:111-120.

Witt JL (1997). National Mitigation Strategy: Partnerships for Building Safer Communities. Federal Emergency Management Agency.

Modelling pan evaporation using mean air temperature and mean pan evaporation relationship in middle south Saurashtra Region

Gundalia Manoj J.[1] and Dholakia Mrugen B.[2]

[1]Gujarat Technological University, Chandkheda, Ahmedabad, Gujarat 382 424, India.
[2]Directorate of Technical Education, Gandhinagar, Gujarat 382 010, India.

Evaporation and evapotranspiration are the major parameters in the hydrologic cycle and water resources research. Most precipitation is lost in the form of evaporation and evapotranspiration with the percentage varying from region to region globally. Therefore, accurate estimate of evapotranspiration is needed for numerous management activities and its application in rainfall-runoff models. The reference evapotranspiration ET_o concept has been gaining significant acceptance by scientists throughout the world since its introduction and it can also be estimated from pan evaporation E_p. E_p provides a measurement of the combined effect of temperature, humidity, wind speed and solar radiation on reference crop evapotranspiration. Hence, it is extensively used to estimate ET_o. A significant relationship exists between mean daily air temperature and observed mean daily pan evaporation in monsoon season in middle South Saurashtra region of Gujarat State (India). The objective of this study is to develop a model for prediction of Ep using mean air temperature. The four quantitative standard statistical performance evaluation measures, (R^2), (MAPE), (RMSE) and Nash-Sutcliffe efficiency coefficient (E) are employed to evaluate the model. These performance measure ratings are found well within the acceptable limits. Prediction of E_p with the help of mean air temperature will lead to minimization of the time, cost, and equipment maintenance necessary for onsite monitoring and will help researchers to use data from other sources too.

Key words: Mean air temperature, pan evaporation, reference evapotranspiration, Saurashtra region

INTRODUCTION

Pan evaporation E_p is one of the most important climatic parameters in the hydrological cycle, and is often applied to estimate terrestrial evaporation and water requirements. Annual $E_p$is an important climatic variable, and it is often used to estimate potential evaporation (Kirono et al., 2009) and reference evapotranspiration (Chen et al., 2005), as well as to forecast agricultural production (Wang et al., 2009). Evaporation can be directly measured from $E_p$and lysimeter. Many methods for estimation of evaporation losses from free water surfaces were reported and it can be divided into several categories including: empirical methods (Kohler et al., 1955), radiation (Priestley and Taylor, 1972), water budget methods (Shuttleworth, 1988; Guitjens, 1982), energy budget methods (Anderson, 1954), mass-transfer methods (Harbeck, 1962), temperature based (Thornthwaite, 1948; Blaney and Criddle, 1950), and combination methods (Penman, 1948). In the direct method of measurement, the observation from Class A Pan evaporimeter and Eddy correlation techniques were used (Linsley et al., 1982), whereas in indirect methods, the evaporation is estimated from other meteorological variables like temperature, wind speed, relative humidity and solar radiation (Benzaghta et al., 2010).

Monitoring evaporation is a great challenge since specific and costly equipment are required. Penman–Monteith method is widely used due to its simplicity and ease of data interpretation and has come to be known as the most precise and accurate measure of ET_o in humid areas (Irmak et al., 2002). Many scientists have attempted to establish relationship between Ep data and other weather parameters and to explain this puzzling phenomenon.

Singh et al. (1992) investigated relationship between evaporation from US Class A open pan evaporimeter and meteorological parameters at Hisar. Wind velocity, sunshine hours, mean air temperature and solar radiation were positively correlated with evaporation and relative humidity was negatively correlated with evaporation. The highest correlation value (r = 0.78) were obtained with relative humidity.

Singh et al. (1995) obtained simple correlations between different meteorological parameters and evaporation measured from US Class A open pan evaporimeter at Hisar. The highest value of correlation coefficient (0.85) was found with maximum temperature followed by wind speed (0.82). The coefficient of determinates for minimum air temperature, relative humidity and bright sunshine hours were 0.70, –0.56 and 0.15, respectively.

Khanikar and Nath (1998) established relationship between meteorological parameters and evaporation from an open pan evaporimeter at Jorhat, Assam. The coefficient of determination for maximum temperature (0.64), minimum temperature (0.65), wind speed (0.53) and soil temperature at 5 cm depth (0.68) was positively correlated.

Shrivastava et al. (2001) developed a statistical relationship between pan evaporation and the meteorological parameters recorded at Jabalpur. The method of regression was adopted as suggested by Mendenhall and Sincich. From the analysis, it was revealed that the morning relative humidity and the maximum temperature have a significant influence on the rate of evaporation. The highest correlation value was obtained with morning relative humidity (R2 = 0.95) followed by the maximum temperature (R2 = 0.94).

Snyder et al. (2005) presented sine-wave approach to estimate ET_o from E_p data according to fetch distances by eliminating the need for relative humidity and wind speed data. Jhajharia et al. (2009) examined the influence of different meteorological parameters on E_p at Agartala and using the linear and exponential methods concluded that the wind speed and mean temperature have positive and significant influence on E_p. Chu et al. (2010) used wind tunnel experiments to investigate the wind effects on the evaporation rate of the Class A evaporation pan and found that the evaporation rate increased as the wind speed increased; It was also discovered that the wind could blow water over the edges of the pan when wind speed exceeded 6.0 m/s. Xiaomang et al. (2011) concluded that increasing air temperature dominated the change in E_p, which offset the effect of wind speed and led to the increase in E_p. Yuhe and Guangsheng (2011) discovered strongest correlation between annual relative humidity and E_p in the Liaohe Delta in the period 1961 to 2005. Xing-Jie et al. (2012) evaluated the linear trends in annual and seasonal E_p and detected significant increasing trends in mean temperature at annual and seasonal time scales, except for summer and the E_p paradox in the lower Yellow River Basin. Fekih et al. (2013) evaluated energy budget equation and the computational fluid dynamics methods for estimation of rate of evaporation from dams in wet, arid and semi-arid areas in Algeria.

Necessity of the study

Quantitative estimation of ET is of great significance in water resource planning, estimation of crop water requirements for irrigation, agricultural production forecasting, in understanding the hydrologic cycle and rainfall-runoff modelling. Unfortunately weather stations are scarce and do not always have the instrumentation to measure relevant variables for its calculation. Therefore standard method PM is not appropriate in many such situations. Saurashtra, a water scarcity-prone area of Gujarat State has only limited number of weather stations facilitated with U.S. Weather Bureau evaporation pans to measure evaporation. It is impractical to place evaporation pans at every point and also highly unlikely to have it in inaccessible areas where accurate instruments cannot be established or maintained. A practical means of estimating the amount of pan evaporation where no pans are available is of considerable significance to the hydrologists, agriculturists, and meteorologists. On the other hand, air temperature is relatively easy to measure and available at most of the weather stations. Therefore, a model is developed with the motivation to bridge between research findings, especially prediction of E_p by air temperature.

METHODOLOGY

Study area and data collection

Geographical areas of middle south Saurashtra region of Gujarat State, India (Figure 1) encompasses Junagadh district (lies between 20.44° to 21.40° North latitude and 69.4° to 71.05° East longitude) and Amreli district (lies between 20.45° to 22.25° North latitude and 70.30° to 71.75° East longitude). The area is situated in semi-arid region with mean annual rainfall of 955 mm, mean maximum temperature 33.70°C and mean minimum temperature 22.70°C.

Mean temperature and mean E_p data are collected from Agro Meteorological Cell, Department of Agronomy, College of Agriculture, Junagadh Agricultural University, Junagadh, Gujarat,

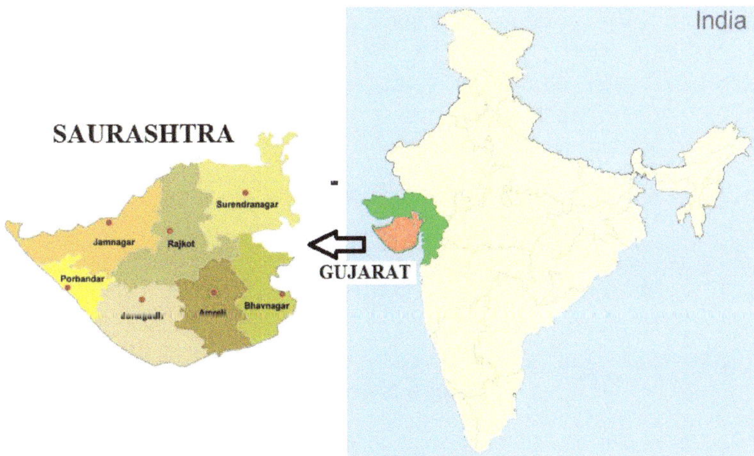

Figure 1. Saurashtra Region of Gujarat State.

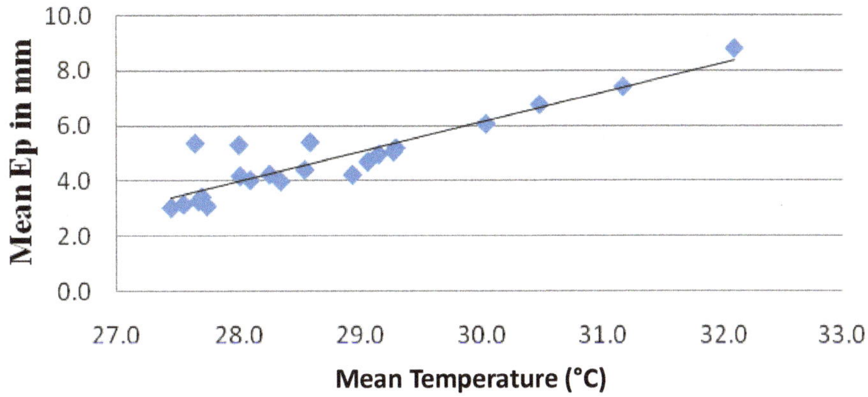

Figure 2. Scatter plot of weekly mean temperature V_s. Mean E_p during monsoon season in Junagadh district.

India.

Model development

Jhajharia et al. (2009) found in his study, wind speed and the mean temperature have positive and significant influence on E_p. Scatter plots of mean weekly temperature v_s, mean weekly E_p for monsoon season and scatter plots of mean weekly wind speed v_s, mean weekly E_p for monsoon season of Junagadh district are shown in Figures 2 and 3 respectively. These scatter plots are offered better relationship between E_p and temperature than between E_p and wind speed during monsoon season in middle south Saurashtra region. A significant relationship has been found between mean daily air temperatures and observed mean daily E_p during monsoon season in Junagadh district of Gujarat State, India through data analysis. Equation 1 is developed based on the relationship between mean E_p and mean air temperature for monsoon season.

$$E_p = \frac{\frac{E_{pm}}{T_m}(T_d - a)}{\text{Cosb}} \quad (\text{Where, } b \neq \frac{\pi}{2}) \qquad (1)$$

Where, E_p is the daily pan evaporation in mm, E_{pm} is the mean daily evaporation, T_m is the mean daily air temperature in °C, T_d is the daily air temperature in °C and 'a' and 'b' are the calibration coefficients. The developed model is calibrated for daily air temperature and daily E_p data using dataset of period 1984 to 1991 and validated using dataset of period 1995 to 2000 for Junagadh district. Model is also tested on weekly time scale for Junagadh and Amrelii districts.

Goodness of fit

In this study, coefficient of determination (R^2) is used as standard regression to describe the degree of collinearity between simulated and measured data. R^2 ranges from 0 to 1, with higher values indicating less error variance, and typically values greater than 0.5 are considered acceptable (Santhi et al., 2001; Van Liew et al., 2003).

Two standard error measures: mean absolute percentage error (MAPE) and root of mean of square of errors (RMSE) are employed to quantify the deviation in the units of the data. A scale to judge the accuracy of the model based on the MAPE measure, developed by

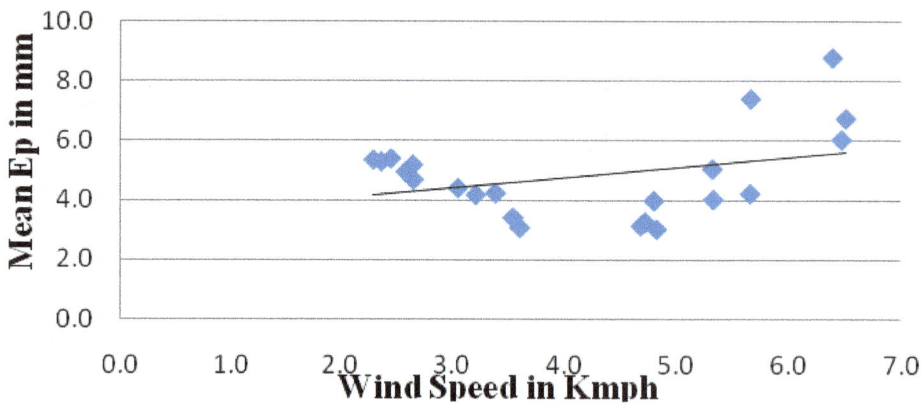

Figure 3. Scatter plot of weekly mean wind speed V_s weekly mean E_p during monsoon season in Junagadh district.

Table 1. A scale of judgment of forecast accuracy (Lewis, 1982).

MAPE	Judgment of Forecast Accuracy
Less than 10%	Highly Accurate
11 to 20%	Good Forecast
21 to 50%	Reasonable Forecast
51% or more	Inaccurate Forecast

Lewis (1982).

The smaller the MAPE value, the better the fit of the model. Using MAPE, and applying Lewis's scale, provides some framework as shown in Table 1 to judge the model.

RMSE is one of the commonly used error index statistics (Chu and Shirmohammadi, 2004; Singh et al., 2004; Vasquez-Amábile and Engel, 2005; Gupta et al., 2009).

The higher the RMSE, the poorer the performance of the model, and vice versa. RMSE = 0 indicates a perfect fit. In addition to (R^2), MAPE and RMSE, accuracy of the developed model is also evaluated by dimensionless statistic, Nash-Sutcliffe criterion (E) (Nash and Sutcliffe, 1970), (E) indicates how well the plot of observed versus simulated data fits the 1:1 line.

Values of (E) between 0.0 and 1.0 are generally viewed as acceptable levels of performance, whereas values less than 0.0 indicates that the mean observed value is a better predictor than the simulated value, which indicates unacceptable performance.

Parameter estimation

In this study, objective function of minimising sum of squares errors between observed and computed E_p (Equation 2) is selected to optimise the values of model parameters 'a' and 'b'.

$$E_{min} = \text{Min} \sum_{i=1}^{n} \left| E_{po} - E_{pc} \right|^2 \qquad (2)$$

Where E_{po} is the observed E_p; E_{pc} is calculated E_p and n is the number of observations. The optimum values of model parameters are computed using Microsoft Excel spread sheet Microsoft Excel built-in optimisation tool Solver (Front Line 2010).

RESULTS AND DISCUSSION

The graphical results of the developed model in calibration and validation for daily time scale are shown in Figures 4 and 5 respectively. The MAPE, RMSE, E and R^2 values of Junagadh station for daily time in calibration and validation are produced 15.79, 0.84 mm, 0.87, 0.75 and 14.17, 0.99 mm, 0.69, 0.71 respectively. The statistical results are presented in Table 2.

The model is also evaluated on weekly time step for Junagadh as well as for Amreli district and presented in plots shown in Figures 6, 7, 8 and 9. The MAPE and RMSE values of Junagadh station are estimated in calibration and validation 10.71, 0.66 mm, 10.49 and 0.62 mm respectively. The E and R^2 values of Junagadh station are estimated in calibration and validation 0.8, 0.81, 0.8 and 0.82 respectively. The MAPE and RMSE values of Amreli station are estimated in calibration and validation 10.3, 0.71 mm, 8.1 and 0.72 mm respectively. The E and R^2 values of Amreli station are estimated in calibration and validation 0.83, 0.71, 0.83 and 0.76 respectively.

In all applications, the initial value of parameter 'a' and 'b' is taken equal to 1. The values of 'a' and 'b' on daily time scale for Junagadh station are found to be 24.0697 and 1.4013 respectively. On weekly time scale, the values of 'a' and 'b' are estimated for Junagadh and Amreli stations 24.3975, 1.4197 and 21.8337, 1.3148 respectively. The values of 'a' and 'b' on daily scale are found to be 1.34 and 1.40% lower than that of the corresponding values on weekly time scale for Junagadh station. While the values of 'a' and 'b' on weekly scale for Amreli station are found to be 10.51 and 7.39% lower than that of the corresponding values for Junagadh station. Model evaluation results and optimized calibrated values of parameters for weekly time scale are presented in Table 3.

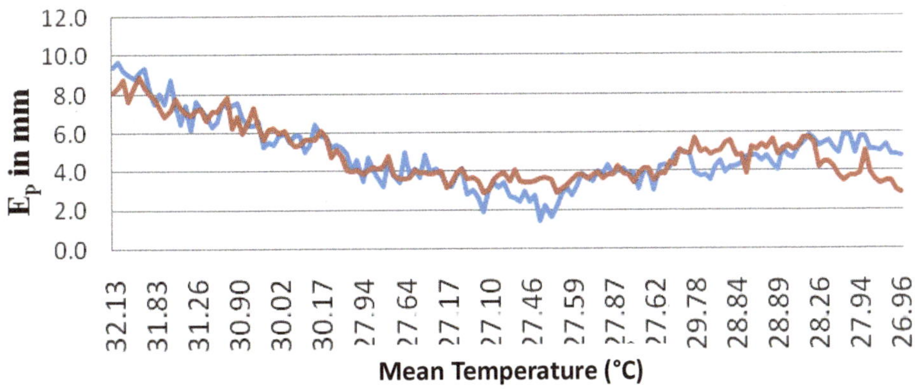

Figure 4. Relationship between mean daily T and mean daily E_p during monsoon season in Junagadh for calibration period 1984 to 1991.

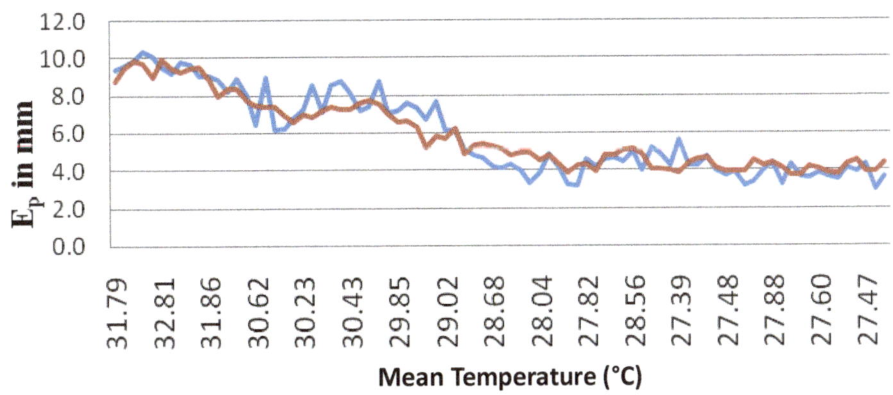

Figure 5. Relationship between mean daily T and mean daily E_p during monsoon season in Junagadh for validation period 1995 to 2000.

Table 2. Results of model on daily time scale for Junagadh Station.

Calibrated model parameters	a = 24.0697, b = 1.4013		
	Calibration (1984 to 1991)	**Validation (1995 to 2000)**	**% difference**
MAPE	15.79	14.17	-10.26
RMSE	0.84	0.99	17.86
E	0.87	0.69	-20.69
R^2	0.75	0.71	-5.33

Conclusions

The minimum values of MAPE and RMSE and values close to 1.0 of (R^2) and (E) show the performance of the model clearly. The results in this study indicated that, the model is found to be more accurate for Junagadh station with RMSE and (E) values 0.66 mm and 0.80 in calibration and 0.62 mm and 0.81 in validation respectively on weekly time scale. Moreover, percentage difference in calibration and validation results of model evolution parameters for Junagadh station was found to be less than that of Amreli station. According to the model evaluation guidelines, the developed model satisfactorily simulated E_p.

The poor performance of model on daily time scale might be because of comparatively higher scatter pattern observed in daily data than that of weekly data. Moreover, averaging over daily data to weekly time scale will not reduce the trend but will reduce the effect of natural variability and thus more detection probability and lead to more robust prediction.

The mean air temperature is observed to be a

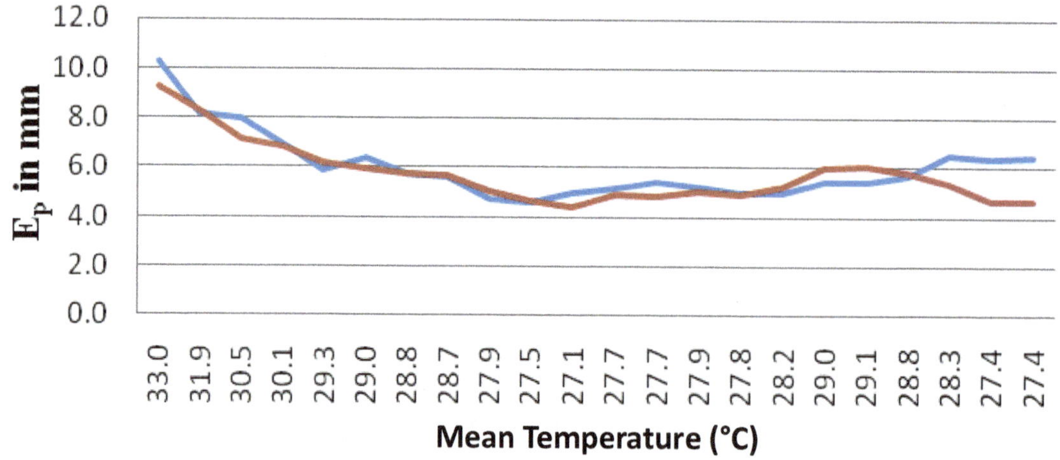

Figure 6. Relationship between mean weekly T and mean weekly E_p during monsoon season in Junagadh for calibration period 1984 to 1991.

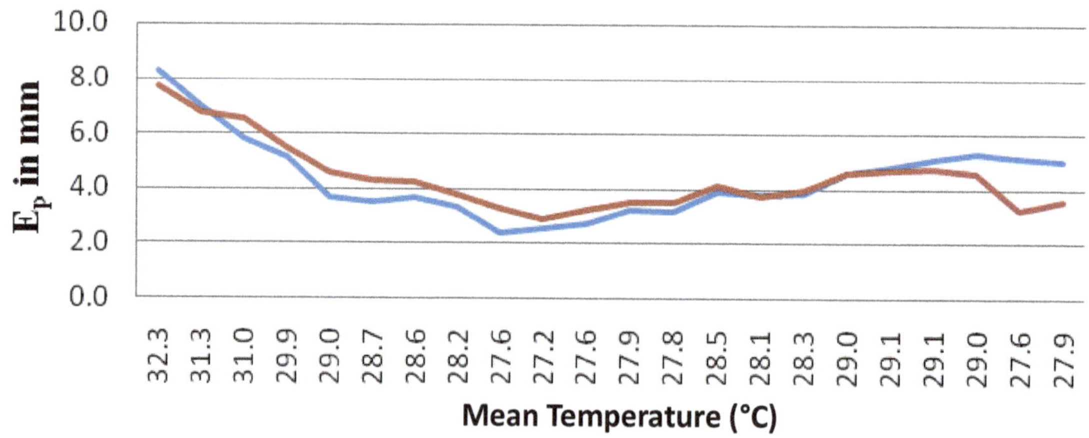

Figure 7. Relationship between mean weekly T and mean weekly E_p during monsoon season in Junagadh for Validation period 1995 to 2011.

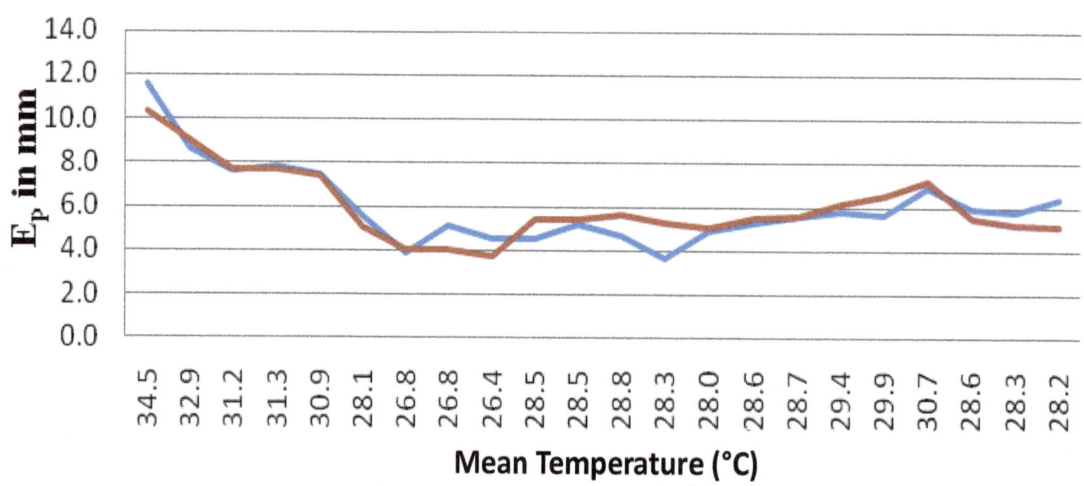

Figure 8. Relationship between mean weekly T and mean weekly E_p during monsoon season in Amreli for calibration period 1992 to 1998.

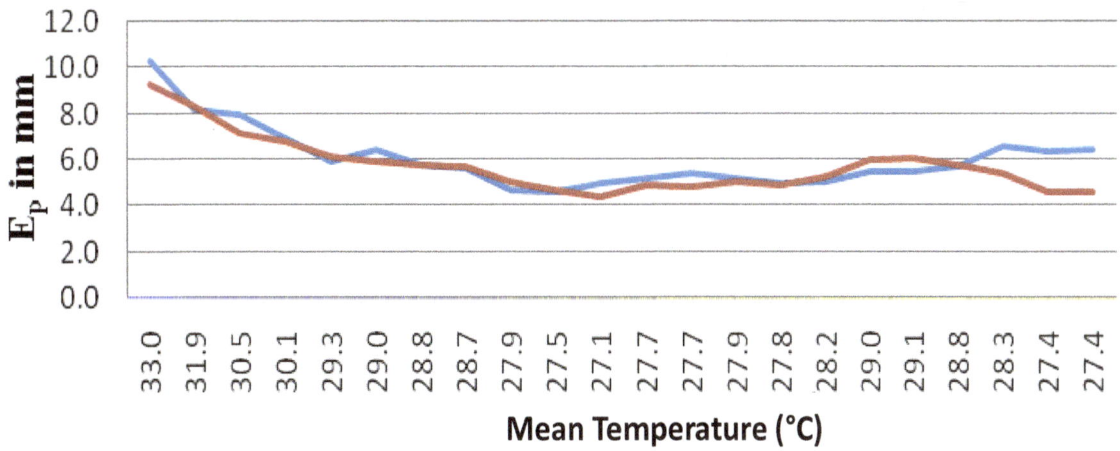

Figure 9. Relationship between mean weekly T and mean weekly E_p during monsoon season in Amreli for validation period 1999 to 2005.

Table 3. Results of model on weekly time scale for Junagadh and Amreli stations.

Junagadh Station			
Calibrated model parameters	a = 24.3975, b = 1.4197		
	Calibration (1984 to 1991)	Validation (1995 to 2011)	% difference
MAPE	10.71	10.49	-2.05
RMSE	0.66	0.62	-6.06
E	0.80	0.81	1.25
R^2	0.80	0.82	2.50

Amreli Station			
Calibrated model parameters	a = 21.8337, b = 1.3148		
	Calibration (1984 to 1991)	Validation (1995 to 2000)	% difference
MAPE	10.30	8.10	-21.36
RMSE	0.71	0.72	1.41
E	0.83	0.71	-14.46
R^2	0.83	0.76	-8.43

significant parameter for the estimation of pan evaporation in middle south Saurashtra region and explicit relationship between mean air temperature and E_p is found during monsoon period. An attempt is made in this study to develop model based on this relationship to predict pan evaporation. The model predictions are comparable with the observations. It is noticed that the variation between observed and estimated E_p increases as the rainfall decreases. Performance of the model is found better during June to September months than in October month. The poor performance of the model might be because of little rainfall in the October month. In the middle south Saurashtra region, only 6 to 8% of total mean annual rainfall occurs in October month.

The model has simplicity advantage as it can estimate evaporation from only one climatological parameter of air temperature. From a practical point of view, this model can be considered suitable to serve as a tool to estimate evaporation when input meteorological variables are insufficient. The developed model can successfully be used in prediction of E_p and will lead to minimization of the time, cost, and equipment maintenance necessary for onsite monitoring. It will also help researchers use data from other sources.

ACKNOWLEDGEMENT

We are grateful to Agro Meteorological Cell, Department of Agronomy, College of Agriculture, Junagadh Agricultural University, Junagadh, Gujarat, India for providing maximum and minimum temperature data, pan evaporation data, wind speed and rainfall data for Junagadh and Amreli Stations.

REFERENCES

Anderson ER (1954). Energy-budget studies. In: Water Loss Investigations: Lake Hefner Studies. U.S. Geol. Surv. Prof. Pap. 269:71-119.

Benzaghta MA, Mohammed TA, Ghazali AH, MohdSoom MA (2010). Study on Simplified Model for Estimating Evaporation from Reservoirs, Austr. J. Basic Appl. Sci. 4(12):6473-6482.

Blaney HF, Criddle WD (1950). Determining Water Requirements in Irrigated Areas from Climatological Irrigation Data. US Department of Agriculture, Soil Conservation Service, Washington, D.C., Techn. Paper No. 96:48.

Chen D, Gao G, Xu CY, Guo J, Ren G (2005). Comparison of the Thornthwaite method and pan data with the standard Penman-Monteith estimates of reference evapotranspiration in China. Clim. Res. 28:123-132.

Chu CR, Li MH, Chen YY, Kau YH (2010). A wind tunnel experiment on the evaporation rate of Class A evaporation pan. J. Hydrol. 381(3–4):221-224.

Chu TW, Shirmohammadi A (2004). Evaluation of the SWAT model's hydrology component in the piedmont physiographic region of Maryland. Trans. ASAE 47(4):1057-1073.

Fekih M, Bourabaa A, Mohamed S (2013). Evaluation of two methods for estimation of evaporation from Dams water in arid and semi-arid areas in Algeria. Int. J. Appl. Innov. Eng. Manage. 2(1).

FrontLine, 8 November 2010, www.solver.org

Guitjens JC (1982). Models of Alfalfa Yield and Evapotranspiration. J. Irrig. Drain. Div. Proc. Am. Soc. Civ. Eng. 108(IR3):212-222.

Gupta HV, Kling H, Yilmaz KK, Martineza GF (2009). Decomposition of the mean squared error and NSE performance criteria: Implications for improving hydrological modeling. J. Hydrol. 377(1-2):80-91.

Harbeck GE (1962). A practical field technique for measuring reservoir evaporation utilizing mass-transfer theory. Geol. Surv. Prof. Pap. 272-E:101-105.

Irmak S, Haman DZ, Jones JW (2002). Evaluation of class A pan coefficients for estimating reference evapotranspiration in humid location. J. Irrig. Drain. Eng. 128(3):153-159.

Jhajharia D, Agrawal G, Sevda MS (2009). Influence Of Meteorological Parameters on Pan Evaporation at Agartala. J. Agric. Eng. 46(1).

Khanikar PG, Nath KK (1998). Relationship of open pan evaporation rate with some important meteorological parameters. J. Agril. Sci. Soc. N - E India 11(1):46-50.

Kirono DGC, Jones RN, Cleugh HA (2009). Pan-evaporation measurements and Morton-point potential evaporation estimates in Australia: are their trends the same. Int. J. Climatol. 29:711-718.

Kohler MA, Nordenson TJ, Fox WE (1955). Evaporation from Pans and Lakes. U.S. Dept. Com.Weather Bur. Res. Paper 38(21).

Lewis CD (2013). Industrial and business forecasting methods: A Radical guide to exponential smoothing and curve fitting." Butterworth Scientific, London; Boston: 1982.

Linsley RK, Kohler MA, Paulhus JLH (1982). Hydrology for Engineers. McGraw-Hill, London.

Nash JE, Sutcliffe JV (1970). River flow forecasting through conceptual models: Part 1. A discussion of principles. J. Hydrol. 10(3):282-290.

Penman HL (1948). Natural evaporation from open water, bare soil and grass. Proc. Royal Soc. Lond. 193:120-145.

Priestley CHB, Taylor RJ (1972). On the assessment of the surface heat flux and evaporation using large-scale parameters. Monthly Weather Rev. 100:81-92.

Santhi C, Arnold JG, Williams JR, Dugas WA, Srinivasan R, Hauck LM (2001). Validation of the SWAT model on a large river basin with point and nonpoint sources. J. Am. Water Res. Assoc. 37(5):1169-1188.

Shrivastava SK, Sahu AK, Dewangan KN, Mishra SK, Upadhyay AP, Dubey AK (2001). Estimating pan evaporation from meteorological data for Jabalpur. Ind. J. Soil Cons. 29(3):224-228.

Shuttleworth WJ (1988). Evaporation from Amazonian rainforest. Proc. Royal Soc. Lond. Ser. B. Biol. Sci. 233(1272):321-346.

Singh R, Bishnoi OP, Ram N (1992). Relationship between evaporation from class 'A' open pan evaporimeter and meteorological parameters at Hisar. Haryana Agri. Univ. J. Res. 22(2):97-98.

Singh R, Prakash O, Khicher ML, Singh R, Prakash O (1995). Estimation of evaporation from different meteorological parameters Ann. Arid-Zone 34(4):263-265.

Snyder RL, Orang M, Matyac S, GrismerMark E (2005). Simplified Estimation of Reference Evapotranspiration from Pan Evaporation Data in California. J. Irrig. Drain. Eng. 131(3):249-253.

Thornthwaite CW (1948). An approach toward a rational classification of climate. Geogr. Rev. 38:55-94.

Van Liew MW, Arnold JG, Garbrecht JD (2003). Hydrologic simulation on agricultural watersheds: Choosing between two models. Trans. ASAE 46(6):1539-1551.

Vazquez-Amábile GG, Engel BA (2005). Use of SWAT to compute groundwater table depth and streamflow in the Muscatatuck River watershed. Trans. ASAE 48(3):991-1003.

Wang ZY, Liu ZX, Zhang ZX, Liu XB (2009). Subsurface drip irrigation scheduling for cucumber (*Cucumis sativus* L.) grown in solar greenhouse based on 20 cm standard pan evaporation in Northeast China. Sci. Hortic. 123:51-57.

Xiaomang L, Yuzhou L, Dan Z, Minghua Z, Changming L (2011). Recent changes in pan-evaporation dynamics in China. Geophys. Res. Lett. 38(13).

Xing-Jie JI, Ji-Jun W, Wan-Long GU, Ye-Yu Z, Feng-Xiu LI (2010). Trends in Annual and Seasonal Pan Evaporation in the Lower Yellow River Basin from 1961 to 2010. Adv. Clim. Change Res. 3(4):195-204.

Yuhe J, Guangsheng Z (2011). Important factors governing the incompatible trends of annual pan evaporation, evidence from a small scale region. Clim. Change 106:303-314.

Permissions

All chapters in this book were first published in IJWREE, by Academic Journals; hereby published with permission under the Creative Commons Attribution License or equivalent. Every chapter published in this book has been scrutinized by our experts. Their significance has been extensively debated. The topics covered herein carry significant findings which will fuel the growth of the discipline. They may even be implemented as practical applications or may be referred to as a beginning point for another development.

The contributors of this book come from diverse backgrounds, making this book a truly international effort. This book will bring forth new frontiers with its revolutionizing research information and detailed analysis of the nascent developments around the world.

We would like to thank all the contributing authors for lending their expertise to make the book truly unique. They have played a crucial role in the development of this book. Without their invaluable contributions this book wouldn't have been possible. They have made vital efforts to compile up to date information on the varied aspects of this subject to make this book a valuable addition to the collection of many professionals and students.

This book was conceptualized with the vision of imparting up-to-date information and advanced data in this field. To ensure the same, a matchless editorial board was set up. Every individual on the board went through rigorous rounds of assessment to prove their worth. After which they invested a large part of their time researching and compiling the most relevant data for our readers.

The editorial board has been involved in producing this book since its inception. They have spent rigorous hours researching and exploring the diverse topics which have resulted in the successful publishing of this book. They have passed on their knowledge of decades through this book. To expedite this challenging task, the publisher supported the team at every step. A small team of assistant editors was also appointed to further simplify the editing procedure and attain best results for the readers.

Apart from the editorial board, the designing team has also invested a significant amount of their time in understanding the subject and creating the most relevant covers. They scrutinized every image to scout for the most suitable representation of the subject and create an appropriate cover for the book.

The publishing team has been an ardent support to the editorial, designing and production team. Their endless efforts to recruit the best for this project, has resulted in the accomplishment of this book. They are a veteran in the field of academics and their pool of knowledge is as vast as their experience in printing. Their expertise and guidance has proved useful at every step. Their uncompromising quality standards have made this book an exceptional effort. Their encouragement from time to time has been an inspiration for everyone.

The publisher and the editorial board hope that this book will prove to be a valuable piece of knowledge for researchers, students, practitioners and scholars across the globe.

List of Contributors

M K Verma
Department of Mechanical and Automation Engineering, Amity School of Engineering and Technology, Amity University, Noida (U.P), India

R. K Tyagi
Department of Mechanical and Automation Engineering, Amity School of Engineering and Technology, Amity University, Noida (U.P), India

H. Taebi
Graduate Student, School of Water Sciences Engineering, Shahid Chamran University, Ahwaz, Iran

M. Fathi-Moghadam
School of Water Sciences Engineering, Shahid Chamran University, Ahwaz, Iran

M. F Ali
Flood-Marine Excellence Center, IIESM, Faculty of Civil Engineering University Teknologi MARA, Malaysia

A. F Kamarudin
Flood-Marine Excellence Center, IIESM, Faculty of Civil Engineering University Teknologi MARA, Malaysia

K. Khalid
Flood-Marine Excellence Center, IIESM, Faculty of Civil Engineering University Teknologi MARA, Malaysia

N. F. Abd Rahman
Flood-Marine Excellence Center, IIESM, Faculty of Civil Engineering University Teknologi MARA, Malaysia

Samane Arvandi
Department of Water Science, Islamic Azad University, Science and Research Branch, Tehran, Iran

Amir Khosrojerdi
Soil Conservation and Watershed Management Research Institute, Tehran, Iran

Mohammad Rostami
Department of Water Engineering, Amirkabir University of Technology, Tehran, Iran

Hosein Baser
Master of Water Engineering, Amirkabir University of Technology

B. F. Sule
Department of Civil Engineering, University of Ilorin, Ilorin, Nigeria

S. A. Alabi
Lower Niger River Basin Development Authority, Ilorin, Nigeria

K. Srinivasa Reddy
ICAR-Central Research Institute for Dryland Agriculture, Hyderabd-59, India

O. E Sami
Faculty of Agriculture, Alzaiem Alazhari University, Khartoum, Sudan

Y. M Mustafa
Faculty of Agriculture, Alzaiem Alazhari University, Khartoum, Sudan

M. A. Shamseddin
Faculty of Agriculture, Alzaiem Alazhari University, Khartoum, Sudan

H. S Hilmi
Faculty of Agriculture, Alzaiem Alazhari University, Khartoum, Sudan

HUSSEIN I. Ahmed
Civil Engineering Department, College of Water Resources and Architectural Engineering, Northwest A & F University, Yangling, Shaanxi Province, P. R. CHINA

Junmin LIU
Civil Engineering Department, College of Water Resources and Architectural Engineering, Northwest A & F University, Yangling, Shaanxi Province, P. R. CHINA

ZHANG -Yinqin
Civil Engineering Department, College of Water Resources and Architectural Engineering, Northwest A & F University, Yangling, Shaanxi Province, P. R. CHINA

M. A. Shamseddin
Water Management and Irrigation Institute, University of Gezira, P. O. Box 20, Wadmedani, Sudan

H.S.M. Hilmi
Faculty of Agriculture, Alzaiem Alazhari University, Khartoum North, P. O. Box 1432, Sudan

E. A. A Oyedele
Physics Department, University of Ado - Ekiti, Nigeria

T Oyedele
Geological Services, Ogun State Ministry of Commerce and Industry, Abeokuta, Nigeria

K Oyedele
Omega - Inlands and Associates, Abeokuta, Nigeria

Ejikeme Ugwoha
Department of Chemical and Environmental Engineering, University of Nottingham, UK

John M. Andresen
Department of Chemical and Environmental Engineering, University of Nottingham, UK

M. O Oke
Department of Geography and Environmental Management, Tai Solarin University of Education, Ijagun, Nigeria

O Martins
Department of Geography and Environmental Management, Tai Solarin University of Education, Ijagun, Nigeria

O. A Idowu
Department of Water Resources Management and Agricultural Meteorology, University of Agriculture, Abeokuta, Ogun State, Nigeria

Nhamo Luxon
International Water Management Institute (IWMI), Pretoria, South Africa

Manyamba Christopher
International Water Management Institute (IWMI), Pretoria, South Africa

Chilonda Pius
International Water Management Institute (IWMI), Pretoria, South Africa

J. V. Tyagi
National Institute of Hydrology, Jal Vigyan Bhawan, Roorkee - 247 667, Uttarakhand, India

S. P. Rai
National Institute of Hydrology, Jal Vigyan Bhawan, Roorkee - 247 667, Uttarakhand, India

Nuzhat Qazi
Forest Research Institute, Dehradun - 248006, Uttarakhand, India

M. P. Singh
Forest Research Institute, Dehradun - 248006, Uttarakhand, India

H. Samadi-Boroujeni
Water Engineering Department, Shahrekord University and Water Recources Research Center, Iran

A. H. Haghiabi
Water Engineering Department, Lorestan University, Iran

E. Ardalan
Water Engineering Department, Shahrekord University, Iran, Msc Water Engineering Graduated from Shahrekord University, Iran

Mohammadhossein Kashefizadeh
Construction Management, Civil Engineering Faculty, University Teknologi Malaysia, Skudai, Johor, Malaysia

Zulkifli bin Yusop
Water Research Alliance, University Teknologi Malaysia, Skudai, Johor, Malaysia

Ali Hekmat
Water Resource Management, Civil Engineering Faculty, University Teknologi Malaysia, Skudai, Johor, Malaysia

Irwin Anthony Akpoborie
Department of Geology, Delta State University, Abraka

Oghenevwede Efobo
Centre for Research in Water and Environment, Abraka

Khosro Nezamkhiavy
Young Researchers and Elite Club, Meshkinshahr Branch, Islamic Azad University, Meshkinshahr, Iran

Sohrab Nezamkhiavy
Young Researchers and Elite Club, Meshkinshahr Branch, Islamic Azad University, Meshkinshahr, Iran

Assem Afify
Khatib and Alami, Salman University, Al Kharj, Saudi Arabia

A. A Adetoyinbo
Department of Physics, University of Ibadan, Ibadan, Nigeria

F. T Adelegan
Department of Physics, University of Ibadan, Ibadan, Nigeria

A. K Bello
Department of Physical Sciences, Bells University of technology, Ota, Nigeria

D. Anonye
Williamson Research Centre for Molecular Environmental Sciences, School of Earth, Atmospheric and Environmental Sciences, University of Manchester, Manchester, UK M13 9PL.

A. N. Nwachukwu
Williamson Research Centre for Molecular Environmental Sciences, School of Earth, Atmospheric and Environmental Sciences, University of Manchester, Manchester, UK M13 9PL.

A. Aldoma
Faculty of agriculture AlzaiemAlazhary University, Sudan

M. Y. Mohamed
Faculty of agriculture AlzaiemAlazhary University, Sudan

Rose E. Daffi
Department of Civil Engineering, Plateau State Polytechnic, Barkin Ladi, Nigeria

Johnson A. Otun
Department of Water Resources and Environmental Engineering, Ahmadu Bello University, Zaria, Nigeria

Abubakar Ismail
Department of Water Resources and Environmental Engineering, Ahmadu Bello University, Zaria, Nigeria

OKOLIE, Emmanuel Chukwuemeka
Department of Physics, Delta State University Abraka, Nigeria

Sudevi Basu
Department of Biotechnology, Sir M Visvesvaraya Institute of Technology, Bangalore - 562157, Karnataka, India

K. S Lokesh
Department of Environmental Engineering, Sri Jayachamarajendra College of Engineering (Autonomous), Mysore - 570006, Karnataka, India

Michael A. Nwachukwu
Department of Environmental Technology, Federal University of Technology, Owerri, Nigeria

Amos T. Kabo-bah
Department of Energy and Environmental Engineering, School of Engineering, University of Energy and Natural Resources, P. O. Box 214, Sunyani, Ghana

G. K. Anornu
Department of Civil Engineering, Kwame Nkrumah University of Science and Technology, P. M. B., Kumasi, Ghana

Eric Ofosu
Department of Energy and Environmental Engineering, School of Engineering, University of Energy and Natural

Resources, P. O. Box 214, Sunyani, Ghana

Robert Andoh
Hydro International, 94 Hutchins Dr., Portland, USA

Kamila Justyna List
Department of Geography, School of Geosciences, University of Energy and Natural Resources, P. O. Box 214, Sunyani, Ghana

Ngo Van Quan
Department of Water Resources Engineering, Water Resource University, 175 Tay Son street, Dong Da district, Hanoi Vietnam

Gwangseob Kim
Department of Civil Engineering, Kyungpook National University, Sankyuk-Dong, Buk-Gu, Daegu, 702-701, South Korea

J. Gundalia Manoj
Gujarat Technological University, Chandkheda, Ahmedabad, Gujarat 382 424, India

B. Dholakia Mrugen
Directorate of Technical Education, Gandhinagar, Gujarat 382 010, India